Wide Bandgap Semicon

Related Titles

The IGBT Device: Physics, Design and Applications of the Insulated Gate Bipolar Transistor (ISBN 978-1-4557-3143-5)
Power Electronics Handbook, Third Edition (ISBN 978-0-12-382036-5)
Wide Bandgap Power Semiconductor Packaging: Materials, Components, and Reliability (ISBN 978-0-08-102094-4)

Woodhead Publishing Series in Electronic and Optical Materials

Wide Bandgap Semiconductor Power Devices

Materials, Physics, Design, and Applications

Edited by

B. Jayant Baliga

ELSEVIER

WP
WOODHEAD
PUBLISHING
An imprint of Elsevier

Woodhead Publishing is an imprint of Elsevier
The Officers' Mess Business Centre, Royston Road, Duxford, CB22 4QH, United Kingdom
50 Hampshire Street, 5th Floor, Cambridge, MA 02139, United States
The Boulevard, Langford Lane, Kidlington, OX5 1GB, United Kingdom

Notices

Knowledge and best practice in this field are constantly changing. As new research and experience broaden our understanding, changes in research methods, professional practices, or medical treatment may become necessary.

Practitioners and researchers must always rely on their own experience and knowledge in evaluating and using any information, methods, compounds, or experiments described herein. In using such information or methods they should be mindful of their own safety and the safety of others, including parties for whom they have a professional responsibility.

To the fullest extent of the law, neither the Publisher nor the authors, contributors, or editors, assume any liability for any injury and/or damage to persons or property as a matter of products liability, negligence or otherwise, or from any use or operation of any methods, products, instructions, or ideas contained in the material herein.

British Library Cataloguing-in-Publication Data
A catalogue record for this book is available from the British Library

Library of Congress Cataloging-in-Publication Data
A catalog record for this book is available from the Library of Congress

ISBN: 978-0-08-102306-8 (print)

ISBN: 978-0-08-102307-5 (online)

For information on all Woodhead Publishing publications
visit our website at https://www.elsevier.com/books-and-journals

Publisher: Matthew Deans
Acquisition Editor: Kayla Dos Santos
Editorial Project Manager: Thomas Van Der Ploeg
Production Project Manager: Debasish Ghosh
Cover Designer: Greg Harris

Typeset by MPS Limited, Chennai, India

Contents

List of Contributors

Cornelius Armbruster Fraunhofer Institute for Solar Energy Systems ISE, Freiburg, Germany

B. Jayant Baliga North Carolina State University, Raleigh, NC, United States

Ishwara Bhat Electrical Computer and Systems Engineering Department, Rensselaer Polytechnic Institute, Troy, NY, United States

Subhashish Bhattacharya North Carolina State University, Raleigh, NC, United States

T. Paul Chow Rensselaer Polytechnic Institute, Troy, NY, United States

Srabanti Chowdhury Electrical and Computer Engineering, UC Davis, Davis, California, United States

Chao Fei Center for Power Electronics Systems, Virginia Tech, Blacksburg, VA, United States

Zhibo Guo Rensselaer Polytechnic Institute, Troy, NY, United States

Andreas Hensel Fraunhofer Institute for Solar Energy Systems ISE, Freiburg, Germany

Patrick Hercegfi Fraunhofer Institute for Solar Energy Systems ISE, Freiburg, Germany

Noriyuki Iwamuro University of Tsukuba, Faculty of Pure and Applied Sciences, Tsukuba, Ibaraki, Japan

Tsunenobu Kimoto Kyoto University, Kyoto, Japan

Dirk Kranzer Fraunhofer Institute for Solar Energy Systems ISE, Freiburg, Germany

Fred C. Lee Center for Power Electronics Systems, Virginia Tech, Blacksburg, VA, United States

Qiang Li Center for Power Electronics Systems, Virginia Tech, Blacksburg, VA, United States

Stefan Schönberger Fraunhofer Institute for Solar Energy Systems ISE, Freiburg, Germany

Jürgen Thoma Fraunhofer Institute for Solar Energy Systems ISE, Freiburg, Germany

Yuchen Yang Center for Power Electronics Systems, Virginia Tech, Blacksburg, VA, United States

Preface

In 1979, I derived a theoretical relationship between the specific on-resistance of unipolar semiconductor power devices and the basic properties of the semiconductor material while employed by the General Electric Company. My theory produced the Baliga's figure-of-merit (BFOM) for power devices that can be used to predict the performance enhancement produced by replacing silicon with wide bandgap semiconductors. The most technologically mature semiconductor after silicon was gallium arsenide (GaAs) at that time due to its applications for infrared lasers and light emitting diodes. The BFOM predicted a 13.6-fold reduction in the specific on-resistance of unipolar power devices by replacing silicon with GaAs extending their applications to higher voltages and power levels. The existing manufacturing infrastructure at GE for GaAs devices prompted its management to assign a team of 10 scientists and technicians to work under my guidance to create a GaAs-based power device technology in the early 1980s. I organized a focused effort to create GaAs epitaxial layers with lower doping levels to make high voltage devices, create a process platform to make high performance ohmic and Schottky contacts, and innovate novel device structures to exploit the material. This effort culminated in the first wide bandgap semiconductor power devices—Schottky rectifiers and vertical metal–semiconductor field-effect transistors—in the 1980s that validated the theoretical predictions.

My equation predicted a reduction in resistance by a factor of 200-times when replacing silicon with silicon carbide using the know properties in the 1980s. By the early 1990s, silicon carbide wafers became commercially available allowing the demonstration of the first high voltage Schottky diodes at the Power Semiconductor Research Center under my leadership in 1992. We were able to demonstrate a high performance SiC power metal-oxide-field-effect transistor (MOSFET) using available 6H-SiC material in 1997. Measurements of the impact ionization coefficients for silicon carbide under my direction provided data that increased the BFOM to 1000 for silicon carbide. These break-throughs encouraged major investments in development of better material and devices in the United States, Europe, and Japan. The first commercial SiC product, a high voltage Junction Barrier Schottky (JBS) diode, became available in the early 2000s. The market for these devices has now grown to over $200 million due to their applications as antiparallel diodes for silicon insulated gate bipolar transistors (IGBTs) in numerous applications.

After many years of effort to improve the interface properties between 4H-SiC and thermally grown oxide, it became feasible to introduce the first SiC power MOSFETs into the market in 2011. The initial concerns of application engineers regarding the reliability of these devices have been overcome by rigorous testing by the industry. The devices are now finding acceptance in applications such as PV inverters and power supplies. The devices must compete against mature silicon power devices—the IGBT and the superjunction FETs. The main impediment to market growth has been the much higher cost of SiC power devices. Considerable effort is underway around the world to drive down the cost of SiC power devices which portends a healthy market projection into the future.

The evolution of GaN power devices took an unusual path with the growth of GaN layers on silicon substrates by using a transition layer to accommodate the lattice mismatch. This breakthrough has made the GaN high electron mobility transistor (HEMT) structure possible with a highly conductive two-dimensional electron gas layer. These lateral devices offer much superior drift region resistance. However, it has been a challenge to create normally-off devices and even the normally-on structures still suffer from dynamic on-resistance issues. Some companies have taken the approach of building normally-on GaN HEMT products using the Baliga-Pair or Cascode configuration. Others have employed structural modifications to obtain a positive threshold voltage. These devices have been shown to enable power circuits to operate at multi-MHz switching frequencies to make very compact electronics possible. The ability to integrate multiple devices on a single chip creates opportunities to make power IC products as well.

This book on wide bandgap semiconductor power devices was motivated by the success of my book "The IGBT Device" published by Elsevier in 2015, which won the prestigious PROSE award for the best book published that year in engineering and technology. The IGBT book provided an extensive description of the applications of the IGBT in all sectors of our society and its social impact during the last 25 years.

For this book on wide bandgap semiconductor power devices, I wanted to cover the entire spectrum from material properties to device structures to their applications. I was pleased that all the people I approached to make contributions to the book enthusiastically accepted my proposal. Unfortunately, some of the authors failed to fulfill their promises due to commitments to their employers. Despite this, the contents of this book provide a comprehensive discussion of the state of the art for wide bandgap semiconductor power devices that is beneficial to the power electronics community.

The book begins with an introductory chapter where I have provided an overview of the benefits of wide bandgap semiconductor materials for power devices. Various types of power device structures are described in the chapter to familiarize the reader with the technology discussed in more depth in the rest of the book.

Chapter 2, SiC material properties, prepared by Professor Kimoto from Kyoto University, provides information on the basic properties of silicon carbide material that is relevant to power device design and analysis. The emphasis is on the 4H-SiC polytype because of its dominance for manufacturing SiC power devices. The discussion includes defects that influence the minority carrier lifetime due to its relevance for bipolar SiC power devices such as very high voltage IGBTs.

Chapter 3, Physical properties of gallium nitride and related III–V nitrides, prepared by Professor Bhat from Rensselaer Polytechnic Institute, provides information on the basic properties of gallium nitride material. The electrical properties of the two-dimensional electron gas in the AlGaN/GaN heterojunction structure are included because of its importance to the lateral GaN HEMT devices that have been commercialized. A discussion of defects produced during the growth of GaN layers on silicon substrates is included here due to its relevance to the reliability of these devices.

Chapter 4, SiC power device design and fabrication, prepared by Professor Iwamuro from the University of Tsukuba, provides a comprehensive discussion of silicon carbide power diodes and transistors. The physics of operation of SiC P–i–N diodes and JBS rectifiers is described and their performance is quantified for various blocking voltages. The design of a robust edge termination is critical to maximizing their performance. An extensive discussion of the SiC power MOSFET structure with either the planar or trench gate approach is provided. Good short-circuit capability for these devices is essential to their acceptance in applications. The potential to develop very high voltage SiC IGBTs is analyzed here as well.

Chapter 5, GaN smart power devices and ICs, prepared by Professor Chow from Rensselaer Polytechnic Institute, provides a thorough discussion of gallium nitride power devices. Lateral power devices based up on GaN-on-Si technology with the HEMT structure are covered in detail here. Prospects for making GaN-based power ICs are included in the chapter.

Chapter 6, GaN-on-GaN power device design and fabrication, prepared by Professor Chowdhury from University of California Davis, describes recent progress with design and fabrication of vertical GaN devices using bulk GaN substrates. The challenges of making the CAVET structure with enhancement-mode operation are described here.

Chapter 7, Gate drivers for wide bandgap power devices, prepared by Professor Bhattacharya from North Carolina State University, highlights the importance of providing adequate gate drive capability for wide bandgap semiconductor power devices. The challenges and solutions for driving these devices to achieve higher operating frequencies are described here. The operation of SiC and GaN devices at higher frequencies offsets the higher cost of the devices due to reduction of size, weight, and cost of passive elements. The chapter also provides insight into designing drivers for very high blocking voltage IGBTs with extremely high dV/dt transients in power circuits.

Chapter 8, Applications of GaN power devices, prepared by a group of professors from Virginia Polytechnic Institute, describes potential applications for GaN power devices. A number of converter designs implemented by the authors for power supply applications are described here. The improvement in efficiency by replacing silicon with GaN devices is quantified here.

Chapter 9, Applications of SiC devices, prepared by authors from the Franhaufer Institute, provides a focused perspective of the benefits of silicon carbide power devices on solar (PV) inverters. The benefits of replacing silicon devices with SiC devices for inverter efficiency are quantified here.

In the final chapter, I have provided a perspective on the history of wide bandgap semiconductor power device development since 1980. Some technology trends that are anticipated are provided here. The growth in the applications for silicon carbide and gallium nitride power devices is prognosticated with system requirements defined to make this happen. The projected market size that has motivated the investments in this technology is provided into the future until 2025. In addition, price targets for silicon carbide power devices with various blocking voltage ratings are defined followed by a manufacturing strategy to achieve them.

It has been my privilege to have initiated the worldwide interest in creating a quantum leap in power device performance by replacing silicon with wide bandgap semiconductors. Although the merits of achieving this goal were immediately recognized, it has taken 35 years of effort to develop the wafer material technology and reengineer the device structures to make commercially viable products.

B. Jayant Baliga
North Carolina State University, Raleigh, NC, United States
June 2018

Introduction

1

B. Jayant Baliga
North Carolina State University, Raleigh, NC, United States

1.1 Silicon power devices

Efficient electrical power generation, distribution, and management has become essential in modern society. These functions were first served by the development of bipolar silicon power devices to displace vacuum tubes in the 1950s. The ratings of silicon bipolar transistors and thyristors grew rapidly to serve an ever broader system need. However, their fundamental limitations in terms of the cumbersome control and protection circuitry led to bulky and costly solutions. A new class of devices evolved in the 1970s for power switching applications with the advent of metal oxide semiconductor (MOS) technology for digital circuits. These silicon power metal oxide semiconductor field effect transistors (MOSFETs) have found extensive use in high frequency applications with relatively low operating voltages (under 100 V). The merger of MOS and bipolar physics enabled creation of yet another class of silicon devices in the 1980s. The most successful innovation in this class of devices has been the insulated gate bipolar transistor (IGBT) [1]. The high power density, simple interface, and ruggedness of the IGBT have made it the technology of choice for all medium and high power applications.

In the 1990s, the concept of two-dimensional charge coupling was introduced using two basic approaches to significantly reduce the on-state resistance of silicon power devices. The first approach utilized a source connected electrode embedded inside a deep vertical trench. This method enabled a new generation of silicon power devices that have been commercialized with voltage ratings of 30–200 V. The second approach utilized alternating vertical columns of P- and N-type silicon regions. Products with blocking voltage around 600 V have been commercialized using this method. Any wide bandgap semiconductor power devices must outperform these advanced silicon power devices now available in the market.

1.2 Silicon power device applications

Power devices are used in all sectors of the economy with systems that operate over a broad spectrum of power levels and frequencies. The applications for power devices are shown as a function of operating frequency in Fig. 1.1. High voltage direct current (HVDC) power distribution and locomotive drives that require the

Wide Bandgap Semiconductor Power Devices. DOI: https://doi.org/10.1016/B978-0-08-102306-8.00001-0

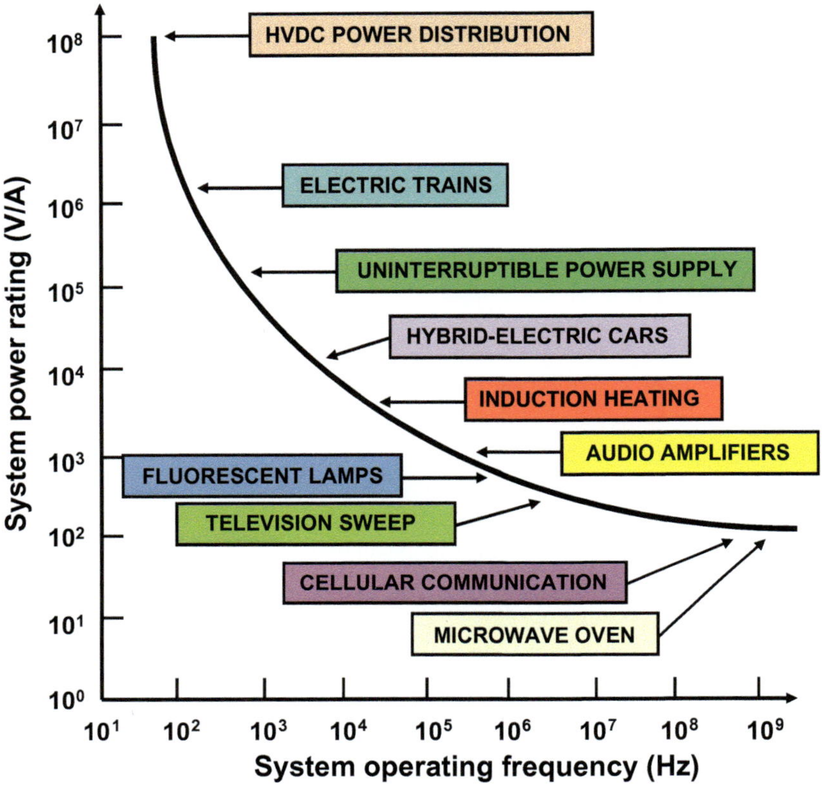

Figure 1.1 Applications for power devices over a broad range of frequencies.

control of megawatts of power operate at relatively low frequencies. The power ratings decrease for the devices when the operating frequency increases with typical microwave devices handling about 100 W. All of these applications are served by silicon devices today. Power MOSFETs are preferred for the high frequency applications operating from low power source voltages. These applications include power supplies for computers and laptops, power management in smart phones, and automotive electronics. Until recently, thyristors were the only devices available with sufficient voltage and current ratings favored for the HVDC power distribution applications. The ratings of IGBTs have now grown to levels where they are now preferred over thyristors for voltage source converters and flexible alternating current transmission (FACTs) designs. The medium frequency and power applications such as electric trains, hybrid-electric cars, home appliances, compact fluorescent lamps, medical equipment, and industrial motor drives also utilize the IGBT.

Silicon power devices can also be classified based on their current and voltage-handling requirements as shown in Fig. 1.2. Thyristors are available that can individually handle over 6000 V and 2000 A enabling the control of over

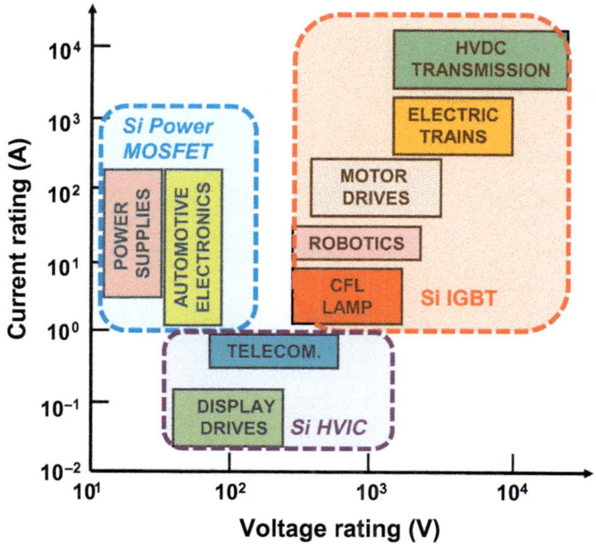

Figure 1.2 Applications for silicon power devices using system voltage and current ratings.

10 MW of power by a single monolithic device. These devices are suitable for the HVDC power transmission applications. During last 10 years, silicon IGBT modules have been developed with blocking voltages of up to 6500 V and current handling capability above 1000 A. This has allowed the silicon IGBT to replace thyristors in HVDC. The silicon IGBT is the optimum solution for a broad range of systems that require operating voltages between 300 V and 3000 V with significant current handling capability. These applications span all sectors of the economy including consumer, industrial, transportation, lighting, medical, defense, and renewable energy generation [1]. It is feasible to integrate multiple silicon devices on a single monolithic chip when the current requirements fall below 1 A to provide greater functionality for systems such as telecommunications and display drives. However, when the current exceeds a few amperes, it is more cost effective to use discrete power MOSFETs with appropriate control ICs to serve applications such as automotive electronics and switch mode power supplies.

1.3 Silicon carbide ideal specific on-resistance

The ideal specific on-resistance for the drift region in a vertical unipolar power devices is given by [2]:

$$R_{\text{on-ideal}} = \frac{4BV^2}{\varepsilon_S \mu_n E_C^3} \tag{1.1}$$

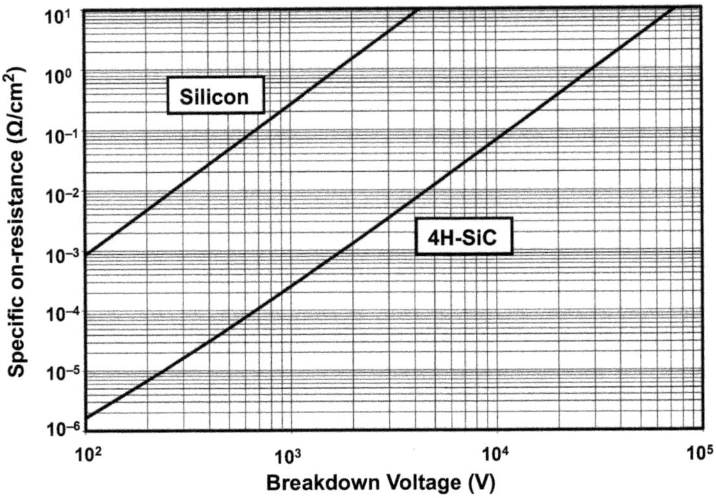

Figure 1.3 Ideal specific on-resistance for silicon carbide power devices.

where ε_S is the dielectric constant of the semiconductor, μ_n is the electron mobility, and E_C is the critical electric field for breakdown in the semiconductor. The denominator is called Baliga's figure-of-merit (BFOM) [3,4]:

$$\text{BFOM} = \varepsilon_S \mu_n E_C^3 = \frac{4BV^2}{R_{\text{on-ideal}}} \tag{1.2}$$

It is a measure of the power-handling capability (W/cm^2) of a power device. Unipolar silicon carbide power devices have low on-state voltage drop due to its large BFOM. This is mainly due to the approximately tenfold increase in critical electric field for breakdown for SiC compared with silicon.

The ideal specific on-resistance for the drift region in 4H-SiC devices is compared with that for silicon in Fig. 1.3 for breakdown voltages from 100 to 100,000 V. A significant reduction in the specific on-resistance of drift regions is predicted by replacing silicon with 4H-SiC. The ratio of the specific on-resistance for silicon to that for 4H-SiC increases from 527 at a breakdown voltage of 100 V to 1280 for breakdown voltages above 40,000 V.

1.4 Silicon carbide power rectifiers

Silicon bipolar power P–i–N diodes operate with the injection of minority carriers during on-state current flow [2]. These carriers must be removed when switching the device from the on-state to the off-state. This is accomplished by the reverse recovery process that produces a large reverse current during turnoff. This current produces significant power losses in the diode and the switches in the circuits.

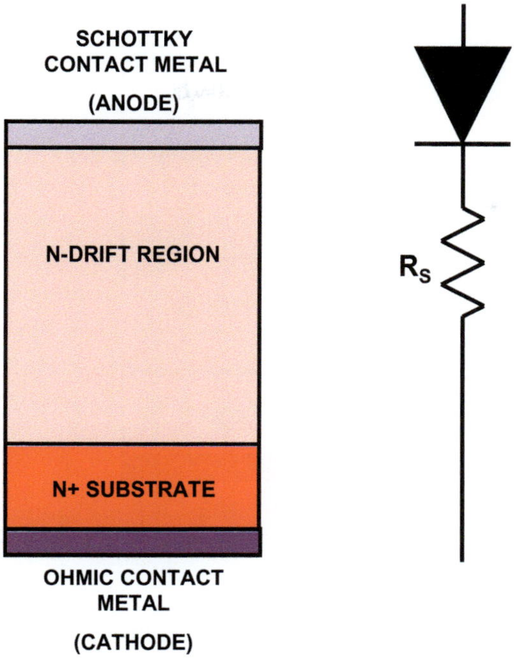

Figure 1.4 The basic Schottky rectifier structure.

It is therefore preferable to utilize unipolar current conduction in a power diode. The commonly used unipolar power diode structure is the Schottky rectifier that utilizes a metal-semiconductor barrier to produce current rectification. The high voltage Schottky rectifier structure contains a drift region, as show in Fig. 1.4, which is designed to support the reverse blocking voltage. The resistance of the drift region increases rapidly with increasing blocking voltage capability. Silicon Schottky rectifiers are commercially available with blocking voltages of up to 150 V. Beyond this value, the on-state voltage drop of silicon Schottky rectifiers becomes too large for practical applications. Silicon P–i–N rectifiers are favored for designs with larger breakdown voltages due to their lower on-state voltage drop despite their slower switching properties.

Silicon carbide Schottky rectifiers have much lower drift region resistance enabling design of very high voltage devices with low on-state voltage drop. These devices are excellent replacements for silicon P–i–N rectifiers used as fly-back or free-wheeling diodes with IGBTs in inverters. However, a major problem observed in SiC Schottky rectifiers is the large increase in the reverse leakage current with increasing reverse bias voltage. An increase in reverse leakage current by five orders of magnitude occurs due to Schottky barrier lowering and tunneling. This is a serious problem for high temperature operation and stability for these rectifiers.

The rapid increase in leakage current for Schottky rectifiers with increasing reverse bias voltage can be mitigated by using the junction-barrier controlled

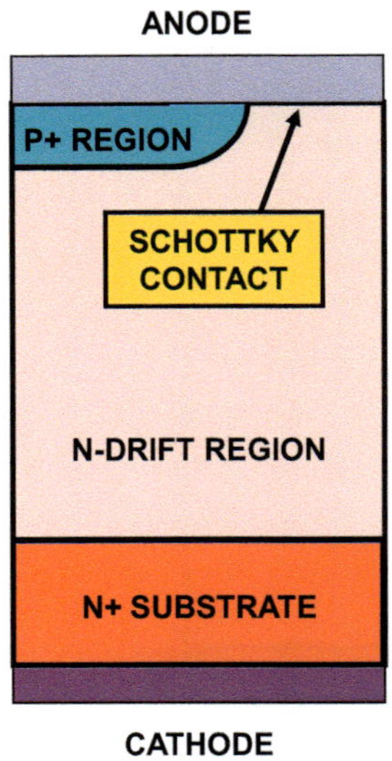

Figure 1.5 The silicon JBS rectifier structure.

Schottky (JBS) structure [5,6] shown in Fig. 1.5. This structure contains P^+ regions surrounding the Schottky contacts. A depletion region extends from the junction and forms a potential barrier under the Schottky contact. This suppresses the electric field at the Schottky contact. The lower electric field at the Schottky contact reduces the Schottky barrier lowering and tunneling at the contact [7].

1.5 Silicon power MOSFETs

The commercially available silicon power MOSFET has been widely used for lower power applications where the supply voltages are below 200 V. The commercially available silicon power MOSFET products are based up on the structures shown in Fig. 1.6. In the D-MOSFET structure, the P-base region and the N^+ source regions are self-aligned to the edge of the polysilicon gate electrode by using ion-implantation of boron and phosphorus with their respective drive-in thermal cycles. The N-type channel is defined by the difference in the lateral extension of the junctions under the gate electrode. The device supports positive voltage applied to the drain across the P-base/N-drift region junction. The voltage blocking capability is determined by the doping and thickness of the drift region. Although low voltage (<100 V) silicon power MOSFET have low on-resistances, the drift region resistance increases rapidly with increasing blocking voltage limiting the performance of silicon power MOSFETs to below 200 V. The silicon U-MOSFET structure has a gate structure embedded within a trench etched into the silicon surface. The N-type channel is formed on the side-wall of the trench at the surface of the P-base region. The channel length is determined by the difference in vertical extension of

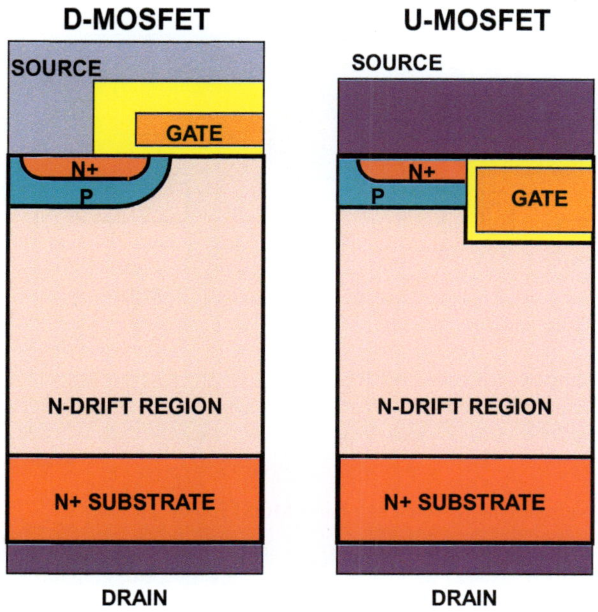

Figure 1.6 The silicon power MOSFET structures.

the P-base and N^+ source regions as controlled by the ion-implant energies and drive times for the dopants. The silicon U-MOSFET structure was developed to reduce the on-state resistance by elimination of the JFET component within the D-MOSFET structure [2].

The charge coupling concept was an important innovation for silicon power MOSFETs. It alters the electric field distribution in the drift region and allows supporting high voltages with large doping concentrations in the drift region [8]. The first charge coupled vertical silicon power MOSFET was the GD-MOSFET structure shown in Fig. 1.7 on the left-hand side [9,10]. The device contains a deep trench region with a source connected electrode. A uniform electric field can be generated in the drift region by using a graded doping profile in the drift region with high doping concentrations [11]. Breakdown voltages well above the parallel-plane breakdown voltage can be achieved using this idea. The specific on-resistance of these devices has been shown to be well below (5 to 25 times) that for the conventional silicon devices for blocking voltages ranging from 50 to 1000 V [6]. Many companies have released products using this approach.

An alternate charge-coupled silicon power MOSFET is the COOLMOS structure shown in Fig. 1.7 on the right-hand side. Here, the charge coupling is accomplished across the vertical P–N junction formed between columns of P and N drift regions [12]. Many studies have been performed to optimize this device structure for blocking voltages of 500–1000 V [6]. It has been demonstrated that the COOLMOS structure has about 3 to 10 times lower specific on-resistance than the conventional

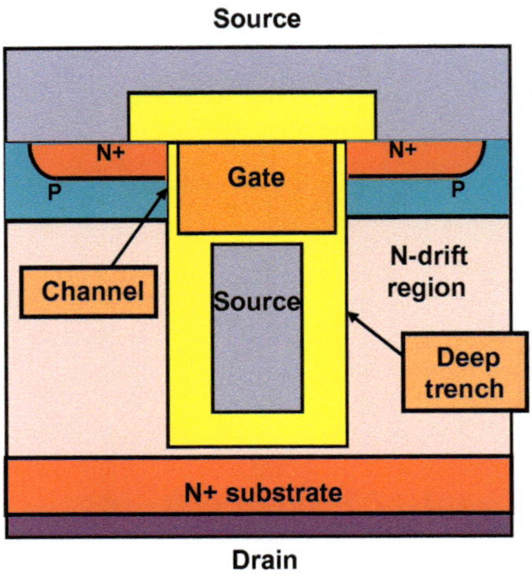

Figure 1.7 The silicon charge-coupled power MOSFET structures.

silicon power MOSFETs at a breakdown voltage of 600 V. Many companies have commercialized this device structure under various names.

Any proposed GaN or SiC power switch technology must compete with not only the conventional silicon power MOSFET structures but also the new charge coupled silicon power MOSFETs. The charge coupled silicon devices offer much better performance but require a more expensive fabrication process. This difference must also be taken into account.

1.6 Silicon carbide power MOSFETs

The silicon D-MOSFET structure cannot be replicated in silicon carbide for several reasons. First, the dopants in silicon carbide do not diffuse even at very high temperatures. Consequently, the channel in silicon carbide planar power MOSFETs is created by staggering the P-base and N^+ source ion implants [13]. This has been called the double-implanted or DI-MOSFET structure [14]. Second, a large amount of depletion occurs in the P-base region of silicon carbide devices due to the large electric field at the blocking junction [15]. This leads to very large channel length resulting in poor on-resistance of devices. This problem can be solved by using the shielded SiC planar power MOSFET structures [16] shown in Fig. 1.8 with either an inversion layer or an accumulation layer channel. A deep P^+ shielding region has been incorporated into the structures to prevent the depletion of the base regions. In the case of the structure with the inversion layer channel, the P^+ shielding region extends under both the N^+ source region as well as under the P-base region. In the case of the structure with the accumulation layer channel, the P^+ shielding region extends under the N^+ source region and the N-base region located under the gate. This N-base region can be formed using an uncompensated portion

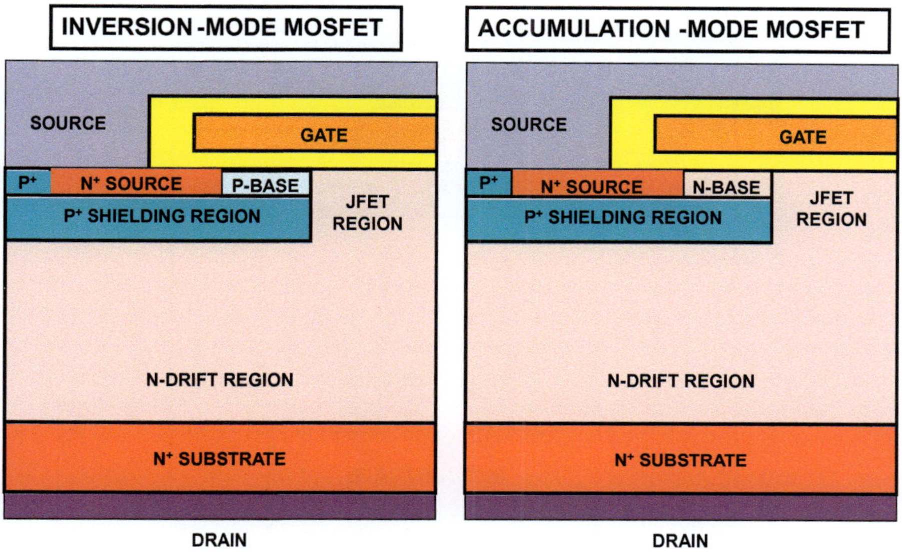

Figure 1.8 The shielded SiC power MOSFET structures.

of the N-type drift region or it can be created by adding N-type dopants near the upper surface with ion implantation or epitaxial growth to independently control its thickness and doping concentration.

Another important problem for silicon carbide power MOSFETs is the large electric field generated in the gate oxide due to the large electric field in the semiconductor when blocking high voltages. This problem can be overcome by using the shielded structures shown in Fig. 1.8. The gap between the P^+ shielding regions is optimized to obtain a low specific on-resistance while simultaneously shielding the gate oxide interface from the high electric field in the drift region.

It has been demonstrated that significantly larger mobility for electrons is observed in the channel for accumulation-mode SiC power MOSFETs when compared with inversion-mode SiC power MOSFETs [17]. This allows reducing the specific on-resistance of SiC power MOSFETs with blocking voltages below 3 kV. For devices with larger blocking voltages, the drift region resistance becomes dominant [18]. At present, most of the SiC power MOSFET commercialization effort is focused on devices with blocking voltages of 1.2 and 1.7 kV.

1.7 Silicon carbide power junction barrier Schottky field effect transistors (JBSFETs)

Most applications for power devices require current flow in not only the first quadrant but also in the third quadrant. One common such application is the

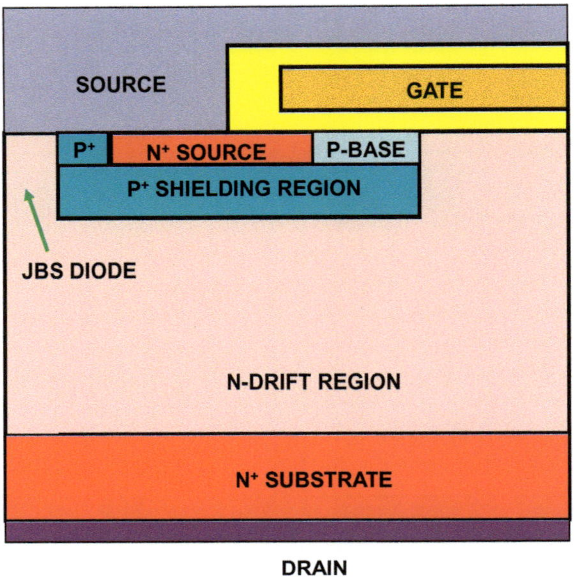

Figure 1.9 The SiC power JBSFET structure.

H-bridge circuit used for motor control. The presence of the body diode in the power MOSFET structure provides a convenient path for current flow in the third quadrant. Alternately, the gate can be turned on even when the drain voltage has a negative potential allowing current flow via the channel. Unfortunately, the perfect synchronization of the gate signal with the switching of the drain voltage into the third quadrant is not possible producing current flow via the body diode. For SiC power MOSFETs, the on-state voltage drop for the body diode exceeds 4 V producing high conduction losses. In addition, the body diode conduction introduces minority carrier injection and stored charge in the drift layer. The removal of the stored charge is accompanied by a reverse recovery current that produces enhanced turn-on switching losses in the MOSFETs. It has also been found that power MOSFET characteristics can be degraded when the body diode is turned-on due to the generation of stacking faults at basal plane dislocations.

The above problems have been overcome by creating the SiC junction barrier controlled Schottky field effect transistor (JBSFET) structure where a JBS diode is integrated into the power MOSFET structure [19,20] as illustrated in Fig. 1.9. It has been demonstrated that the JBS diode inside the JBSFET can be fabricated using a single metal to make the ohmic contacts to the N^+ source region and the P-base region while producing a Schottky contact to the N-drain region. The JBS diode has a voltage drop of only 2 V when conducting current in the third quadrant which prevents turn-on of the MOSFET P–N junction body diode.

1.8 Silicon carbide power MOSFETs with improved high frequency performance

In order to maximize the benefits of replacing the silicon IGBT with SiC power MOSFETs, it is necessary to increase the circuit operating frequency to reduce the size and cost of passive elements. SiC power MOSFETs must be optimized to reduce their switching losses. This can be achieved by reducing the reverse transfer capacitance (C_{GD}) with innovations in the gate structure. Two device structures that produce a significantly improved (smaller) high-frequency figures-of-merit (HFFOM), defined by [R_{on}*C_{GD}] and [R_{on}*Q_{GD}], are shown in Fig. 1.10.

The split-gate (SG)-MOSFET structure has a polysilicon gate with an opening in the middle of the gate electrode where it overlaps the drift region. This structure can be fabricated using the same process used to make the conventional SiC planar-gate power MOSFET. The measured HFFOM [R_{on}*C_{GD}] for the SG-MOSFET has been found to be 1.3 times smaller than for the conventional MOSFET while its HFFOM [R_{on}*Q_{GD}] is 2.4 times smaller than for the conventional MOSFET [21].

The buffered-gate (BG)-MOSFET structure has a polysilicon gate with an opening in the middle of the gate electrode where it overlaps the drift region. In addition, the P^+ shielding region is extended beyond the edge of the gate electrode to completely screen it from the drain. To prevent complete depletion of the N-type region above the P^+ shielding region, it is necessary to add a second junction field effect transistor (JFET) 2 region with higher doping concentration than in the JFET 1 region. The measured HFFOM [R_{on}*C_{GD}] for the BG-MOSFET has been found

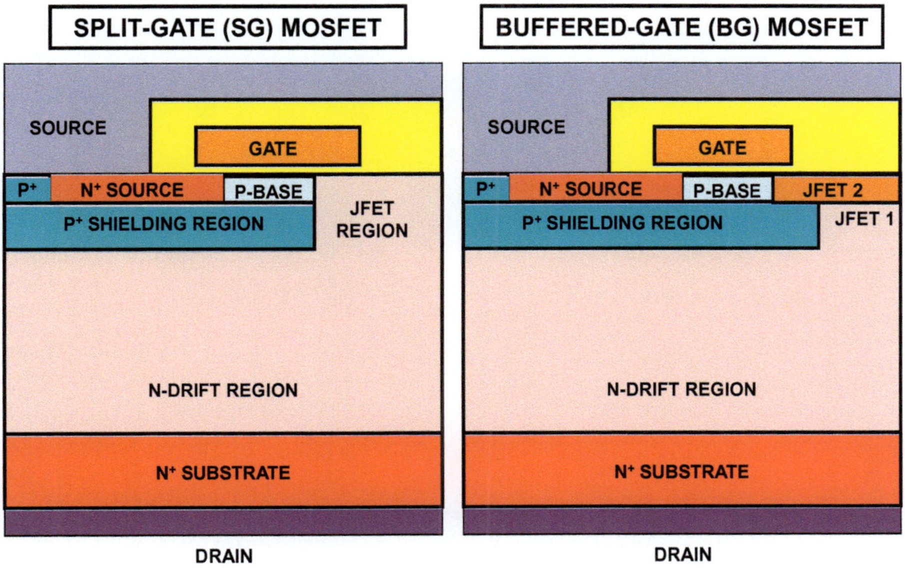

Figure 1.10 The SiC power MOSFET structures with improved HFFOM.

to be 3.6 times smaller than for the conventional MOSFET while its HFFOM [R_{on}*Q_{GD}] is 4.0 times smaller than for the conventional MOSFET [22].

1.9 Silicon carbide bidirectional field effect transistor

Matrix converters require power devices that can block high voltage in the first and third quadrants and carry gate controlled current in both quadrants [23]. The development and commercialization of these types of converters has been hindered by the lack of availability of a cost-effective bidirectional switch with low on-state voltage drop and switching losses. Many approaches to creating a bidirectional switch have been proposed using silicon IGBTs and SiC power MOSFETs. They require multiple independently packaged devices with a high net on-state voltage drop.

The SiC bidirectional field effect transistor (BiDFET) was proposed to address this application [24]. It consists of two SiC JBSFETs connected in series as shown in Fig. 1.11. These devices can also be monolithically integrated by building them adjacent to each other on the same SiC wafer. Terminal T1 serves as the reference terminal of the BiDFET with high AC voltages applied to terminal T2. Both gates G1 and G2 are used to control the operation of the device in the first and third quadrant. It is worth emphasizing that, unlike conventional power MOSFETs, no external electrical connection is performed to the N^+ substrate (drain) in the proposed BiDFET.

High blocking voltage capability in the first quadrant is achieved in the BiDFET with zero bias applied to gate G1 with respect to terminal T1. Under these

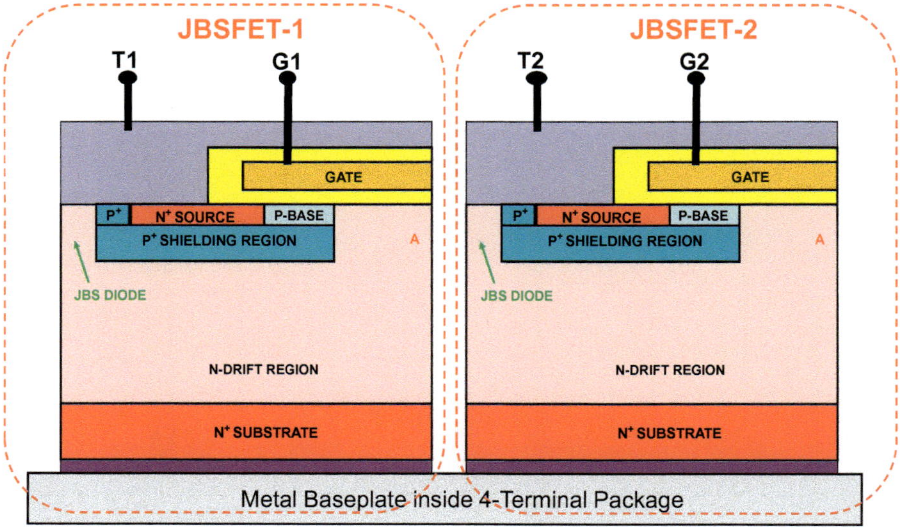

Figure 1.11 The SiC BiDFET.

conditions, the body diode of power JBSFET 2 is forward biased and the high voltage is supported across power JBSFET-1 and its edge termination. High blocking voltage capability in the third quadrant is achieved with zero bias applied to gate G2 with respect to terminal T2. Under these conditions, the body diode of power JBSFET 1 is forward biased and the high voltage is supported across power JBSFET-2 and its edge termination.

Current conduction in the first quadrant is achieved in the BiDFET by the application of a positive gate drive voltage to both gates G1 and G2 with reference to the corresponding terminals T1 and T2. This turns-on the channel for both SiC power JBSFETs. The on-resistance of the BiDFET is then the sum of the on-resistance of both SiC power JBSFETs 1 and 2. It can be reduced as desired by scaling the area of both power JBSFETs to achieve a low on-state voltage drop.

Gate voltage-controlled current saturation with excellent output characteristics is achieved in the BiDFET as previously demonstrated for SiC power MOSFETs. The BiDFET also has fast switching capability as previously demonstrated for SiC power MOSFETs. These features make it well suited for matrix converters operating at high frequencies to achieve high power density.

One of the problems that can be encountered in bidirectional switches used in matrix converters occurs during the dead-time or commutation-time. This can lead to current conduction via the body diode in the MOSFETs. It has been documented that current flow via the body diode of SiC power MOSFETs can lead to bipolar degradation of the devices. This problem is circumvented in the BiDFET by implementing it using SiC JBSFETs. In this case, the current flow occurs via the integrated JBS diodes within the JBSFET structure preventing current follow via the P–N body diode to completely suppress the bipolar degradation phenomenon.

The BiDFET has been experimentally demonstrated with 1.2 kV rated JBSFETs [25]. The devices were shown to exhibit gate voltage-controlled output characteristics in the first and third quadrant with blocking voltages of up to 1650 V. The on-resistance of the devices in both quadrants is the sum of the resistance of each of the JBSFETs with the on-state gate bias of 20 V. The on-state characteristics exhibit a knee of about 1 V if the gate voltage is not applied to the device with the forward biased JBS diode. The BiDFET requires only a single package in contrast to previous bidirectional switches that need 4–6 separately packaged devices. Its on-state voltage drop can be reduced to only 0.5 V by scaling the on-resistance of the JBSFETs compared with more than 1.25 V for the prior devices.

1.10 Silicon carbide power device applications

In principle, silicon carbide power MOSFETs are excellent candidates to replace silicon IGBTs in all their applications because of their low on-resistance and reduced switching losses. However, the cost of the SiC power MOSFETs is substantially larger than that of the silicon IGBT. As a consequence, SiC power MOSFETs have been successfully used in selected applications as shown in

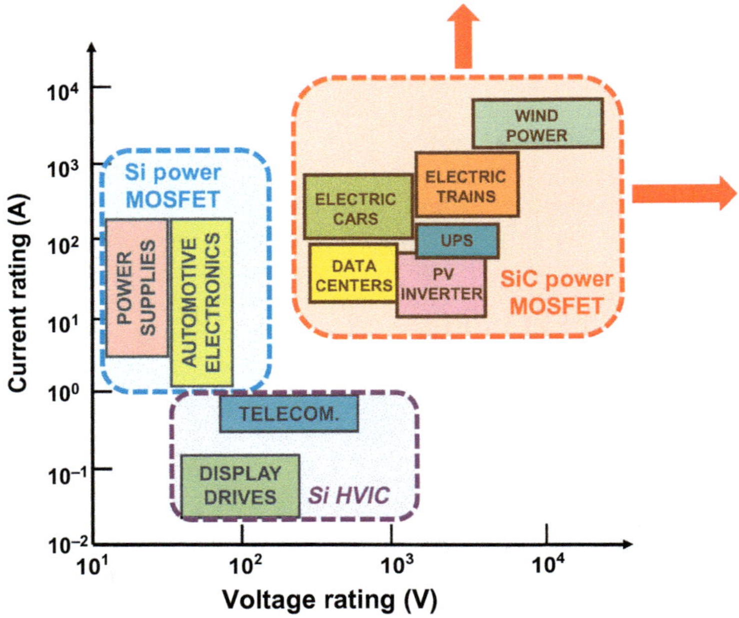

Figure 1.12 Applications for SiC power MOSFET structures.

Fig. 1.12. One of these applications is inverters for solar power generation. The replacement of silicon IGBTs with SiC power MOSFETs has been shown to increase the efficiency by 1%–2%. This modest gain in energy generation offsets the larger initial cost of the SiC devices.

Another promising application for SiC power MOSFETs is in inverters for electric and hybrid-electric vehicles. The improved inverter efficiency allows extending the range of the vehicles. In addition, the SiC inverters can be operated at higher frequencies than those based on silicon IGBTs. This allows reduction of passive component size and weight, an important benefit for in electric vehicles.

As the cost of SiC power MOSFET is reduced in the future, it is anticipated that their applications will expand as indicated by the arrows in Fig. 1.12. The penetration of SiC power MOSFETs into applications served by the Si IGBT will be accelerated by making products with superior HF-FOMs, such as the SG-MOSFET and BG-MOSFET.

1.11 Gallium nitride power devices

The commercialization of gallium–nitride devices was accelerated by the successful growth of high quality gallium nitride layers on silicon substrates by using a transition layer as illustrated in Fig. 1.13 to ameliorate the lattice mismatch between the materials. The cost of the wafers is greatly reduced in comparison with SiC or GaN

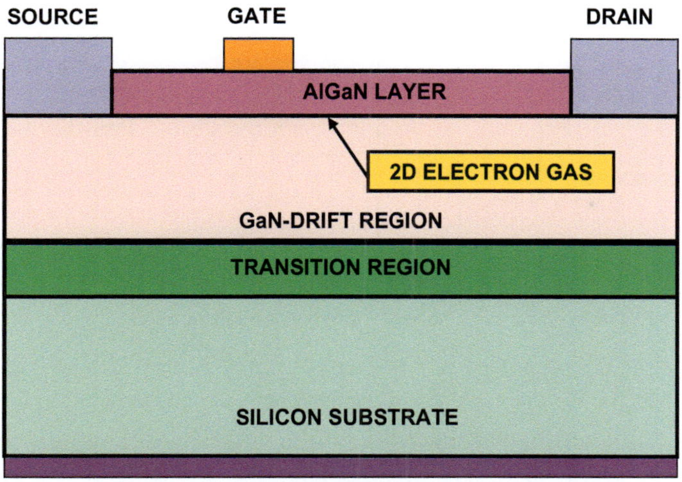

Figure 1.13 The GaN HEMT structure formed on a silicon substrate.

substrates by the use of large diameter silicon substrates. A thin aluminum–gallium nitride layer is formed on top of the gallium nitride layer to produce the two-dimensional electron gas in the gallium nitride. The electrons in the two-dimensional electron gas have a high mobility ($\sim$2000 cm^2/Vs) and charge ($\sim 10^{13}$ cm^{-2}). This creates a high conductivity current path through the GaN drift region with low specific on-resistance despite the lateral device structure. This allows creating lateral devices with high-blocking voltages and low-specific on-resistance.

GaN HEMT structures have been demonstrated with very high blocking voltages. However, commercial devices have blocking voltages in the range of 600–900 V. The interdigitated cell structure for the lateral GaN HEMT device is difficult to scale to higher current levels. This is restricted the current ratings between 50 A and 100 A. Furthermore, it has been difficult to achieve an enhancement-mode lateral GaN HEMT device. Unfortunately, normally-on devices are unacceptable for power electronics applications. Normally-off GaN devices have been realized by using the Baliga-Pair or cascode configuration. In the Baliga-Pair [26], a normally-on silicon carbide high voltage JFET/MESFET device or a GaN HEMT device is used together with a low voltage silicon MOSFET to create a configuration with the desired features for a high-quality power switch. The basic idea is illustrated in Fig. 1.14. It consists of a high voltage silicon carbide JFET or MESFET structure with its source electrode connected to the drain electrode of a low voltage silicon power MOSFET. An important feature of configuration is that the gate of the silicon carbide device is connected to the source of the silicon power MOSFET which serves as the ground or reference terminal in circuits. Gate signals are exclusively applied to the gate of the silicon power MOSFET. The drain of the silicon carbide device is connected to the load in power circuits as would be done with the drain of silicon power MOSFETs. The Baliga-Pair configuration is a three-terminal power switch with an MOS-input interface provided by the silicon power

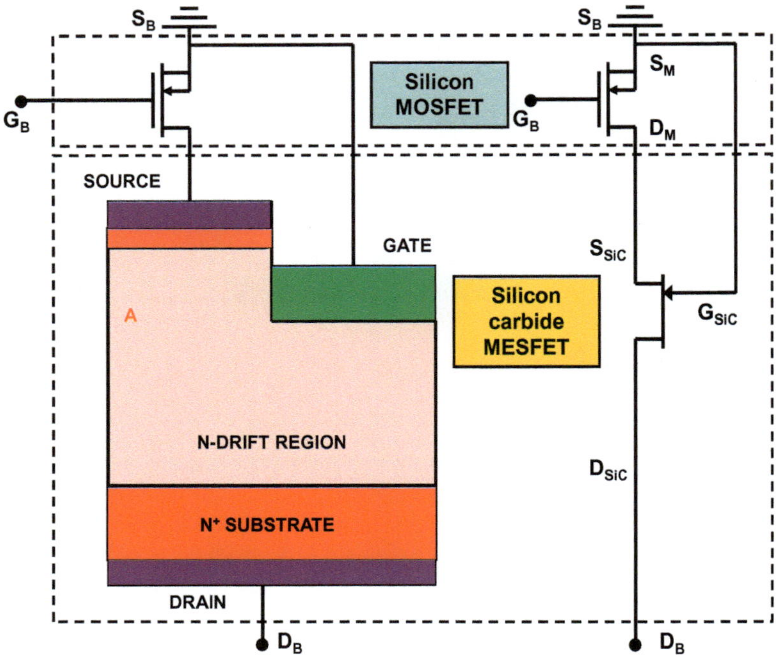

Figure 1.14 The Baliga-Pair configuration with SiC MESFET.

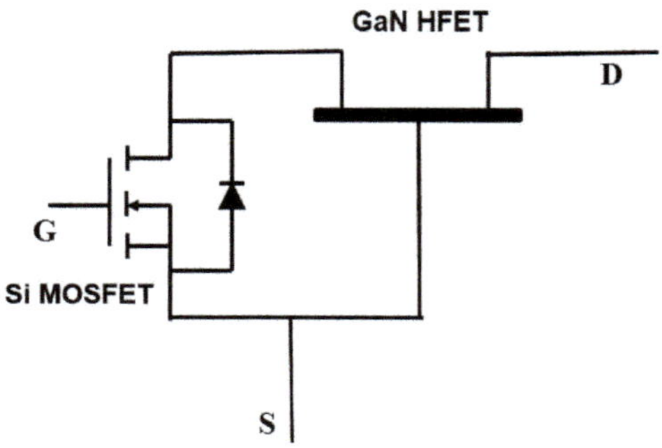

Figure 1.15 The Baliga-Pair configuration with GaN HEMT.

MOSFET and high blocking voltage capability provided by the SiC JFET/MESFET device or GaN HEMT device. The Baliga-Pair implemented with a GaN HEMT device is shown in Fig. 1.15.

1.12 Gallium nitride power device applications

GaN HEMT devices have been commercialized by using the Baliga-Pair configuration. This approach was used by the industry due to the high specific on-resistance

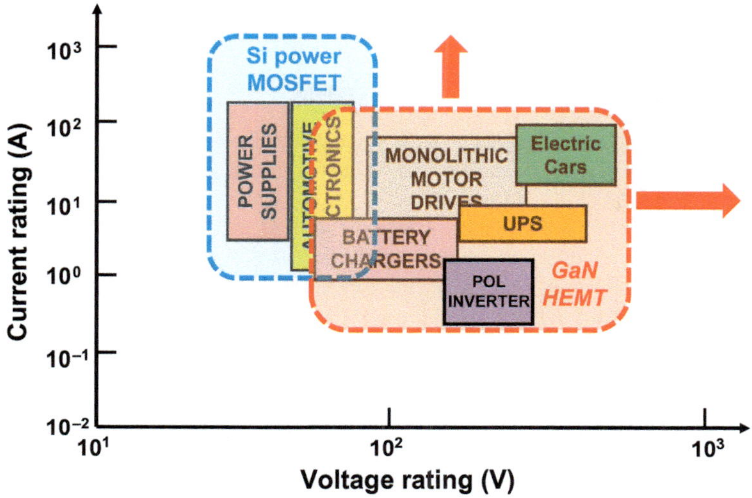

Figure 1.16 Applications for GaN HEMT devices.

and performance degradation of normally-off structures. Their applications have been focused on low power (<100 W) applications such as battery chargers for cell phones. The fast switching capability of the GaN devices has enabled increasing the circuit operating frequency to above 1 MHz. This allows shrinking the passive components to create a small form factor that is attractive for mobile consumer devices.

The lateral configuration of the GaN HET structure leads to all the device terminals being accessible on the top surface of the chip. This is convenient for interconnecting multiple devices together to create power ICs. This is suitable for applications such as fractional hp motor-drives. The applications for GaN devices is shown in Fig. 1.16.

1.13 Summary

The development of silicon carbide and gallium nitride power devices has been motivated by the opportunity to create unipolar devices with high blocking voltage capability. Silicon carbide JBS rectifiers and power MOSFETs with excellent on-state voltage drop have been demonstrated with blocking voltages up to 5000 V. These devices offer low switching losses when compared with the silicon IGBT enabling increasing the circuit operating frequency which reduces the size of passive components and filters in applications. Gallium–nitride HEMT devices have been commercialized with blocking voltages of 100–600 V. The applications of the SiC and GaN devices is constrained by their higher cost when compared with silicon devices. The lower switching losses of these devices can to utilized to increase the circuit operating frequency leading to smaller, less expensive passive elements to offset the higher device cost.

References

[1] B.J. Baliga, The IGBT Device: Physics, Design, and Applications of the Insulated Gate Bipolar Transistor, Elsevier Press, 2015.

[2] B.J. Baliga, Fundamentals of Power Semiconductor Devices, Springer-Science, 2008.

[3] B.J. Baliga, Semiconductors for high voltage vertical channel field effect transistors, J. Appl. Phys. 53 (1982) 1759–1764.

[4] B.J. Baliga, Power semiconductor device figure of merit for high frequency applications, IEEE Electron Device Lett. 10 (1989) 455–457.

[5] B.J. Baliga, The pinch rectifier: a low forward voltage drop high speed power diode, IEEE Electron Device Lett. 5 (1984) 194–196.

[6] B.J. Baliga, Pinch rectifier, U.S. Patent 4,641,174, Issued February 3, 1987.

[7] B.J. Baliga, Advanced Power Rectifier Concepts, Springer-Science, New York, 2010.

[8] B.J. Baliga, Advanced Power MOSFET Concepts, Springer-Science, New York, 2010.

[9] B.J. Baliga, Vertical field effect transistors having improved breakdown voltage capability and low on-state resistance, U.S. Patent 5,637,898, Issued June 10, 1997.

[10] B.J. Baliga, Power semiconductor devices having improved high frequency switching and breakdown characteristics, U.S. Patent 5,998,833, Issued December 7, 1999.

[11] B.J. Baliga, Trends in power discrete devices, IEEE International Symposium on Power Semiconductor Devices and ICs, Abstract P-2 (1997) 5–10.

[12] L. Lorenz, et al., COOLMOS – A new milestone in high voltage power MOS, IEEE International Symposium on Power Semiconductor Devices and ICs (1999) 3–10.

[13] B.J. Baliga, M. Bhatnagar, Method of fabricating silicon carbide field effect transistor, U.S. Patent 5,322,802, Issued June 21, 1994.

[14] J.N. Shenoy, J.A. Cooper, M.R. Melloch, High voltage double-implanted power MOSFETs in 6H-SiC, IEEE Electron Device Lett. 18 (1997) 93–95.

[15] B.J. Baliga, Gallium Nitride and Silicon Carbide Power Devices, World Scientific Publishers, Singapore, 2016.

[16] B.J. Baliga, Silicon carbide semiconductor devices having buried silicon carbide conduction barrier layers therein, U.S. Patent 5,543,637, Issued August 6, 1996.

[17] W. Sung, K. Han, B.J. Baliga, A comparative study of channel designs for SiC power MOSFETs: Accumulation-mode vs inversion-mode channel, IEEE International Symposium on Power Semiconductor Devices and ICs (2017) 375–378.

[18] J.W. Palmour, Silicon carbide power device development for industrial markets, IEEE International Electron Devices Meeting, Abstract 1.1.1 (2014) 1–8.

[19] W. Sung, B.J. Baliga, Monolithically integrated 4H-SiC MOSFET and JBS diode (JBSFET) using a single Ohmic/Schottky process scheme, IEEE Electron Device Lett. 37 (2016) 1605–1608.

[20] W. Sung, B.J. Baliga, On developing one-chip integration of 1.2 kV SiC MOSFET and JBS diode (JBSFET), IEEE Trans. Ind. Electron. 64 (2017) 8206–8212.

[21] K. Han, B.J. Baliga, W. Sung, Split-gate 1.2 kV 4H-SiC MOSFET: analysis and experimental results, IEEE Electron Device Lett. 38 (2017) 1437–1440.

[22] K. Han, B.J. Baliga, W. Sung, A novel 1.2 kV 4H-SiC buffered-gate (BG) MOSFET: analysis and experimental results, IEEE Electron Device Lett. 39 (2018) 248–251.

[23] P.W. Wheeler, et al., Matrix converters: a technology review, IEEE Trans. Ind. Electron. 49 (2002) 276–288.

[24] B.J. Baliga, Monolithically integrated AC switch having JBSFETs therein with commonly connected drain and cathode electrodes, U.S. Provisional Patent Application No. 62/526,192, Filed June 28, 2017.
[25] B.J. Baliga, K. Han, Monolithic SiC bi-directional field effect transistor (BiDFET): concept, implementation, and electrical characteristics, GOMACTech 2018, Paper 3.2, pp. 32−35, March 13, 2018.
[26] B.J. Baliga, Silicon carbide switching device with rectifying gate, U.S. Patent 5,396,085, Issued March 7, 1995.

SiC material properties

2

Tsunenobu Kimoto
Kyoto University, Kyoto, Japan

2.1 Crystal and band structures

SiC is a IV–IV compound semiconductor, where only a rigid stoichiometry (Si:C = 1:1) is allowed. The large SiC bond energy (about 4.6 eV) gives this material a wide bandgap, high critical electric field strength, and high phonon energies [1–3]. Compared with other wide bandgap semiconductors such as III-nitrides, p- and n-type doping control in the wide range (10^{14}–10^{20} cm^{-3}) by either in situ doping or ion implantation is easy in SiC. Thermal oxidation of SiC yields SiO_2 on the surface as in the case of Si, which provides an opportunity of easy fabrication of metal-oxide-semiconductor (MOS)-based devices (though the interface quality needs much improvement), surface passivation, and sacrificial oxidation to remove a surface damage. Availability of large-diameter single crystalline wafers with a reasonable quality is another advantage of SiC. Since 2012, 150-mm SiC wafers with a low resistivity of 0.02 Ω cm are commercially available and the dislocation density of the commercial wafers has been reduced to about 3000 cm^{-2} (as in 2017). In SiC, however, formation of a heterojunction is extremely difficult. A low stacking-fault energy is another weakness of this material.

SiC crystallizes in many different crystal structures which vary in one dimension (that is, in stacking sequence). This phenomenon is known as polytypism and each SiC structure is called "polytype." Regarding the detailed description, see books on SiC or polytypism [1–4]. Each SiC polytype is often represented by using Ramsdell's notation, where polytypes are expressed by the number of Si–C bilayers in the unit cell and the crystal system (C for cubic, H for hexagonal, and R for rhombohedral). Among more than 200 SiC polytypes, 3C-SiC, 4H-SiC, and 6H-SiC are popular and have been grown and studied. Although it is not fully understood why so many SiC polytypes appear, the intermediate ionicity of SiC (11% according to Pauling's definition) may be a possible reason for the occurrence of SiC polytypism.

Individual SiC polytypes exhibit different bandgap and mobility, and other electronic properties are usually very different in the different polytypes. For power device applications, 4H-SiC is the most suitable and has exclusively been developed [3,5]. The major reasons include its high electron mobility [6], high critical electric field strength [7], and the availability of single crystalline wafers. The small anisotropies in the mobility [8] and critical electric field strength [9] of 4H-SiC are also advantages, because 6H-SiC exhibits a very small electron mobility along ⟨0001⟩ and a relatively low critical electric field strength perpendicular to ⟨0001⟩.

Wide Bandgap Semiconductor Power Devices. DOI: https://doi.org/10.1016/B978-0-08-102306-8.00002-2

Device-quality 3C-SiC has never been grown and more importantly, the critical electric field strength of 3C-SiC is too low to compete with Si power devices. Thus, "SiC" means 4H-SiC hereafter in this chapter unless specified.

Fig. 2.1 schematically illustrates (A) the crystal structure, (B) the stacking structure of SiC (4H-SiC), where the *open* and *closed circles* denote Si and C atoms, respectively, and (C) the definition of several major planes in a hexagonal structure with fundamental translation vectors $\boldsymbol{a}_1$, $\boldsymbol{a}_2$, $\boldsymbol{a}_3$, and $\boldsymbol{c}$. The (0001) face, where one bond from a tetrahedrally bonded Si atom is directed along the *c*-axis (⟨0001⟩), is called the "Si face," while the $(000\bar{1})$ face, where one bond from a tetrahedrally bonded C atom is directed along the *c*-axis, is called the "C face." The $\{11\bar{2}0\}$ face is called "A face (or *a*-face)", and the $\{1\bar{1}00\}$ face is "M face (or *m*-face)." The surface of standard SiC wafers is (0001) or the Si face. Note that an off-angle of several degrees (typically 4 degrees) toward $[11\bar{2}0]$ is introduced to ensure perfect polytype replication in an epitaxial layer on the wafer [10]. In SiC, the valence electrons are slightly localized near C atoms, due to a higher electronegativity of carbon than that of silicon (C: 2.5, Si: 1.8). This ionicity gives this material polarity, which induces different electronic and chemical properties on different faces (e.g., Si face vs C face).

Table 2.1 shows the lattice constants and several mechanical properties of SiC at room temperature. The atom density (sum of Si and C atoms) in SiC is approximately 9.6×10^{22} cm^3 at room temperature. Note that the lattice constants slightly change when the temperature, dopant, and doping density are varied, as is the case of other semiconductors. Nitrogen (donor) doping results in lattice contraction and aluminum (acceptors) doping causes lattice expansion [3,11]. Therefore, it is crucially important to control a stress caused by the doping-induced lattice-mismatch near a junction (e.g., n^-/n^+, p^+/n^-, p^+/n^+) in device fabrication.

Fig. 2.2 depicts (A) the first Brillouin zone and (B) electron energy–wavenumber dispersions of SiC [12]. Here the bandgap is underestimated, because of a limitation of theoretical calculation (density functional theory), but the

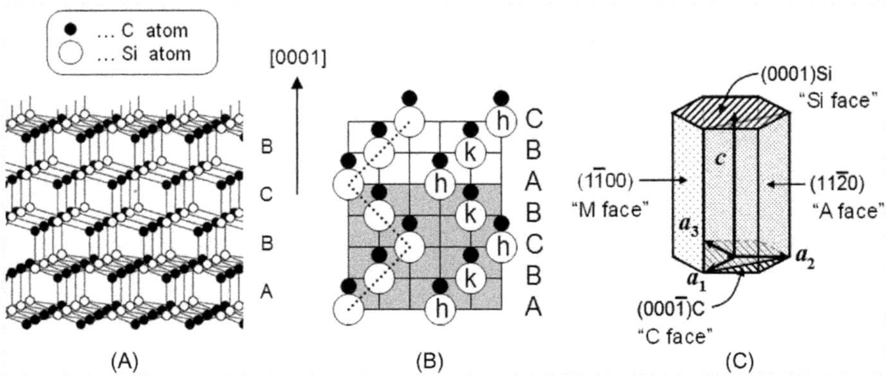

Figure 2.1 (A) Crystal structure, (B) schematic stacking structure of SiC (4H-SiC), where the *open* and *closed circles* denote Si and C atoms, respectively. (C) Definition of several major crystal planes in a hexagonal structure with fundamental translation vectors $\boldsymbol{a}_1$, $\boldsymbol{a}_2$, $\boldsymbol{a}_3$, and $\boldsymbol{c}$.

Table 2.1 Lattice constants and several mechanical properties of SiC at room temperature

Properties	SiC (4H)
Lattice constant a (Å) c (Å)	 3.0798 10.0820
Density (g/cm^3)	3.21
Young modulus (GPa)	390–690
Fracture strength (GPa)	21
Poisson's ratio	0.21
Elastic constant (GPa) c_{11} c_{12} c_{13} c_{33} c_{44}	 501 111 52 553 163
Specific heat (J/g K)	0.69

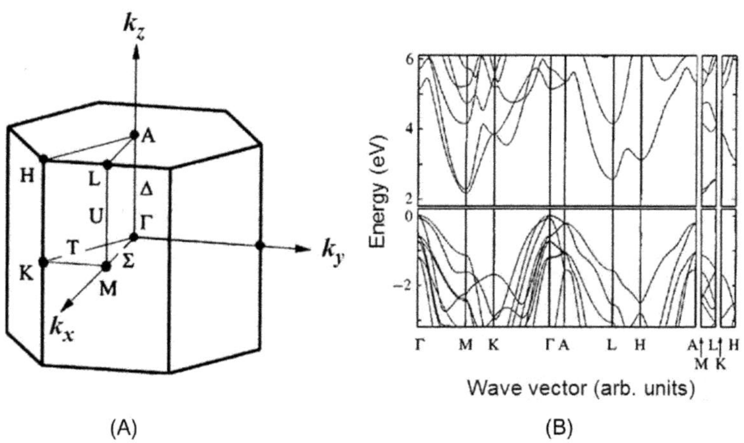

Figure 2.2 (A) First Brillouin zone and (B) electron energy–wavenumber dispersions of SiC [12]. Here the bandgap is underestimated, because of a limitation of theoretical calculation (density functional theory). The conduction band minima are located at the M point.

dispersions are mostly correct. The conduction band minima are located at the M point, and thus the number of conduction band minima in the first Brillouin zone (M_c) is 3. The indirect transition in SiC leads to a relatively long carrier lifetime, which is beneficial for developing bipolar devices.

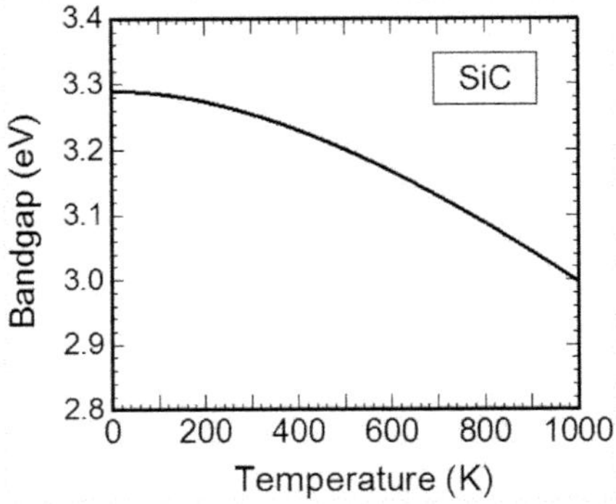

Figure 2.3 Temperature dependence of the bandgap for SiC.

The bandgap (E_g) of SiC is 3.26 eV at room temperature and it decreases at elevated temperature because of thermal expansion. The temperature dependence of the bandgap for SiC can be semiempirically expressed as the following equation (Fig. 2.3): [3,13]

$$E_g(T) = E_{g0} - \frac{\alpha T^2}{T + \beta}, \tag{2.1}$$

where E_{g0} is the bandgap at 0K (3.292 eV), T is the absolute temperature, and α and β are fitting parameters ($\alpha = 8.2 \times 10^{-4}$ eV K^{-1}, $\beta = 1.8 \times 10^3$K). The bandgap narrowing caused by heavy impurity doping becomes significant when the doping density exceeds 10^{19} cm^3, and the detail has been reported in literature [14].

2.2 Electrical properties

2.2.1 Impurity doping and carrier density

Nitrogen or phosphorus are employed for n-type doping and aluminum for p-type doping. Nitrogen substitutes at the C sublattice site, while phosphorus, aluminum, and boron substitute at the Si sublattice site. Table 2.2 shows the ionization energies and the solubility limits of nitrogen, phosphorus, and aluminum in SiC [3,15,16]. The ionization energies of the donors are relatively small (though larger than those in Si), and the ionization ratio of donors in lightly doped SiC is reasonably high, ranging from 80% to nearly 100% at room temperature. In contrast, the ionization energy of aluminum is large (200 meV), leading to incomplete ionization (10%–60%, depending on the doping density) at room temperature. The ionization energy decreases when the doping density is increased, and this decrease becomes very sharp in the range of 10^{19}–10^{20} cm^{-3}. In spite of the relatively large ionization energy of aluminum, nearly-perfect ionization is observed in heavily aluminum-doped SiC ($> 2 \times 10^{20}$ cm^{-3}) [17].

Table 2.2 Ionization energies and solubility limits of nitrogen, phosphorus, and aluminum in SiC [3,15,16]

	Nitrogen	Phosphorus	Aluminum
Ionization energy (meV) (hexagonal/cubic sites)	61/126	60/120	198/201
Solubility limit (cm^{-3})	2×10^{20}	$(\sim 1 \times 10^{21})$	1×10^{21}

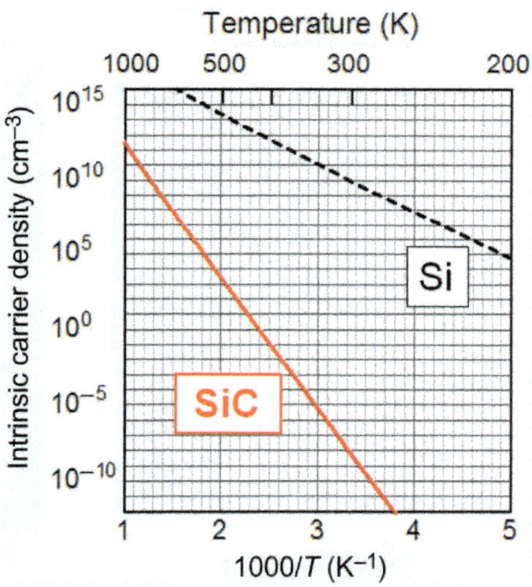

Figure 2.4 Temperature dependence of the intrinsic carrier density for SiC and Si. The intrinsic carrier density of SiC is extremely low (5×10^{-9} cm^{-3} at 300K) because of the wide bandgap, whereas the density is about 1×10^{10} cm^{-3} in Si.

Using the density-of-state effective mass and the number of conduction band minima, the effective densities of states in the conduction band (N_C) and valence band (N_V) are determined to be 1.8×10^{19} cm^{-3} and 7.6×10^{18} cm^{-3}, respectively, at 300K. The temperature dependence of the intrinsic carrier density for SiC and Si is plotted in Fig. 2.4. The intrinsic carrier density of SiC is extremely low because of the wide bandgap, being 5×10^{-9} cm^{-3} at 300K, whereas the density is about 1×10^{10} cm^{-3} in Si. This extremely low intrinsic carrier density enables high-temperature operation of SiC electronic devices. Furthermore, the leakage current caused by carrier generation via crystalline defects (both point and extended defects) is very low in SiC, indicating superior tolerance of SiC devices against defects.

In general, the position of the Fermi level E_F in nondegenerate semiconductors is calculated by [18]:

$$E_F = E_C - kT \ln\left(\frac{N_C}{n}\right), \tag{2.2}$$

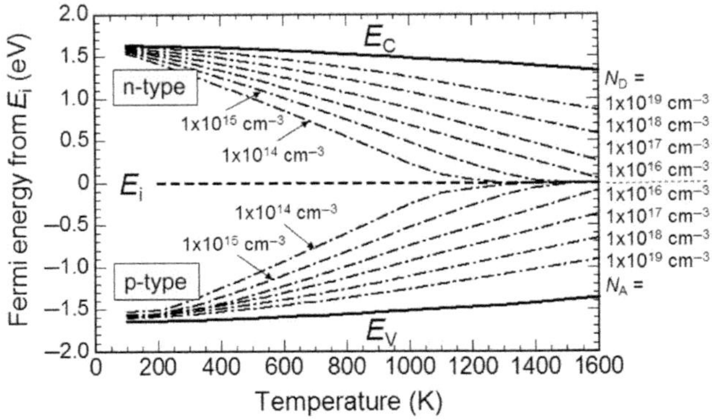

Figure 2.5 Fermi level for nitrogen- or aluminum-doped SiC as a function of temperature and impurity concentration, taking into account the temperature dependence of the bandgap and the incomplete ionization of dopants at low temperature.

$$E_F = E_V + kT \ln\left(\frac{N_V}{p}\right). \tag{2.3}$$

Here E_C (E_V) is the energy of the conduction (valence) band edge. Fig. 2.5 shows the Fermi level for nitrogen- or aluminum-doped SiC as a function of temperature and impurity concentration, taking into account the temperature dependence of the bandgap and the incomplete ionization of dopants at low temperature. Because of the wide bandgap, the Fermi level does not approach the intrinsic level even at a high temperature of 800–1000K, keeping the n- or p-type conductivity.

2.2.2 Mobility

Fig. 2.6 shows the low-field electron and hole mobilities (mobilities perpendicular to the *c*-axis) versus doping density for SiC at room temperature. The low-field mobilities can be expressed by Caughey–Thomas equations as follows [3]:

$$\mu_e = \frac{1020}{1 + \left(\frac{N_D + N_A}{1.8 \times 10^{17}}\right)^{0.6}} \text{cm}^2/\text{Vs}, \tag{2.4}$$

$$\mu_h = \frac{118}{1 + \left(\frac{N_D + N_A}{2.2 \times 10^{18}}\right)^{0.7}} \text{cm}^2/\text{Vs}. \tag{2.5}$$

Here N_D and N_A are given in the unit of cm^{-3}. Note that SiC (4H-SiC) exhibits small anisotropy in mobility [8]. The electron mobility along the *c*-axis direction is about 20% higher than that perpendicular to the *c*-axis, and thus the electron

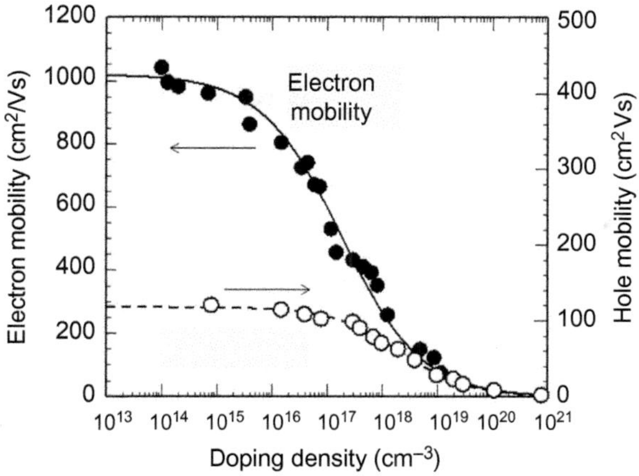

Figure 2.6 Low-field electron and hole mobilities (mobilities perpendicular to the *c*-axis) versus doping density for SiC at room temperature.

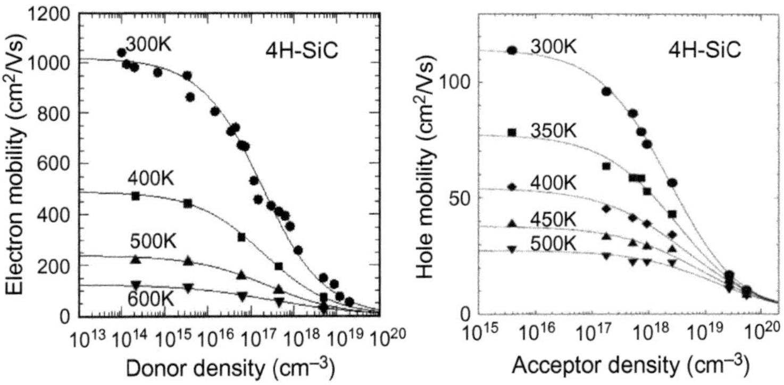

Figure 2.7 Doping density dependence of (A) electron mobility and (B) hole mobility for SiC at different temperatures [3,20,21].

mobility in high-purity SiC is approximately 1200 cm^2/Vs at room temperature. On the other hand, the hole mobility along the *c*-axis is about 15% lower than that perpendicular to the *c*-axis [19].

Fig. 2.7 depicts the doping density dependence of (A) electron mobility and (B) hole mobility for SiC at different temperatures [3,20,21]. The doping dependence of mobility becomes small at elevated temperature, because the contribution of impurity scattering becomes less dominant. In general, the temperature dependence of mobility is discussed by using a relationship of $\mu \sim T^{-n}$, where μ is the mobility and T is the absolute temperature. The n value is strongly dependent on the doping density, because the dominant scattering mechanism is changed for SiC with different doping density. For example, the n value is 2.6 for lightly doped and 1.5 for heavily doped n-type SiC. In lightly doped n-type SiC, the electron mobility is mainly determined by acoustic phonon scattering at low temperature (70–200K)

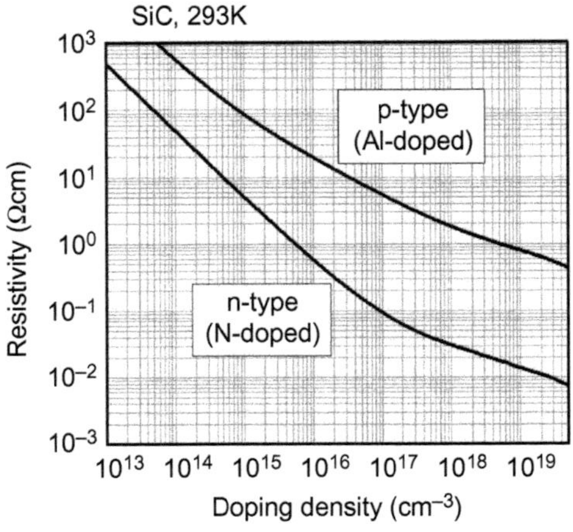

Figure 2.8 Resistivity of nitrogen- or aluminum-doped SiC epitaxial layers at 300K as a function of the doping density.

and by intervalley scattering at temperatures higher than 300K. In lightly doped p-type SiC, the hole mobility is mainly determined by acoustic phonon scattering at or below room temperature, and by nonpolar optical phonon scattering at high temperature (>400K).

The resistivity of nitrogen- or aluminum-doped SiC epitaxial layers at 300K is plotted as a function of the doping density in Fig. 2.8. In very heavily doped materials, the resistivity decreases to 0.003 Ω cm for n-type and 0.016 Ω cm for p-type. Substrates grown by sublimation (or other techniques) exhibit higher resistivities than those shown in Fig. 2.8 because of a higher density of unwanted impurities and point defects.

2.2.3 Drift velocity

There have been only limited investigations on drift velocity in SiC. Fig. 2.9 shows the drift velocity of electrons versus the applied electric field for n-type SiC experimentally obtained [22]. A low-field mobility of 450 cm^2/Vs was determined from the slope at low electric fields ($<10^4$ V/cm) at room temperature, being in good agreement with the data shown in Fig. 2.6 for the donor density (1×10^{17} cm^{-3}). The saturated drift velocity is estimated as 2.2×10^7 cm/s.

The saturated drift velocity (v_{sat}) is approximately given by [18]:

$$v_{sat} = \sqrt{\frac{8\hbar\omega}{3\pi m*}}, \tag{2.6}$$

where $\hbar\omega$ is the energy of the optical phonon (LO phonon). The value experimentally obtained agrees well with that estimated from Eq. (2.6). Though the drift

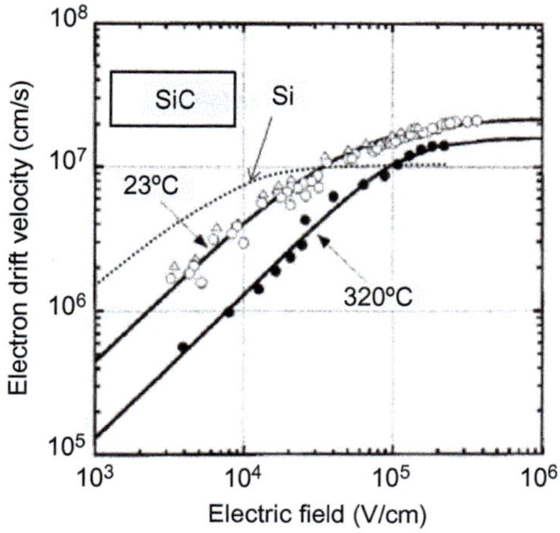

Figure 2.9 Drift velocity of electrons versus the applied electric field strength for n-type SiC experimentally obtained [22].

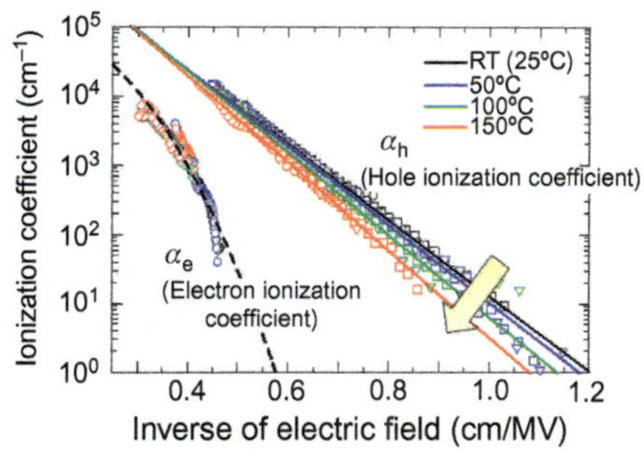

Figure 2.10 Impact ionization coefficients for electrons and holes in SiC along the *c*-axis at different temperatures versus the inverse of electric field strength [24].

velocity of holes in SiC has not been experimentally studied, the saturated value can be estimated as 1.3×10^7 cm/s from Eq. (2.6).

2.2.4 Impact ionization coefficients and critical electric field strength

Avalanche breakdown of a junction is well described by using the impact ionization coefficients of electrons and holes and the profile of the electric field strength [23]. The impact ionization coefficients for electrons (α_e) and holes (α_h) are usually determined by photomultiplication experiments. Fig. 2.10 shows the impact ionization coefficients for electrons and holes in SiC along the *c*-axis at different temperatures versus the inverse of electric field strength [24]. At room temperature, the impact ionization coefficients are approximately expressed by the following equations:

$$\alpha_e = 1.43 \times 10^5 \exp\left\{-\left(\frac{4.93 \times 10^6}{E}\right)^{2.37}\right\} \text{cm}^{-1}, \tag{2.7}$$

$$\alpha_\text{h} = 3.12 \times 10^6 \exp\left\{-\left(\frac{1.18 \times 10^7}{E}\right)^{1.02}\right\} \text{cm}^{-1}. \tag{2.8}$$

Here, E is the electric field in V/cm. The ionization coefficients for SiC are remarkably lower than those for Si at a given electric field strength, due to the wide bandgap. It should be noted that the ionization coefficient for holes is much larger than that for electrons ($\alpha_\text{h} > \alpha_\text{e}$) in SiC, which is completely opposite to the case of Si ($\alpha_\text{e} > \alpha_\text{h}$). As shown in Fig. 2.10, the hole impact ionization coefficient decreases at elevated temperature because of enhanced phonon scattering, which is naturally expected. In contrast, the electron impact ionization coefficient is almost independent of the temperature. This unusual temperature dependence and the smaller ionization coefficient of electrons than that of holes may originate from the peculiar conduction band structure of SiC [25].

The critical electric field strength (or breakdown electric field strength) E_B can be determined by calculation of the ionization integral using the impact ionization coefficients described above. Fig. 2.11 depicts the critical electric field strength versus doping density for SiC⟨0001⟩ [24]. The data for Si at room temperature are also shown for comparison. SiC exhibits about eight times higher critical electric field strength than Si at a given doping density. Note that the critical field strength is dependent on the doping density in any semiconductor materials. This should be considered when one estimates a breakdown voltage or compares the critical

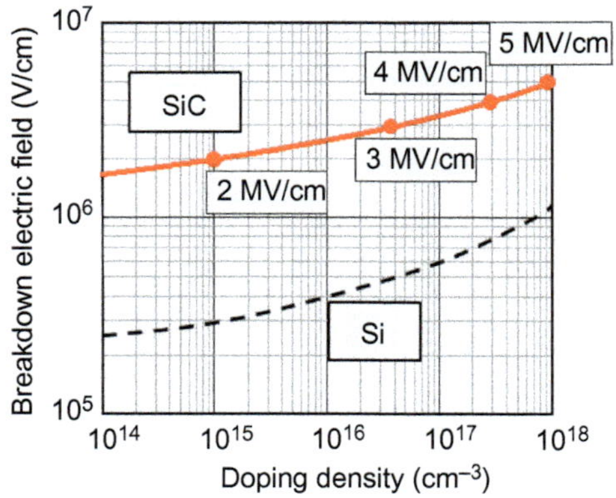

Figure 2.11 Critical electric field strength versus doping density for SiC⟨0001⟩ [24]. The data for Si at room temperature are also shown for comparison.

electric field strength of different materials. One must be also aware that the critical field strength is valid only for junctions with nonpunchthrough structures. The critical field strength slightly increases at elevated temperature, especially in the low-doping range, and can be approximately expressed by the following equation [24]:

$$E_B = \frac{2.653 - 1.166 \times 10^{-3}T + 2.222 \times 10^{-6}T^2}{1 - \frac{1}{4}\log_{10}\left(\frac{N_D}{10^{16}}\right)} \text{MV/cm.} \tag{2.9}$$

Here, T is the temperature in Kelvin, N_D is the reduced doping density in cm^{-3}. Since the carrier acceleration and scattering are strongly influenced by the energy band structure, the impact ionization coefficients and thereby the critical field strength depend on the crystallographic orientation. In fact, SiC (4H-SiC) exhibits some anisotropy in impact ionization and breakdown characteristics [9,26]. The critical field strength of SiC$\langle 11\bar{2}0\rangle$ is about 20%–25% lower than that of SiC$\langle 0001\rangle$.

2.3 Other physical properties

The optical absorption coefficient versus wavelength for SiC is shown in Fig. 2.12 [27]. Because of the indirect band structure, the absorption coefficient gradually increases, when the photon energy exceeds the bandgap. The absorption coefficient of SiC at room temperature is 69 cm^{-1} at 365 nm (3.397 eV, Hg emission line), 1350 cm^{-1} at 325 nm (3.815 eV, He-Cd laser), and 14200 cm^{-1} at 244 nm (5.082 eV, 2HG Ar ion laser). Therefore, the penetration depth, as defined by the

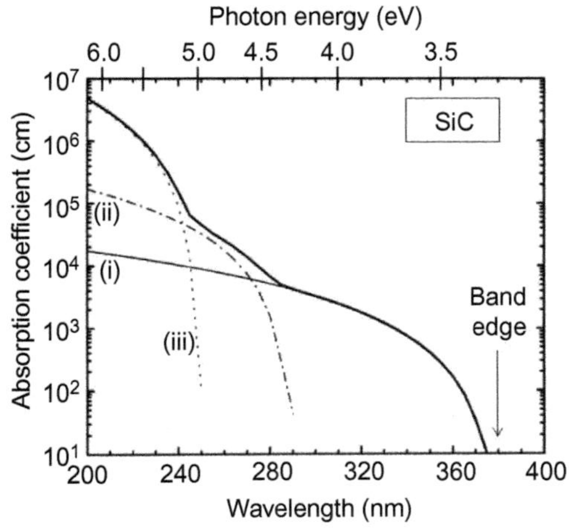

Figure 2.12 Optical absorption coefficient versus wavelength for SiC.

inverse of absorption coefficient, is 145 μm at 365 nm, 7.4 μm at 325 nm, and 0.7 μm at 244 nm.

The relative dielectric constants in the high-frequency (100 kHz–1 MHz) region for SiC at room temperature are 9.76 perpendicular to the *c*-axis and 10.32 parallel to the *c*-axis [28]. The refractive index at a wavelength of 600 nm is 2.64.

High thermal conductivity is an important feature of SiC, in which contribution of phonons is significant. Fig. 2.13 depicts the temperature dependence of thermal conductivity for SiC [29]. The thermal conductivity depends on the doping density and the crystal direction, and it is 4.9 W/cm K for high-purity SiC at room temperature [29,30]. In heavily nitrogen-doped 4H-SiC, which is usually employed as n^+-substrates for vertical power devices, the thermal conductivity along ⟨0001⟩ is about 4.1 W/cm K at room temperature.

The phonon energies in SiC are high owing to the large bonding energy. The energies of major phonons which create phonon replicas in luminescence from SiC are 36 (TA), 46, 51, 77 (LA), 95, 96 (TO), 104, and 107 meV (LO) [31].

The major physical properties of SiC are summarized in Table 2.3.

2.4 Defects and carrier lifetimes

2.4.1 Extended defects

Commercial SiC wafers contain various types of extended defects [dislocations and stacking faults (SFs)], though the defect density has remarkably been reduced in the last decade. Most extended defects in a homoepitaxial SiC layer, which works as an active layer in any SiC power devices, are replicated from the underlying substrate (bulk wafer), but some types of extended defects are generated during the epitaxial

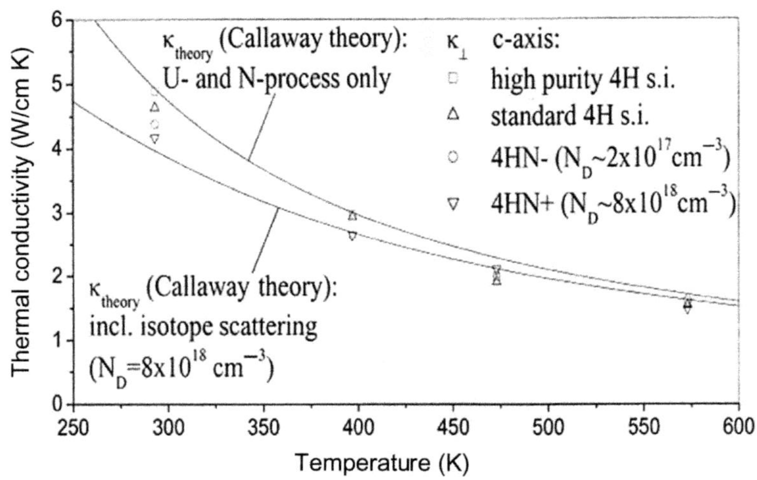

Figure 2.13 Temperature dependence of thermal conductivity for SiC [29].

Table 2.3 Major physical properties of SiC (4H-SiC)

Properties	**SiC (4H)**
Bandgap (eV)	3.26
Electron mobility (cm^2/Vs)	
μ perpendicular to c-axis	1020
μ parallel to c-axis	**1200**
Hole mobility (cm^2/Vs)	120
Electron saturated drift velocity (cm/s)	2.2×10^7
Hole saturated drift velocity (cm/s)	$(\sim 1.3 \times 10^7)$
Breakdown electric field (MV/cm)	
E_B perpendicular to c-axis	2.0
E_B parallel to c-axis	**2.5**
Relative dielectric constant	
ε_s perpendicular to c-axis	9.76
ε_s parallel to c-axis	10.32

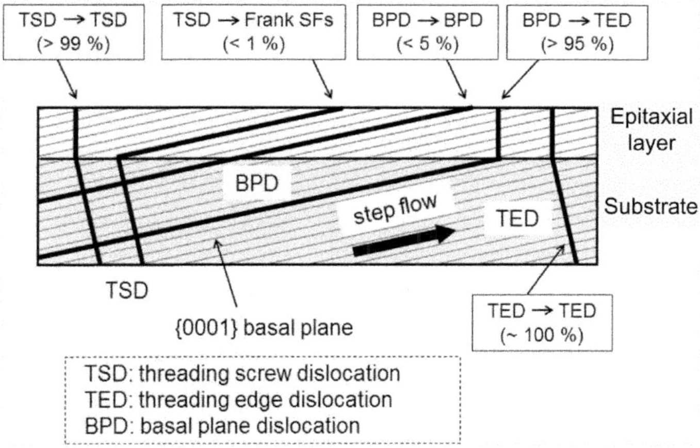

Figure 2.14 Schematic illustration of replication and conversion of various dislocations typically observed in SiC epitaxial layers grown on off-axis {0001} by chemical vapor deposition.

growth process [32]. Major dislocations present in SiC are threading screw dislocations (TSDs), threading edge dislocations (TEDs), and basal plane dislocations (BPDs), the detail of which is summarized in ref. [3]. Micropipe (MP) defects have been mostly eliminated (density $< 0.1\ cm^{-2}$) and are no longer a primary concern.

Fig. 2.14 schematically illustrates replication and conversion of various dislocations typically observed in SiC epitaxial layers grown on off-axis {0001} by chemical vapor deposition [32,33]. Almost all the TSDs and TEDs in a substrate are replicated in an epitaxial layer. Behavior of BPDs during epitaxial growth is more complicated. Most BPDs in a substrate are converted to TEDs in the initial stage of epitaxial growth [34], because this conversion results in a substantial decrease of the dislocation length in an epitaxial layer, leading to reduction of total dislocation energy. The BPD–TED conversion ratio is usually high (>95%) but the ratio depends on the epitaxial growth condition and process (including the in situ etching). The ratio has been improved to 99.9% or higher in recent years and the density of BPDs replicated from a substrate is currently 0.01–0.2 cm^{-2}.

Although it is hard to observe SFs in SiC wafers, SFs may nucleate during epitaxial growth (especially in the initial stage), which are referred to "in-grown SFs" (IGSFs) [35,36]. Several IGSFs having different structures (stacking sequences) have been identified, but the nucleation mechanism is still an open question. The IGSF density of high-quality SiC epitaxial layers is below 0.1 cm^{-2}.

In addition to common dislocations and SFs, SiC epitaxial layers contain several types of surface morphological defects, most of which accompany some sort of extended defect(s). These morphological defects are readily discernable by optical microscopy. Such macroscopic defects include so-called "carrot" defects, "comet" defects, "triangular" defect, and "downfalls" [3]. The density of these defects is also dependent on the epitaxial process and the substrate quality, and it is typically 0.1–0.8 cm^{-2}.

Table 2.4 shows the impacts of major extended defects on SiC devices [33]. Though it was believed that threading dislocations (especially TSDs) severely degrade the blocking performance of SiC devices in the early stage, it turned out that threading dislocations in the state-of-the-art SiC wafers are not killing defects when the devices are fabricated by an adequate process [37]. It is crucially important to suppress pit formation at the location where a dislocation appears on the surface. The superior tolerance of SiC devices against dislocations is most likely due to the extremely low intrinsic carrier density, because any carrier-generation current is proportional to the intrinsic carrier density [18]. Since the dislocation density of the state-of-the-art SiC wafers is relatively low (~3000 cm^{-2}), the contribution of dislocation-induced leakage is almost hidden by other leakage components. The failure of a gate oxide evaluated by time-dependent dielectric breakdown tests is also not influenced much by dislocations as far as the surface pits are not formed [38].

A BPD is a detrimental defect for any kinds of SiC bipolar devices, because a BPD can work as the source of a single Shockley-type SF upon carrier injection [39], and such an SF induces local reduction of carrier lifetimes (increase of on-resistance) and increase in leakage current. This is called "bipolar degradation" and has been a severe obstacle for developing highly reliable SiC bipolar devices. Although bipolar degradation is not induced during on-state and off-state operation of SiC power metal-oxide-semiconductor field effect transistors (MOSFETs), current conduction through a body diode embedded in the power MOSFET can also initiate the degradation phenomenon. In recent years, this degradation has substantially been suppressed by combination of enhanced BPD–TED conversion during epitaxial growth and introduction of a "recombination-enhancing layer" between

Table 2.4 Current understanding on impacts of major extended defects on SiC devices [33]

Defects/devices	SBD	MOSFET, JFET	PiN, BJT, Thyristor, IGBT
TSD (without pit)	No	No	No, but causes local reduction of carrier lifetime
TED (without pit)	No	No	No, but causes local reduction of carrier lifetime
BPD (including interface dislocation, half-loop array)	No	No[a], but can cause degradation of body diode	Bipolar degradation (increase of on-resistance and leakage current)
In-grown SF	V_B[b] reduction (20%–50%)	V_B reduction (20%–50%)	V_B reduction (20%–50%)
Carrot, triangular defects	V_B reduction (30%–70%)	V_B reduction (30%–70%)	V_B reduction (30%–70%)
Downfall	V_B reduction (50%–90%)	V_B reduction (50%–90%)	V_B reduction (50%–90%)

TSD: Threading Screw Dislocation, TED: Threading Edge Dislocation,
BPD: Basal Plane Dislocation, SF: Stacking Fault.
[a]"No" means "negligibly small"
[b]"VB" means "breakdown voltage"

the voltage-blocking layer and the substrate [40]. In the case of power MOSFETs, embedding a Schottky barrier diode in the body diode yields a very promising result toward complete elimination of bipolar degradation in SiC power MOSFETs [41].

As summarized in Table 2.4, BPDs, IGSFs, and macroscopic (morphological) defects have been identified as device-killing defects in SiC devices. The total density of these killing defects is typically 0.2–0.8 cm^{-2}, which determines a maximum chip area of SiC power devices. In order to detect and identify these defects in large-diameter SiC wafers in a nondestructive manner, a photoluminescence (PL) mapping or imaging technique has been developed [42,43]. Since each dislocation or SF in SiC epitaxial layers yields PL at a peculiar wavelength, the location and types of individual defects can be easily identified by PL mapping/imaging with an appropriate band-pass filter. With an advanced measurement system, mapping of major defects in a 150 mm SiC wafer can be completed in 30 min or shorter.

During device processing such as ion implantation followed by activation annealing at high temperature, various extended defects are created. Bombardment of high-energy particles in the case of ion implantation naturally causes formation of extended (and point) defects inside and near the implanted region. Furthermore, when the thermal stress caused by an inhomogeneous temperature distribution (and doping-induced lattice mismatch) exceeds a critical value, dislocation glide is activated. Or formation of a BPD half-loop can be introduced from the surface under a large stress. Thus, stress control becomes more and more stringent as the diameter of SiC wafers increases.

2.4.2 Point defects

In contrast to the extended defects, all the point defects, which generally create deep level(s), in SiC epitaxial layers are independent of the substrate quality and are dominated by the epitaxial growth conditions. The total density of deep levels in lightly doped as-grown SiC epitaxial layers is typically 3×10^{12} to 2×10^{13} cm^{-3}. This value is fairly low as a compound semiconductor and is well acceptable for fabricating SiC unipolar devices, because the density is much lower than the doping density of a drift layer. For bipolar device applications, a long carrier lifetime is required, which can be achieved by reduction of deep levels (point defects). Of course, additional point defects are created during device processing such as ion implantation and dry etching. For example, a high density of point defects (mainly intrinsic type defects) is generated inside the implanted region as well as the implant-tail region [44]. Most of these generated point defects are persistent and survive after activation annealing at high temperature.

The energy levels of major deep levels observed in as-grown n-type and p-type SiC epitaxial layers are shown in Fig. 2.15. Among these levels, the $Z_{1/2}$ (E_c−0.63 eV) [45] and EH6/7 (E_c−1.55 eV) [46] centers are commonly observed with a highest density in all as-grown epitaxial layers. Both the deep levels are very stable against high-temperature (~1700°C) annealing. In the lower half of the bandgap, several deep levels such as HK4 (E_v + 1.44 eV) are observed, but these

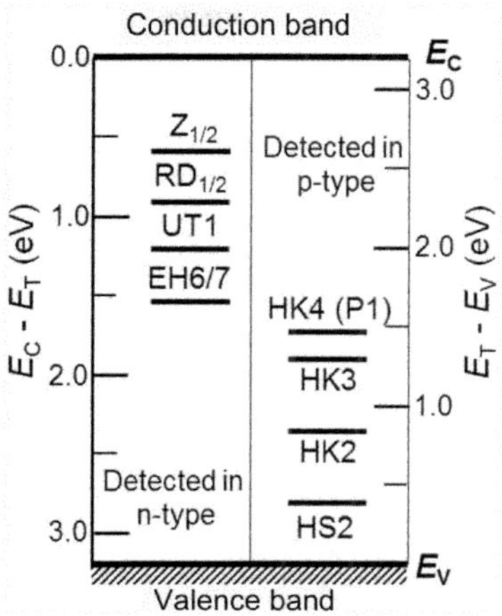

Figure 2.15 Energy levels of major deep levels observed in as-grown n-type and p-type SiC epitaxial layers.

deep levels disappear after annealing at 1450–1550°C. As described in the next subsection, the $Z_{1/2}$ center is the dominant carrier-lifetime killer, being the most important deep level in SiC.

The $Z_{1/2}$ center and the EH6/7 center are observed with almost identical density in all the SiC samples, and recently the origin of these centers was identified as a carbon monovacancy (V_C) with different charge states [47]: The $Z_{1/2}$ and EH6/7 centers correspond to the acceptor-type (negatively charged) and donor-type (positively charged) levels, respectively. The densities of $Z_{1/2}$ and EH6/7 centers can be reduced when epitaxial growth is performed under a C-rich condition, as expected.

Apart from these intrinsic point defects, a few impurity-related levels are often observed in SiC epitaxial layers. Boron (B) can be unintentionally incorporated, which results in formation of the boron acceptor level (E_v + 0.28–0.35 eV) and the boron-related "D center" (E_v + 0.55 eV) [48]. Another common impurity is titanium (Ti), which creates very shallow electron traps (E_c–0.11/0.17 eV) and probably other deep levels in SiC. The typical B and Ti densities are approximately $(1-20) \times 10^{12}$ cm^{-3}. The density of other metallic impurities such as iron and vanadium is usually very low.

2.4.3 Carrier lifetimes

A carrier lifetime is a key physical property, which determines the on-state and switching characteristics of bipolar devices, though it does not matter in unipolar devices such as power MOSFETs. The carrier lifetimes in SiC are usually characterized by time-resolved photoluminescence or microwave-detected photoconductance decay (μ-PCD) measurements. The carrier lifetime killer in SiC has been identified as the $Z_{1/2}$ center (acceptor-type levels of carbon vacancy) [49]. Although several other deep levels are present in SiC, the impacts of those levels other than the $Z_{1/2}$ center on the carrier lifetimes are very small. Fig. 2.16 depicts the inverse of the measured carrier lifetime versus the density of the carbon vacancy defects (half of the $Z_{1/2}$ center density) for 50 μm or 230 μm thick lightly doped n-type SiC epitaxial layers. The inverse of the carrier lifetime is proportional to the carbon vacancy density, when the carbon vacancy density is relatively high (higher than about 1×10^{13} cm^{-3} in the case of 50 μm-thick epitaxial layers). However, the carrier lifetime is governed by other recombination paths when the carbon vacancy density is relatively low. Because of relatively long carrier lifetimes in lightly doped SiC epitaxial layers, the diffusion length of carriers can reach 30−100 μm. Note that the measured carrier lifetimes are severely affected (underestimated) by surface recombination and recombination in the underlying substrate (or recombination near the interface between the epitaxial layer and substrate). This is one of the major reasons why the linear relationship is lost when the carbon vacancy density is low in Fig. 2.16.

Enhancement of carrier lifetimes can be achieved by reduction of carbon vacancy defects. The carbon vacancy density in as-grown epitaxial layers decreases when the growth is performed under carbon-rich condition. As a postgrowth process for enhancement of carrier lifetimes, a few successful approaches have been established. In the first technique, excess carbon is introduced by carbon implantation and diffusion of carbon interstitials is promoted by post-implantation annealing [50]. In the second approach, thermal oxidation at high temperature (1300−1400°C) is effective for carbon vacancy reduction, because some carbon atoms are emitted into the SiC bulk side and diffuse toward a deep region during oxidation [51]. By these techniques, the carbon vacancy defects can be almost eliminated (the density: below 3×10^{10} cm^{-3}) from the surface to very deep (100−200 μm) region. It has been reported that annealing with a carbon cap in Ar or annealing in a hydrocarbon ambient is also effective for reduction of carbon vacancy defects in SiC. The carrier lifetime of commercial n-type SiC epitaxial

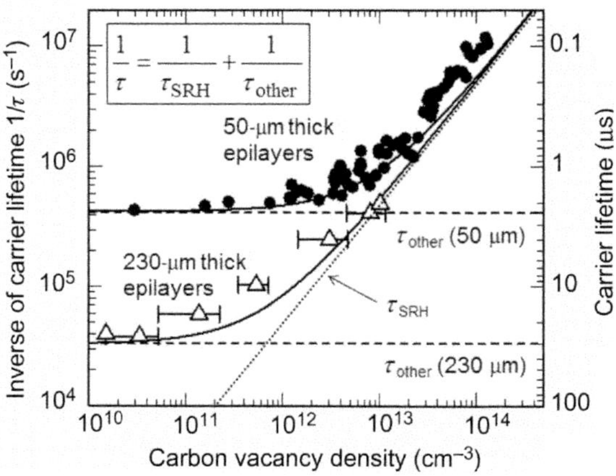

Figure 2.16 Inverse of the measured carrier lifetime versus the density of the carbon vacancy defects for 50 or 230 μm thick lightly doped n-type SiC epitaxial layers.

wafers is about 1–2 μs at room temperature, and it exhibits a gradual increase with elevating the temperature. The carrier lifetime can be increased to 30 μs or longer by the carbon-vacancy-reduction process [52]. Although the carrier lifetimes in p-type SiC have been less systematically studied so far, the lifetimes can substantially be enhanced by reduction of carbon vacancy defects as well. Defect passivation by annealing in H_2 at 800–1000°C is also effective for p-type SiC, leading to a long lifetime over 10 μs [53].

When the doping density or the excess carrier density becomes high, band-to-band recombination and Auger recombination are involved, limiting the bulk carrier lifetime. Recombination of excess carriers in an n-type semiconductor is governed by the following equation [54].

$$\begin{aligned}\frac{d\Delta n}{\mathrm{d}t} &= -\frac{\Delta n}{\tau_{\mathrm{SRH}}} - B(n_0 + p_0 + \Delta n)\Delta n - C_{\mathrm{n}}({n_0}^2 + 2n_0\Delta n + \Delta n^2)\Delta n \\ &\quad - C_{\mathrm{p}}({p_0}^2 + 2p_0\Delta n + \Delta n^2)\Delta n. \end{aligned} \tag{2.10}$$

Here, Δn, n_0, and p_0 are the excess carrier density, the equilibrium electron density, and the equilibrium hole density, respectively. SRH recombination (τ_{SRH}), band-to-band (direct) recombination (coefficient: B), and Auger recombination (coefficients: C_{n}, C_{p}) are considered. By using the band-to-band and Auger recombination coefficients (B, C_{n}, C_{p}) reported in literature [55], the decay of the excess carrier density can be calculated and the effective carrier lifetime is extracted. Fig. 2.17 shows the effective carrier lifetime as a function of the excess carrier

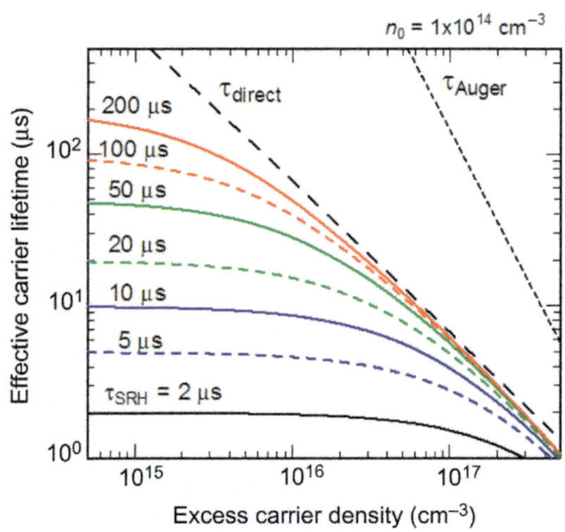

Figure 2.17 Effective carrier lifetime as a function of the excess carrier density calculated by using Eq. (2.10). In this particular case, the equilibrium electron density (n_0) is 1×10^{14} cm^{-3}, and the results for SRH lifetimes varied from 2 to 200 μs are shown. In the figure, the lifetimes limited purely by band-to-band and Auger recombinations are indicated by *dashed* and *dotted lines*, respectively.

density calculated by using Eq. (2.10). In this particular case, the equilibrium electron density (n_0) was assumed to be 1×10^{14} cm^{-3}, and the results for SRH lifetimes varied from 2 to 200 μs are shown. In the figure, the lifetimes limited purely by band-to-band and Auger recombinations are indicated by *dashed* and *dotted lines*, respectively.

As shown in Fig. 2.17, the carrier lifetimes are limited by Auger recombination when the excess carrier density is very high ($>10^{18}$ cm^{-3}), and a similar trend is observed in heavily doped SiC. The band-to-band recombination affects the effective carrier lifetime even at a relatively low excess carrier concentration of 1×10^{16} cm^{-3} when the SRH lifetime is long ($>$50 μs). In such a case, the effective carrier lifetime is mainly limited not by SRH recombination via defects but by band-to-band recombination. Since the excess carrier density in a voltage-blocking layer of bipolar devices exceeds 1×10^{16} cm^{-3} at high current density, the lifetimes limited by band-to-band recombination can be one intrinsic performance-limiting factor in high-voltage SiC bipolar devices [56].

References

[1] G.L. Harris, Properties of Silicon Carbide, INSPEC, London, 1995.

[2] W.J. Choyke, H. Matsunami, G. Pensl (Eds.), Silicon Carbide: A Review of Fundamental Questions and Applications to Current Device Technology, Vols. I & II, Akademie Verlag, Berlin, 1997.

[3] T. Kimoto, J.A. Cooper, Fundamentals of Silicon Carbide Technology, John Wiley & Sons, Singapore, 2014.

[4] A.R. Verma, P. Krishna (Eds.), Polymorphism and Polytypism in Crystals, John Wiley & Sons, Inc, New York, 1966.

[5] B.J. Baliga, Silicon Carbide Power Devices, World Scientific, Singapore, 2006.

[6] A. Itoh, H. Akita, T. Kimoto, H. Matsunami, High-quality 4H-SiC homoepitaxial layers grown by step-controlled epitaxy, Appl. Phys. Lett. 65 (1994) 1400.

[7] A.O. Konstantinov, Q. Wahab, N. Nordell, U. Lindefelt, Ionization rates and critical fields in 4H silicon carbide, Appl. Phys. Lett. 71 (1997) 90.

[8] W.J. Schaffer, G.H. Negley, K.G. Irvin, J.W. Palmour, Conductivity anisotropy in epitaxial 6H and 4H SiC, Mater. Res. Soc. Symp. Proc. 339 (1994) 595.

[9] S. Nakamura, H. Kumagai, T. Kimoto, H. Matsunami, Anisotropy in breakdown field of 4H-SiC, Appl. Phys. Lett. 80 (2002) 3355.

[10] H. Matsunami, T. Kimoto, Step-controlled epitaxial growth of SiC: high quality homoepitaxy, Mater. Sci. Eng., R 20 (1997) 125.

[11] M. Stockmeier, R. Muller, S.A. Sakwe, P.J. Wellmann, A. Magerl, On the lattice parameters of silicon carbide, J. Appl. Phys. 105 (2009) 033511.

[12] C. Persson, U. Lindefelt, Relativistic band structure calculation of cubic and hexagonal SiC polytypes, J. Appl. Phys. 82 (1997) 5496.

[13] W.J. Choyke, Optical properties of polytypes of SiC: interband absorption, and luminescence of nitrogen-exciton complexes, Mater. Res. Bull. 4 (1969) 141.

[14] C. Persson, U. Lindefelt, B.E. Sernelius, Band gap narrowing in n-type and p-type 3C-, 2H-, 4H-, 6H-SiC, and Si, J. Appl. Phys. 86 (1999) 4419.

[15] Yu. A. Vodakov, E.N. Mokhov, Diffusion and solubility of impurities in silicon carbide, in: R.C. Marshall, J.W. Faust Jr., C.E. Ryan (Eds.), Silicon Carbide, University of South Carolina Press, Columbia, 1974, p. 508.
[16] I.G. Ivanov, A. Henry, E. Janzén, Ionization energies of phosphorus and nitrogen donors and aluminum acceptors in 4H silicon carbide from the donor−acceptor pair emission, Phys. Rev. B 71 (2005) 241201.
[17] S. Ji, K. Kojima, Y. Ishida, S. Saito, T. Kato, H. Tsuchida, et al., The growth of low resistivity, heavily Al-doped 4H−SiC thick epilayers by hot-wall chemical vapor deposition, J. Crystal Growth 380 (2013) 85.
[18] S.M. Sze, K.K. Ng, Physics of Semiconductor Devices, 3rd ed., Wiley, New York, 2007.
[19] T. Hatakeyama, T. Watanabe, M. Kushibe, K. Kojima, S. Imai, T. Suzuki, et al., Measurement of Hall mobility in 4H-SiC for improvement of the accuracy of the mobility model in device simulation, Mater. Sci. Forum 433−436 (2003) 443.
[20] S. Kagamihara, H. Matsuura, T. Hatakeyama, T. Watanabe, M. Kushibe, T. Shinohe, et al., Parameters required to simulate electric characteristics of SiC devices for n-type 4H-SiC, J. Appl. Phys. 96 (2004) 5601.
[21] A. Koizumi, J. Suda, T. Kimoto, Temperature and doping dependencies of electrical properties in Al-doped 4H-SiC epitaxial layers, J. Appl. Phys. 106 (2009) 013716.
[22] I.A. Khan, J.A. Cooper Jr., Measurement of high-field electron transport in silicon carbide, IEEE Trans. Electron Devices 47 (2000) 269.
[23] B.J. Baliga, Fundamentals of Semiconductor Power Devices, Springer, Berlin, 2008.
[24] H. Niwa, J. Suda, T. Kimoto, Impact ionization coefficients in 4H-SiC toward ultrahigh-voltage power devices, IEEE Trans. Electron Devices 62 (2015) 3326.
[25] A.P. Dmitriev, A.O. Konstantinov, D.P. Litvin, V.I. Sankin, Impact ionization and super-lattice in 6H-SiC, Soviet Phys. Semicond. 17 (1983) 686.
[26] T. Hatakeyama, Measurements of impact ionization coefficients of electrons and holes in 4H-SiC and their application to device simulation, Phys. Status Solidi A 206 (2009) 2284.
[27] H.-Y. Cha, P.M. Sandvik, Electrical and optical modeling of 4H-SiC avalanche photodiodes, Jpn. J. Appl. Phys. 47 (2008) 5423.
[28] S. Ninomiya, S. Adachi, Optical constants of 6H−SiC single crystals, Jpn. J. Appl. Phys. 33 (1994) 2479.
[29] J.R. Jenny, St. G. Müller, A. Powell, V. Tsvetkov, H.M. Hobgood, R.C. Glass, et al., High-purity semi-insulating 4H-SiC grown by the seeded sublimation method, J. Electron. Mater. 31 (2002) 366.
[30] H.M. Hobgood, M. Brady, W. Brixius, G. Fechko, R. Glass, D. Henshall, et al., Status of large diameter SiC crystal growth for electronic and optical applications, Mater. Sci. Forum 338−342 (2000) 3.
[31] W.J. Choyke, Optical and Electronic Properties of SiC, in: R. Freer (Ed.), The Physics and Chemistry of Carbides, Nitrides, and Borides, Kluwer Academic Publishers, Dordrecht, 1990, p. 563.
[32] H. Tsuchida, M. Ito, I. Kamata, M. Nagano, Formation of extended defects in 4H-SiC epitaxial growth and development of a fast growth technique, Phys. Status Solidi B 246 (2009) 1553.
[33] T. Kimoto, Material science and device physics in SiC technology for high-voltage power devices, Jpn. J. Appl. Phys. 54 (2015) 040103.
[34] S. Ha, P. Mieszkowski, M. Skowronski, L.B. Rowland, Dislocation conversion in 4H silicon carbide epitaxy, J. Cryst. Growth 244 (2002) 257.
[35] G. Feng, J. Suda, T. Kimoto, Characterization of major in-grown stacking faults in 4H-SiC epilayers, Phys. B 23−24 (2009) 4745.

[36] I. Kamata, X. Zhang, H. Tsuchida, Photoluminescence of Frank-type defects on the basal plane in 4H-SiC epilayers, Appl. Phys. Lett. 97 (2010) 172107.
[37] H. Fujiwara, H. Naruoka, M. Konishi, K. Hamada, T. Katsuno, T. Ishikawa, et al., Impact of surface morphology above threading dislocations on leakage current in 4H-SiC diodes, Appl. Phys. Lett. 101 (2012) 042104.
[38] J. Senzaki, A. Shimozato, K. Kojima, T. Kato, Y. Tanaka, K. Fukuda, et al., Challenges of high-performance and high-reliablity in SiC MOS structures, Mater. Sci. Forum 717–720 (2012) 703.
[39] M. Skowronski, S. Ha, Degradation of hexagonal silicon-carbide-based bipolar devices, J. Appl. Phys. 99 (2006) 011101.
[40] T. Tawara, T. Miyazawa, M. Ryo, M. Miyazato, T. Fujimoto, K. Takenaka, et al., Short minority carrier lifetimes in highly nitrogen-doped 4H-SiC epilayers for suppression of the stacking fault formation in PiN diodes, J. Appl. Phys. 120 (2016) 115101.
[41] S. Hino, H. Hatta, K. Sadamatsu, Y. Nagahisa, S. Yamamoto, T. Iwamatsu, et al., Demonstration of SiC-MOSFET embedding Schottky barrier diode for inactivation of parasitic body diode, Mater. Sci. Forum 897 (2017) 477.
[42] M. Tajima, M. Tanaka, N. Hoshino, Characterization of SiC epitaxial wafers by photoluminescence under deep UV excitation, Mater. Sci. Forum 389–393 (2002) 597.
[43] R.E. Stahlbush, K.X. Liu, Q. Zhang, J.J. Sumakeris, Whole-wafer mapping of dislocations in 4H-SiC epitaxy, Mater. Sci. Forum 556–557 (2007) 295.
[44] K. Kawahara, G. Alfieri, T. Kimoto, Detection and depth analyses of deep levels generated by ion implantation in n- and p-type 4H-SiC, J. Appl. Phys. 106 (2009) 013719.
[45] T. Dalibor, G. Pensl, H. Matsunami, T. Kimoto, W.J. Choyke, A. Schöner, et al., Deep defect centers in silicon carbide monitored with deep level transient spectroscopy, Phys. Status Solidi A 162 (1997) 199.
[46] C. Hemmingsson, N.T. Son, O. Kordina, J.P. Bergman, E. Janzén, J.L. Lindstrom, et al., Deep level defects in electron-irradiated 4H SiC epitaxial layers, J. Appl. Phys. 81 (1997) 6155.
[47] N.T. Son, X.T. Trinh, L.S. Løvlie, B.G. Svensson, K. Kawahara, J. Suda, et al., Negative-U system of carbon vacancy in 4H-SiC, Phys. Rev. Lett. 109 (2012) 187603.
[48] T. Troffer, M. Schadt, T. Frank, H. Itoh, G. Pensl, J. Heindl, et al., Doping of SiC by implantation of boron and aluminum, Phys. Status Solidi A 162 (1997) 277.
[49] K. Danno, D. Nakamura, T. Kimoto, Investigation of carrier lifetime in 4H-SiC epilayers and lifetime control by electron irradiation, Appl. Phys. Lett. 90 (2007) 202109.
[50] L. Storasta, H. Tsuchida, Reduction of traps and improvement of carrier lifetime in 4H-SiC epilayers by ion implantation, Appl. Phys. Lett. 90 (2007) 062116.
[51] T. Hiyoshi, T. Kimoto, Reduction of deep levels and improvement of carrier lifetime in n-type 4H-SiC by thermal oxidation, Appl. Phys. Express 2 (2009) 041101.
[52] T. Kimoto, K. Kawahara, B. Zippelius, E. Saito, J. Suda, Control of carbon vacancy in SiC toward ultrahigh-voltage power devices, Superlattices Microstruct. 99 (2016) 151.
[53] T. Okuda, T. Miyazawa, H. Tsuchida, T. Kimoto, J. Suda, Carrier lifetimes in lightly-doped p-type 4H-SiC epitaxial layers enhanced by post-growth processes and surface passivation, J. Electron. Mater. 46 (2017) 6411.
[54] D.K. Schroder, Semiconductor Material and Device Characterization, third ed., Wiley-IEEE, New York, 2006.
[55] A. Galeckas, J. Linnros, V. Grivickas, U. Lindefelt, C. Hallin, Auger recombination in 4H-SiC: Unusual temperature behavior, Appl. Phys. Lett. 71 (1997) 3269.
[56] T. Kimoto, K. Yamada, H. Niwa, J. Suda, Promise and challenges of high-voltage SiC bipolar power devices, Energies 9 (2016) 908.

Physical properties of gallium nitride and related III–V nitrides

3

Ishwara Bhat
Electrical Computer and Systems Engineering Department, Rensselaer Polytechnic Institute, Troy, NY, United States

In this chapter, a concise review of the material and transport properties of gallium nitride and other III–V nitrides are given with respect to power device application. Some discussions on the use of different substrates for the epitaxial growth and their advantages are presented. Nature and role of defects in nitrides and some methods to alleviate the effect of defects on GaN on silicon substrate are presented at the end. This is by no means complete. Further information can be found in the references listed.

3.1 Crystal structure and related properties

The most commonly observed crystal structure for GaN, AIN, and InN is wurtzite 2H polytype [1]. In the structure, each group Ill atom is surrounded by four equidistant nearest group V atoms as the corners of a tetrahedron, and vice versa. The wurtzite crystal structure is a member of the hexagonal crystal system and thus has two lattice constants: a (the length of the sides of the base) and c (the height of the unit cell). In an ideal case the ratio of c/a is equal to sqrt(8/3) = 1.633. Each unit cell has six atoms and consists of two interpenetrating hexagonal close-packed lattices. It can be viewed as two interpenetrating hexagonal close-packed lattices, with a basis consisting of a group Ill atom at (0,0,0,0) and a group V atom shifted along c-axis to (0,0,0,0.375) position [2]. Another way to visualize it as an hexagonal close packed structure where each lattice site consists of two atoms, one from group III and one from group V. Each anion is surrounded by four cations at the corners of a tetrahedron, and vice versa. The point group and space group of the wurtzite structure are 6 mm and $P6_3mc$, respectively [3]. Due to the lack of inversion symmetry, which means that (0001) basal planes differ from $(000\bar{1})$ planes, wurtzite structure materials exhibit crystallographic polarity.

All the above nitrides may also crystallize in a metastable zinc-blende 3C structure when a cubic substrate is used [4–7]. The structure is metastable and can convert to wurtzite on high temperature treatment. Zinc-blende structure has a face centered cubic unit cell and consist of two interpenetrating face centered cubic sublattices, one of them shifted ¼ of the distance along the body diagonal. The crystal can also be thought of as a face centered cubic lattice with each atom site consisting

Wide Bandgap Semiconductor Power Devices. DOI: https://doi.org/10.1016/B978-0-08-102306-8.00003-4

of two atoms, a group III and a group V atom. The space group of zinc-blende structure is *F43m* [8].

An important difference between zinc-blende structures and wurtzite structures is the bond angle of the second nearest neighbor. In wurtzite structures, atoms in the first layer and third layer are directly aligned together, which means there is mirror symmetry along the c-axis, but in zinc-blende structures the third layer is rotated by 60 degrees. As a result, the stacking sequence of close-packed planes is different. The wurtzite structure has triangularly arranged alternating biatomic close-packed (0001) planes in ABABAB sequence in $\langle 0001 \rangle$ direction, while the zinc-blende structures have a close-packed (111) plane stacking sequence of ABCABC in the $\langle 111 \rangle$ direction, where the letter A, B, or C represents an anion–cation bond. The possible crystal structures are shown in Fig. 3.1.

The bond length of GaN is 1.94 Å at $T = 300$K and the cohesive energy per bond is 2.24 eV [9]. Due to the strong bonding between Ga and N atoms, GaN is a very hard semiconductor material and thermally stable. The density of GaN is 6.1 g/cm^3 at room temperature [10]. These data are also listed in Table 3.1, together with the information about AlN and InN.

The commonly accepted lattice constant values of wurtzite GaN are $a = 3.189$ Å and $c = 5.185$ Å at room temperature [11], and their temperature dependence is also provided in the same literature. There are wide variation in the data obtained early on since these data were obtained from the heteroepitaxially grown films. More accurate data are made available from bulk GaN single crystal. The lattice constants with better accuracy are $a = 3.1880$ Å and $c = 5.1856$ Å at 294K [12], and these values vary slightly from sample to sample depending on the size, polarity, and the free carrier concentration. The thermal expansion coefficients (TECs) in the temperature ranges of 300–700K are $\Delta a = 5.59 \times 10^{-6}\text{K}^{-1}$ and $\Delta c = 3.17 \times 10^{-6}\text{K}^{-1}$ [11]. The temperature dependence of the lattice constants of GaN bulk crystal is shown in Fig. 3.2 [13]. The temperature dependence of the TECs of GaN within

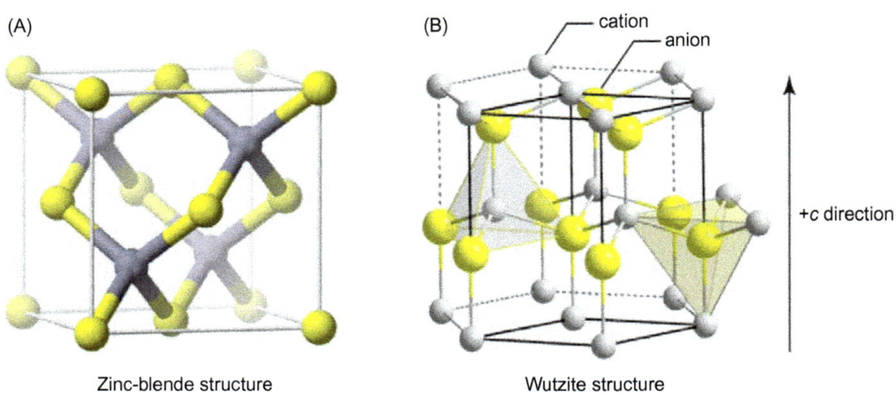

Figure 3.1 Schematic diagram showing (A) zinc-blende and (B) wurtzite structure of GaN.

Table 3.1 Some properties of GaN, AlN, and InN

Property	Wurtzite GaN	Wurtzite AlN	Wurtzite InN
Lattice constant a (Å)	3.1880 ± 0.0001	3.1127 ± 0.0003	3.53–3.548
Lattice constant c (Å)	5.1856 ± 0.0005	34.9816 ± 0.0005	5.69–5.76
TEC of a-direction at 300K (1/K)	3.1×10^{-6}	2.56×10^{-6}	3.1×10^{-6}
TEC of c-direction at 300K (1/K)	2.8×10^{-6}	5.3×10^{-6}	
Density (g/cm^3)	6.1	3.255	6.88
Bond length (Å)	1.94	1.89	2.15
Cohesive energy per bond (eV)	2.24	2.88	1.93
Thermal conductivity κ (W/cm K)	2.3	3.19 (exp)	0.8
Elastic constants (GPa)			
C11	390	345	190 ± 7
C12	145	125	104 ± 3
C13	106	120	121 ± 7
C33	398	395	182 ± 6
C44	105	118	10 ± 1
C66	123	110	–

294 to 753K is shown in Fig. 3.3 [14]. The lattice constants and TECs of AlN and InN are included in Table 3.1 [15–17]. For the III–V nitride ternary alloys like AlGaN and InGaN, usually a linear extrapolation between GaN and AlN (or InN) is assumed.

The Young's modulus of GaN is 150 GPa for (0001) orientation and 196 GPa for (1210) orientation. The corresponding Poisson's ratio for these two orientations are 0.37 and 0.33, respectively [18]. The elastic constants of GaN have also been reported by several authors but the values differ a lot depending on the techniques they used [19,20]. The elastic constants of GaN determined by Polian using Brillouin scattering measurement [20] looks more reliable because the measurement was conducted on a GaN single bulk crystal. These values are: $c11 = 390$ GPa, $c12 = 145$ GPa, $c13 = 106$ GPa, $c33 = 398$ GPa, $C44 = 105$ GPa, and $C66 = 123$ GPa. The elastic constants of AlN and InN are available in refs. [21] and [22] and are also included in Table 3.1.

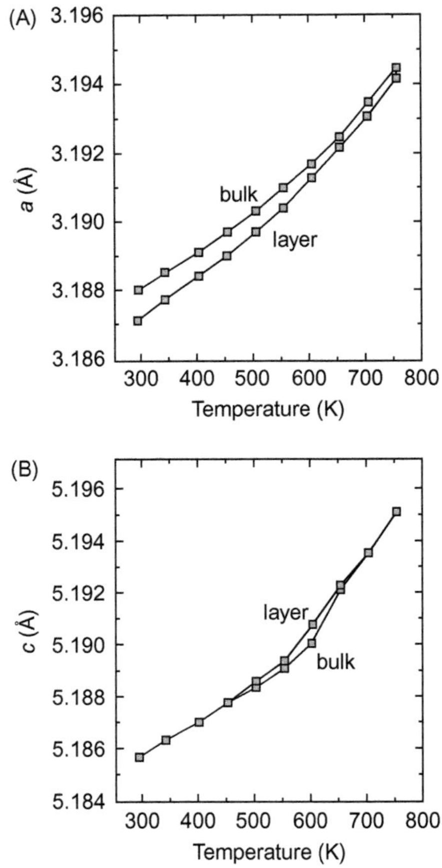

Figure 3.2 Temperature dependence of GaN lattice constants. (A) Lattice constant *a*, parallel to the interface and (B) lattice constant *c*, perpendicular to the interface [13].

GaN is a relatively good thermal conductor and its room temperature thermal conductivity $\kappa = 2.3$ W/cm K [23]. Since the thermal conductivity is very sensitive to the dislocation density and decreases with the increase in dislocation density for density $>10^6\ cm^{-2}$, early reports quote a value of 1.3 W/cm K along *c*-direction [24]. At low temperature of around 30K, this value decreases to 0.4 W/cm K. AlN shows a better thermal conductivity (3.19 W/cm K obtained on single crystal AlN bulk, and the theoretical prediction is 3.2 W/cm K [25]. On the other hand, InN has a thermal conductivity of only 0.8 W/cm K at 300K [26].

GaN begins to dissociate at a much lower temperature than its possible melting temperature. The decomposition process can be described by a reversible reaction as:

$$GaN \leftrightarrow Ga + 1/2\ N_2 \qquad (3.1)$$

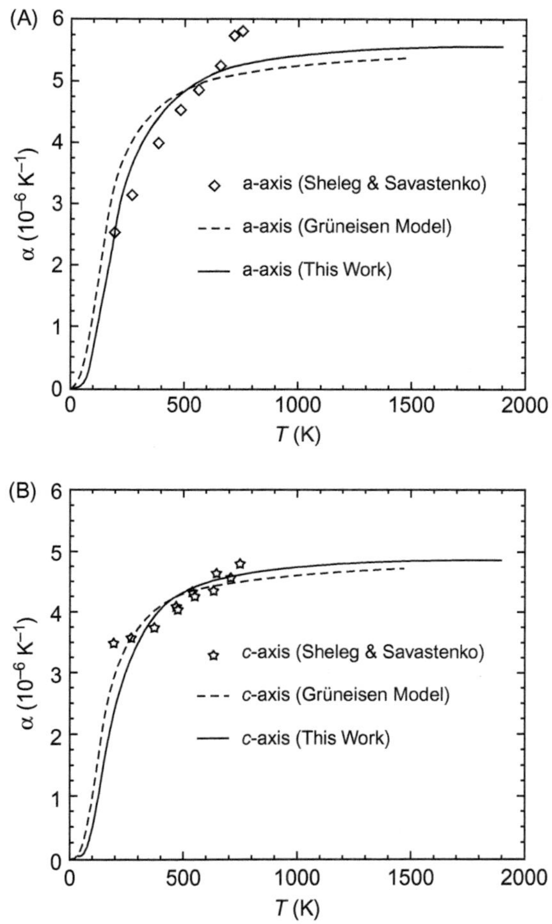

Figure 3.3 Comparison of the coefficient of thermal expansion for (A) *a* and (B) *c* axes [14].

In Fig. 3.4 the equilibrium pressure of N_2 over GaN is shown as the *solid line* [27,28]. Along this curve, GaN, N_2, and Ga can coexist. Below this line, GaN is not stable and only Ga + N_2 can exist in the final equilibrium stage. On the other hand, in the region above this line all Ga will synthesize to GaN in N_2 ambient. From this figure, both the standard heat of formation (ΔH°) and the standard entropy of formation (ΔS°) of GaN can be extracted as −37.7 and −32.43 kcal/mol K, respectively [29]. This graph illustrates the difficulty of growing bulk GaN material from the liquid phase since very high pressure of nitrogen will be needed for the process to be successful.

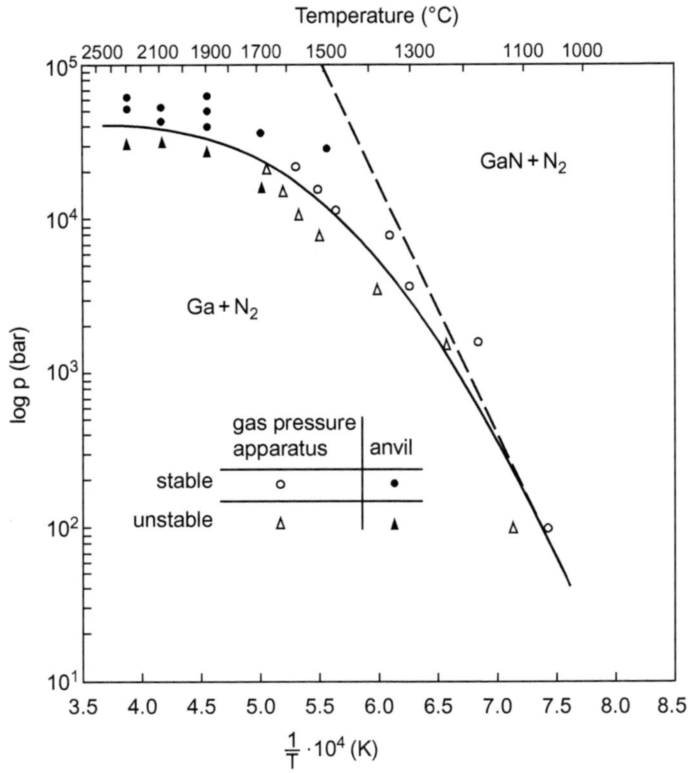

Figure 3.4 Equilibrium pressure of nitrogen over GaN.

3.2 Polarization charges

The bonds in compound semiconductors have some ionic character, which leads to a net electrical dipole between—for example—adjacent gallium and nitrogen atoms in the case of III-nitrides, or gallium and phosphorus atoms in the case of III-phosphides. Zinc-blende GaAs has a Phillips iconicity f_i of 0.310, whereas zinc-blende GaN has f_i of 0.5, indicating a more ionic bond [30,31]. Ionic bonds and the resultant dipoles are present in compound semiconductors but not in elemental semiconductors like Si or Ge, because only in the case of compound semiconductors is there a difference of electronegativity between adjacent atoms.

Unlike the zinc-blende lattice, wurtzite is non-centrosymmetric and lacks inversion symmetry. The centrosymmetric character of zinc-blende means that there is no unique direction in the crystal; for any direction in the crystal, there are other equivalent directions in which the atomic arrangement is identical. The effect of this symmetry is such that each dipole between an adjacent pair of atoms is canceled by the dipoles of other bonded atoms, leaving no net dipole moment. For non-centrosymmetric wurtzite, however, the crystal has a unique direction—the *c*-direction. Dipoles along the *c*-direction are not canceled and along it there is a net dipole moment, and therefore net polarization [32]. The material polarization is determined by the magnitude of dipoles between the adjacent bonded atoms, and is therefore influenced by the material composition (be it AlN, GaN, InN, or any alloy thereof). This is the reason why the actual *c*/*a* ratio of GaN or AlN is not exactly equal to sqrt(8/3) as is ideal for an hexagonal close packed (hcp) structure, but slightly smaller than this value. The polarization will also change with any strain in the layer. Strain acts to distort the lattice and change the interatomic spacing, and therefore also affects the dipole strength. The strain can be present when the layer is grown on lattice mismatched substrate.

Typically III-nitride films are grown epitaxially on the *c*-plane—that is, perpendicular to the *c*-direction. Growth along this direction makes high material quality possible, and is favored in the industry, although growth on other planes—such as the *m*-plane or *a*-plane—is a topic of active research [33,34]. However, due to the material polarization, growth in the *c*-plane has profound consequences which will be discussed later.

In the presence of built-in material polarization, the electric displacement field D is given by

$$D = \varepsilon_o \varepsilon_r E + P_m \tag{3.2}$$

Here, E is the external electric field, P_m is the material polarization field, ε_o is the permittivity of free space, and ε_r is the relative dielectric constant of the material. From Gauss' law we have

$$\Delta.D = \rho \tag{3.3}$$

where ρ is the charge density. Now consider an interface between two materials with different relative dielectric constants ε_1 and ε_2 and material polarization P_1 and P_2. Gauss' law states that there must be a sheet charge σ at the interface with magnitude equal to the difference in electric displacement on the two sides of the interface:

$$\varepsilon_o\, \varepsilon_2\, E_2 + P_2 - (\varepsilon_o\, \varepsilon_1\, E_1 + P_1) = \sigma \tag{3.4}$$

This implies that for *c*-plane III-nitride films, sheet charges will exist at material interfaces whenever there is a polarization mismatch and is termed spontaneous polarization. This can occur at the interfaces between bulk n-type or p-type material, heterointerfaces or quantum well edges, or even at the surfaces and the contacts. It has been found that sheet charges at the surfaces of III-nitride semiconductors make it difficult to achieve ohmic contact behavior [35–37].

Due to the spontaneous polarization, a negative charge results on the Ga face and a positive charge on the N face. In the absence of strain, the spontaneous polarization charge for GaN is $2.1 \times 10^{13}\ \text{cm}^{-2}$ [32,38]. In the absence of compensating charge, an internal electric field of 4 MV/cm would be present in the [0001] direction. However, these high fields are typically screened by other free or bound charges in the semiconductor or surface traps. AlGaN is expected to have even higher polarization charge. Piezoelectric polarization charge is induced by a

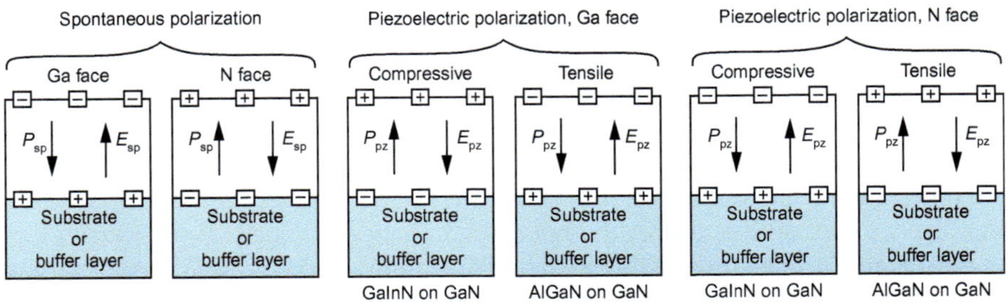

Figure 3.5 Surface charges and direction of internal electric field and polarization field for spontaneous and piezoelectric polarization in III-nitrides for Ga- and N-face orientation [41].

mechanical perturbation such as crystal strain between AlGaN and GaN. AlGaN/GaN high electron mobility transistor (HEMT) has made use of the spontaneous polarization charge difference and piezoelectric polarization due to crystal stain, which results in high electron density in 2DEG. With thicker AlGaN (>30 nm for 35% Al content), the layer is expected to be relaxed but still there will be spontaneous polarization charge at the interface due to the band discontinuity. In an metal oxide semiconductor field effect transistor (MOSFET) structure, the polarization charge is treated similar as fixed oxide charge. However, this spontaneous charge density varies with temperature (referred to as pyroelectric polarization) [39,40], which can significantly affect the flatband voltage of the MOS structure and cause accumulation or depletion on GaN surface.

Since the sign of piezoelectric polarization depends on the crystal strain, this can either aid or oppose the spontaneous polarization. Fig. 3.5 shows the sign of different polarization and the direction of electric field for GaN and GaAlN grown on a substrate [41]. These properties are used for the fabrication of HEMT device structures as shown in Fig. 3.6. Here, the 2DEG is not induced by doping but instead facilitated by spontaneous and piezoelectric electric field in the AlGaN layer. The 2DEG density in GaN-based HEMTs is therefore a strong function of the barrier layer (AlGaN) thickness. By applying a negative voltage to the gate, the device can be turned off.

3.3 Substrates for GaN epitaxial growth

Since GaN bulk material growth is met with difficulties, no commercially available large area bulk III-nitride substrates exist, despite considerable improvement in GaN and AlN bulk growth in recent years. Epitaxial growth mostly has to be performed on foreign substrates, with all the problems associated with this form of deposition. The lattice constant mismatch is a primary criterion for choosing substrates, as well as other important issues such as difference in the TECs, thermal conductivities, and polarity. Usually, lateral lattice constant (a) mismatch leads to high misfit dislocation density. These misfit dislocations, when they are extending

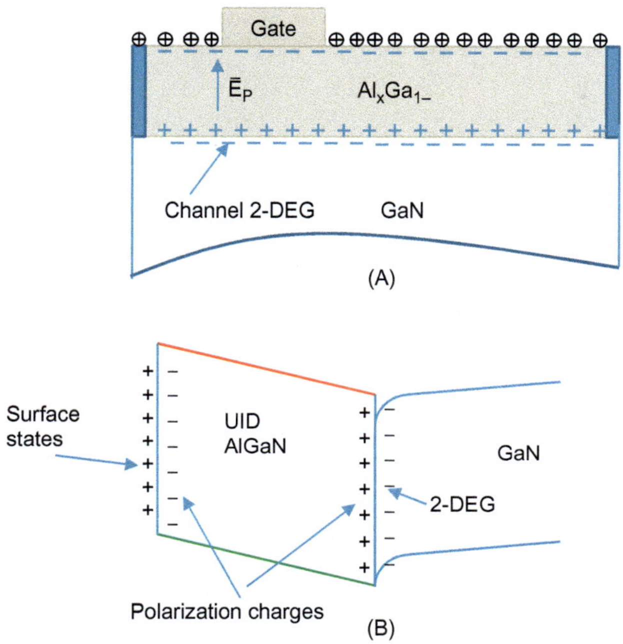

Figure 3.6 (A) Device structure schematic (B) Band diagram of heterostructure showing different types of charges.

into the layers above, form threading dislocations that degrade electronic and optical properties [42]. Vertical lattice constant (*c*) mismatch can form antiphase boundaries. Mismatch of TECs induces thermal stress in the film and the substrate during cooling down to room temperature, which can bow and even crack the epitaxial films. In addition, the strain will cause a large built in electric field in the film by piezoelectric polarization because of the high piezoelectric constant of III-nitride.

3.3.1 Sapphire substrates

Sapphire (β-Al_2O_3) is the most popular and extensively used substrate for the growth of III-nitride. It is optically transparent from the visible into deep UV. It is stable at high temperatures and pre-growth cleaning is well established due to its use in SOI wafer fabrication in Si technology. The technology for the growth of bulk sapphire is relatively mature and good quality sapphire wafers of large diameter are manufactured at relatively low cost. Sapphire has a hexagonal unit cell with lattice constants $a = 0.4765$ nm and $c = 1.2982$ nm [43], resulting in lattice constant mismatch between sapphire and GaN as high as 30%. When GaN grows on *c*-plane (0001) sapphire, the crystal orientations of sapphire and GaN are parallel, but the unit cell of GaN is rotated by 30 degrees about the *c*-axis with respect to the sapphire unit cell. This reduces the lattice constant mismatch between them. However, the lattice mismatch of sapphire substrate after this rotation is still as high as 16%, which generates large density of threading dislocations ($\sim 10^{10}$ cm^{-2}) in as-grown nitride layer above the substrate. A schematic illustration of lattice mismatch and in-plane orientation relation of (0001)GaN with (0001) sapphire substrates is shown in Fig. 3.7. The mismatch of TEC between *c*-plane sapphire and GaN is about 39%. As a result, a large biaxial compressive stress exists in the GaN epitaxial layer during cooling from the deposition temperature to room temperature. Based on the lattice mismatch, it is expected that GaN should be under tension, but as grown layer at the growth temperature is relaxed with the presence of high density of dislocations so that the TEC mismatch is the dominant factor here. The mismatch of lattice constants and TECs of several popular substrates are shown in the Table 3.2. It should be noted that more than the lattice mismatch, TEC mismatch is more important once the lattice mismatch is more than a few percent. Since these values are temperature dependent, one has to be careful in using these values to estimate the residual stress in the materials.

GaN is grown on sapphire with a low temperature buffer of AlN or GaN, followed by the growth of thicker GaN at higher temperature. In such a case, the layer is generally be of Ga-face GaN, resulting in an electric field going from the interface to the surface. This will have important influence on the design of device structures.

3.3.2 SiC substrates

Besides sapphire, SiC is another popular substrate for III-nitride. Compared with sapphire there are several advantages of SiC. It has a smaller lattice mismatch (3%)

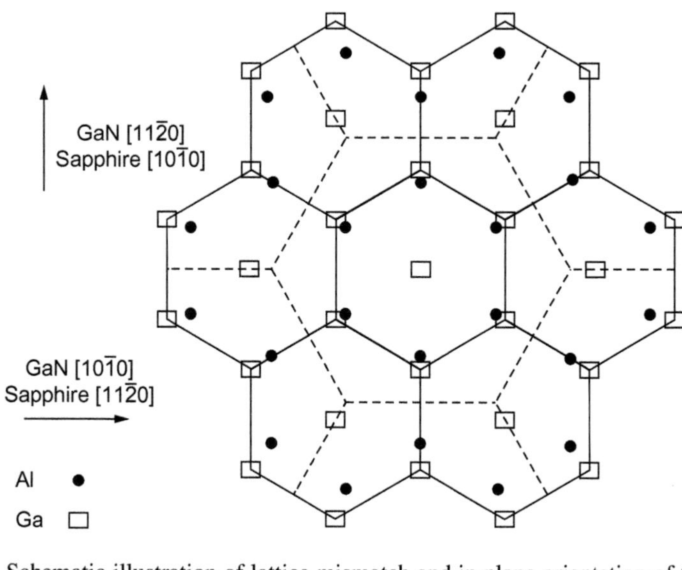

Figure 3.7 Schematic illustration of lattice mismatch and in plane orientation of GaN and sapphire.

Table 3.2 Lattice constant and TEC of several substrates for GaN

	Lattice constants (nm)	% mismatch with GaN along a	TEC values (K^{-1})	TEC % mismatch with GaN along a
GaN	$a = 0.318843$ $c = 0.518524$	0	$a = 3.1 \times 10^{-6}$ $c = 2.8 \times 10^{-6}$	0%
AlN	$a = 0.3112$ $c = 0.4982$	2.4%	$a = 2.8 \times 10^{-6}$ $c = 2.7 \times 10^{-6}$	− 10.7%
Sapphire	$a = 0.4765$ $c = 1.2982$	33%	$a = 6.06210^{-6}$ $c = 6.66 \times 10^{-6}$	+ 48%
6H-SiC	$a = 0.30806$ $c = 1.51173$	3.4%	$a = 4.46 \times 10^{-6}$	+ 30%
Si	0.543102		$a = 2.61 \times 10^{-6}$	−18%

for [0001] orientation growth and higher thermal conductivity (3.8 W/cm K). Conductive SiC makes possible backside substrate electrical contacts. SiC does have its drawbacks, main one being that the SiC substrates are still quite expensive and of relatively small size compared to sapphire or silicon. So, in addition to sapphire and SiC, III-nitride have been grown on silicon and is being the primary substrate used for power device applications [44].

3.3.3 Silicon substrates

Of all the substrates, epitaxial growth on silicon substrates is of particular interest since silicon substrates are inexpensive and processing of layers grown on silicon substrate can be easier compared to layers grown on any other substrates. There are several challenges in growing high quality GaN and its alloys on silicon substrates. There is a large lattice mismatch of ~17% between silicon and GaN ($a_{Si} = 3.84$ Å on (111) plane, $a_{GaN} = 3.189$ Å). Also there is a large thermal-expansion-coefficient mismatch of ~54% between GaN and Si(111). The large lattice mismatch results in high density of dislocations in the active layer of GaN on silicon. The greater TEC of GaN as compared to that of Si (5.6×10^{-6}K versus 2.6×10^{-6}K) puts the film in tension, hence cracks may be generated in the GaN during the cool-down cycle or results in significant amount of bowing of the wafer, especially large diameter wafers. In addition, there is a tendency for Si to react with nitrogen source available in the growth system forming SiN_x at the interface before any films can be grown, and this results in significant amount of dislocations in the grown layer. The reason for this is that the direct growth of GaN on silicon starts with localized growth since GaN does not wet silicon [44] and ammonia in the system reacts with exposed silicon. Hence a buffer layer of AlN is introduced first since AlN is found to wet the surface completely and this technique also prevents the direct reaction of ammonia with silicon substrate. For example, the start of AlN in molecular beam epitaxy (MBE) at least is started by the deposition of Al first in order to avoid the formation of SiN_x. To overcome the above problems, many researchers have used various buffer layer schemes such as AlN nucleation layer grown at low temperature, followed by step grading from AlN to GaN, or AlN nucleation layer followed by superlattice of AlN/GaN to control the dislocation followed by the thick GaN growth or growth of thick AlN followed by the growth of GaN. Such multilayer schemes are used not only to control the defect density in the active layer but also to engineer the strain in the layer to prevent cracking or bowing of the wafer.

The most successful transition layer scheme to get high quality active GaN layers seems to be the one shown in Fig. 3.8 [45]. The AlN initial layer is used to prevent the undesired reaction between Si substrate and the reactor ambient and also relaxes the stress caused by large lattice mismatch between the epitaxial layers and the Si substrate. This layer is 160-nm-thick and is expected to be fully relaxed. A 40-nm-thick $Al_{0.3}Ga_{0.7}N$ is then grown on this layer that is believed to improve the surface roughness of the initial AlN layer and provides a flat surface to grow the AlN/GaN superlattices (SLs) structure and the AlGaN/GaN channel layer with smooth heterointerfaces. The AlN/GaN SLs structure is introduced to create a compressive stress to cancel a tensile stress caused by the thermal expansion mismatches between III−V nitride and Si. The stress created by the AlN/GaN SLs structure need to be accurately controlled to cancel out the tensile stress caused by the thermal expansion mismatch to reduce the bowing. This can be accomplished by optimizing the thickness of individual AlN and GaN thickness such that each layer is only partially relaxed. Ishida et al. [45] have calculated the optimum thickness of individual layers so that crack free large area GaN can be grown on silicon. For example, 100 pairs AlN/GaN (5/20 nm) SLs structure and an AlGaN/GaN heterojunction consist of a 50-nm-thick $Al_{0.2}Ga_{0.8}N$ layer and a 2-μm-thick GaN layer

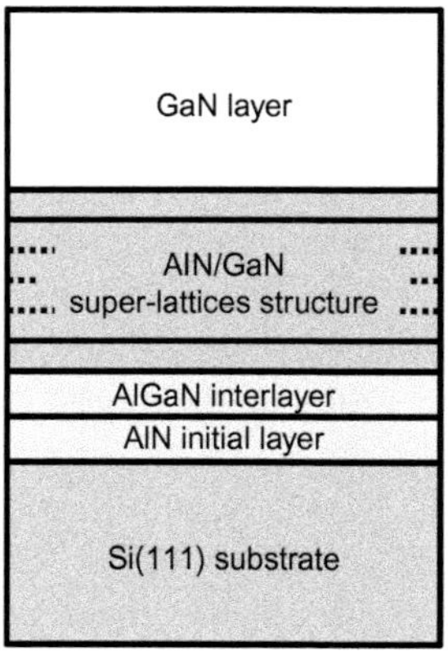

Figure 3.8 Schematic illustration of transition layer used to grow high quality GaN on Si substrates [45].

can result in wafers with bow less than 50 μm over a 6-in. diameter substrate. The thickness of individual layer in the SL structure is chosen such that AlN layer will be pseudomorphic where as GaN layer will be partially relaxed, thus providing a compressive stress for the overall structure. The extent of GaN relaxation can be controlled by the growth conditions such as the temperature so that optimum bow after the growth can be obtained. Other variations on the transition layers include the growth of thick AlN layer or graded layer from AlN at the interface to $Al_{0.2}Ga_{0.8}N$ buffer [46–50].

In addition to obtaining high quality GaN on silicon for the HEMT structure fabrication, it is also important to get the thickness of the GaN suitable for the high voltage operation since the substrate is highly conducting. Hence undoped or highly resistive GaN buffer layer of sufficient thickness is necessary for high voltage operation. Even though the breakdown voltage of the HEMT device depends on the gate to drain distance and optimum field profile near the surface, for large gate-to-drain distances, the substrate conduction will also be a factor. Hence buffer layer thickness will be critical to obtain high breakdown voltage. Fig. 3.9 shows the vertical breakdown voltage across the GaN buffer layer as a function of the buffer layer thickness for undoped GaN buffer [45]. It is clear that at least 5.4 μm thick buffer layer is needed to hold 1200 V across the buffer layer. The buffer layer is made semi-insulating by the introduction of unknown intrinsic deep levels. However, such buffers have shown that charge trapping in compensated intrinsic deep donors and acceptors can result in "current collapse" which varies with geometry, buffer doping, and trap energy level. So, improved buffer layers are needed to reduce the charging and detrapping effect in power devices when operated at high frequency.

Recent GaN/AlGaN HFET devices have often preferred to use extrinsic deep-level dopants to make the buffer insulating instead of using intrinsic defects, partly due to the ease of control during growth since growth conditions can be optimized

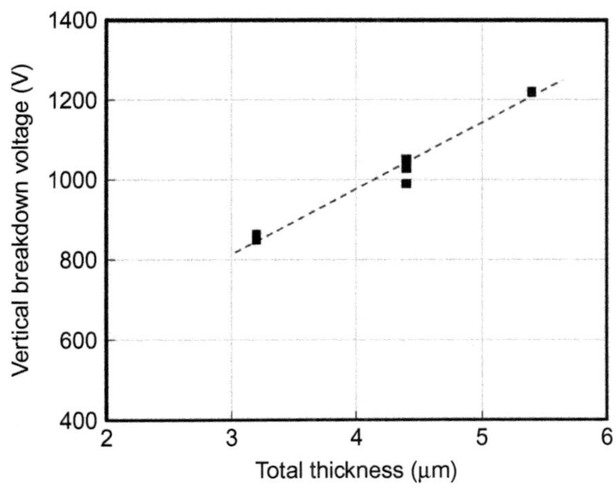

Figure 3.9 Vertical breakdown voltage as a function of the GaN buffer layer thickness. Layers are grown on (111)Si substrates [45].

based on other requirements such as morphology of layers. Two widely used dopants are iron (Fe) and carbon (C). They lead to two different deep acceptor centers, one in the upper and one in the lower half of the GaN bandgap, respectively. Iron has a trap level at 0.7 eV below the conduction band and carbon has a trap levels 0.9 eV above the valence band, respectively [51,52]. These extrinsically doped buffer layers provide excellent isolation and power device performance. For example, since carbon has an acceptor level at 0.9 eV above the valence band, introducing carbon at a level higher than any donor states will result in Fermi level being pinned at the carbon energy level. This is easier to control than depending on the intrinsic levels based on the growth conditions.

3.4 Band structure and relevant properties

Electronic band structure of wurtzite GaN has been calculated using various methods [53–55] and a detailed picture was given in [53]. Here a simplified band structure near the Γ point is presented in Fig. 3.10 [56]. The Γ_9 state is always termed heavy-hole (hh) state and the energy below (Γ_{7+}) forms the light-hole state (lh). The third state (Γ_{7-}) is defined as crystal field split-off state (cs). The energy differences between these states can be calculated using two parameters—the spin splitting energy Δ_{so} and the crystal field splitting energy Δ_{cr}, which are

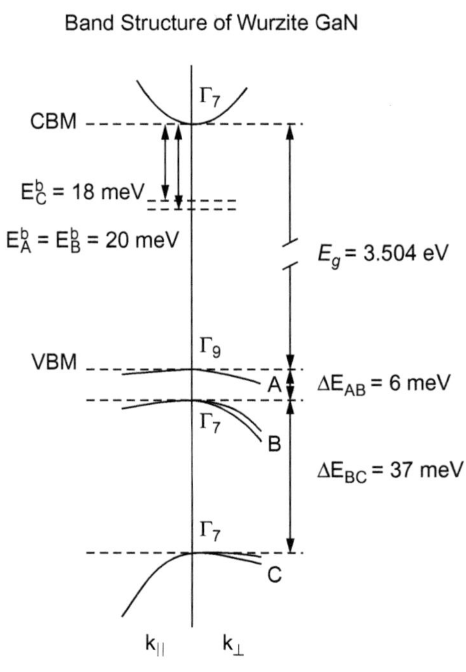

Figure 3.10 Calculated band structure near the Γ point of WZ GaN.

12 and 37.5 meV in case of GaN. The valence bands show considerable nonparabolic behavior. For GaN, both the conduction band minimum (Γ_7) and the valence band maximum (Γ_9) are at Γ point, therefore it has a direct bandgap with an energy of 3.39 eV at room temperature, and this value increases to 3.503 eV at 1.6K [57]. The temperature dependence of GaN bandgap is shown in Fig. 3.11 and the relation can also be expressed empirically as:

$$E_g(T) = 3.503 - 5.08 \times 10^{-4}\, T^2/(996 - T)(\text{eV});\quad \text{for } T < 295 \tag{3.5}$$

At temperature higher than 180K, the value of $dE_g(dT)$ is approximately -6.0×10^{-4} eV/K [58]. The bandgap energy will shift to a higher energy under hydrostatic pressure. The pressure dependence of bandgap energy is shown in Fig. 3.12 and a $dE_g/(dP)$ of 3.7 ± 0.4 meV/kbar is approximated [59]. But more commonly, we will encounter the problem of biaxial strain in the epitaxial layer on heterogeneous substrate. The biaxial strain effect on the energy gaps between the conduction band minimum and three valence band maximums has been theoretically calculated and shown in Fig. 3.12, together with some experimental data [60].

Electronic band structures of wurtzite AlN and InN are very similar to the GaN band structure except the detail energy values [61,62]. The bandgap energies for

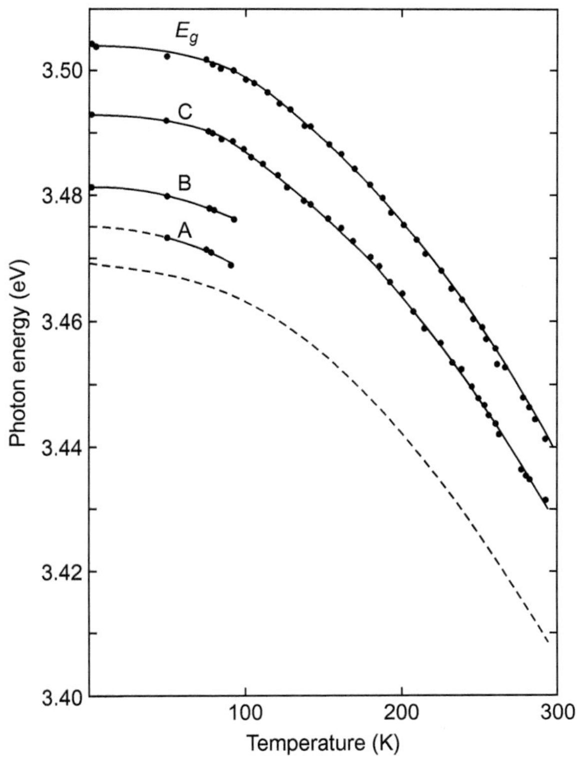

Figure 3.11 Temperature dependence of band diagram.

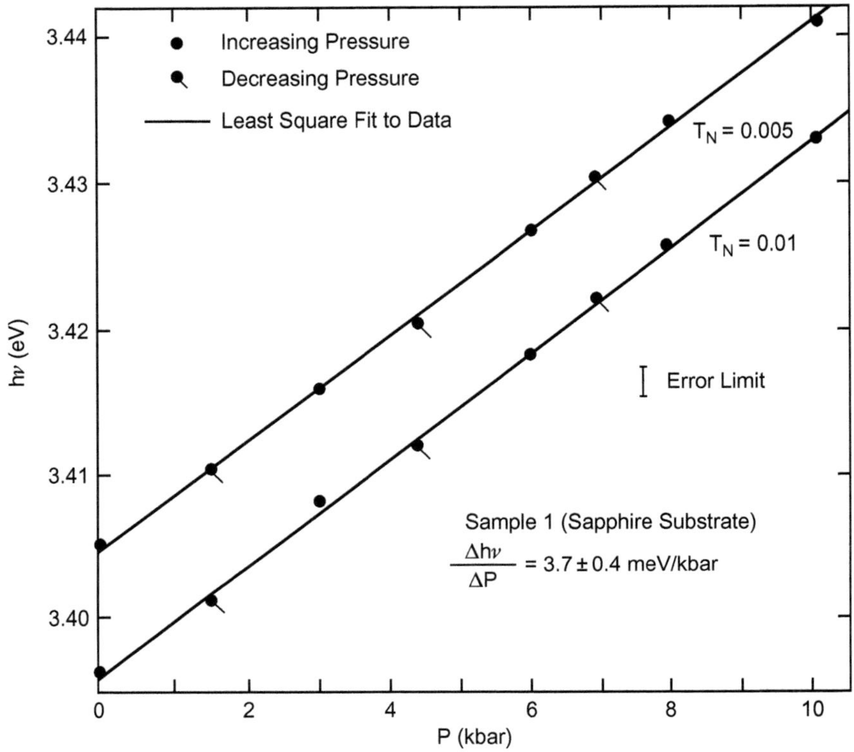

Figure 3.12 Effect of pressure on the band diagram.

AlN and InN at 300K are 6.2 and 1.89 eV, respectively. The InN bandgap is now believed to be even lower, at 0.7 eV range. Earlier data for InN materials gave higher bandgap values. This may be due to the fact that optical absorption measurement can give higher values if the films are heavily doped and Burstein–Moss effect is present. The temperature dependence is available in Ref. [63]. For AlGaN ternary alloys, the composition dependence of their bandgap has been determined experimentally by the optical absorption edge method and is presented in Fig. 3.13 [64,65]. The *solid line* is the best fit to the experimental data with the equation $E(x) = E(0) + ax + bx^2$ where $E(0) = 3.43$ eV is the bandgap energy of GaN determined by the absorption measurements at room temperature, x is the AlN alloy fraction, and energy is given in eV. With the bandgap energy of AlN fixed at 6.2 eV, the best fit using least-square fitting yields $a = 1.44$ and $b = 1.33$ eV. The parameter b is termed as the bowing parameter.

3.4.1 Effective mass of carries

The effective mass of electrons and holes in a band is important for the transport property and also for describing various electrical and optical properties of the material. The value of the effective mass is dependent on the local structure of the

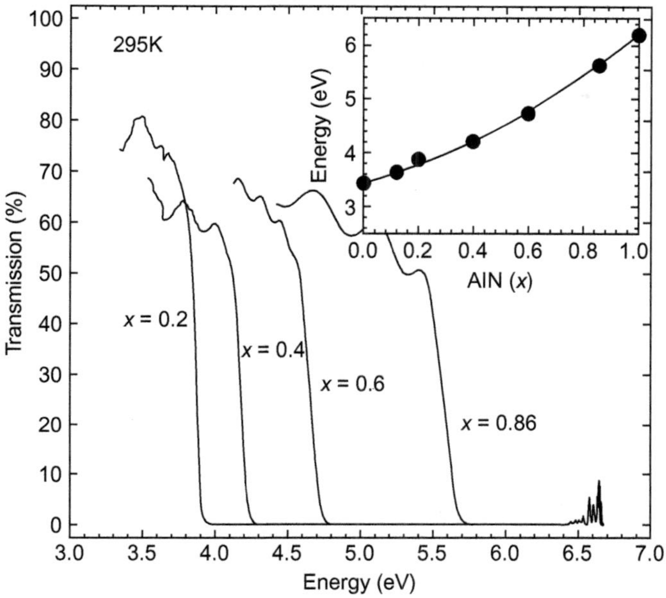

Figure 3.13 Bandgap energy versus composition.

band near the band extremum and is defined by the local curvature of the band. Since the valence bands are split into light and heavy holes, the effective masses will also be different for different types of carriers, as well as whether the carriers are moving parallel to the growth plane or perpendicular to the growth plane. The overall effective masses for electrons and holes are tabulated in Table 3.1.

3.4.2 Effective density of states

Knowledge of density of states (DOS) is important to study the electronic and optical properties of GaN. However, more important parameter which is frequently used to describe the band property is the effective DOS at the conduction and valence band edge. It is defined as:

$$N = 2\left(2\pi m * kT/h^2\right)^{3/2} \tag{3.6}$$

where h is the Plank's constant, k is the Boltzmann's constant, T is the temperature, and m^* is the DOS effective mass. In the case of GaN, $m_e^*$of electrons at the conduction band edge is $0.19m_o$ and the m_h^* of holes at the valence band edge (Γ_9 point) is $0.66m_o$ [66] and is given in Table 3.1. Putting this values back into Eq. (3.6) yields:

$$N_e = 0.4 \times 10^{15}\ T^{3/2}\ \text{cm}^{-3} = 2.08 \times 10^{18}\ \text{cm}^{-3}(\text{at 300K}) \tag{3.7a}$$

$$N_V = 2.6 \times 10^{15}\ T^{3/2}\ \text{cm}^{-3} = 1.35 \times 10^{19}\ \text{cm}^{-3}(\text{at 300K}) \tag{3.7b}$$

Therefore the intrinsic carrier concentration can be obtained as:

$$n_i = 5.3 \times 10^{18} \exp\left(-E_g/2kT\right) \tag{3.8}$$

The value of n_i at room temperature is only $\sim 1.65 \times 10^{-10}$ cm^{-3} and a temperature of ~ 1500K is needed to obtain an intrinsic carrier concentration level of 10^{13} cm^{-3}. So, at a typical growth temperature, GaN is generally extrinsic.

3.5 Transport properties

Transport of charges in the semiconductor material takes place by the motion of holes in the valence band and the electrons in the conduction band. At low electric field, the drift velocity (v_d) of these charges is proportional to the electric field strength (E) and the proportionality constant is described as the mobility μ. The mobility of the carriers in semiconductor material is determined by the different scattering mechanisms which are defect scattering, carrier–carrier scattering and lattice scattering. Defect scattering includes neutral and ionized impurity scattering, alloy scattering and scattering due to crystal imperfection such as dislocations. Lattice scattering includes acoustic phonon scattering (deformation potential and piezoelectric) and optical phonon scattering. The detail of each process can be found in standard texts.

The transport behavior of electrons and holes is governed by the Boltzmann equation and can be solved using relaxation-time approximation as long as all the scattering processes can be considered as elastic, i.e., no energy transfer involved. In this case, the mobility can be expressed as:

$$\mu = q\tau/m^* \tag{3.9}$$

where m^* is the effective mass of the electrons and $\langle \tau \rangle$ is the average relaxation time and is the sum of the relaxation time of all individual scattering process τ_i. This solution can be further simplified by assuming that for each scattering process there is a corresponding mobility μ_i which can be calculated individually. The overall mobility μ_i is given by:

$$\frac{1}{\mu} = \sum_i \frac{1}{\mu_i} \tag{3.10}$$

At low electric field, the electron mobility is related to the drift current density:

$$J = qnv = qn\mu E \tag{3.11}$$

where n, v, μ, and E are the carrier concentration, drift velocity, mobility, and electric field, respectively. The drift velocity increases almost linearly with the field and the mobility has a constant value μ_0, the low-field mobility. In GaN, the lattice and impurity scattering are dominant and hence the low-field mobility is primarily a function of the doping concentration and the temperature. At very low doping, mobility is mainly dependent on the lattice scattering, whereas at high doping both lattice scattering and impurity scattering should be considered. The well-known Caughey–Thomas [67, 68] model is widely used for low-field mobility dependence on doping and temperature and can be described as:

$$\mu_o = \mu_{\min} + \frac{\mu_{\max}(T/300)^{\nu} - \mu_{\min}}{1 + \left(\frac{T}{300}\right)^{\xi}\left(\frac{N}{Nref}\right)^{\alpha}} \tag{3.12}$$

where N is the ionized doping concentration, T is the absolute temperature, and $\mu_{\min}$, $\mu_{\max}$, N_{ref}, ξ, and α are fitting parameters. The parameter $\mu_{\max}$ represents the mobility of undoped or unintentionally doped samples, where lattice scattering is the main scattering mechanism, $\mu_{\min}$ is the mobility in the heavily doped material, where impurity scattering is dominant but lattice scattering is also present. N_{ref} is the doping concentration at which the mobility is halfway between $\mu_{\max}$ and $\mu_{\min}$ and α is a measure of how fast the mobility changes from $\mu_{\min}$ to $\mu_{\max}$. The temperature dependence of the mobility is modeled in the parameters ν and ξ.

At high electric fields, the carrier drift velocity ν saturates due to the increasing optical phonon scattering, finally reaching the material saturation velocity ν_{sat}. As the carrier velocity approaches ν_{sat}, the relationship with electric field is no longer linear and the mobility is given by Eq. (3.12). The reduction in mobility at high electric fields can be modeled as:

$$\mu = \frac{\mu_o}{\left[1 + \frac{\mu_o E^{\beta}}{\nu_{\text{sat}}}\right]^{1/\beta}} \tag{3.13}$$

where ν_{sat} is the saturation velocity and β is an empirical constant specifying how abruptly the velocity goes into saturation. The saturation temperature dependence is as follows [69]:

$$\nu_{\text{sat}} = \frac{2.7 \times 10^7}{1 + 0.8\exp\left(\frac{T}{600}\right)} \tag{3.14}$$

Based on the experimental work, Mnatsakanov et al. [70] have summarized the values of the parameters above for electrons and holes. The values found for electrons are listed in Table 3.3. The parameters commonly used for silicon and 4H-SiC [71] are also given for comparison. The low-field mobility in the materials for different doping concentration (at 300K) and at different temperatures (at various doping concentration) are plotted in Figs. 3.14 and 3.15, respectively. The mobility plotted and the parameters tabulated are in the c-plane. The dependence of the low-field electron mobility on doping level is given and compared to many

Table 3.3 Parameters for electron mobility models in wurtzite GaN (*c*-plane)

Parameter	Wurtzite GaN	Si
μ_{max} (cm^2/Vs)	1000	1430
μ_{min} (cm^2/Vs)	55	55
N_{ref} (cm^{-3})	2.0×10^{17}	1.0×10^{17}
α	1	0.73
ν	−2.0	−2.3
ξ	−2.7	−3.8
v_{sat} (cm/s)	2.5×10^7	1.0×10^7
β	2	2

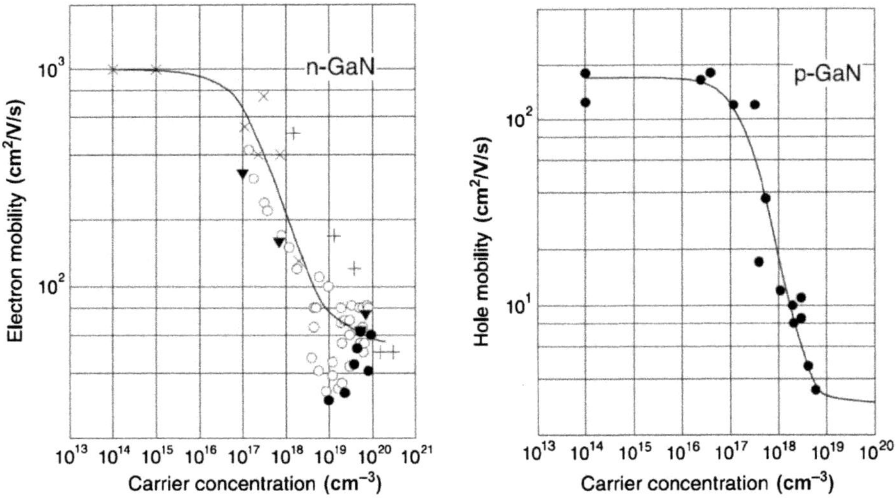

Figure 3.14 Low-field electron and hole mobility as a function of doping concentration in GaN at room temperature.

available experimental data on room temperature electron mobility in GaN. In Fig. 3.16, an analogous comparison is made for hole mobility. It is clear that there are appreciable scatter in the experimental data for high doping concentration for electrons which is presumably caused by different compensation in many experimental data.

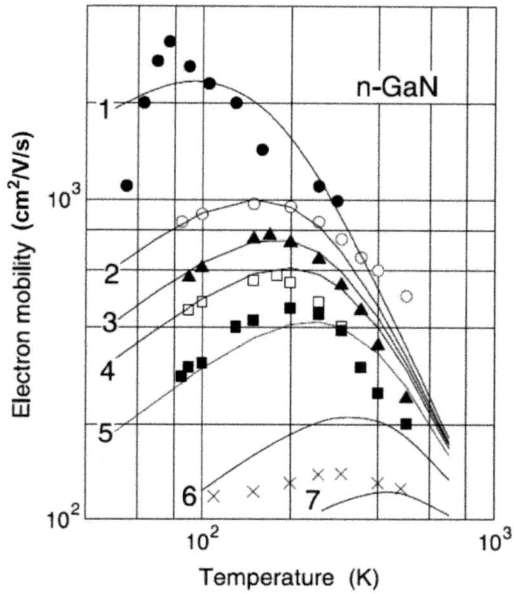

Figure 3.15 Temperature dependencies of low-field mobility in wurtzite GaN at different values of doping concentration.

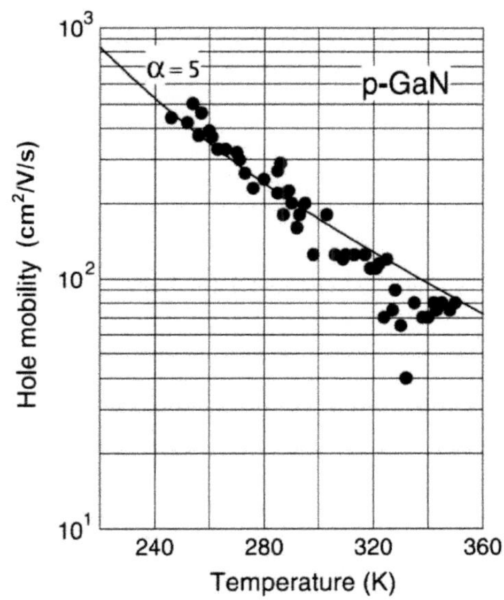

Figure 3.16 Temperature dependencies of low-field hole mobility in wurtzite GaN.

3.5.1 2D mobility in GaN/AlGaN structures

In AlGaN/GaN HEMT structures, in which the electrons derive from the effect of spontaneous and piezoelectric polarization in the AlGaN barrier, the electron drift mobility in the channel at room temperature is important. There have been many different studies on the electron and hole mobilities of 2D carriers both as a function of the interface roughness, temperature, and sheet carrier concentration [38,71–73]. The 2D carrier concentration is a strong function of the barrier layer composition and thickness. The residual stress in the barrier layer has a strong influence on the density of 2D gas and common values are in the $10^{13}\ cm^{-2}$ range. The 2D gas mobilities have been measured as a function of the carrier concentration and it was found that the mobility increases with the concentration which is attributed to the shielding of surface roughness at high concentration. Typical mobilities for electrons are measured in the 2000 cm^2/Vs range. Nakajima et al. [74] have studied the hole and electron 2D gas mobilities using a structure as shown in Fig. 3.17. By selectively contacting the appropriate carriers, they have measured the mobilities of carriers as a function of temperature. For comparison, a bulk Mg-doped GaN carrier concentration and hole mobilities are also measured. Fig. 3.18 shows the sheet carrier density and the hole and electron mobilities of carriers as a function of temperature.

3.6 Impact ionization coefficients

One of the most attractive properties of GaN is its ability to withstand higher electric field than Si, which allow semiconductor to achieve the same blocking with thinner and less resistive layer. There are three main breakdown mechanisms: Thermal instability, tunneling effect, and avalanche multiplication, the latter typically represents the physical limit to the design of power devices. Under high

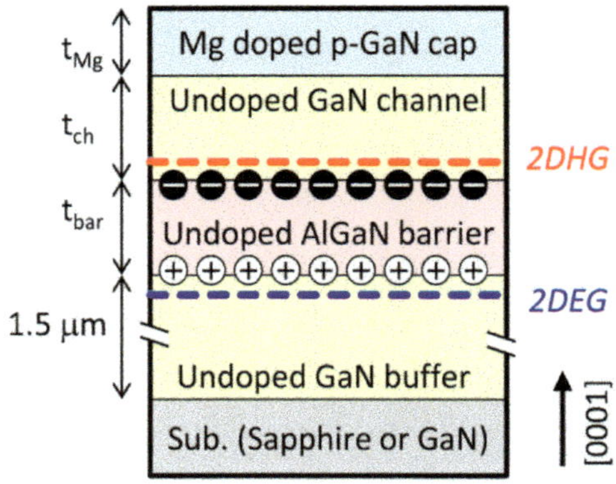

Figure 3.17 Schematic of structure used to measure 2D hole and electron mobilities in GaN/AlGaN structure.

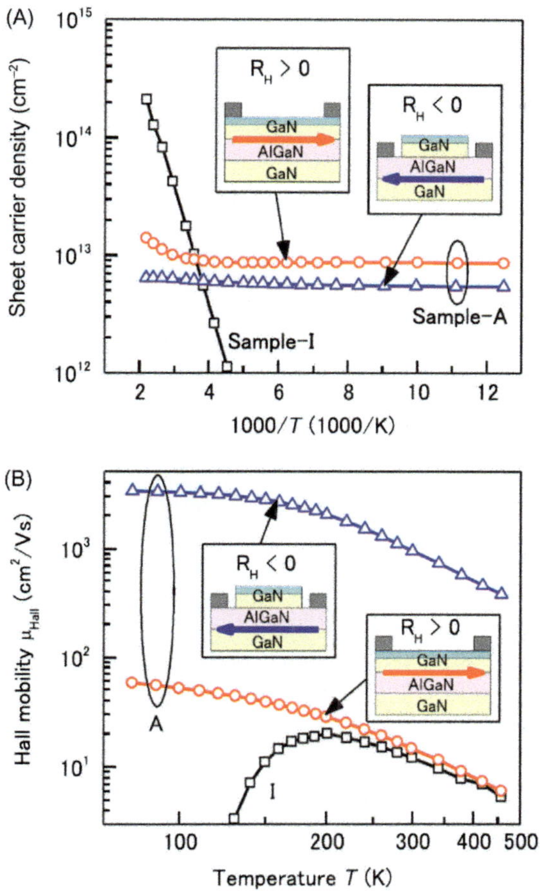

Figure 3.18 Sheet carrier concentration and Hall mobility as a function of temperature for 2D hole and electron gas. Also shown are the data for bulk Mg-doped GaN.

electric field, high-energy electrons and holes generate electron–hole pairs by collisions with the atoms in the lattice and exciting electrons from the valence band to conduction band. This process for the generation of electron–hole pairs is called impact ionization. Impact ionization is a multiplicative phenomenon that produces a cascade of electron–hole pairs, which are swept out through the depletion region, resulting in an increase of current flow. The avalanche breakdown occurs when the impact ionization process attains an infinite rate. The impact ionization coefficient is defined as the numbers of electron–hole pairs generated per carrier (electron or hole) per centimeter traveled in the direction of the electric field. The lower the ionization coefficient higher the breakdown voltage. It is also important that the breakdown voltage of a device increase with temperature for it to have stable reliable behavior. This again is determined by the variation of impact ionization coefficients with temperature. The impact ionization coefficients not only determine the reverse bias characteristics of a device but they also influence the forward characteristics. Hence, it is important to have a good understanding of the impact ionization coefficients and their variation with temperature. The impact ionization coefficients are strong functions of electric field and are typically modeled by [75,76]:

$$\alpha_n = a_n \exp\left(-\frac{b_n}{E}\right)^{c_n} \quad \text{or} \quad \alpha_p = a_p \exp\left(-\frac{b_p}{E}\right)^{c_p} \tag{3.15}$$

where $a_{n,p}$, $b_{n,p}$, and $c_{n,p}$ are model parameters and E is the electric field.

Impact ionization coefficients are usually extracted by applying a high reverse voltage to an illuminated junction and measuring the reverse current flow. However the leakage current in the reverse biased junction depends strongly on the impurities and defects in the material, which still remain relatively high in GaN. The device design also contributes to excess leakage current such as poor edge termination.

In GaN, the impact ionization coefficients are lower than that of Si, which results in higher breakdown with the same design. The critical electric field is defined as the electric field when junction breakdown happens. By fitting to the ionization coefficients, a single expression may be derived to describe the breakdown of one-dimensional junction dependence on the doping concentration and other dependences.

Since the defect density is rather high in GaN materials, traditional method of measuring the impact ionization may result in rather high coefficients. Use of a localized method may be better and was used in Ref. [77]. A pulsed electron beam technique was used here to localize the measurements to avoid defects detected in the material using electron beam induced current (EBIC) scans [77,78]. Schottky barrier diodes were fabricated on n-type GaN epitaxial layers with low doping concentration grown on highly doped N^+ GaN substrates. The problem of high electric fields at the edges of the diodes was solved by using argon ion implantation which raised the breakdown voltage from 300 to 1650 V. Pulsing the electron beam during the impact ionization measurements greatly improved the signal to noise ratio by using a lock-in amplifier.

From the measured data for electrons in GaN, a_n has a value of 1.5×10^5 (+ or − 0.2×10^5) cm^{-1} and b_n has a value of 1.41×10^7 (+ or − 0.03×10^7) V cm^{-1} at room temperature. From the measured data for holes in GaN, a_p has a value of 6.4×10^5 (+ or − 0.1×10^5) cm^{-1} and b_p has a value of 1.46×10^7 (+ or − 0.01×10^7) V cm^{-1} at room temperature. This is shown in Fig. 3.19. The temperature dependence of these parameters for GaN has also been measured by using a heating stage installed inside the scanning electron microscope used for the pulsed electron beam experiments: $a_n = 4.4 \times 10^5 - 9.73 \times 10^2 T$ and $a_p = 1.68 \times 10^6 - 3.44 \times 10^3 T$. The coefficients b_n and b_p were found to be independent of temperature within experimental tolerances.

For measurements for GaN epitaxial layers grown on Sapphire substrates gave an a_n value of 9.17×10^5 cm^{-1} and a b_n value of 1.7×10^7 V/cm for the impact ionization coefficient for electrons at room temperature. For the impact ionization

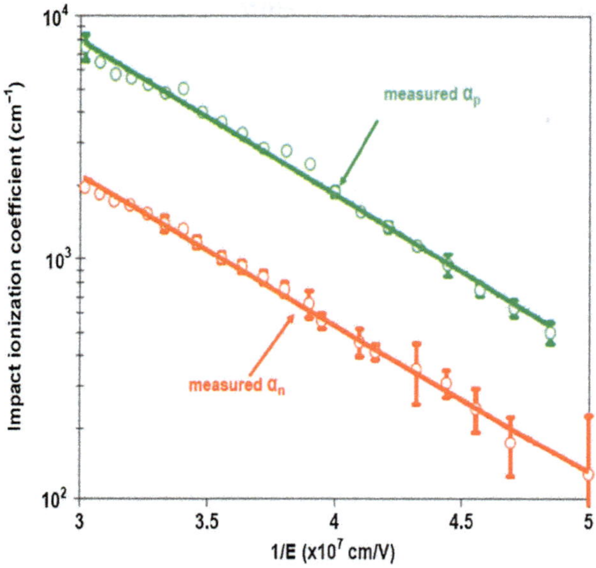

Figure 3.19 Measured impact ionization coefficients of electrons and holes in GaN measured on epitaxial layer grown on bulk GaN.

coefficients for holes at room temperature, the values of a_p and b_p were found to be 8.7×10^5 cm^{-1} and 1.46×10^7 V/cm, respectively. The values for both coefficients are larger than those measured for GaN grown on GaN substrates. The temperature dependence measurement results are $a_n = 2.82 \times 10^6 - 6.34 \times 10^3 T$ and $a_p = 2.98 \times 10^6 - 7.02 \times 10^3 T$. The coefficients b_n and b_p were found to be independent of temperature within experimental tolerances. The higher values indicates that the defect density is still too high and localized e-beam excitation still has some limitations.

3.7 Defects in GaN

Due to the lack of lattice-matched substrate and nonoptimized growth conditions, GaN epitaxial layers always contain a large number of defects, including point defects, threading dislocations, and grain boundaries. Even in the nitride-based, superbright LEDs, it is surprising to find that the dislocation density in the active region is still on the order of 10^{-8}–10^{-10} cm^{-2} [79] which is definitely unacceptable in other Ill–V compound semiconductors. Therefore, defects may play some unique roles in III–V nitrides. Understanding the nature and location of energy levels associated with these defects is important to improve both the material and device properties. In this section, we will first review the intrinsic point defects, then discuss the planar and line defects found in GaN resulting from the lattice-mismatched heteroepitaxial growth. We will also review the defects in GaN grown on silicon substrates.

3.7.1 Intrinsic point defects

Native point defects in GaN include Ga vacancy (V_{Ga}), Ga interstitial (G_I), Ga antisite (Ga_N), N vacancy (V_N), N interstitial (N1), and N antisite (N_{Ga}). These native defects can also form complexes with extrinsic impurities. The appearance of these defects depends on the growth condition (such as N partial pressure) and Fermi level during growth in the material. The energy levels of these defects have been calculated extensively using different methods [80–83], and the theoretical results have been compared with the deep levels observed in experiments. Fig. 3.20 shows the distribution of defect levels in the bandgap of GaN [84]. Column 1 are those reported from experimental studies while column 3 are the electron states of point defects calculated by Jenkins and Dow [81]. The N vacancy V_N, calculated as a shallow donor in GaN [81], have widely been used to explain the high electron background concentration in unintentionally doped films. It is suggested that V_N may have three electron states in the bandgap which are located at about 30 meV, 100 meV, and 0.4 eV below the conduction band. Correspondingly, three groups of energy levels are detected in experiment, marked as A, B, and C in Fig. 3.18. However the theory of this V_N as shallow donor in unintentionally doped GaN was challenged by Neugebauer and Van de Walle's calculation [83] in which they argued that the formation energy of V_N in n-type GaN is too high to let V_N be the source for the n-type conductivity. In MOCVD growth of GaN, under some growth

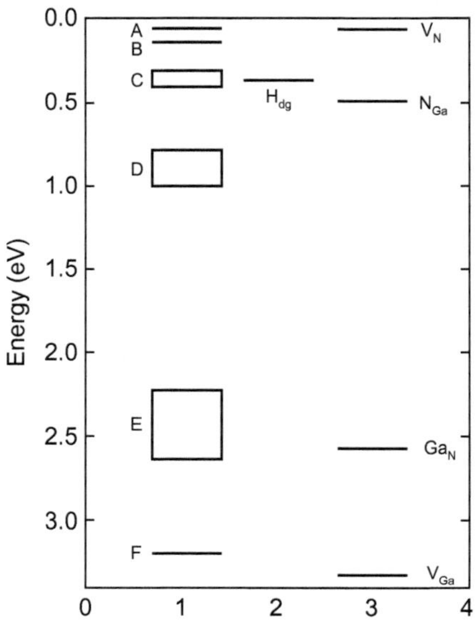

Figure 3.20 Distribution of defect levels in the bandgap of GaN.

conditions, the background carrier concentration now can be reduced two or three magnitude by depositing a thin buffer layer before the epilayer growth. This is hard to explain using just the nitrogen vacancy theory.

Therefore, Neugebauer and Van de Walle's calculation may be right, and high background concentration in poor quality GaN material is probably due to some other structural related defects or extrinsic contamination. A deep level around 0.8–1.1 eV below the conduction-band edge (group D in column 1 of Fig. 3.20) was observed in the thermal activation experiment and is interpreted by Tansley and Egan [80] as the nitrogen antisite defect. If this argument is true, then the yellow band emission located at around 2.2 eV in PL can be attributed to the recombination between the N_{Ga} level and the valence band. The complementary antisite defect GaN, was thought to appear as an absorption edge tail in GaN and have an energy level about 2.3–2.7 eV below the conduction band edge (corresponding to group E level in Fig. 3.20) Ga vacancy, which has an energy level of 3.26 eV below the conduction band edge [82], has been positively identified as a deep level acceptor which plays a role of compensation center in n-type GaN.

3.7.2 Other defects

Besides these point defects, high-density planar and line defects also appear in epitaxially grown GaN due to the large lattice and TEC mismatch between the substrate and the GaN epilayer. These defects include threading dislocations [86], nanopipes [87,88], stacking faults, etc. Among them, threading dislocation is the major component which usually have a density of 10^8–10^{10} cm^{-2} in typical epitaxially grown GaN films. The main sources of threading dislocations are the low angle grain boundaries which are formed during coalescence of islands at the initial stage of GaN growth. Most dislocations lie along the c-axis and have been identified as pure edge, with Burgers vectors of the 1/3 (11–20) type [85]. Another type of defect in GaN is nanopipe which is oriented along the [0001] growth direction and exhibit a funnel-like shape with its wider crater ended at the GaN crystal free surface. Both the crater and the pipe are generally hexagonal shape, with some pipes empty inside while others filled with amorphous material. The origin of nanopipes is postulated to be most likely the open-cores of screw dislocations formed under local thermodynamic equilibrium. Stacking faults and microtwins are also detected by TEM, usually more dominate in p-GaN grown on zinc-blende substrates. Up to now, the best crystal quality is reported on MOCVD-grown GaN film on bulk GaN substrate.

3.7.3 Impurities in GaN

Extrinsic impurities could be incorporated into samples intentionally or unintentionally during the growth. Unintentional impurities usually come from the source gas and the growth environment and should be reduced to a minimum. On the other hand, some other impurities are brought into reactor intentionally to control the electrical and optical properties of the film. In this section, the properties of films doped with different impurities are discussed [89].

3.7.4 Group-II impurities

Group II elements (including Be, Zn, Mg, Cd, Hg) usually occupy the Ga site and therefore behave like acceptors. Among these impurities, Mg is the most extensively studied. Study of Mg doping in GaN began in the early 70 s. However, prior to 90 s, all Mg-doped GaN showed highly compensated behavior and no p-type conductivity was observed. P-type Mg doped GaN was first achieved in 1989 by low energy electron beam irradiation (LEEBI) of highly compensated Mg-doped GaN samples [90]. Later, p-type GaN:Mg was obtained by several other groups using LEEBI or other postgrowth treatments [91–94]. Soon after, p-type behavior was realized in Mg-doped GaN grown by MBE without any postgrowth treatment [95]. The formation of Mg–H complexes under the abundant presence of hydrogen during the metalorganic vapor phase epitaxy (MOVPE) growth has been suggested as the origin for the passivation of Mg in GaN [96]. Another reason is that early undoped GaN was generally heavily doped n-type, and addition of p-type dopants will results in significant compensation. The breakthrough came when undoped GaN could be grown with sufficiently low n-type doping by optimizing the growth process on sapphire. Currently the highest hole concentration up to 3×10^{18} cm^{-3} with a room temperature hole mobility of 9 cm^2/Vs could be obtained (check). As to MBE-grown samples, hole concentrations as high as 10^{19} cm^{-3} have been reported [97]. The activation energy of Mg acceptors is between 120 meV and 160 meV depending on the doping level. Due to such a high activation energy, less than 1% of Mg acceptors are ionized at room temperature which results in the low hole concentration in GaN at room temperature. Other column II elements have been used for p-type doping but were not as successful as that of Mg.

3.7.5 Group-IV impurities

Group-IV elements (including C, Si, and Ge) are called amphoteric dopants in III–V semiconductors because they could either take a cation site to behave like a donor, or take an anion site to be an acceptor. This amphoteric behavior will result in compensation in films doped with these impurities. However, in most cases, these impurities will prefer to occupy one site depending on the detailed properties of impurities and host lattice, such as size of the atoms and electron negativity of the elements.

Among group-IV elements, carbon is expected to be an acceptor in GaN [97]. Carbon can be incorporated as a contamination that comes from the MO source materials, and also from the graphite susceptor in MOVPE. It is generally believed that the chemistry of the source material plays a critical role in the carbon level of the growing layer [98]. For example, layers grown by $(C_2H5)_3Ga$ will contain less carbon than those grown by $(CH_3)_3Ga$ because decomposition mechanisms of these two compounds are different. This kind of carbon is undesired because they are uncontrollable and will cause compensation in n-type material. Hence external sources such as CCl_4 can be used as carbon source.

Silicon is the most commonly used n-type dopant in GaN. As to the wurtzite GaN, the highest electron concentration obtained is $6 \times 10^{19}\ cm^{-3}$ [99]. However, the stress induced by high Si doping was found to cause degradation of the film morphology by forming V-shaped grooves and cracks [100]. The highest doping concentration free from V-grooves and cracks is $2 \times 10^{19}\ cm^{-3}$ reported by Nakamura [99]. It was found that the carrier concentration has a linear dependence on the silane flow rate. The activation energy of Si donor in wurtzite GaN was 26.2 meV. Ge will also behave as n-type dopant and [99] doping concentration in the range from 7×10^{16} to $1 \times 10^{19}\ cm^{-3}$ was achieved using GeH_4 as the dopant source.

3.7.6 Group VI impurities

Group-VI elements include O, S, Se, and Te. When they are incorporated into III–V nitride films, they will occupy the N sites and behave like donors. However, due to their comparatively large atomic radii, they are not expected to dope easily in GaN except oxygen. Among them, oxygen has been studied in detail because it could be one of the main contaminants in the source gases and in the reactor. Seifert et al. [101] postulated that oxygen, substitutionally incorporated onto nitrogen sites, could be the cause for high levels of electron concentration in unintentionally doped GaN layers. They have seen a significant reduction in electron concentration when special care was taken to remove H_2O from the ammonia gas. The donor behavior of oxygen was confirmed in this study and an activation energy of only 28.7 meV was extracted from their variable temperature resistivity measurements.

Besides oxygen, selenium (Se) is another group-VI element that has been studied for its potential use as an n-type dopant in GaN. However, these dopants are not common now since Si is a well behaved and easy to use n-type dopant. Some common impurities and their levels are shown in Fig. 3.21.

3.7.7 Deep levels

Growth of GaN on (111)Si is often used to fabricate HEMT devices it is essential to grow an insulating buffer layer. It is especially problematic on Si since residual Si is often an n-type dopant in GaN. Iron and carbon are the most commonly used impurities to get insulating buffer layer. They lead to two different deep acceptor centers, one in the upper and one in the lower half of the GaN bandgap, respectively. Iron has a trap level at 0.7 eV below the conduction band and carbon has a trap levels 0.9 eV above the valence band, respectively [52, 102]. These dopants are used to get insulating buffer layers of GaN on (111)Si substrates.

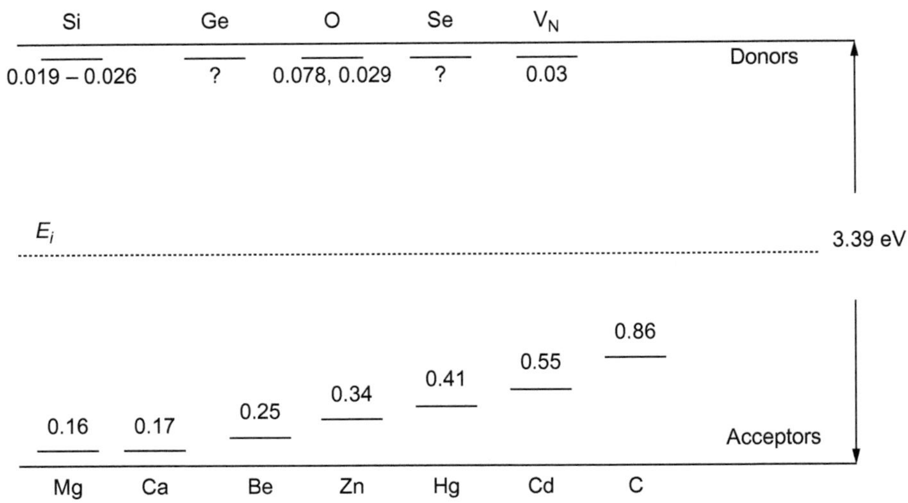

Figure 3.21 Ionization energies of substitutional impurities in GaN.

3.8 Conclusions

This chapter provides a brief review of the properties of GaN and related alloys relevant to power devices. More detailed information are available from the references cited as well as many other review articles that appeared in open literature.

References

[1] S. Strite, H. Morkoc, GaN, AIN, and InN: a review, J. Vac. Sci. Technol. B 10 (1992) 1237.

[2] P. Kung, C.J. Sun, A. Saxler, H. Ohsato, M. Razeghi, Crystallographyof epitaxial growth of wurtzite-type thin films on sapphire substrates, J. Appl. Phys. 75 (1994) 4515.

[3] S. Adachi, Properties of Group-IV, III-V and II-VI Semiconductors, John Wiley and Sons, 2005.

[4] M. Mizuta, S. Fujieda, Y. Matsumoto, T. Kawamura, Low temperature growth of GaN and AIN on GaAs utilizing metalorganics and hydrazine, Jpn. J. Appl. Phys. 25 (1986) L945.

[5] H. Okumura, K. Ohta, G. Feuillet, K. Balakrishnan, S. Chichibu, H. Hamaguchi, et al., Growth and characterization of cubic GaN, J. Cryst. Growth 178 (1997) 113.

[6] I. Petrov, E. Mojab, R.C. Powell, J.E. Greene, L. Hultman, J.-E. Sundgren, Synthesis of metastable epitaxial zinc-b]ende-structure AIN by solid-state reaction, Appl. Phys. Lett. 60 (1992) 2491.

[7] S. Strite, D. Chandrasekhar, D.J. Smith, J. Sariel, H. Chen, N. Teraguchi, et al., Structural properties of InN films grown on GaAs substrates: observation of the zincblende polytype, J. Cryst. Growth 127 (1993) 204.

[8] P.Y. Yu, M. Cardona, Fundamentals of Semiconductors physics and Materials properties, third ed., Springer, New York, 2001.
[9] W.A. Harrison, Electronic strncture and the properties of solids, Dover Publications, New York, 1989.
[10] D.R. Lide, CRC Handbook of Chemistry and Physics, CRC Press, 1992–1993.
[11] H.P. Maruska, J.J. Tietjen, The preparation and properties of vapor-deposited single-crystalline GaN, Appl. Phys. Lett. 15 (1969) 327.
[12] M. Leszczynski, H. Teisseyre, T. Suski, I. Grzegory, M. Bockowski, J. Jun, et al., Lattice parameters of gallium nitride, Appl. Phys. Lett. 69 (1996) 73.
[13] M. Leszczynski, T. Suski, H. Teisseyre, P. Perlin, I. Grzegory, J. Jun, et al., Thermal expansion of gallium nitride, J. Appl. Phys. 76 (1994) 4909.
[14] R.R. Reeber, K. Wang, Lattice parameters and thermal expansion of GaN, J. Mater. Res. 15 (2000) 40.
[15] G.A. Slack, T.F. McNelly, Growth of high purity AlN crystals, J. Cryst. Growth 34 (1976) 263.
[16] G.A. Slack, S.F. Bartram, Thermal expansion of some diamondlike crystals, J. Appl. Phys. 46 (1975) 89.
[17] W.M. Yim, R.J. Paff, Thermal expansion of AlN, sapphire, and silicon, J. Appl. Phys. 45 (1974) 1456.
[18] I.F. Chetverikova, M.V. Chukichev, L.N. Rastorguev, X-ray phase analysis and the elastic properties of gallium nitride, Inorg. Mater. 22 (1980) 53.
[19] V.A. Savastenko, A.U. Sheleg, Study of the elastic properties of gallium nitride, Phys. Status Solidi A 48 (1978) Kl35.
[20] A. Polian, M. Grimsditch, I. Grzegory, Elastic constants of gallium nitride, J. Appl. Phys. 79 (1996) 3343.
[21] K. Tsubouchi, N. Mikoshiba, Zero-temperature-coefficient SAW devices on AlN epitaxial films, IEEE Trans. Sonics Ultrason SU-32 (1985) 634.
[22] U. Sheleg, V.A. Savastenko, Determination of elastic constants of hexagonal crystals from measured values of dynamic atomic displacements, Inorg. Mater. 15 (1979) 1257.
[23] C. Mion, J.F. Mutha, E.A. Preble, D. Hanser, Accurate dependence of gallium nitride thermal conductivity on dislocation density, Appl. Phys. Lett. 89 (2006) 092123.
[24] E.K. Sichel, J.I. Pankove, Thermal conductivity of GaN, 25–360 K, J. Phys. Chem. Solids 38 (1977) 330.
[25] G.A. Slack, T.F. McNelly, AJN single crystals, J. Cryst. Growth 42 (1977) 560.
[26] H. Morkoc, S. Strite, G.B. Gao, M.E. Lin, B. Sverdlov, M. Bums, Large-band-gap SiC, III-V nitride, and II-VI ZnSe-based semiconductor device technologies, J. Appl. Phys. 76 (1994) 1363.
[27] J. Karpinski, J. Jun, S. Porowski, Equilibrium pressure of N2 over GaN and high pressure solution growth of GaN, J. Cryst. Growth 66 (1984) 1.
[28] S. Porowski, I. Grzegory, Thermodynamic properties of III-V nitrides and crystal growth of GaNat high N2 pressure, J. Cryst. Growth 178 (1997) 174.
[29] J. Karpinski, S. Porowski, High pressure thermodynamics of GaN, J. Cryst. Growth 66 (1984) 11.
[30] A. Garcia, M.L. Cohen, First-principles iconicity scales. I. Charge asymmetry in the solid state, Phys. Rev. B 47 (1993) 4215.
[31] M. Schubert, Overcoming the efficiency droop in GaInN light-emitting diodes and novel technologies for *c*-plane GaInN polarized emitters, Ph.D. Thesis, Rensselaer Polytechnic institute, 2009.

[32] F. Bernardini, V. Fiorentini, D. Vanderbilt, Spontaneous polarization and piezoelectric constants of III-V nitrides, Phys. Rev. B 56 (1997). R10 024–R10 027.

[33] P. Waltereit, O. Brandt, M. Ramsteiner, A. Trampert, H.T. Grahn, J. Menniger, et al., Growth of m-plane GaN (11-00): a way to evade electrical polarization in nitrides, Phys. Status Solidi A 180 (2000) 133.

[34] A. Chakraborty, B.A. Haskell, S. Keller, J.S. Speck, S.P. Denbaars, J. Nakamura, et al., Demonstration of nonpolar m-plane InGaN/GaN light-emitting diodes on free-standing m-plane GaN substrates, Jpn. J. Appl. Phys. 44 (2005) L173.

[35] Th Gessmann, J.W. Graff, Y.-L. Li, E.L. Waldron, E.F. Schubert, Ohmic contact technology in III nitrides using polarization effects of cap layers, J. Appl. Phys. 92 (2002) 3740.

[36] Y.-L. Li, J.W. Graff, E.L. Waldron, Th Gessmann, E.F. Schubert, Novel polarization enhanced ohmic contacts to n-type GaN, Phys. Status Solidi A 188 (2001) 359.

[37] E.T. Yu, X.Z. Dang, L.S. Yu, D. Qiao, P.M. Asbeck, S.S. Lau, et al., Schottky barrier engineering in III-V nitrides via the piezoelectric effect, Appl. Phys. Lett. 73 (1998) 1880.

[38] O. Ambacher, J. Smart, J.R. Shealy, N.G. Weimann, K. Chu, M. Murphy, et al., Two-dimensional electron gases induced by spontaneous and piezoelectric polarization charges in N- and Ga-face AlGaN/GaN heterostructures, J. Appl. Phys. 85 (1999) 3222–3233.

[39] M.S. Shur, A.D. Bykhovski, R. Gaska, Pyroelectric and piezoelectric properties of GaN-based materials, MRS Internet J. 4S1 (1999). G1.6 1–G1.6 12.

[40] V. Fuflygin, E. Salley, A. Osinsky, P. Norris, Pyroelectric properties of AlN, Appl. Phys. Lett. 77 (2000) 3075–3077.

[41] E.F. Schubert, Light-Emitting Diodes, second ed., Cambridge University Press, Cambridge, 2006.

[42] V. Narayanan, K. Lorenz, W. Kim, S. Mahajan, Origins of threading dislocations in GaN epitaxial layers grown on sapphire by metalorganic chemical vapor deposition, Appl. Phys. Lett. 78 (2001) 1544.

[43] L.M. Belyaev, Ruby and Sapphire, Amerind Publishing Co., New Delhi, 1980.

[44] L. Liu, J.H. Edgar, Substrates for gallium nitride epitaxy, Mater. Sci. Eng., R 37 (2002) 61.

[45] M. Ishida, T. Ueda, T. Tanaka, D. Ueda, GaN on Si technologies for power switching devices, IEEE Trans. Electron Devices 60 (2013) 3053.

[46] H.-P. Lee, J. Perozek, L.D. Rosario, C. Bayram, Investigation of AlGaN/GaN high electron mobility transistor structures on 200-mm silicon (111) substrates employing different buffer layer configurations, Sci. Rep. 6 (2016) 37588.

[47] N. Ikeda, Y. Niiyama, H. Kambayashi, Y. Sato, T. Nomura, S. Kato, GaN power transistors on Si substrates for switching applications, Proc. IEEE 98 (2010) 1151.

[48] M.J. Uren, J. Möreke, M. Kuball, Buffer design to minimize current collapse in GaN/AlGaN HFETs, IEEE Trans. Electron Devices 59 (2012) 3327.

[49] K.J. Chen, O. Häberlen, A. Lidow, C.L. Tsai, T. Ueda, Y. Uemoto, et al., GaN-on-Si power technology: devices and applications, IEEE Trans. Electron Devices 64 (2017) 779.

[50] C. Zhou, Q. Jiang, S. Huang, K.J. Chen, Vertical leakage/breakdown mechanisms in AlGaN/GaN-on-Si devices, IEEE Electron Device Lett 33 (2012) 1132.

[51] A. Tzou, D. Hsieh, S. Chen, Y. Liao, Z. Li, C. Chang, et al., An investigation of carbon-doping-induced current collapse in GaN-on-Si high electron mobility transistors, Electronics 5 (2016) 28.

[52] M. Silvestria, M.J. Urena, D. Marconb, M. Kuballa, Gan Buffer design: electrical characterization and prediction of the effect of deep level centers in GaN/AlGaN HEMTs, CS MANTECH Conference, 2013, New Orleans, Louisiana, USA, p. 195.

[53] S. Bloom, G. Harbeke, E. Meier, I.B. Ortenburger, Band structure and reflectivity of GaN, Phys. Status Solidi B 66 (1974) 161.

[54] W.C. Lu, K.M. Zhang, X.D. Xie, Band structures and pressure dependence of the bandgap ofGaN, J. Phys. C 5 (1993) 875.

[55] M. Palunmo, C.M. Bertoni, L. Reining, et al., The electronic structure of gallium nitride, Phys. B 185 (1993) 404.

[56] G.D. Chen, M. Smith, J.Y. Lin, H.X. Jiang, S. Wei, M.A. Khan, et al., Fundamental optical transitions in GaN, Appl. Phys. Lett. 68 (1996) 2784.

[57] B. Monemar, Fundamental energy gap of GaN from photoluminescence excitation spectra, Phys. Rev. B 10 (1974) 676.

[58] J.I. Pankove, J.E. Berkeyheiser, H.P. Maruska, J. Wittke, Luminescent properties of GaN, Solid State Commun. 8 (1970) 1051.

[59] D.L. Camphausen, G.A.N. Cornell, Pressure and temperature dependence of the absorption edge in GaN, J. Appl. Phys. 42 (1971) 4438.

[60] J.A. Majewks, M. Stadele, P. Vogl, Electronic structure of biaxially strained wurtzite crystals GaN and AIN, Mater. Res. Soc. Symp. 449 (1997) 887.

[61] A. Kobayaski, O.F. Sankey, S.M. Volz, J.D. Dow, Semiempirical tight binding band structures of wurtzite semiconductors: AIN, CdS, CdSe, ZnS and ZnO, Phys. Rev. B 28 (1983) 935.

[62] C.P. Foley, T.L. Tansley, Pseudo potential band structure of indium nitride, Phys. Rev. B 33 (1986) 1430.

[63] Q. Guo, A. Yoshida, Temperature dependence of bandgap change in InN and AIN, Jpn. J. Appl. Phys. 35 (1994) 2453.

[64] S. Yoshida, S. Miawa, S. Gonda, Properties of AlxGa 1.xN films prepareed by reactive molecular beam epitaxy, J. Appl. Phys. 53 (1982) 6844.

[65] K. Osamura, K. Nakajima, Y. Murakami, P.H. Shingu, A. Ohtsuki, Fundamental absorption edge in GaN, InN and their alloys, Solid State Commun. 11 (1972) 617.

[66] M. Drechsler, B.K. Meyer, D.M. Hofinann, D. Detchprohm, H. Amano, I. Akasaki, Determination of the conduction band electron effective mass in hexagonal GaN, Jpn. J. Appl. Phys. 34 (1995) Ll178.

[67] W. Huang, High-voltage lateral MOS-gated FETs in gallium nitride, Ph.D. Thesis, Rensselaer Polytechnic Institute, 2008.

[68] D.M. Caughey, R.E. Thomas, Carrier mobilities in silicon empirically related doping and field, Proc. IEEE 52 (1967) 2192–2193.

[69] S.J. Pearton, C.R. Abernathy, F. Ren, Gallium Nitride Processing for Electronics, Sensors and Spintronics, Spring-Verlag, London, 2006.

[70] T.T. Mnatsakanov, M.E. Levinshtein, L.I. Pomortseva, S.N. Yurkov, G.S. Simin, M.A. Khan, Carrier mobilities model for GaN, Solid-State Electron. 47 (2003) 111–115.

[71] D. Zanato, S. Gokden, N. Balkan, B.K. Ridley, W.J. Schaff, The effect of interface-roughness and dislocation scattering on low temperature mobility of 2D electron gas in GaN/AlGaN, Semicond. Sci. Technol. 19 (2004) 427–432.

[72] R. Oberhuber, G. Zandler, P. Vogl, Mobility of two-dimensional electrons in AlGaN/GaN modulation-doped field-effect transistors, Appl. Phys. Lett. 73 (1998) 818.

[73] E. Ahmadi, S. Keller, U.K. Mishra, Model to explain the behavior of 2DEG mobility with respect to charge density in N-polar and Ga-polar AlGaN-GaN heterostructures, J. Appl. Phys. 120 (2016) 115302.

[74] A. Nakajima, P. Liu, M. Ogura, T. Makino, K. Kakushima, S. Nishizawa, et al., Generation and transportation mechanisms for two-dimensional hole gases in GaN/AlGaN/GaN double heterostructures, J. Appl. Phys. 115 (2014) 153707.
[75] J. Kolnik, I.H. Oguzman, K.F. Brennan, R. Wang, P. Ruden, Monte Carlo calculation of electron initiated impact ionization in bulk zinc-blende and wurtzite GaN, J. Appl. Phys. 81 (1997) 726–733.
[76] H. Oguzman, E. Belloti, K.F. Brennan, J. Kolnik, R. Wang, P. Ruden, Theory of hole initiated impact ionization in bulk zinc blende and wurtzite GaN, J. Appl. Phys. 81 (1997) 7827–7834.
[77] M. Ozbek, Measurement of impact ionization coefficients in GaN, Ph.D. Thesis, North Carolina State University.
[78] B.J. Baliga, Gallium nitride devices for power electronic applications, Semicond. Sci. Technol. 28 (2013) 074011.
[79] S.D. Lester, F.A. Ponce, M.G. Crawford, D.A. Steigerwald, High dislocation densities in high efficiency GaN-based light-emitting diodes, Appl. Phys. Lett. 66 (1995) 1249.
[80] T.L. Tansley, R.J. Egan, Point-defects energies in the nitrides of aluminum, gallium, and indium, Phys. Rev. B 45 (1992) 10941.
[81] D.W. Jenkins, J.D. Dow, M.H. Tsai, N vacancies in AlxGa1x-N, J. Appl. Phys. 72 (1992) 4130.
[82] P. Boguslawski, E.L. Briggs, J. Beernholc, Native defects in gallium nitride, Phys. Rev. B 51 (1995) 17255.
[83] J. Neugebauer, C.G. Van de Walle, Atomic geometry and electronic structure of native defects in GaN, Phys. Rev. B 50 (1994) 8067.
[84] O. Lagerstedt, B. Monemar, Luminescence in epitaxial GaN:Cd, J. Appl. Phys. 45 (1974) 2266.
[85] W. Qian, M. Skowronski, M. De Graef, K. Doverspike, L.B. Rowland, D.K. Gaskill, Microstructural characterization of a-GaN films grown on sapphire by organometallic vapor phase epitaxy, Appl. Phys. Lett. 66 (1995) 1252.
[86] W. Qian, M. Skowronski, K. Doverspike, L.B. Rowland, D.K. Gaskill, Observation of nanopipes in a-GaN crystals, J. Cryst. Growth 151 (1995) 396.
[87] F.A. Ponce, W.T. Young, D. Chems, J.W. Steeds, S. Nakamura, Nanopipes and inversion domains in high-quality GaN epitaxial layers, Mater. Res. Soc. Symp. Proc. 449 (1997) 405.
[88] Z.L. Weber, S. Ruvimov, C.H. Kisielowski, Y. Chen, W. Swideer, J. Washburn, et al., Structural defects in heteroepitaxial and homoepitaxial GaN, MRS Symp. Proc. 395 (1996) 351.
[89] H.-Q. Lu, Studies on the growth and processing of III-V nitrides for light emitting diode applications, Ph.D. Thesis, Rensselaer Polytechnic Institute, 1997.
[90] H. Amano, M. Kito, K. Hiramatsu, I. Akasaki, P-type conduction in Mg-doped GaN treated with low-energy electron beam irradiation, Jpn. J. Appl. Phys. 28 (1989) L2112.
[91] S. Nakamura, M. Senoh, T. Mukai, Highly p-type Mg-doped GaN films grown with GaN buffer layer, Jpn. J. Appl. Phys. 30 (1991) LI 708.
[92] S. Nakamura, T. Mukai, M. Senoh, N. Iwasa, Thermal annealing effects on p-type Mg-doped GaN films, Jpn. J. Appl. Phys. 31 (1992) L139.
[93] H. Lu, I. Bhat, Magnesium doping of GaN by metalorganic chemical vapor deposition, MRS Proc 395 (1995) 497.
[94] J.W. Huang, H. Lu, I. Bhat, T.F. Kuech, Electrical characterization of Mg doped GaN grown by metalorganic vapor phase epitaxy, Appl. Phys. Lett. 68 (1996) 2392.

[95] T.D. Moustakas, Epitaxial growth of GaN films produced by ECR-assisted MBE, Mater. Res. Soc. Symp. Proc. 395 (1996) 111.
[96] S. Nakamura, N. Iwasa, M. Senoh, T. Mukai, Hole compensation mechanism of p-type GaN films, Jpn. J. Appl. Phys. 31 (1992) 1258.
[97] P. Boguslawski, E.L. Briggs, J. Bemholc, Amphoteric properties of substitutional carbon impurity in GaN and AIN, Appl. Phys. Lett. 69 (1996) 233.
[98] T.F. Kuech, J.M. Redwing, Carbon doping in metalorganic vapor phase epitaxy, J. Cryst. Growth 145 (1994) 382.
[99] S. Nakamura, T. Mukai, M. Senoh, Si- and Ge-doped GaN films grown with GaN buffer layer, Jpn. J. Appl. Phys. 31 (1992) 2883.
[100] B. Rowland, K. Doverspike, D.K. Gaskill, Silicon doping of GaN using disilane, Appl. Phys. Lett. 66 (1995) 1495.
[101] W. Seifert, R. Franzheld, E. Butter, H. Sobotta, V. Riede, Cryst. Res. Technol. 18 (1983) 383.
[102] M. Meneghini, I. Rossetto, D. Bisi, A. Stocco, A. Chini, A. Pantellini, et al., Buffer traps in Fe-doped AlGaN/GaN HEMTs, IEEE Trans. Electron Devices 61 (2014) 4070.

SiC power device design and fabrication

4

Noriyuki Iwamuro
University of Tsukuba, Faculty of Pure and Applied Sciences, Tsukuba, Ibaraki, Japan

4.1 Introduction

The "Internet of Things (IoT)" is one of the most important technology trends during coming next ten years because it has the potential to impact on connecting the consumers and business each other and the social infrastructures. The IoT connects all physical things of people, places, computers, home appliances, cars, and production machinery through the internet. And, all things are equipped with embedded electronics systems, software, and sensors. It is said that 12.5 billion electric apparatus are connected to the network at this time and this numbers is growing rapidly. According to a research company, 50 billion apparatus will be connected throughout the world in 2020. At the same time, the worldwide market for IoT solutions is expected to grow at a 20% CAGR and to seven trillion US$ in 2020. As the IoT technology is growing up, many people immigrate to the urban area and administrative offices will look for ways to make the megacities, their infrastructure and energy systems smarter, more secure and more power-efficient with the help of embedded systems intelligence and modern ICT technology. This is the so-called Smart City and semiconductor devices, especially power semiconductor devices, will play an important role in achieving these objectives. For example, a large number of EVs and Plug-in HEVs (PHEVs) with some intelligent functions will be operated in the smart city, so that an efficient power management system for EVs and PHEVs has to be built to achieve the most efficient powertrain solutions. For a secure data center operation which is essential for the IoT, the highest reliable uninterruptable power supplies (UPSs) are indispensable. The power-efficient generation, storage, management, and distribution of energy are strongly required for smart business/industry and home/consumer. For these applications, medium, and high-power semiconductors like silicon insulated gate bipolar transistor (Si-IGBT) and silicon carbide metal-oxide-field effect transistor (SiC-MOFET) are suitable.

Fig. 4.1 shows a field of power semiconductor devices and their applications. It is clear from Fig. 4.1 that the Si-IGBTs are now applied for many applications such as EV/PHEV/HEV inverter, UPS, and power control unit (PCS) for solar cell. It should be noted that suitable applications field of SiC metal-oxide-field-effect transistors (SiC-MOSFETs) are almost same with the Si-IGBTs described earlier; this means that the SiC-MOSFETs are could be competitor or superior alternatives

Wide Bandgap Semiconductor Power Devices. DOI: https://doi.org/10.1016/B978-0-08-102306-8.00004-6

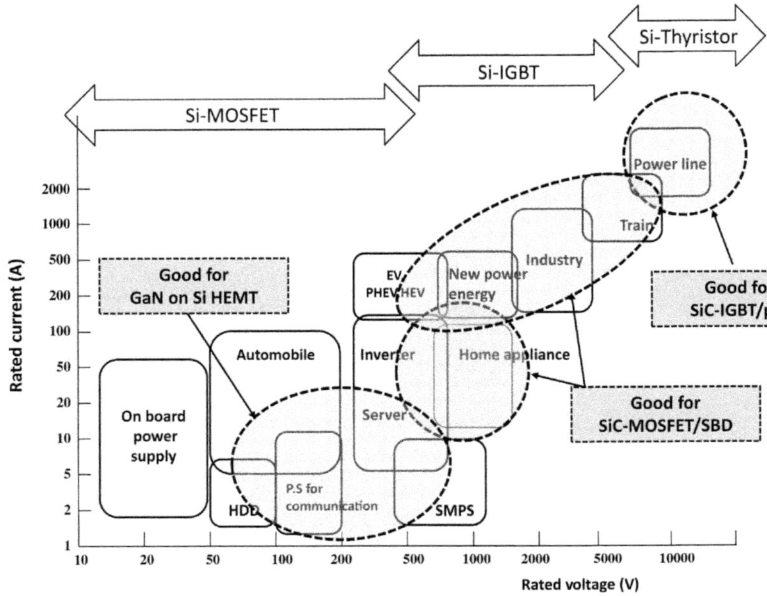

Figure 4.1 Power semiconductor devices and their applications.

to the IGBTs. In the range of application field over 10 kV, Si thyristors are still applied because low on-state voltage drop cannot be achieved by the Si-IGBT structure, so that there are few IGBT devices applied in this ultrahigh-voltage application field. However, since thyristor devices have characteristics of current source operation with large gate current leading to a complex gate control circuit and require passive protection elements (fuses) in case of a short-circuit accident because this device does not have a current saturation capability, the SiC-IGBTs are considered to be superior alternatives to the Si thyristors.

Si-IGBTs have improved significantly over the past 30 years and the progress is not stopping at all. It has been said that the Si-IGBTs are now about to approach their performance limits imposed by the fundamental material properties of silicon; however, their on-state voltage drops and switching characteristics have been still improved by the precise physical analysis and sophisticated fabrication processes, so that their high cost performance is being kept. One example is that state-of-the-art Si-IGBT today is a punch-through construction with a lowly doped field stop layer between drift zone and low efficient back emitter without any carrier life time reduction needs, often called as field stop structure or also as soft punch-through approach [1]. With respect to the well-known theoretical limit of IGBTs on-state performance [2], lot of activities and investigations are running with submicrometer mesas between adjacent deep trenches down to mesas below 0.5 μm [3–5]. Out of that, there are indications that the forward voltage drop can indeed be reduced close to 1 V for a 1200 V blocking rated device. Also within the vertical structure of the IGBTs, there is still room to improve as it is valid. As in the case for every vertical bipolar power device, the thinner the drift region, the lower both on-state and switching losses can be adjusted. Today 1200 V IGBTs with a chip thickness about 100–130 μm are common, 600 V IGBTs with around 70 μm. From theoretical limits for 1200 V blocking capability, chip thicknesses at about 80–90 μm should seem feasible, and for 600 V blocking capability it is about 45–50 μm.

The further aspects within state-of-the-art IGBTs are integration of additional functions within the chip. One example is the implementation of the free-wheeling diode (FWD) function within the IGBT, called as reverse conducting IGBT (RC-IGBT). In 2004, the fabricated results of RC-IGBT were reported at first [6]. Approaches fitting well only for special applications such as inductive heating [7] could be optimized meanwhile in terms of the integrated diode function also for typical consumer and general-purpose drives application in the 600 to 1200 V voltage class range with its needs for diode hard commutation [8,9]. The challenge is mainly to integrate technologically very differently optimized features (lifetime killing of diode, high p-dose in IGBT cell) within one single die. But meanwhile even first inverter and traction applications in the voltage area of 1.2 and 3.3 kV were discussed [10–14].

SiC is one of the wide bandgap semiconductor materials and its wide bandgap and high thermal stability make it possible to operate the SiC devices at very high junction temperatures of over 200°C. Furthermore, SiC is the only compound semiconductor whose native oxide is silicon dioxide (SiO_2), the same insulator as Si. The SiO_2 has good performance as passivation film in Si power devices, so that this makes it possible to fabricate the entire MOS-based devices in SiC. The first SiC Schottky barrier diodes (SBDs) were released in the market in early 2000s, and SiC junction field effect transistors (SiC-JFETs) and MOSFETs were in mid- or late-2000s. At this time, many industries are produced and distributed SiC power devices, especially SiC-SBDs and SiC-MOSFETs, and are taking advantage of the benefits of SiC power devices like the volume and weight reduction of a power supply or an inverter. This means that the real competition between SiC-MOSFETs and Si-IGBTs has just started.

The emphasis in this chapter is on the device processing, design concept of SiC rectifiers and switching devices of MOSFETs and IGBT, features of the unipolar and bipolar devices operations. Further, state-of-the-art SiC device structure and its fabrication process and the characteristics are presented.

4.2 SiC diode

4.2.1 Introduction

SiC is IV–IV compound semiconductor material with a bandgap of 2.3–3.3 eV. One knows that there are many polytypes in the SiC. Owing to the high critical field strength and high electron mobility along the *c*-axis, a 4H-SiC exhibits a significantly higher Baliga's figure-of-merit (BFOM) [15] than other SiC polytypes. This is the main reason why the 4H-SiC has been almost exclusively employed for power device application. Therefore, the discussion is constrained to the 4H-SiC in this chapter.

When realizing SiC power devices that will operate in applications with lower power dissipation compared to Si counterpart, it is very important to design and fabricate devices that make use of the better electrical properties of SiC. The improved device performance should result in less power dissipation for the same blocking voltages for Si devices and higher operating temperatures, so that the benefit is less cooling equipment and fewer numbers of components.

A SBD having a rectification characteristic can be fabricated by depositing a metal on the surface of semiconductor material. The important advantage of the SBD over the pin diode is a very fast recovery characteristic because of the lack of minority carrier injection; therefore, Si SBDs are now applied in many high frequency applications such as low power supplies, for example. When reverse voltage is applied to a typically fabricated SBD, a very small amount of current (leakage current) flows. When forward voltage is applied to the diode, the diode has a small on-resistance. When the SBD has a high Schottky barrier height, the device can restrain a leakage current to increase a breakdown voltage but its on-state voltage drop becomes large. On the contrary, when the Schottky barrier height is low, the on-resistance becomes small but the leakage current becomes large. Therefore, there is a trade-off relation between the leakage current in the reverse electric characteristic and the on-state voltage drop in the forward electric characteristic. From these reasons, the metal should be selected in accordance with a purpose for fabricating SBD. However, since the Schottky barrier height of the SBD is characterized by electron affinity of the semiconductor and the work function of the metal, an optimum SBD for the purpose cannot be always fabricated. If the Schottky barrier height is low, the leakage current increases even in a wide bandgap semiconductor SBD. Also, Baliga mentioned in Ref. [16] that the Schottky barrier lowering in SiC-SBD can be expected to be significantly larger than that in Si-SBD and this leads to a more rapid increase in leakage current with increasing reverse bias. For example, the Schottky barrier lowering is found to be three times larger in SiC SBD for the case of drift region doping level of $1 \times 10^{16}\ \mathrm{cm}^{-3}$. Furthermore, it should be noted that the measured leakage current with applied reverse voltage for the high-voltage SiC-SBD devices is much larger than that can be accounted for with the Schottky barrier lowering model [17–19]. In order to investigate the reasons of this larger leakage current in SiC SBD, Hatakeyama and Shinohe suggested that it is necessary to include the field emission component (or tunneling effect) of the leakage current [20]. According to this literature, the tunneling process dominates the reverse leakage current of the SiC SBD when a high reverse voltage is applied. Thus, the inclusion of the field emission model enhances the leakage current; therefore, the dominant mechanisms of the measured large leakage current of SiC SBD are identified as not only the barrier lowering model but also the field emission one. A diode using a junction barrier Schottky structure (JBS structure) is the structure that combines pin and Schottky diode, and is used as a method for solving these problems mentioned earlier.

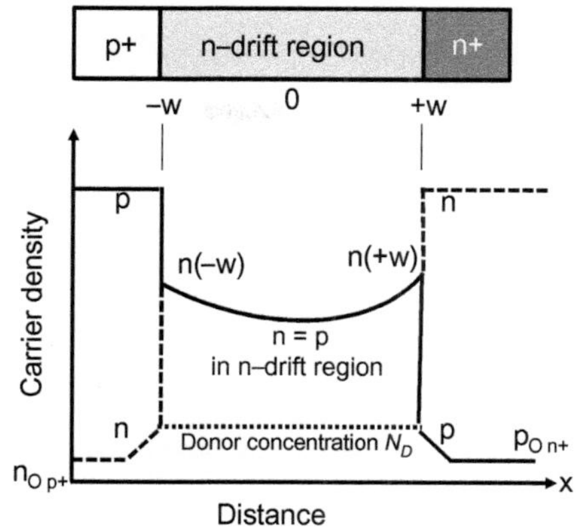

Figure 4.2 Schematic carrier distribution under high-level injection condition for pin diode.

SiC pin diodes have a much thinner drift region when compared with Si devices due to the higher critical electric field for breakdown. This implies that the stored charge in the SiC pin diode is much smaller than the Si pin diode providing an improvement in switching behavior. However, the improved switching performance is accompanied by a substantial increase in the on-state voltage drop associated with the larger energy bandgap for the SiC. Though, the physics of operation of the SiC pin diode is the same with Si one, the parameters for the SiC device defer from the Si devices. For example, this has a strong impact on the voltage drop with the pn junction. The junction voltage drop is given by

$$V_p + V_n = \frac{kT}{q}\ln\left[\frac{n(-w)n(+w)}{n_i^2}\right] \tag{4.1}$$

where k is Boltzmann's constant, q is the electron charge, T is the absolute temperature, and n_i is the intrinsic carrier concentration. The other parameters in this equation are shown in Fig. 4.2. Since the intrinsic carrier concentration for 4H-SiC is very small of 6.7×10^{-11} cm^{-3} at 300K due to its larger energy band cap when compared with 1.4×10^{10} cm^{-3} for Si. Therefore, the junction voltage drop of SiC is 3.24 V whereas that of Si is only 0.82 V if the carrier density in the drift region is assumed to be 1.0×10^{17} cm^{-3}. Therefore, the power dissipation during the on-state in SiC pin diode becomes about four times higher than the Si pin diode. As a result, it is preferable to develop the SiC pin diode for ultrahigh-voltage application above 10 kV.

4.2.2 SiC-JBS device design for low on-state loss

Fig. 4.3 shows a schematic cross section of Schottky, pin, and JBS diode. The JBS diode structure was first demonstrated in Si [21] and is a Schottky structure with a p+ n junction grid integrated into its drift region. The metal layer on top forms ohmic

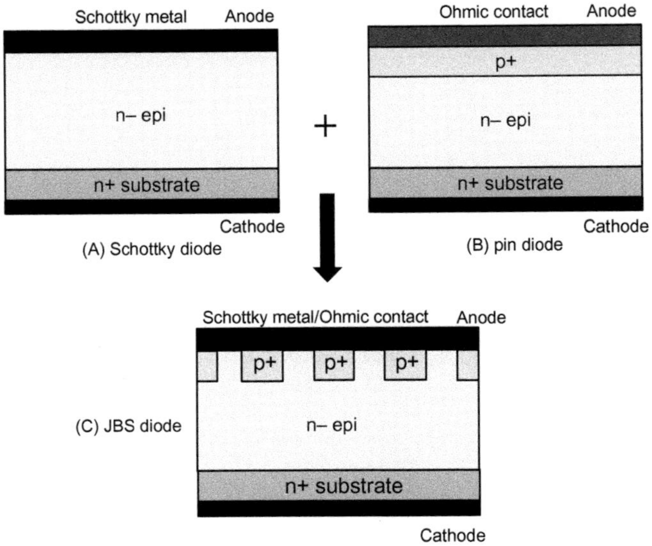

Figure 4.3 Schematic cross section of the Schottky, pin, and JBS diodes.

contacts to the p+ regions and Schottky contacts to the n− regions, so the overall device consists of interdigitated Schottky and pin diodes connected in parallel.

For JBS diodes, the switching power losses are very low and therefore, the design strategy is to minimize the on-state losses for a rated voltage. The static on-state losses consist of forward voltage drop over the Schottky junction and the on-resistance of n− drift layer. The most important design parameters are the n− drift resistance which was determined by n− epitaxial doping concentration and its thickness, the Schottky contact properties such as barrier height, current ideality and the p+ grid dimensions. An SiC-JBS diode was usually fabricated on Si-face (0001) 4H-SiC wafers. An n-type semiconductor region (n− drift layer) under a Schottky electrode is sandwiched between p-type semiconductor regions to deplete the n− drift layer in a Schottky interface portion to thereby restrain a leakage current. In addition to this, when the thickness of a depletion layer (the width of a depletion layer extending from the Schottky interface toward a semiconductor substrate) becomes larger, the leakage current is restrained more greatly. In reverse blocking mode, the p+ n junctions become reverse biased and the depletion layers spread into the channel and pinch off the Schottky barrier. After pinching off, a potential barrier is formed which limits the electric field at the Schottky contact while the drift region supports further increase in voltage. The spacing between the p+ regions should be designed so that the pinch off is reached before the electric field at the Schottky contact increases to the point where excessive leakage currents occur due to tunneling currents. In SiC devices, the electric field strength at the Schottky contact becomes very high and approximately ten times higher than that for Si device. Therefore, the band diagram at the contact becomes so sharp that the potential barrier can be very thin, and the leakage current can be described by a thermionic field emission model [20,22]. This means that the effective method to reduce the leakage current through the Schottky contact is to lower the electric field strength at the Schottky barrier interface so that the potential barrier does not become too thin. For this purpose, it is clear that the JBS structure is suitable.

In forward conduction mode, the current flows in unipolar mode through the multiple conductive channels between the p+ regions under the Schottky contact with a voltage drop determined by the metal semiconductor Schottky barrier height and the drift region resistance. The channel regions have to be spaced far enough apart that their depletion regions do not touch under zero or forward bias condition. This leaves a conductive path through the n− drift region between each Schottky contact and the n+ substrate. Since the forward drop leads to static losses in the SiC-JBS diodes, the thickness and doping concentration of the n− drift layer are carefully designed in order to achieve higher blocking voltage compared to the device rated voltage and lower resistance at the same time. The relation between the forward voltage and the current density is same as that of Schottky diode, except that the Schottky barrier current density needs to take the area taken up by the p+ region into account. Fig. 4.4 exhibits a unit cell of JBS diode with geometrical parameters. The modified current density across the Schottky barrier for a stripe p+ grid design can be written as [23]:

$$J_{F,\mathrm{JBS}} = \frac{s+w}{s-2d} J_F \tag{4.2}$$

where w is the width of the p+ regions and parameter s is the spacing in between the p+ region. The parameter d is the junction depletion width from the p+ regions. By using Schottky barrier theory based upon thermionic emission model the total forward voltage drop of a JBS diode at defined current density can be written as:

$$V_{F,\mathrm{JBS}} = \frac{kT}{q}\ln\left(\frac{J_{F,\mathrm{JBS}}}{A^{**}T^2}\right) + \Phi_B + R_{\mathrm{grid}} \times J_F + R_{\mathrm{drift,JBS}} \times J_F \tag{4.3}$$

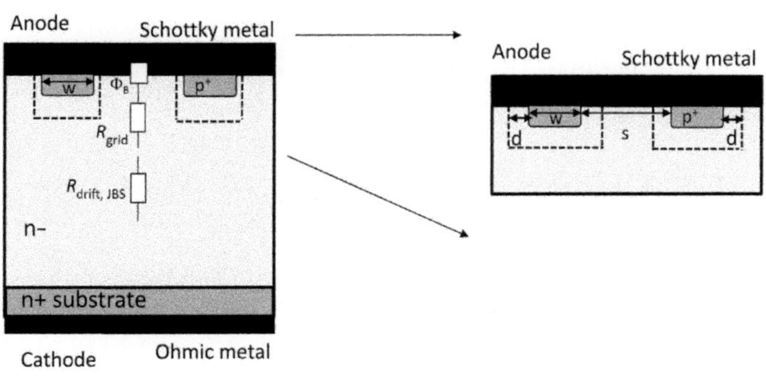

(A) JBS diodes structure (B) Upper part of JBS diode

Figure 4.4 Unit cell of JBS rectifier with geometrical parameters defined.

where Φ_B is the Schottky barrier height, k is Boltzmann's constant, q is the electron charge, T is the absolute temperature and J_F is the forward current density at $V_{F,\mathrm{JBS}}$. Parameter A^{**} is the Richardson's constant. The parameter of R_{grid} is the sum of resistance from the p+ grid and current spreading below the grid. From Eq. (4.3), it is clear that the main parameters to optimize are the drift region resistance, the Schottky barrier height and the p+ grid design, controlling the trade-off in forward voltage drop and leakage current. Lowering of the leakage current, i.e., shielding of the Schottky barrier, without too much increase in the on-resistance can be obtained if an optimized spacing and width of the p+ grid is used. For a nonpunch-through design, i.e., the depletion layer at the blocking voltage is not exceeding the epi thickness, the drift resistance increases quadratically with blocking voltage as shown in Eq. (4.4) [16]. Since the drift resistance is a major part of the JBS on-resistance, it is also important to consider an optimization of the drift region resistance, which is determined by epi thickness and doping according to Eq. (4.5).

$$R_{\mathrm{drift},npt} = \frac{4V_B^2}{\varepsilon \mu_n E_c^3} \tag{4.4}$$

$$R_{\mathrm{drift},npt} = \frac{t_{\mathrm{epi}}}{q \mu_n N_d} \tag{4.5}$$

By introducing a punch-through factor z, defined as the ratio between the electric field at the substrate and the junction field, Eq. (4.4) can be modified for a punch-through epi design:

$$R_{\mathrm{drift},pt} = \frac{4V_B^2}{\varepsilon \mu_n E_c^3} \times \frac{1}{(1 - z^2)(1 + z)} \tag{4.6}$$

Minimizing this expression gives a minimum drift resistance for $z = 1/3$. This reduces the drift resistance by 16% compared to the non-punch-through case for the appropriate epi doping and thickness combinations [24].

A stripe or square/hexagonal grid geometry for the JBS diodes were designed with an optimum p+ region width. The spacing between the p+ regions was optimized mainly dependent upon the n− drift layer concentration; however, for 1200 V device design, the spacing was usually set at about a few micrometers. A punch-though design was usually applied, for example, giving an n− epi thickness of ~10−15 μm and a doping of about $5.0 \times 10^{15}\ \mathrm{cm}^{-3}$ for 1200 V devices [25]. The fabrication sequence of this diode is as follows:

1. After the preparation of SiC wafers, the p+ regions in active area and p+ guard rings in edge termination regions were ion implanted at the same time with aluminum (Al^+) to a depth of ~0.5 μm.
2. Followed by annealing process at 1600−1800°C.
3. A thick SiO_2 deposition.

4. In case of making the junction termination extension (JTE) structure in the JBS device, an optimum dose of Al^+ is usually a smaller value and not the same one of p+ regions mentioned earlier, was used to implement the edge termination.
5. For the Schottky and combined ohmic contact process, titanium (Ti) or molybdenum (Mo) were usually chosen on the top side.
6. On the bottom of the wafer, nickel (Ni) was widely used as ohmic metal layer.
7. And finally, thick aluminum (Al) layer was deposited on top and Ti/Ni/Au layer was on bottom of the wafer, respectively, then the devices were covered with a passivation film such as polyimide.

Though the Ni can efficiently react with SiC and demonstrate ohmic behavior after annealing just below 1000°C, the Ni/SiC interface becomes rough with a large amount of voids and there is a large amount of carbon accumulating on the surface [26]. The accumulated carbon on the surface could lead to practical problems in device such as peeling off the top metal of aluminum and/or bottom metal of Ti/Ni/Au layer; therefore, it should be noted that the ohmic metal process has to be carefully designed.

Fig. 4.5 shows a trench JBS diode structure [27]. Trench structure extended in the p-type regions and in the reverse bias condition, the electric field at the Schottky interface can be successfully reduced. Fig. 4.6 exhibits the simulated electric field of the trench and the standard Schottky diode based upon the structure in this literature. It is clear that the electric field can be successfully reduced by ~60% (1.66 MV/cm and 0.68 MV/cm); therefore, the trench JBS diode can achieve lower on-state voltage drop by applying the Schottky metal with low barrier height Φ_B while keeping its reverse blocking characteristics. Fig. 4.7 exhibits the comparison of calculated forward current–voltage curves between them. Barrier heights Φ_B are set at 1.31 eV for the standard Schottky diode and at 0.85 eV for the trench JBS diode, so that the built-in voltage of trench structure is 0.46 V and it is smaller by about 0.45 V than the standard Schottky structure. This lower built-in voltage contributes to reduce the on-state voltage drop.

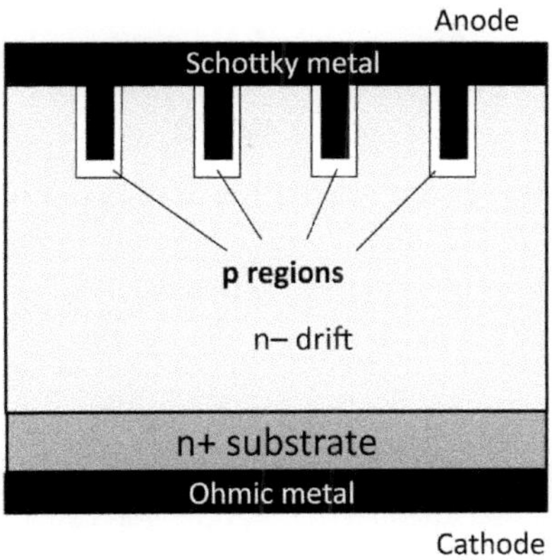

Figure 4.5 Schematic cross section of the trench structure Schottky diodes.

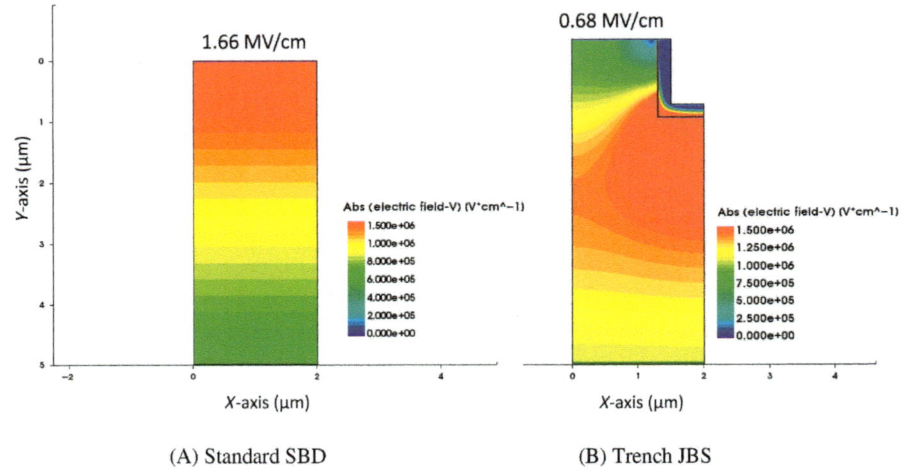

Figure 4.6 Simulated electric field distribution on the drift layer at Vak = −600 V. This result was calculated based upon the device structure described in Ref. [27] by the author of this chapter.

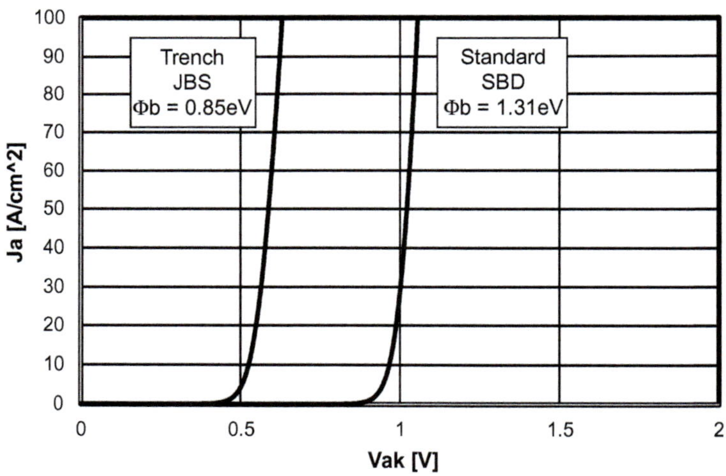

Figure 4.7 Simulated forward IV characteristics of the standard SBD structure and the trench JBS structure.
This result was calculated based upon the device structure described in Ref. [27] by the author of this chapter.

These calculated results are almost same as the measured results reported in Ref. [27], so that this means that the barrier height Φ_B can be lowered in the trench structure, resulting in the improvement of trade-off characteristics between the leakage current and on-state voltage drop.

4.2.3 Edge terminations for SiC-JBS device

The planar junction edge termination with some floating field rings is commonly used in Si power devices. Fig. 4.8 shows this edge structure [16]. One of the advantages of this structure is that the floating field rings can be fabricated simultaneously with the main junction without adding process steps. The breakdown voltage of the termination structure is strongly dependent upon the spacing (Ws_1) of the first floating ring from the main junction and the spacing of each floating rings (Wsn). For example, if the spacing Ws_1 is too large, the electric field at the edge of main junction remains high (point A), resulting in a breakdown voltage similar to the planar junction without the edge structure. When the spacing is too small, a high electric field concentrates on the next-outer edge of the floating ring leading to the possibility of breakdown voltage deterioration. Therefore, an optimum spacing is required to achieve an increase in the breakdown voltage. In Ref. [28], the breakdown voltage of a planar junction with a single floating ring can be calculated using an analytical method. According to this method, optimum field ring spacing from the main junction was calculated 4.5 μm in case of device drift doping concentration of $1 \times 10^{16}\ \mathrm{cm}^{-3}$ and junction depth of 0.9 μm [29]. This is the device design example for 1200 V SiC-JBS diode. However, at this time, in order to achieve more reliability in the edge structure, the multiple field-limiting structures are applied for SiC-JBS devices. A more precise optimization of the spacing can be performed by two-dimensional numerical simulations of the structure. For example, according to 1200 V SiC-JBS devices [30], five field limiting rings are implemented. Since the doping density of n− drift layer in SiC JBS have to be set of about $1 \times 10^{16}\ \mathrm{cm}^{-3}$ for 1200 V device, which is about 100 times higher than the Si devices, the depletion layer is not easy to expand; therefore, the spacings of Ws_1, Ws_2, and Wsn have to be set very small of a few micrometers or less. As a result, the multiple guard ring structures have a drawback that their breakdown characteristics might change easily for the spacing unevenness of Ws_1, Ws_2 to Wsn during its fabrication process.

The other methodology for the junction edge termination is the JTE structure. For silicon devices, the breakdown voltage has been shown to be greatly improved

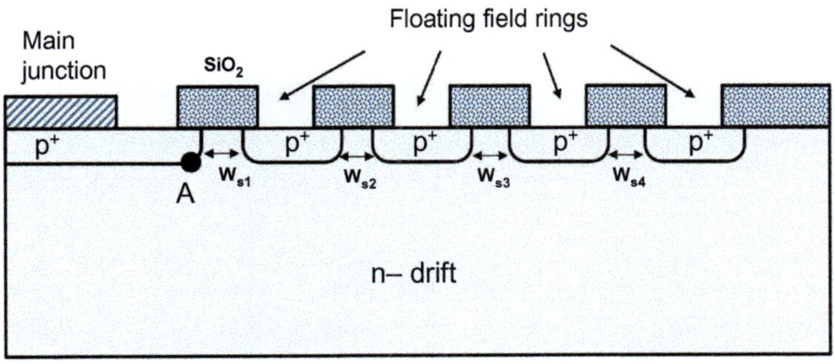

Figure 4.8 Schematic cross section of multiple floating field ring terminations.

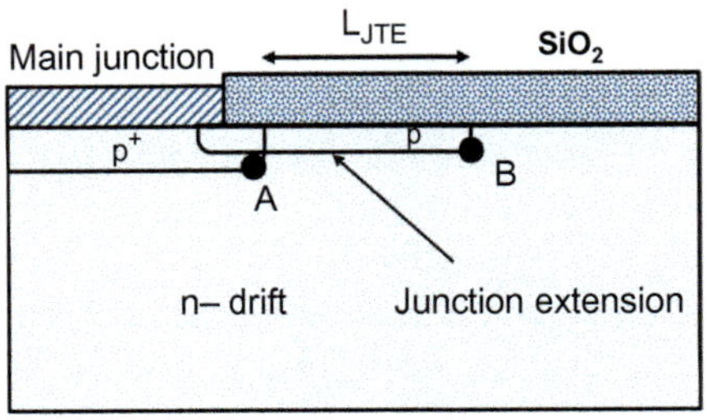

Figure 4.9 Schematic cross section of junction termination extension.

by using this structure. This structure, illustrated in Fig. 4.9, contains a p-type region formed at the periphery of p/n junction [16]. This lightly doped p-type region is usually formed by using ion implantation to precisely control the dopant charge within the layer. When the doping concentration in the p-type region is too high, the breakdown occurs at the edge (point B) at a lower breakdown voltage than the main junction due to its smaller radius of curvature. When the doping is too low in the p-type region, it becomes completely depleted at low reverse bias voltage resulting in breakdown at the main junction (point A) at the same voltage as unterminated junction.

The optimum charge in the p-type region is

$$Q = \varepsilon_s \times E_c \tag{4.7}$$

which amounts to 2.58×10^{-6} C/cm^2 for a critical electric field Ec of 3.0×10^6 V/cm for 4H-SiC. Parameter ε_s is the permittivity of the SiC. This is equivalent to a dopant dose of 1.6×10^{13} cm^{-2}, which is about ten times larger than that used in silicon devices. As in the case of the field limiting structure, precise optimization of the dose can be realized by the numerical simulations. For example, according to Ref. [31], 40 to 50 μm width two zone JTE structures with no spacing are applied for SiC devices. Although the breakdown characteristics of the JTE structure are so sensitive to the lightly doped p-type region, the total dose of this region can be precisely controlled by the ion implantation process, their breakdown characteristics might not change easily for the doping unevenness in the p-type regions.

4.2.4 SiC-JBS device design for higher ruggedness

By operating the diode at a higher forward voltage where the holes start to inject from p+ regions but at the same time having current conduction through the Schottky contact the reverse recovery current is still low with only a little sacrifice in forward voltage and leakage currents. When the JBS diode is operated in this mode it is usually referred to as the merged PiN Schottky (MPS) rectifier [32]. A pn junction in SiC has a large built-in voltage of ~2.5 V and more at room temperature because of the wide bandgap energy; therefore, the forward voltage drop becomes in excess of 2.5 V and looks worse compared to the

Si diode counterpart, so that it is suitable to apply not the MPS but the JBS diode in low and medium voltage applications such as 600–3300 V. However, there are sometimes in accidental operation mode like a surge current flowing through the diode. If this surge current flows through a pure SiC Schottky diode, there would be very high forward voltage drop exceeding 10 V between the anode and the cathode electrode, resulting in device destructive failure due to the huge power dissipation. On the contrary, when the SiC-JBS diode is operated in such mode, the device operates as the MPS diode and the forward voltage drop becomes so small of several volts due to the bipolar operation. As a result of this, the power dissipation becomes much less when compared to the pure Schottky diode and the JBS diode can be successfully withstand such a harsh environment. According to Ref. [33], the forward behavior of SiC-MPS diode allows a two to three times higher surge current density rating than a pure Schottky diode, leading to a I_{FSM} value (surge nonrepetitive forward current, sine half wave, 10 ms) of eight to nine times rated nominal current. Accordingly, the I^2t rating can be increased by roughly a factor of 5.

In an actual power electronics application, it is necessary for the SiC-SBD to exhibit high withstand capabilities, especially avalanche capability. This design was chosen to optimize the electrical field reduction at the Schottky interface, the ratio between p+ regions and Schottky areas and the surge current robustness of the diodes. As a side effect the maximum electrical field in this structure always occurs at the bottom of the implanted p+ regions. If it is ensured that the breakdown voltage of the edge termination is higher than the breakdown voltage of the cell structure, the avalanche occurs only in the cell field. Those power devices are avalanche rugged and therefore very good candidates for studying the avalanche behavior of SiC. Tsuji et al. exhibited measured results of avalanche withstand capability of 1200 V SiC-JBS device [34]. This device was designed the maximum electric field always occurred in the p+ region of the active region. The results exhibit the extremely high avalanche withstanding capability of more than 5000 mJ/cm^2 (@ 25°C) and this characteristic was more than one order of magnitude higher than those of Si-pin diodes. Also, it should be noted that the destruction points of the SiC-JBS device in this measurement was recognized within the active areas. Rupp et al. and Harada et al. reported that 1200 V SiC-JBS device exhibits the extremely high avalanche withstanding capability by optimized design of cells and edge structures combination [35,36]. Rupp et al. also reported that, in order to prove this, two different junction termination concentrations have been tested by simulation and experiment. It is noted that the optimum design is such a way that the termination dose is set close to dose corresponding to the maximum breakdown voltage of the edge termination. In this case, the avalanche current is located in the p-well area of the JBS cell, whereas for an SiC JBS with not-optimized termination design, the avalanche location is the p+ /p− junction of the JTE region. This means that every p well of the JBS structure is contributing to the avalanche current allowing a good stable long term avalanche, on the other hand being a good precondition for well distributed power dissipation in high current single pulse avalanche events by the optimized design.

4.2.5 SiC-JBS and Si-IGBT hybrid type module

Because of the successful achievement of high withstand capability described in Section 4.2.4, the SiC-JBS diodes started to be applied as FWDs in the state-of-the-art Si-IGBT modules. Hybrid type module is a power module using Si-IGBTs as transistors and SiC-JBSs as FWDs. According to Ref. [34], the power loss in the reverse recovery of the SiC-JBS was extremely low compared to that of the Si pin diode and no temperature dependence could be recognized, so that the total power dissipation of the inverter system using the 1200 V SiC-JBS and Si-IGBT hybrid modules was successfully reduced. Thanks to the unipolar operation of SiC-JBS, the reverse-recovery loss and turn-on loss of the hybrid module were successfully reduced by 77% and 51%, respectively. It should be noted that no significant difference could be identified between the both modules in the forward conduction loss of the diodes and Si-IGBTs, and the turn-off loss of the IGBTs. This means that extremely low recovery loss of SiC-JBS helped the reduction of power dissipation when the IGBTs were turned on and leads to a significant improvement of the inverter efficiency; as a result, the total power loss of the inverter system was reduced by 35% compared with that of the standard Si-IGBT module, in which the Si pin diodes were utilized as FWDs(@Vbus = 600 V, fc = 20 kHz). This result demonstrated that the impact of SiC hybrid type modules is quite significant and further impact is realized by all-SiC MOSFET modules, in which SiC-MOSFETs are utilized as transistors and SiC-JBSs are as FWDs, on motor drive and power supplies and so on.

4.2.6 PiN diode

As mentioned earlier, the main advantage of SiC semiconductor is the very low resistance of the drift region even it is designed to support higher voltage. This favors the development of high voltage unipolar devices which have much superior switching speed than bipolar devices. Also, since the built-in potential of SiC pn junction is as high as ~2.5 V in room temperature, the on-state voltage drop of bipolar device becomes high. Therefore, the development of ultrahigh-voltage pin diode has been given significant attention among the researches for very high voltage devices [37]. For example, ultrahigh-voltage and low loss power devices are key components for future smart grids and high-voltage power supplies. A typical voltage of power distribution is 6.5–7.2 kV, where 13–15 kV power devices are required for constructing single-phase converters. Solid-state transformers are one of attractive applications [38]. Since the specific on-resistance of ultrahigh-voltage unipolar devices becomes very high ($>100\ \mathrm{m\Omega\ cm^2}$) even with SiC, bipolar devices will be attractive, owing to the conductivity modulation effect [39,40]. Kimoto et al. presented the device structure and characteristics of ultrahigh-voltage pin diode [41]. In order to support the high voltage, a very thick and low dose n− epitaxial layer was grown on the n+ substrates. For example, the thickness and donor concentration of the n− epitaxial layer was designed 268 μm thick and $(1-2)\times10^{14}\ \mathrm{cm^{-3}}$, respectively. Also, to alleviate electric field crowding, space-modulated JTE (SM-JTE) [42], which is a combination of mesa and the JTE and guard rings structures, was demonstrated. The topside ohmic contact is made using Al/Ti because it has a low hole barrier height to p-type 4H-SiC. The cathode ohmic contact is made using Ni which is same as the JBS diodes. Also, it is important that the p+ anode region was made not by ion implantation but the epitaxial method for such an ultrahigh-voltage pin diode fabrication. This is because the induced defects by the ion implantation for fabrication of the p+ anode region could act as lifetime killer at the vicinity of pn junction, resulting in higher on-state voltage drop [43]. Cross sections of the pin diode and the JTE structure are shown in Fig. 4.10. This pin diode exhibited a breakdown voltage in excess of 26.9 kV. Furthermore, in this ultrahigh-voltage pin diode fabrication process, after growing a very thick and low dose n− epitaxial layer, the carrier lifetime was enhanced by utilizing an elimination process of carbon vacancy (thermal oxidation at high temperature) [44], which has been identified as the lifetime killer after growing a very thick and low dose n− epitaxial layer in this ultrahigh-voltage pin diode fabrication process. By the application of the elimination process of carbon vacancy, minority carrier lifetime was successfully enhanced from 1.8 to 21.6 μs; as a result of this, an extremely low differential on-resistance of 9.72 mΩ cm^2 and on-state voltage drop of 4.72 V (@J = 100 A/cm^2) were achieved. Since the on-resistance of an SBD fabricated on the same wafer was as high as 460 mΩ cm^2, the lower on-resistance of this pin diode indicates remarkable effect of conductivity modulation owing to the higher minority carrier injection as well as the enhanced carrier lifetime. The physics of operation of SiC pin (p+ /n− /n+) diode has been shown to be similar to that used to describe Si devices. High level injection in the drift region enables modulating its conductivity to reduce the on-state voltage drop. Singh et al.

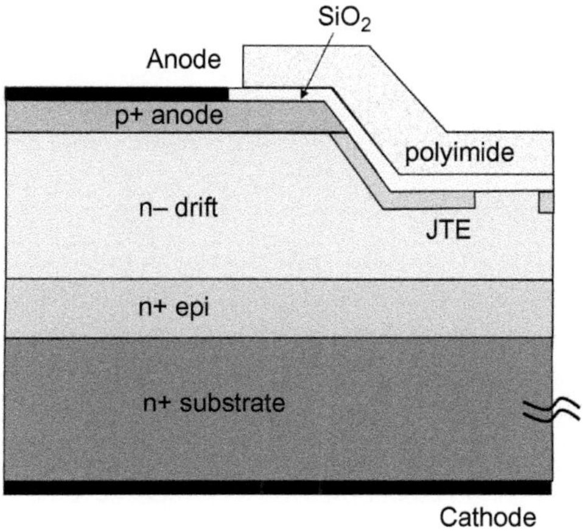

Figure 4.10 Schematic structure of an ultrahigh-voltage SiC pin diode. The pn junction was epitaxially grown. The edge termination was achieved by combination of mesa and Al-implanted. The SM-JTE structure is not shown in this figure.

explained the effect of carrier lifetime on the on-state voltage drop in SiC pin diode [40]. The dominant component in the on-state voltage drop in a pin diode are given by

$$V_F = V_{\mathrm{p+cont}} + V_M + V_{\mathrm{p+n-}} + V_{\mathrm{n+n-}} + V_{\mathrm{subs}} \tag{4.8}$$

where V_F is the on-state voltage drop in a pin diode, $V_{\mathrm{p+cont}}$ is the p contact resistance, V_M is the "middle region" (n− drift region) voltage drop, and $V_{\mathrm{p+n-}}$ and $V_{\mathrm{n+n-}}$ are the junction drops at $\mathrm{p^+n^-}$ and $\mathrm{n^+n^-}$ junctions, and V_{subs} is the resistive voltage drop in the substrate. In case of the ultrahigh-voltage devices, the V_{M} is the largest values among the components mentioned earlier and is determined by the extent of carrier modulation in the n− drift of the pin diode and is dependent upon the carrier lifetime by the following relationship

$$V_M = \frac{3kT}{q}\left(\frac{W}{2L_a}\right)^2 \quad \text{for } W \le 2L_a \tag{4.9}$$

$$V_M = \frac{3\pi kT}{8q} e^{W/L_a} \quad \text{for } W \ge 2L_a \tag{4.10}$$

where k and T are Boltzmann's constant and absolute temperature, L_a is the ambipolar diffusion length, which is given by $L_a = \sqrt{D_a \times \tau_{HL}}$, D_a is the ambipolar diffusion constant given by $D_a = \mu_a \times kT/q$, and τ_{HL} is the high level injection carrier lifetime. The ambipolar carrier mobility μ_a is given by $\mu_a = \mu_{\mathrm{n}}\mu_{\mathrm{p}}/(\mu_{\mathrm{n}} + \mu_{\mathrm{p}})$. Here, μ_{n} and μ_{p} are the electron and hole mobility in the voltage blocking drift layer. $V_{\mathrm{p+n-}}$ and $V_{\mathrm{n+n-}}$ are strongly dependent upon the minority carrier concentration at the two-end regions of the n− drift layer. It is clear from Eqs. (4.9) and (4.10) that the V_M is a function of ambipolar diffusion constant (or carrier lifetime) and n− drift layer thickness W, and a higher carrier lifetime results in better conductivity modulation. In case of the ultrahigh voltage of 26.9 kV pin diode with 268 μm epi layer mentioned earlier, the carrier lifetime of 21.6 μs is so enough to achieve conductivity modulation in the n− drift layer; therefore, in addition to high minority carrier injection from p+ epi anode layer, this device achieved its lower on-state voltage drop of 4.72 V.

To switch the pin diode from its on-state operation to the reverse blocking mode, it is necessary to remove the stored carriers to enable the formation of a depletion region in n− drift layer. In power electronics circuits, power rectifiers are usually used with an inductive load and the improvement of their reverse recovery characteristics is sometimes strongly required. However, in case of the power electronics application for ultrahigh-voltage devices of in excess of 10 kV, total power dissipation of the pin diode is mainly occupied by not their reverse recovery loss but its on-state conduction one, so that the achievement of their lower on-state voltage drop is essential for the SiC pin diodes. The theory and basic operating principle of reverse recovery of SiC pin diode are reviewed and described in many papers,

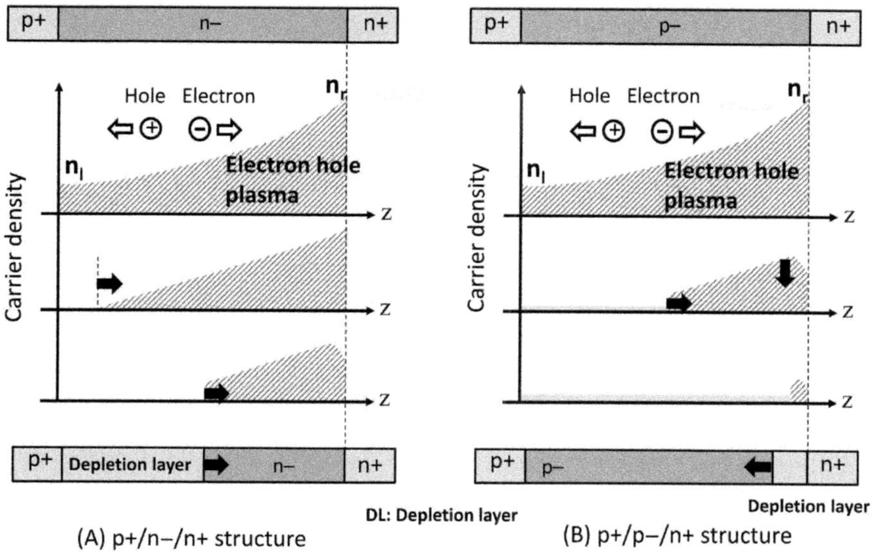

Figure 4.11 Qualitative distributions of injected EHP during the recovery process [46].

e.g., in Refs. [22,45] and in several SiC semiconductor books [16,29], so the details of its operations are not derived in this chapter.

For pulse power applications, a high-voltage pin diode which shows extremely high speed voltage pulse is strongly required. In order to achieve the high breakdown voltage in excess of 10 kV and high speed voltage pulse in the range of a few nanoseconds or less at the same time, Si-drift step recovery diode (Si-DSRD) structure was reported in [46,47]. And, SiC-DSRD with not the p+/n−/n+ structure but the specially-designed p+/p−/n+ one was fabricated and demonstrated with single die [48]. This device was designed to exhibit an extremely hard reverse recovery characteristic with very fast voltage rise time. The reverse recovery behavior of p+/p−/n+ structure was drastically different from that of the p+/n−/n+ one, which means that the d*I*/d*t* of the recovery current was very sharp in the p+/p−/n+ structure. Fig. 4.11 shows the qualitative distribution of injected electron–hole plasma (EHP) during the recovery process. After finishing the initial stage of recovery process when the hole and electron concentrations at the left and right junctions are reduced to their thermal equilibrium values "sweeping-out" boundaries are formed. These "sweeping-out" boundaries moves toward each other with different velocities. The velocity of the left boundary is

$$v_l = \frac{\mu_n}{\mu_n + \mu_p} \times \frac{J_R}{q n_l} \tag{4.11}$$

where μ_n, μ_p are the mobilities of electrons and holes, respectively, q is the electron charge, n_l is the EHP density at the left boundary and J_R is the reverse current density. Correspondingly, the velocity of the right boundary is

$$v_r = \frac{\mu_p}{\mu_n + \mu_p} \times \frac{J_R}{qn_r} \tag{4.12}$$

where n_r is the EHP density at the right boundary.

It is well known that in 4H-SiC the electron mobility, μ_n is approximately seven times higher than the hole one, μ_p. Besides, the density of injected EHP at the left junction, n_l, is lowered as compared to that at the right junction, n_r. For p+ /n− /n+ diodes, owing to $\mu_n/n_l >> \mu_p/n_r$, the region near blocking p+ /n− junction becomes free of minority carriers very soon, so that restoration of depletion layer occurs in presence of the EHP thus predetermining soft recovery behavior. For p+ /p− /n+ diodes, the EHP wave front is first formed near nonblocking p+ /p− region and moves toward the p− /n+ junction with high velocity due to high μ_n/n_l ratio. Besides, the EHP density at the blocking p− /n+ junction starts to reduce very slowly due to low μ_p/n_r ratio. Certain conditions can be selected providing simultaneous events for the wave front to reach the p− /n+ junction and for EHP density to become zero here. So, all excess carriers are sustained away from the base by the instant for depletion layer to start restoration. Then the depletion layer boundary will move to the left with extremely high velocity giving abrupt break of the reverse current.

As mentioned earlier, the SiC DSRD with p+ /p− /n+ structure could show extremely high speed voltage pulse because of sharp decrease of recovery current; however, it is very tough to fabricate and to obtain a high breakdown voltage over 10 kV as a single die because of the difficulty of making a thick p− epi layer and applying a dry etching process for the thick p-epi layer to fabricate a mesa edge structure. Also, there are hardly good p-type SiC substrate with both 4 in. or larger diameter and low resistivity. Therefore, one can fabricate the p+ /p− /n+ structure by utilizing a thin p- epitaxial layer on n+ SiC substrate as shown in Fig. 4.12. Since the thin p-epitaxial layer (below 10 μm) was applicable on the n+ substrate for fabrication of this diode structure, its blocking voltage was attained as low as 1000 V and more. In order to make high voltage characteristics of more than 10 kV and nano/subnanoseconds fast SiC diode for the pulse power application, a large number of diode stacks with dies connected in series were required, resulting in degradation of on-state characteristic and extremely high chip cost. To solve these issues, a newly developed 4H-SiC DSRD was proposed and successfully fabricated [49]. A schematic cross section of this device is shown in Fig. 4.13. The thickness of p-epi layer was optimized as thin as 9 μm in order to achieve the stable and reliable breakdown characteristic by the application of dry etching to make a 10 μm depth mesa edge structure with JTE layers. Furthermore, an extremely low dose of 1.0×10^{15} cm^{-3} and 8 μm thick n− drift layer was grown between the p-drift and n+ substrate in order to attain the higher breakdown voltage and higher dV/dt values by the punch-through of depletion layer during switching operation. The breakdown voltage was achieved 2.8 kV as a single die. The DSRD with smaller

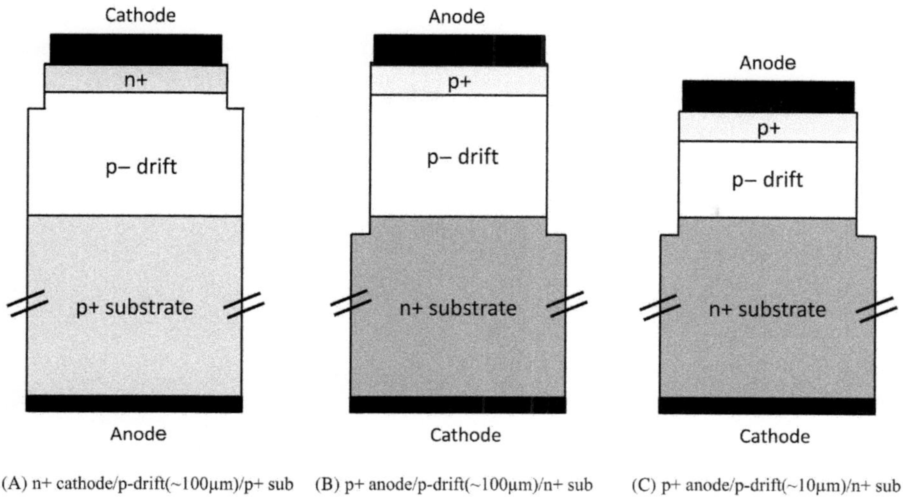

Figure 4.12 Candidate for a ultrahigh-voltage (>10 kV) SiC p+ /p− /n+ pin diode structure.
(A) n+ cathode/p− drift/p+ sub. Not applicable due to lack of p+ substrate
(B) p+ anode/p-drift (∼100 μm thick)/n+ sub.
Not applicable due to requirement of a deep dry etching of in excess of 100 μm
(C) p+ anode/p− drift (∼10 μm thick)/n+ sub. Applicable structure.

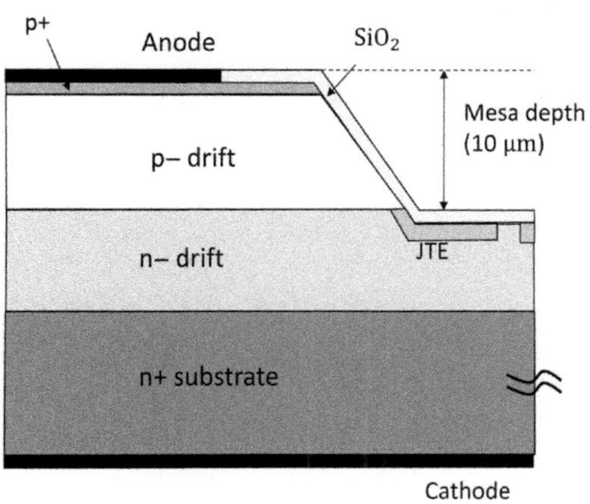

Figure 4.13 Schematic cross section of the newly developed SiC p+ /p− /n− /n+ DSRD.

numbers of five die stacks, which are just a half numbers when compared with the Si DSRD, was fabricated and packaged to exhibit the higher breakdown voltage of 14 kV as shown in Fig. 4.14 and its measured switching waveforms were demonstrated in Fig. 4.15. It should be noted from this figure that the higher peak voltage value of 11.0 kV and the leading edge of voltage rise time as small as 2.3 ns was successfully achieved at the same time.

Figure 4.14 Packaged 14 kV SiC DSRD. Five dies connected in series are succesfully assembled.

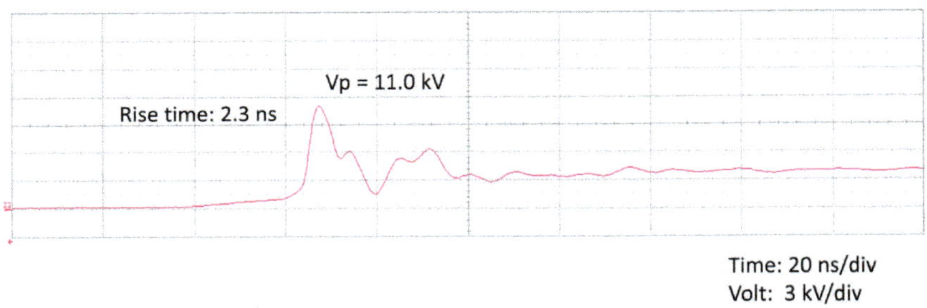

Figure 4.15 Measured reverse recovery voltage waveform of the packaged 14 kV SiC DSRD. V_{peak} = 11.0 kV and rise time of 2.3 ns were successfully acheived.

4.2.7 Bipolar degradation

In the case of 4H-SiC bipolar devices, the forward bipolar degradation is still a problem. In 2001, Bergman et al. [50] reported a correlation between this forward bipolar degradation and the expansion of stacking faults (SFs) coming from the basal plane dislocations (BPDs) in the substrate. In Ref. [51], mechanism of the bipolar degradation was explained as follows:

1. After minority carrier injection followed by recombination, nucleation and expansion of a single Shockley stacking fault (SSF) take place at the location of a BPD or at a basal plane segment of other dislocations replicated in the n− epi layer.
2. The expanded SSFs cause a significant reduction in the carrier lifetimes and probably in the formation of potential barriers for carrier transport.
3. This leads to an increase in the forward voltage drop in SiC bipolar devices such as pin diodes.

The SSF expansion mechanism has been studied and it seems to be intrinsic to the 4H-SiC material. Therefore, complete elimination of the nucleation sites is essential for the development of SiC pin diodes. In order to prevent the SSFs expansion from the BPDs replicated in the n− epi layer, enhanced conversion from the BPDs to threading edge dislocations (TEDs) during epitaxial growth is strongly required. Kimoto et al. reported [52] that the BPD to TED conversion is enhanced under slightly C-rich growth condition. Appropriate in situ hydrogen etching prior to epitaxial growth is also helpful. Further, the BPD density in epi layers tends to decrease with increasing the growth rate. Thus, the BPD density in n− epi layer is estimated to be 0.1 cm^{-2} or less. In addition to this, it should be noted that the new device structure to suppress the forward voltage degradation has been reported from the view point of SiC pin diode designing. Tawara et al. [53] demonstrated that the nitrogen and boron (N + B)-doped n− buffer layer with shorter minority carrier lifetime was very effective to suppress the SSF expansion and the bipolar degradation. This "recombination enhanced layer" that is inserted between the n+ substrate and n− drift layer successfully decreases hole density near the n+ substrate/n− epi interface to prevent the expansion of SSF from BPD segments as shown in Fig. 4.16. The pin diode with ~2 μm thick (N + B)-doped buffer layer was fabricated and its characteristic was measured. The N and B concentrations were determined to be 4×10^{18} and 7×10^{17} cm^{-3}, respectively, and this buffer layer is estimated to have a very short lifetime of less than 30 ns (@250°C). The 16 pin diodes with (N + B)-doped buffer layer were subjected to 1 h current stress test at 600 A/cm^2 and the all diodes exhibited the negligible change of forward voltage drop and no SSF expansion.

The bipolar degradation is the most serious problem in the development of SiC bipolar devices, therefore, not only the reduction of BPDs in bulk growth, enhanced conversion from BPDs to TEDs during epitaxial growth, and elimination of BPD nucleation during device processing but also the application of recombination

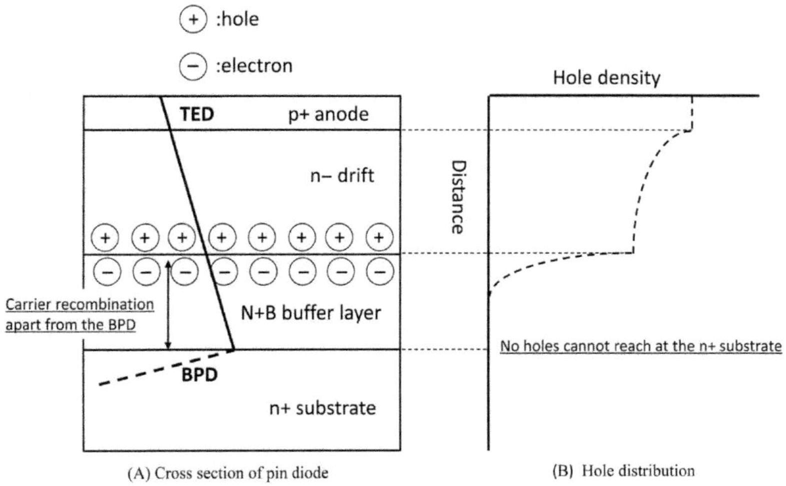

Figure 4.16 (A) Schematic cross section of pin diode with the (N + B)-doped buffer layer. (B) Hole distribution in pin diode with the (N + B)-doped buffer layer in forward conduction state.

enhanced buffer layer to decrease the hole density near the n+ substrate/n− epi interface are the most important remaining issues in SiC bipolar device technologies.

4.2.8 Summary

The SiC-JBS diode structure can achieve a significant improvement in leakage current by shielding the Schottky contact against high electric field. Switching power losses are very low for the SiC-JBS diodes; therefore, the design strategy is to minimize the on-state losses for a rated voltage. The static on-state losses consist of forward voltage drop over the Schottky junction and the on-resistance of n− drift layer, so that the most important design parameters are the n− drift doping concentration and its thickness, Schottky contact properties such as barrier height, ideality factor and the p+ grid dimensions. Also, since the higher surge current and avalanche capabilities are strongly required in many power electronics applications, it is necessary for the SiC-JBS diodes to design to operate as MPS diodes in very large forward current regions and the breakdown voltage of main cells and edge termination properly.

The SiC pin diode structures have much thinner drift layer when compared with Si devices due to the higher critical electric field for breakdown. This favors a faster switching speed with reduced reverse recovery current. However, the larger bandgap for SiC produces four times higher on-state voltage drop than the Si diode. Therefore, SiC pin diodes are of strong interest for ultrahigh-voltage application above 10 kV. The SiC pin diodes can achieve a low on-state voltage drop due to the high level injection in the drift region while showing the very high blocking voltage. For example, a breakdown voltage over 26.9 kV and a differential on-resistance of 9.7 mΩ cm^2 were achieved for a pin diode having a 268 μm-thick n− layer through enhancement of carrier lifetime and optimization of the junction termination. And, interestingly, the reverse recovery behavior was drastically different between the p+/n−/n+ structure and the p+/p−/n+ one in the pin diode structure. The p+/n−/n+ structure, which is popular for many power electronics applications, shows the soft recovery behavior, however, the p+/p−/n+ structure exhibits the hard recovery of the very sharp dI/dt characteristics, resulting in extremely high speed voltage pulse of 11.0 kV and 2.3 ns as its rise time, so that this characteristic is suitable for the extremely high voltage pulse power applications.

The bipolar degradation is still a big problem in the pin diode. It was well known that this was caused by the nucleation and expansion of SSF and its complete elimination was essential. Recently, in order to prevent the SSFs expansion from the BPDs replicated in the n− epi layer, the new epi growth technologies were proposed to attain the enhanced conversion from the BPDs to TEDs. On the other hands, the N + B buffer application between n− drift and n+ substrate was proposed and it was clear that this method was so effective to solve this problem. However, this N + B buffer layer was grown epitaxially, so that its additional fabrication cost might not be negligible.

4.3 SiC-MOSFET

4.3.1 Introduction

The vertical power MOSFET structure was developed using in Si technology in the mid-1970s to obtain improved performance when compared with the existing Si power bipolar transistor. One of the major issues with the power bipolar transistor structure was low current gain when designed to support high voltage. In addition, the power bipolar transistor could not be operated at high frequency due to the large storage time related to the injected minority carriers in their drift regions. The replacement of these current-controlled devices with a voltage-controlled device was attractive from the application point view. In Si power MOSFETs, the on-resistance becomes dominant by the resistance of the drift region when the breakdown voltage exceeds about 200 V. At high breakdown voltages, the specific on-resistance for these devices becomes greater than 10 mΩ cm^2, leading to an on-state voltage drop of more than 1 V. For this reason, the IGBT was developed in the 1980s to serve medium- and high-power systems. The superior performance of the IGBT in high voltage applications makes the Si-MOSFET go to the applications with operating voltage below about 200 V. Superjunction MOSFET structure that utilize the charge-coupling concept has allowed extending the breakdown voltage of power MOSFET to the 600 V range [54,55]. However, their specific on-resistance is still large limiting their use to high frequencies where switching loss dominates.

As discussed in Section 4.2, the main advantage of SiC semiconductor device is the very low resistance of the drift region even it is designed to support higher voltage. This favors the development of high voltage unipolar devices which have much superior switching speed than bipolar devices. Therefore, much lower resistance of the drift region in SiC should enable the development of power MOSFETs with high breakdown voltage. These devices offer not only fast switching speed but also lower on-resistance when compared with Si-IGBT. Fig. 4.17 shows schematic cross section of two types of power MOSFETs in the form of the planar DMOSFET and trench UMOSFET. The name of DMOSFET derives from the Si device of the same name, where the n+ source and p-base regions are formed by diffusion of n-type and p-type impurities through the same photomask opening. In SiC, the same structure is formed by double implantation. The name of UMOSFET derives from the U-shaped gate geometry, but the term "trench MOSFET" is also used. As shown in Fig. 4.17, both DMOSFETs and UMOSFETs consist of an MOSFET formed above a thick n− drift region, with the n+ substrate serving as the drain terminal. The blocking voltage in the off state is supported by the reverse-biased junction between the p-base and the n− drift region, but it is also supported by the MOS capacitor formed by the gate and the drift region. The on-resistance of the power MOSFET is the sum of the on-resistance of the MOSFET channel plus the on-resistance of the drain region and the substrate. Fig. 4.18 exhibits the DMOSFET structure with its internal resistance components [16]. There are eight resistances that must be analyzed to obtain the total on-resistance between the source and the drain electrodes when the device is in the on-state. The total

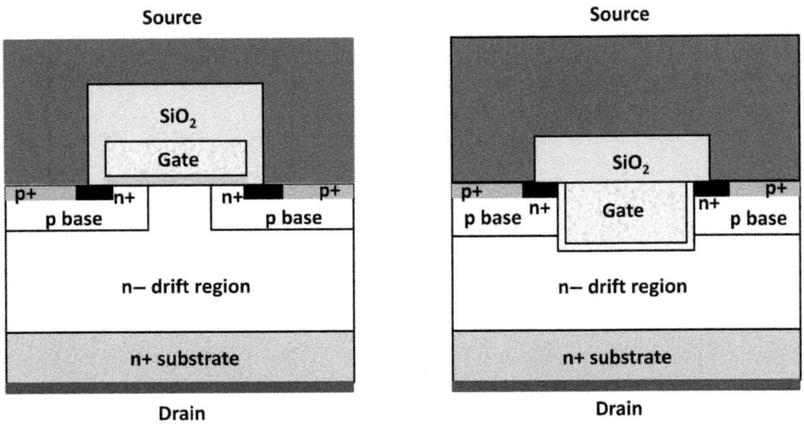

(A) Planar DMOSFET (B) Trench UMOSFET

Figure 4.17 The basic structure of the planar DMOSFET and trench UMOSFET.

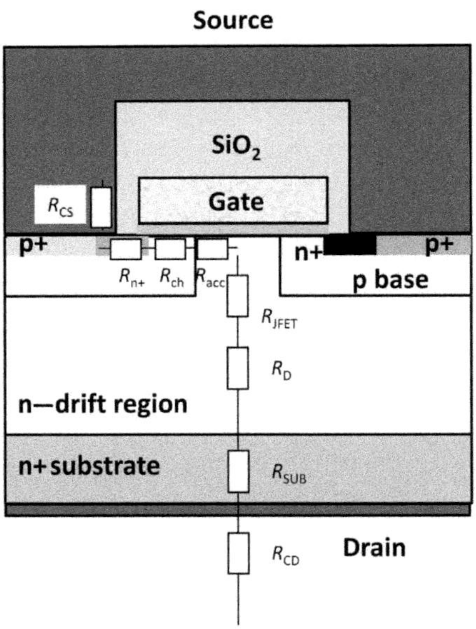

Figure 4.18 DMOSFET structure with its internal resistance.

on-resistance for the power DMOSFET structure is obtained by the addition of the all resistance because they are considered to be in series the current path between the source and the drain electrodes:

$$R_{ON} = R_{CS} + R_{n+} + R_{ch} + R_{acc} + R_{JFET} + R_D + R_{SUB} + R_{CD} \quad (4.13)$$

The goal is to minimize the total on-resistance for a given blocking voltage. This involves adjusting device parameters (doping density, thickness, and lateral dimensions of each layer/regions) to minimize the total resistance. A comprehensive analysis must also include the two-dimensional effects due to current crowding and field crowding. Current crowding occurs at the end of the MOS channel as current spreads into the JFET region, and again at the bottom of the JFET region as current spreads into the wider drift region. In the blocking state, field crowding occurs at the corners of the p-base regions within the cell and at the periphery of the DMOSFET. These effects are usually analyzed by device simulation. The goals of minimizing resistance, minimizing area, and maintaining blocking voltage have resulted in several innovations. For example, a variety of technologies such as a self-align submicrometer MOS channel, a more heavily doped JFET region, p-base contacts that are segmented along the length of the fingers, and the trench gate structure were developed and applied to a modern Si power MOSFETs. However, unfortunately, the power MOSFET structures developed in Si mentioned earlier cannot be directly utilized to form high performance SiC-MOSFET devices. First, the lack of significant diffusion of dopants in SiC materials prevents the use of the Si DMOSFET process. Second, a high electric field occurs in the gate oxide of the SiC-MOSFET exceeding its rupture strength leading to catastrophic failure of devices in the blocking mode at high voltages. Third, the quality of the SiO_2/SiC must be improved to allow good control over the threshold voltage and the channel mobility. These factors pose challenges in the fabrication of SiC devices and limit the types of device structures that can be realized in the material.

The SiC-MOSFETs were first reported in Refs. [56]and [57], and the first SiC power MOSFETs were vertical trench MOSFET structure [58]. The Si trench-gate power MOSFET was developed in the 1990s by applying the trench technology originally developed for dynamic random access memories (DRAMs). The trench gate or UMOSFET structure enabled significant increase in the channel density and elimination of the JFET resistance contribution resulting in a major enhancement in low voltage (<100 V) Si-MOSFET performance. This has motivated the development of the trench-gate device in the SiC. Also, it should be noted that the SiC UMOSFET structure is attractive because the base and the source regions can be formed epitaxially without the need for ion implantation and associated high temperature annealing. Therefore, the first SiC power MOSFETs were UMOSFET structure as described earlier. A disadvantage of the UMOSFET structure becomes apparent when one considers the peak electric fields in the device. At the onset of avalanche breakdown, the electric field at the p–n junction in SiC is almost ten times higher than in Si, owing to the higher critical field of the SiC. At the MOS interface, the field in the oxide exceeds the semiconductor field by the ratio of the dielectric constants, a factor of 2.5. This places the field in the oxide at $\sim$4 MV/cm, a value dangerously close to oxide breakdown. The geometry of the trench corner further increases the oxide field due to field crowding, resulting in local oxide failure. In 1995, SiC UMOSFETs fabricated on the carbon face of SiC had achieved blocking voltages about 260 V [59], limited by oxide breakdown at the trench corners. An obvious way to avoid the problem with oxide breakdown at the trench corners is to eliminate the trench structures. This was accomplished with the introduction of planar double implanted DMOSFET [60].

4.3.2 Device structure and its fabrication process

4.3.2.1 Planar MOSFET structure

The basic structure of the planar power MOSFET is shown in Fig. 4.17, which is same with the Si-power MOSFET. However, in recognition of some difficulties of fabrication process such as high temperature ion implantation and low diffusion coefficients for dopants in SiC, the design of SiC power MOSFET have to be totally different from the Si-MOSFET counterpart, so that the fabricated DMOSFET [60] showed a high specific on-resistance of 125 mΩ cm^2 whereas the device could sustain a blocking voltage of 760 V.

Fabrication process

Fig. 4.19 exhibits a schematic fabrication process procedure for planar SiC-MOSFET. Fabrication process consists of 16 major steps shown as follows:

1. Preparation of SiC wafers with n− epi/n+ substrate.
2. SiO_2 deposition on top of the SiC wafer to protect the wafer surface from the following ion implantation damages.
3. Photolithography and ion implantation with aluminum (Al^+) of the p-base regions. The p+ guard rings in the edge termination regions are also implanted at the same time.
4. Photolithography and ion implantation with phosphorus (P −) or nitrogen (N −) of the n+ source regions.

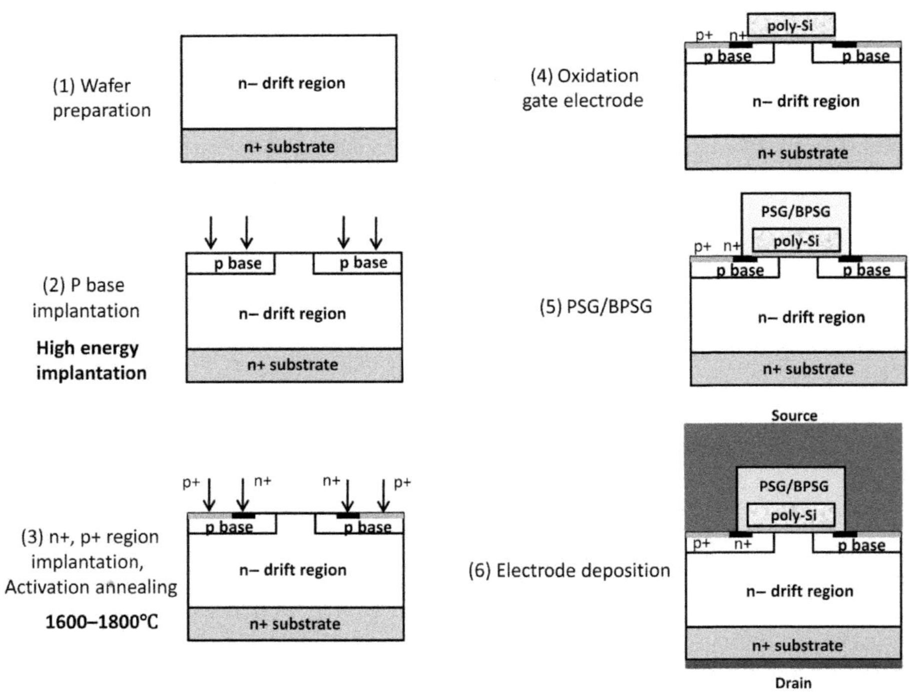

Figure 4.19 Schematic fabrication process procedure for planar SiC MOSFET.

5. Photolithography and ion implantation with aluminum (Al^+) of the p+ regions.
6. After cleaning of the wafer, activation of the dopants at 1600–1800°C.
7. Gate oxidation process. 50–100 nm SiO_2 was thermally grown or deposited by CVD and postoxidation annealing is performed, for example, in wet oxygen and NO/N_2O.
8. Deposition of the gate electrode of doped polycrystalline Si (poly-Si) with the thickness of 500–800 nm. Subsequent annealing of the poly-Si is performed.
9. Dry etching of the gate poly-Si layers.
10. Deposition of phosphorus silicate glass (PSG)/boro-phosphorus silicate glass (BPSG) with ~1 μm thick for the top passivation layer.
11. Dry etching of the top passivation layer and gate oxide to open the source contact.
12. Deposition of the source and the drain ohmic contacts composed of Ni.
13. Silicidation of the source and the drain contacts (~1000°C). Rapid thermal annealing (RTA) method is usually used.
14. Deposition of the source pad metal of aluminum with annealed ohmic contacts.
15. Etching of the source pad metal pad. On the bottom of the wafer, Ti/Ni/Au layer was usually deposited.
16. Polyimide layer was deposited as a passivation film.

This device has been called DIMOSFET because of the double-implant process used for the fabrication. Since the wide-range and precise doping control of both n-type and p-type regions can be achieved by ion implantation, this process is a key process in fabrication of SiC-MOSFET as well as Si-MOSFET. However, there are major differences between ion implantation technologies for SiC and Si as follows [51]:

- Because of the extremely low diffusion constants of dopants within SiC, selective doping by a diffusion process is not realistic. Diffusion of most implanted impurities during post-implantation annealing is negligibly small.
- If the as-implanted lattice damage is of near-amorphous level, lattice recovery (that is, repairing this damage) is very difficult. Therefore, implantation at elevated temperature is often employed, especially when the implant dose is very high. On the other hand, implantation at room temperature is sufficient when the implant dose is not very high, as long as appropriate annealing is carried out.
- Irrespective of the implant dose and the implantation temperature, postimplantation annealing at very high temperature (>1500–1600°C) is required to attain lattice recovery and high electrical activation ratios. Such high-temperature annealing may cause incongruent evaporation of silicon and surface roughening.

The diffusion rate for dopants is extremely small in SiC even at very high temperature such as 1600–1800°C, so that it is not feasible to fabricate the power MOSFET structure like Si-MOSFET. For example, it was proposed that the p-base and n+ source ion implantation can be operated separately by using photomasks [51]. In Si-MOSFET fabrication, so-called self-aligned process was usually applied, that is p-base and n+ source regions were ion implanted by using the photomask defined by the gate edge, followed by the thermal diffusion of implanted regions. However, in case of SiC-MOSFET, the p-base and n+ source regions made by ion implantation and thermal annealing processes have to be fabricated before the gate oxidation/gate poly-Si electrode process because the gate oxidation layer deteriorates and fails due to the extremely high annealing temperature of 1600–1800°C. Therefore, the self-aligned implantation process for p-base and n+ source layer cannot be applied in SiC-MOSFET, and realignment tolerances must be allowed between the p-base, n+ source, and gate features. This might result in a little sacrifice for achievement of short channel length. Furthermore, in Si power MOSFET, photoresist masks are usually applied for the selective formation of n-type and p-type region by ion implantation. A photoresist is easy to coat on the surface of Si wafer, and this method is a very simple and a cost-effective one. However, in SiC-MOSFET case, this photoresist method cannot be applied because the high temperature ion implantation of ~500°C was usually required and this could result in harsh damage of the coated photoresist layer. As a result, silicon dioxide layer (SiO_2) has to be deposited, which is a relatively time-consuming process compared with coating photoresist method, as photomasks for making the selective n-type and p-type regions by ion implantation. However, the recent progress of new photoresist materials which has a feature of higher heat resistance more than 300°C was announced and leads to reduce the complexity of the SiC ion implantation process [61].

Blocking characteristics

The applied drain voltage is supported by the n− drift region and the p-base region with a triangular electric field distribution as shown in Fig. 4.20 if the doping is uniform on both sides. The maximum electric field occurs at the p-base/n− drift junction. The depletion width on the p-base side is related to the maximum electric field by

$$W_p = \frac{\varepsilon_s E_m}{q N_A} \tag{4.14}$$

where N_A is the doping concentration in the p-base region. The minimum p-base thickness (t_p) required to prevent reach-through limited breakdown can be obtained

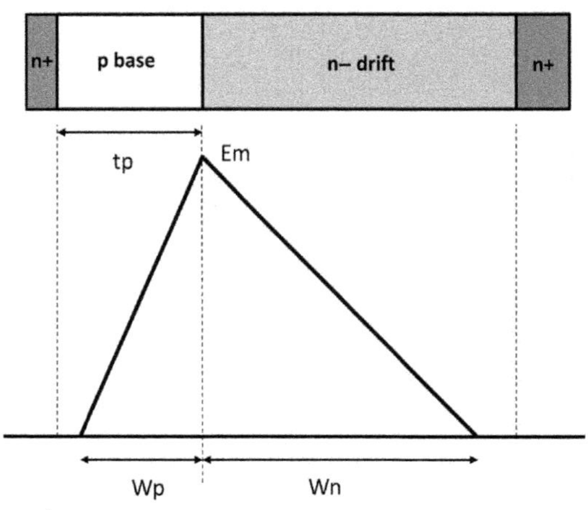

Figure 4.20 Reach-through in a power MOSFET structure.

by assuming that the maximum electric field at p-base/n− drift junction reaches the critical electric field for breakdown when the p-base region is completely depleted:

$$t_p = \frac{\varepsilon_s E_c}{qN_A} \tag{4.15}$$

where E_c is the critical electric field for breakdown in the SiC. In Ref. [16], the calculated minimum p-base thickness for 4H-SiC power MOSFETs is compared with that for Si as shown in Fig. 4.21. According to this calculation results, the thickness for 4H-SiC is about six times larger than for Si. This implies that the minimum channel length required for SiC devices is much larger than the silicon devices, resulting in a big increase in the on-resistance. The minimum thickness of the p-base region required to prevent reach-through breakdown decreases with increasing doping concentration as shown in Fig. 4.21. The typical p-base doping concentration for Si power MOSFET is 1.0×10^{17} cm^{-3} to obtain a threshold voltage between 1 V and 3 V for a gate oxide thickness of 50−100 nm. At this doping level, the p-base thickness can be reduced to 0.3 μm without reach-through limiting the breakdown voltage. In contrast, for the SiC, it is necessary to increase the p-base doping concentration to above 4.0×10^{17} cm^{-3} to prevent reach-through with a 0.4 μm p-base thickness. This higher doping concentration increases the gate threshold voltage.

The electric field profiles of the planar MOSFET in the blocking state are shown in Fig. 4.22. The electric field in the oxide layer (E_{ox}) becomes higher

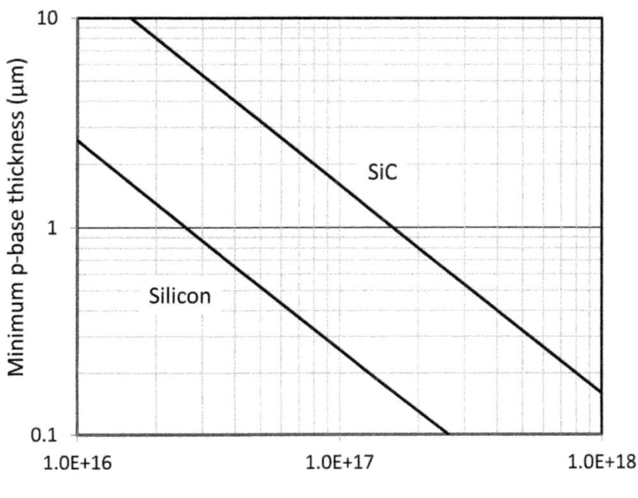

Figure 4.21 Comparison of minimum p-base thickness to prevent reach-through breakdown in SiC and Si.

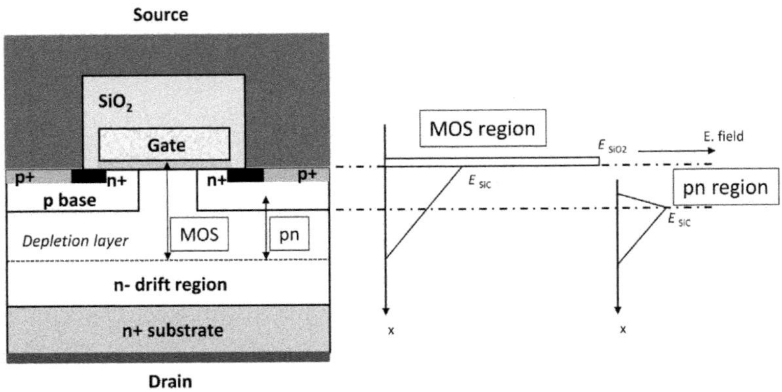

Figure 4.22 Electric fields distribution in the blocking state of DMOSFET.

than the peak field E_s in the SiC by the ratio of dielectric constants, as required by Gauss' law:

$$E_{ox} = \frac{\varepsilon_s}{\varepsilon_{ox}} E_s \tag{4.16}$$

For SiO_2/SiC interface, $\frac{\varepsilon_s}{\varepsilon_{ox}}$ becomes ~2.5 and this means that the oxide field is about 2.5 times higher than the peak field in the SiC. Since the peak electric field of SiC can reach ~1.5 MV/cm due to the high critical field of ~3.0 MV/cm, the corresponding SiO_2 field becomes 3.8 MV/cm. This value is about 5.5 times higher than that in the Si-MOSFET and is close to the maximum allowable field for SiO_2 layer [51]. Therefore, in order to achieve a superior long-term reliability of gate oxide later, it is necessary to design the oxide field below ~4 MV/cm.

Exclusive process technologies for SiC-MOSFET

The specific on-resistance (Ron · A) of the planar MOSFET has been still higher than the theoretical limit of SiC, and this is due to the low channel mobility originate from the poor SiO_2/SiC interfacial quality.

In this device structure, p-base region in contact with the gate oxide is formed by ion implantation followed by activation annealing of 1600–1800°C. This process makes the flat surface of SiC rough, decreasing the channel mobility [62]. There are at least two mechanisms which cause the surface degradation during high-temperature annealing. One is the Si desorption from the SiC surface, and the other is the migration of surface atoms. According to Refs. [63,64], a several capping materials have been investigated; as a result, a carbon cap achieves the most successful results [51]. This technology is granted a patent [65] and is now applied for production of SiC power devices in industry.

To create low on-resistance and high reliable SiC-MOSFETs, the ohmic contact on n-type and p-type ion-implanted SiC layers is one of the key processes. Nickel silicide is a popular contact metal, since it can be formed on n-type and p-type SiC in the same process. In SiC-MOSFET, high-temperature annealing around 1000°C was required to form ohmic contact with low contact resistance [51]. On the other hand, it was also reported that high-temperature annealing degraded adhesion and the thermal stability of the silicide/metal interface owing to the precipitation of graphite, and it increased the oxide/semiconductor interface state density owing to the decomposition of interface passivation [66]. Since both the precipitation of graphite and the decomposition of interface passivation occur at temperature higher than 800°C, the metallization process for ohmic contact is required to be at the lower temperature. The nickel silicide process is said not-well controllable and stable, owing to the thickness nonuniformity and large roughness at the silicide/barrier metal interface. Therefore, a low-temperature (lower than 750°C) metallization process for forming ohmic contact without nickel silicide was tried for an SiC-MOSFET with good performance and high reliability [67]. Low contact resistance of $2.1 \times 10^{-6}\ \Omega\ cm^2$ on the n-type region and $1.3 \times 10^{-3}\ \Omega\ cm^2$ on p-type regions were obtained by using the titanium with high dose implantation at room temperature.

It has been over a several years since SiC-MOSFETs came into the market. Nitridation of gate oxides is known as a key process to improve the SiO_2/SiC interface properties [68] and a lot of SiC-MOSFETs in the market seems to be applied to this kind of process. However, harmful defects that affect electrical properties such as lower channel mobility and the threshold voltage instability by capturing electrons or holes in the channel still exist at and near the interface [69,70]. In order to solve these problems, the gate nitridation process has been eagerly studied and improved very much. As a result, it was reported that the gate threshold voltage shift of 13 kV MOSFET stabilized within ± 0.06 V under the condition of gate oxide field of -3 MV/cm and 200°C [71]. Furthermore, instead of the nitridation process, the other technologies of phosphorus and boron incorporation into the interface were proposed to reduce the interface traps further [72–74]. The improvement of the gate oxidation process has been studied actively and the SiO_2/SiC interface property will be improved so much in near future.

Cell design

In 2007, Saha and Cooper reported measured results of the 1 kV SiC-DMOSFET structure with low on-resistance of $6.95\ m\Omega\ cm^2$ [75]. Fig. 4.23 exhibits the cross section of the DMOSFET cell and layout of the device with p+ contacts. This device structure includes a very short channel of 0.5 μm or less, source ohmic contacts that are self-aligned to the polysilicon gate, a current spreading layer beneath the p-base and p-base contacts that are segmented along the length of the fingers and so on. In addition to these technologies, this device incorporated the other fabrication process/design technologies such as the carbon cap and short JFET width of 1 μm for the gate oxide field reduction of below 4 MV/cm. Therefore, this DMOSFET successfully exhibits a specific on-resistance of as low as $6.95\ m\Omega\ cm^2$ with the blocking voltage of 1050 V. The blocking voltage is found to be limited by the avalanche breakdown, with a gate oxide field of less than 4 MV/cm

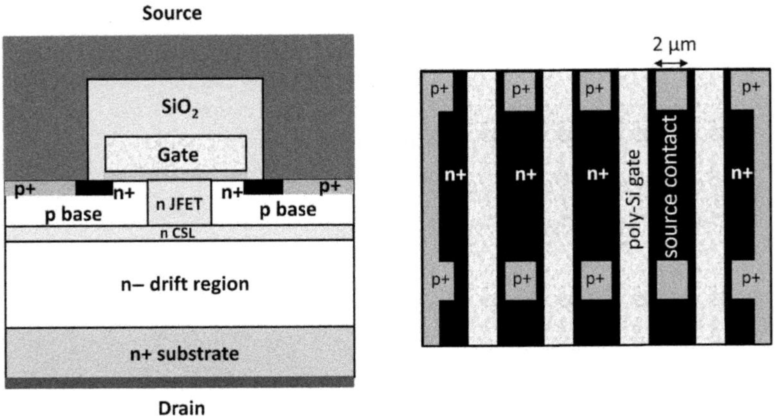

(A) DMOSFET (B) p+ contact and n+ source layout

Figure 4.23 Cross section of DMOSFET and layout of with small p+ contacts.

described earlier. Also, it is noted that the peak value of the measured inversion channel mobility is achieved as high as 15 cm^2/Vs.

In Si-DMOSFETs, many topologies of the cell design at the top surface of MOSFET structure were proposed and examined from the point of view of reducing the on-resistance [76]. Needless to say, these technologies can be applied to design the cell structure of SiC-MOSFETs. As described earlier, Saha and Cooper applied the linear cell structure with short channel length, fine source ohmic, and p-base contacts to achieve the low on-resistance. However, in Ref. [76], the cellular type structure such as square wells in square cell and hexagonal or square wells in hexagonal cell was attractive to reduce the on-resistance when the smaller p-base width was adopted; therefore, SiC-MOSFETs with the cellular type structure with short p-base width could achieve a superior on-resistance. In 2014, Palmour et al. demonstrated superior characteristics of the advanced SiC-DMOSFET devices with the wide range of blocking voltage from 900 V to 15 kV [77]. This paper reported that the new breakthrough performance for voltage ratings from 900 V up to 15 kV with a specific on-resistance as low as 2.3 mΩ cm^2 for a breakdown voltage of 900 V-rating and 208 mΩ cm^2 for a breakdown voltage of 15 kV rating. These characteristics are successfully achieved by not only the improvement of substrate and epitaxial material quality but also further optimizing and improving device design and fabrication processes described earlier. As a result, the simple planar SiC-DMOSFET structures show an extremely low specific on-resistance with the improved blocking performance.

IEMOSFET devices

As the other planar SiC-MOSFET structures, SiC-MOSFETs named implantation, and epitaxial MOSFETs (IEMOSFETs) were proposed and fabricated as shown in Fig. 4.24 [36,78]. In this structure, the upper half of the p-base is formed by

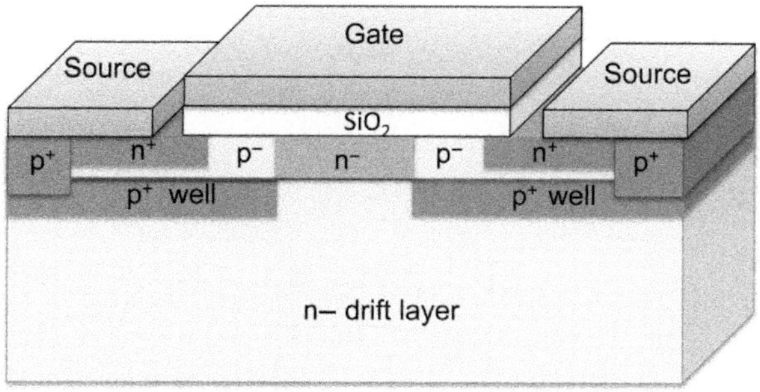

Figure 4.24 Surface structure of implantation and epitaxial MOSFETs (IEMOSFETs).

epitaxial growth without activation annealing. The smooth surface prevents the degradation of the channel mobility as the double implantation method induces. Also, this device utilized the thin n-type buried channel layer in order to increase the channel mobility. In the edge structure of IEMOSFETs, the p+ guard ring process described earlier cannot be used because the p-type epitaxial layer was already formed on an SiC wafer surface. To fabricate the edge structure, the p-type epitaxial layer (p-base) was partially etched by dry etching process, followed by making the JTE or guard rings structure by photolithography and ion implantation with aluminum (Al^+). For example, 600 and 1200 V IEMOSFETs were fabricated. An n− epi layer was doped to 2.1×10^{16} cm^{-3} and had a thickness of 6 μm for 600 V device, and 1.0×10^{16} cm^{-3} and had a thickness of 10 μm for 1200 V device, respectively. Aluminum ions (Al^+) with the concentration of as high as 3.0×10^{18} cm^{-3} were selectively implanted to form the bottom of p-base as p+ well region. This high acceptor concentration of this region shields the channel (p-base region) from the high electric field and prevents the punch-through between the n+ source and n− drift layer [29]. Then the p-epi layer with as low as 5.0×10^{15} cm^{-3} and the thickness of 0.5 μm was grown to form the top p-layer. The smooth surface and low doped p-epi layer improves the channel mobility and the reliability of the gate oxide layer. The JFET region was formed by selective nitrogen ion (N^+) implantation into the p-layer. Its resistance is successfully decreased since the concentration and width of the JFET region is optimized independently of the drift layer. By the hexagonal cell structure for the SiC-IEMOSFET is adopted to maximize the packing density, the superior Ron · A characteristics of 1.8 mΩ cm^2 for 600 V device and 5.0 mΩ cm^2 for 1200 V device were successfully achieved.

4.3.2.2 UMOSFET structure

The UMOSFET structure is shown in Fig. 4.25 with its internal resistance components [16]. The same internal resistances encountered in the DMOSFET structure are present in the UMOSFET structure with the exception of the JFET region resistance. The elimination of the JFET region allows a significant reduction of the overall specific on-resistance for the UMOSFET structure, not only because its resistance is excluded but also because the cell pitch can be made much smaller

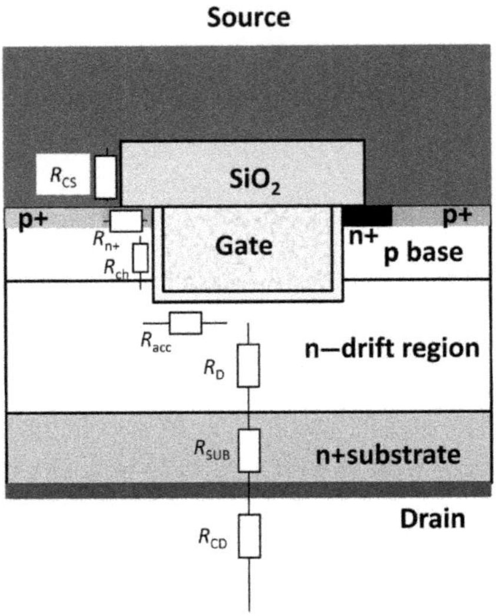

Figure 4.25 UMOSFET structure with its internal resistance.

than that of the DMOSFET. A smaller cell pitch reduces the specific on-resistance contributions from the channel, accumulation, and drift regions. The total on-resistance for the UMOSFET structure is obtained by the addition of all resistances because they are considered to be in series in the current path between the source and the drain electrodes:

$$R_{ON} = R_{CS} + R_{n+} + R_{ch} + R_{acc} + R_D + R_{SUB} + R_{CD} \quad (4.17)$$

It is customary to utilize the linear cell structure surface topology for the UMOSFET structure because the trench surface can be oriented in the preferred direction of most favorable for producing high quality surface to achieve the higher channel mobility and SiO_2/SiC interface reliability. However, there are some exception such as square cells and hexagonal cells applications. The UMOSFET usually utilized the $\{11\bar{2}0\}$ or $\{1\bar{1}00\}$ planes (a-face or m-face) for trench sidewall because these faces exhibit higher channel mobility than the Si-face and C-face for a planer gate structure [79–81]; therefore, a further reduction in on-resistance is expected.

Fabrication process

SiC UMOSFET requires its own unique process sequence. Although a variety of process options have been proposed, a typical case is briefly described here. The fabrication process consists of 17 major steps as shown below:

1. Preparation of SiC wafers with n– epi/n+ substrate.
2. SiO_2 deposition on top of the SiC wafer to protect the wafer surface from the following ion implantation damages.
3. Photolithography and ion implantation with aluminum (Al^+) of the p-base regions. The p+ guard rings in the edge termination regions are also implanted at the same time.

4. Photolithography and ion implantation with phosphorus (P −) or nitrogen (N −) of the n+ source regions.
5. Photolithography and ion implantation with aluminum (Al^+) of the p+ regions.
6. After cleaning of the wafer, activation of the dopants at 1600−1800°C.
7. Formation of SiO_2 mask and dry etching of SiC to make the trench gates.
8. Gate oxidation process. 50−100 nm SiO_2 was thermally grown or deposited by CVD and postoxidation annealing is performed, for example, in wet oxygen, NO/N_2O, etc. In order to make the gate oxide thickness uniform, the CVD method is usually applied.
9. Deposition of the gate electrode of doped polycrystalline Si (poly-Si) to fill the trenches. Subsequent annealing of the poly-Si is performed.
10. Dry etching of the gate poly-Si layers.
11. Deposition of PSG/BPSG with ~1 μm thick for the top passivation layer.
12. Dry etching of the top passivation layer and gate oxide to open the source contact.
13. Deposition of the source and the drain ohmic contacts composed of Ni.
14. Silicidation of the source and the drain contacts (just below 1000°C). RTA method is usually used.
15. Deposition of the source pad metal of aluminum with annealed ohmic contacts.
16. Etching of the source pad metal pad. On the bottom of the wafer, Ti/Ni/Au layer was deposited.
17. Polyimide layer was deposited as a passivation film.

Blocking characteristics

As discussion in Section 4.3.1, it is noted that their blocking voltage of the early UMOSFET was limited by oxide breakdown at the trench corners. Fig. 4.26 illustrates the schematic electric field distribution in the UMOSFET in the blocking state, along with field profiles along two vertical slices. As in the DMOSFET, the oxide field is 2.5 times higher than the peak semiconductor field, but there is significant field crowding at the trench corners, producing much higher local fields at these points. A number of UMOSFET structures were investigated to alleviate this problem by optimizing the p-base layer, tapering the sidewalls or rounding the corners using proprietary processing techniques. In 1997, Hara presented a noble

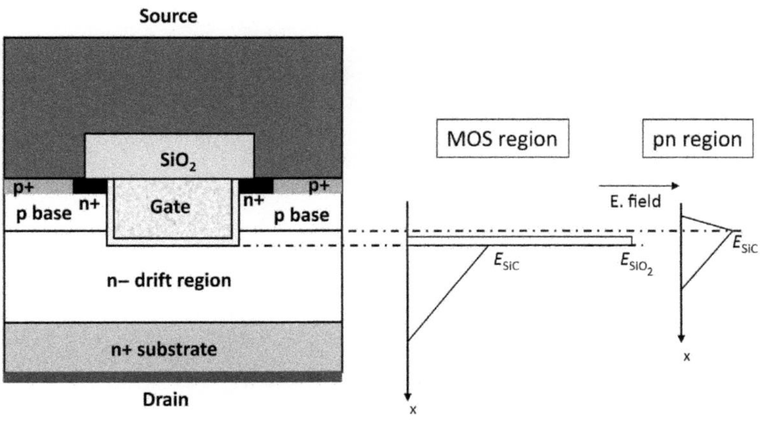

Figure 4.26 Electric fields distribution in the blocking state of UMOSFET.

UMOSFET structure named epi-channel FET [82]. This is the accumulation-channel UMOSFET structure and n− epitaxial layer was grown subsequent to trench etch. A key point of the device design is the electric field relaxation near the trench corner by tailoring the impurity concentration of p-base layer. This accumulation-mode UMOSFET exhibited a blocking voltage of 450 V, and a specific resistance of 10.9 mΩ cm^2. Also, Tan et al. reported the other type of SiC accumulation-channel UMOSFET with new structural features that shielded the trench oxide from high electric fields in the blocking state [83]. A cross section of this device is shown in Fig. 4.27. The new features consist of a p region formed in the trench bottom by self-aligned ion implantation, and a thin n− epitaxial layer incorporated between the n− drift region and the p-base one. The p region formed in the trench bottom is connected to the ground and shields the trench oxide from high electric fields successfully in the blocking state. The thin n− epitaxial layer prevents JFET pinch-off between the p-region in the trench bottom and the p-base and promotes lateral current spreading in the on state, eliminating current crowding at the trench corners. The critical design parameters of the new UMOSFET are the implant doping and thickness of p-region formed in the trench bottom, the depth of trench below the p-base, and the doping and thickness of the thin n− epitaxial layer for current spreading. The doping of this thin n-type epitaxial layer is about a 100 times higher than that of the n− drift layer. The device also incorporates an n-type epitaxial layer grown after the trench etch and implant steps as in the UMOSFET structure Hara presented. This SiC accumulation-channel UMOSFET had an n− drift layer with the doping of 2.5×10^{15} cm^{-3} and the thickness of 10 μm and exhibited a blocking voltage of 1400 V, specific resistance of 15.7 mΩ cm^2 at a gate voltage of 40 V (oxide field of 3.1 MV/cm).

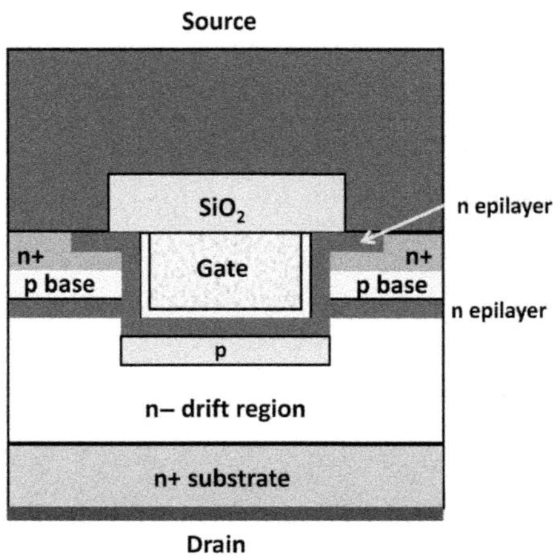

Figure 4.27 Cross section of accumulation-channel UMOSFET.

UMOSFET with gate shielding structure

In 2011, Nakamura et al. presented a new type double-trench MOSFET structure as shown in Fig. 4.28 [27]. This is an inversion-mode UMOSFET and the structure has both source trenches and gate trenches. In order to shield the trench oxide from high electric fields in the blocking state, the double trench structure with p-region at the bottom of source trench, which is deeper than the bottom of the gate trench, are fabricated. By the enhancement of total channel width by the application of small unit cell size of $4 \times 4\ \mu m^2$ and thinner n+ substrate of 100 μm, superior on-resistances of as low as 0.79 mΩ cm^2 (blocking voltage: 630 V) and 1.41 mΩ cm^2 (blocking voltage: 1260 V) were achieved, respectively. Fig. 4.29 exhibits the calculation examples of electric field between the double-trench and the standard single-trench structures at the blocking mode of 600 V when the gate-source voltage is 0 V. This calculation results were derived based upon the device parameters shown in Ref. [27]. The highest electric field at bottom of the trench gate in the double-trench structure is successfully reduced of 1.35 MV/cm. This result is 29% less compared with that of the standard single-trench structure of 1.90 MV/cm and the double-trench structure can prevent the destruction of gate oxide layer while showing extremely on-resistance.

With increasing blocking voltage of 1.7 and 3.3 kV, for example, contribution of JFET resistance becomes larger and could be a serious problem in the SiC UMOSFET with the deep p-regions because the additional pn junction results in the parasitic JFET resistance with lower doping concentration of n− drift layer. Thus, the critical issues for the SiC UMOSFET with higher blocking voltage are the development of a shielding structure for the gate oxide at the trench bottom without an increase in the JFET resistance. Fig. 4.30 shows a new type of SiC UMOSFET structure with 3.3 kV blocking capability [84]. The buried p-region, functioning as the gate shielding structure, was formed beside the gate trench, using high-energy

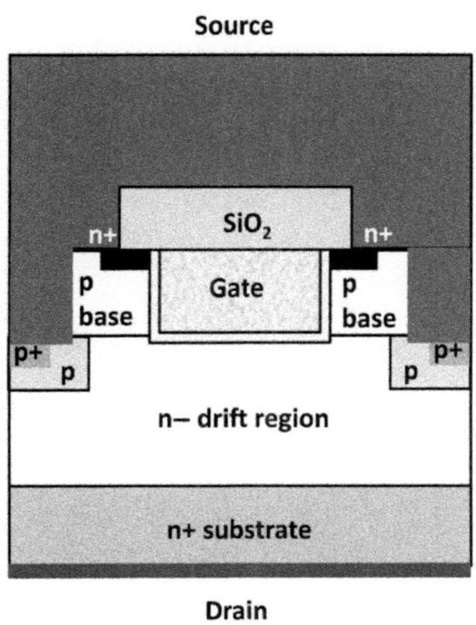

Figure 4.28 Cross section of double trench MOSFET structure.

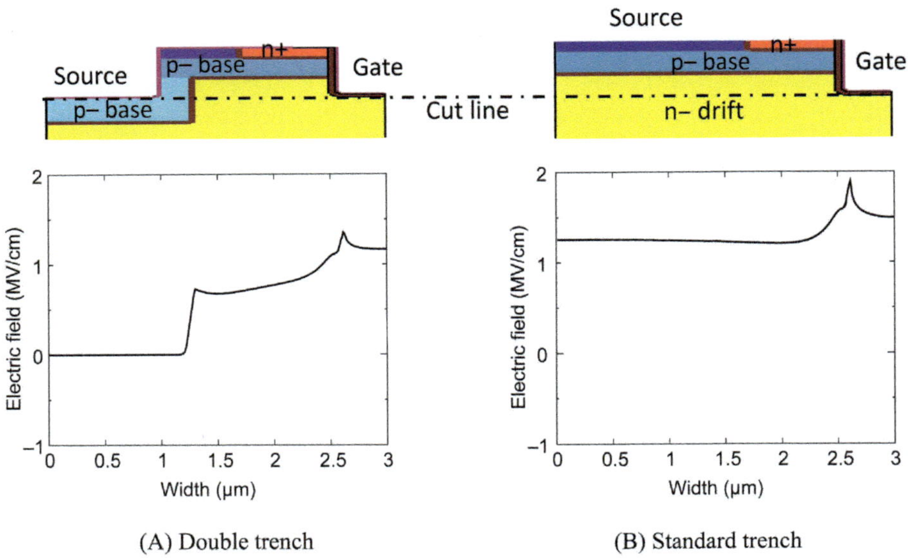

Figure 4.29 Comparison of calculated electric field between (A) double trench and (B) standard UMOSFET (@ DC bus voltage = 600 V, V_g = 0 V). Calculated electric field was picked up along the "cut line" shown in this figure. This result was calculated based upon the device structure described in Ref. [27] by the author of this chapter.

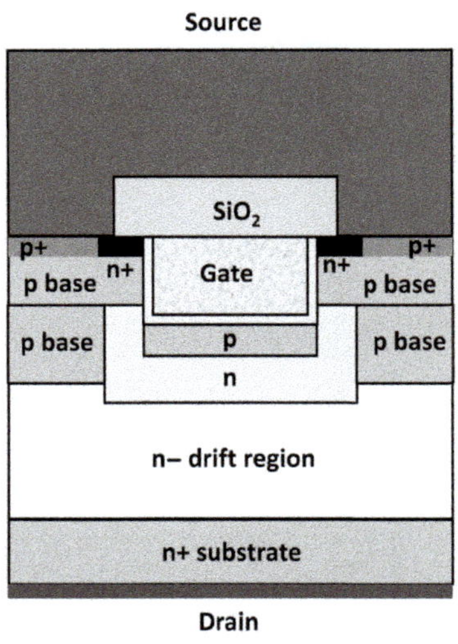

Figure 4.30 Cross section of the another type UMOSFET structure. Trench bottom shielding region and deep p-base are implemented.

MeV aluminum (Al^+) implantation. The p-base regions were also implanted, followed by high-temperature activation annealing. It should be noted for designing this type of structure that the n− epitaxial layer had a two-layer structure with different doping concentrations. The bottom drift layer was doped low value to

3×10^{15} cm^{-3} with a thickness of 29 μm to mainly support high blocking voltage, and the highly doped top layer was doped to 2×10^{16} cm^{-3} with a thickness of 2.5 μm which suppresses the JFET resistance generated between the buried p and p-base regions. Therefore, in order to further enhance the shielding effect due to the application of highly doped top n layer, Al^+ ions were also implanted under the gate trench as a trench bottom shielding region. By the enhancement of total channel width by the application of hexagonal cell, superior on-resistance of as low as 9.4 mΩ cm^2 with blocking voltage higher than 3.3 kV and no gate oxidation rapture were achieved.

It should be noted that the trench MOSFET has a disadvantage of its large reverse transfer capacitance (Crss), which results in degraded switching performance. The buried p-region under the trench gate has been proposed to address the above issues [83]. Typically, the buried p-region is grounded to the source contact. Such the p-region termination demands a more complicated process, and a significant portion of chip area has to be sacrificed for contact holes. Recent studies revealed that although a buried and floating p-region reduces the oxide field in trench MOSFET without degrades its static performance of on-resistance and breakdown voltage, it creates high on-state oxide field after having encountered a high drain bias stress. Further, the floating p-region also results in slower switching speed because of the high Crss. Therefore, the buried p-region in SiC UMOSFET should be grounded despite the more complicated fabrication process [85,86].

IE-UMOSFET and V-groove trench devices

As in the case of planar MOSFET, SiC-UMOSFETs named implantation and epitaxial UMOSFETs (IE-UMOSFETs) can be designed and fabricated [87]. The p-base region was formed using epitaxial growth; therefore, the on-state characteristics such as on-resistance and threshold voltage are not affected by implantation damage. The p-base regions and the buried p-region under the trench bottom, which functions as the gate shielding structure, were formed using high-dose aluminum-ion (Al^+) implantation. By the stripe cell structure with 5 μm pitch and {1 − 100} trench planes for the SiC IE-UMOSFET is adopted, the superior Ron · A characteristics of 2.8 mΩ cm^2 with $V_{th} = 3.1$ V, and 4.4 mΩ cm^2 with $V_{th} = 5.9$ V were successfully achieved.

In 2002, Yano et al. reported that the 4H-SiC (03-38) face has the lower interface state density than the 4H-SiC (0001) face [88]. The 4H-SiC (03-38) face is semiequivalent to cubic (100), tilted by 54.7 degrees toward $\langle 01\bar{1}0 \rangle$from (0001). It is well known that the interface properties of MOS structure in Si strongly depend on the surface orientations, that is, Si (001) has the smallest interface state density. Therefore, 4H-SiC (03-38) face is considered to show the lower interface state density. Based upon these results, Hiyoshi et al. reported that a high channel mobility of 80 cm^2 V^{-1} s^{-1} was attained by using 4H-SiC (0-33-8) face [89]. Subsequently, 1200 V V-groove trench MOSFET structure with the {0-33-8} face as the trench sidewalls for the channel region was proposed and fabricated [90,91]. The 4H-SiC (0-33-8) face is titled by 54.7 degrees toward $\langle 1\bar{1}00 \rangle$ from (0001), so that the name of V-groove MOSFET derives from the V-shaped gate geometry as shown in Fig. 4.31. The deep buried p+ region was implemented as the gate shielding

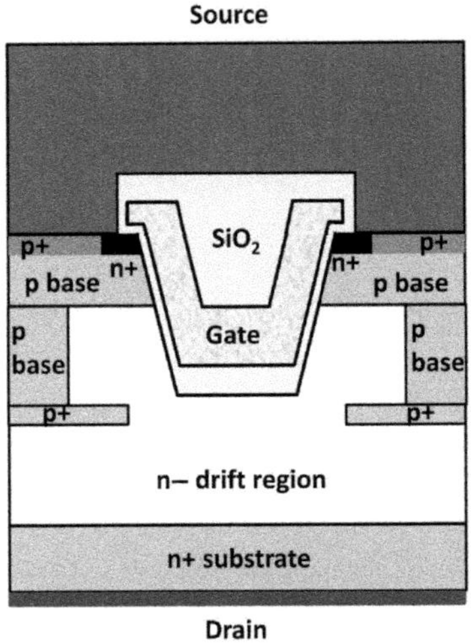

Figure 4.31 Cross section of the V-groove MOSFET structure. Deep buried p+ regions are implemented for trench bottom shielding.

structure, and the trench bottom oxide layer thickness was as thick as 200 nm whereas the gate oxide layer was set of 50 nm. This V-groove trench MOSFET also exhibited a low Ron · A characteristics of 2.0 mΩ cm^2 with high gate oxide reliability.

4.3.2.3 SiC-MOSFET loss estimation due to its high drain–source capacitance

The main advantage of SiC semiconductor device is the very low resistance of the drift region by increasing the doping density and reducing the thickness of n− drift layer to block the same reverse voltage because the SiC has a high critical electric field. However, this lower on-resistance leads to an increase of drain–source capacitance and drain stored charge. The output capacitance of the MOSFET is dissipated during its turn on in case of hard switching [92]. When considering a one-dimensional, one-sided junction with constant doping as a function of depth, the stored energy in p+ n-junction, E_{ds}, is given by as follows:

$$E_{ds} = \frac{q^2 N_D^2 x_n^3}{6\varepsilon_s} \tag{4.18}$$

where N_D is the doping concentration in the n− drift layer, and x_n is the length of depletion layer, respectively. Since the doping of the n− drift layer in SiC-MOSFET is ~100 times higher than that of Si device (e.g., about 1.0×10^{16} cm^{-3} in SiC and 1.0×10^{14} cm^{-3} in Si for 1200 V devices) and the depletion length is one-tenth (e.g., about 10 μm in SiC and 100 μm in Si for 1200 V devices), the stored energy of the SiC output capacitance is ten times larger than that of Si.

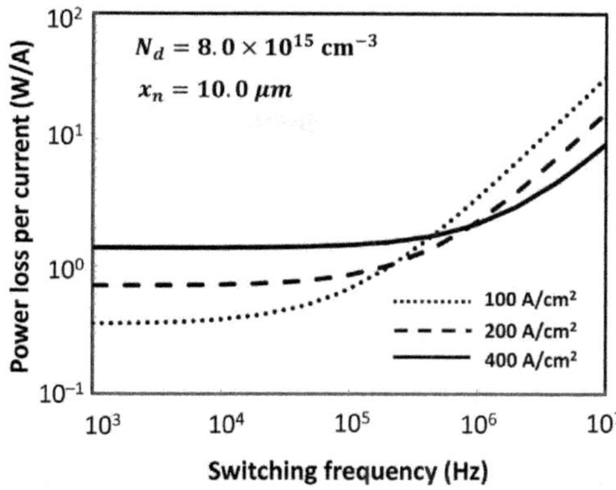

Figure 4.32 Power loss in a 1200 V SiC-MOSFET as a function of switching frequency and current density.

The hard-switching converter, the efficiency is fundamentally limited by the conduction loss and switching loss. The conduction loss P_{on} is shown as:

$$P_{\text{on}} = J^2 \times (R_{\text{on}} \cdot A) \tag{4.19}$$

where $R_{on} \cdot A$ is the specific on-resistance and J is the current density.

The switching loss, $P_{\text{switching}}$, is induces solely by charging the output capacitance during the each switching cycle at frequency f, given by

$$P_{\text{switching}} = E_{\text{ds}} \times f = \frac{q^2 N_D^2 x_n^3}{6\varepsilon_s} \times f \tag{4.20}$$

When designing a converter, it is important to estimate the conduction and switching losses by selecting the proper device size, meaning to define the current density. When normalizing the two loss components to the current density, the total power loss as a function of switching frequency and current density are calculated. Fig. 4.32 exhibits the calculated results of power loss when the specific on-resistances of 1200 V SiC-MOSFET are set of 3.5 mΩ cm^2. For example, it is clear from this figure that conduction loss dominates at the frequency below about 200 kHz, so it is more efficient to operate at low current densities. As switching frequency increases, the switching loss also become dominate, requiring high current densities to reduce the total loss.

4.3.2.4 Short-circuit safe operating area

SiC-MOSFETs are used in various power electronics circuits such as motor control and power supply to regulate the energy delivered to various types of load. A short-circuit state could take place accidentally in the power electronics circuit and one serious problem associated with the SiC-MOSFETs is the destructive failure at this

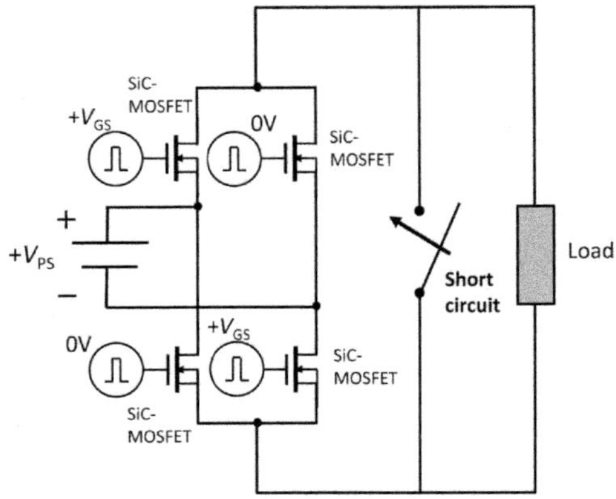

Figure 4.33 Inverter circuit in short-circuit operation of the SiC MOSFET.

short-circuit state. Fig. 4.33 shows an inverter circuit in the short-circuit operation of the SiC-MOSFET. High voltage dc power source directly connected to the drain of the SiC-MOSFET while its gate bias is still turn on. It would be desirable to detect the short-circuit condition and turn off the SiC-MOSFET safely using feedback to the SiC-MOSFET control circuit before the device is in destructive failure. This feedback time is $\sim 10\,\mu s$; therefore, it is strongly required for the SiC-MOSFET to withstand the high-current flow under short-circuit state while supporting the high voltage applied to the drain electrode during this feedback time duration. The withstand capability under this simultaneous application of high current and high voltage for short-circuit durations is referred to as the short-circuit safe operating area (SCSOA). It is well known that the SCSOA and the on-state voltage drop are in trade-off relation for Si-IGBT and SiC-MOSFET. So, the SCSOA of Si-IGBT has to be set at more than $10\,\mu s$ and state-of-the-art IGBT was designed and fabricated to meet this requirement while its on-state voltage drop keeps as low as possible. As a superior alternative to the Si-IGBT, SiC-MOSFET has to obtain lower on-state voltage drop while keeping the SCSOA with the Si-IGBT.

Due to the absence of the load inductance during short-circuit operations, the current flow in the SiC-MOSFET becomes limited only by its saturation current. The power being dissipated by the SiC-MOSFET is given by

$$P_D = J_{D,\mathrm{SAT}} \times V_{\mathrm{PS}} \times \frac{1}{2} \tag{4.21}$$

where $J_{D,\mathrm{SAT}}$ is the saturation current density determined by the gate bias V_{GS} and V_{PS} is the DC power supply voltage. According to Fig. 4.33, each MOSFET drain voltage is a half of V_{PS}. This dissipated power lead to the temperature rise in the SiC-MOSFET. The SiC-MOSFET can withstand the increase in temperature until it reaches a critical one when the built-in potential of the pn junction becomes zero [93]. Fig. 4.34 exhibits the calculated temperature dependence of built-in potential

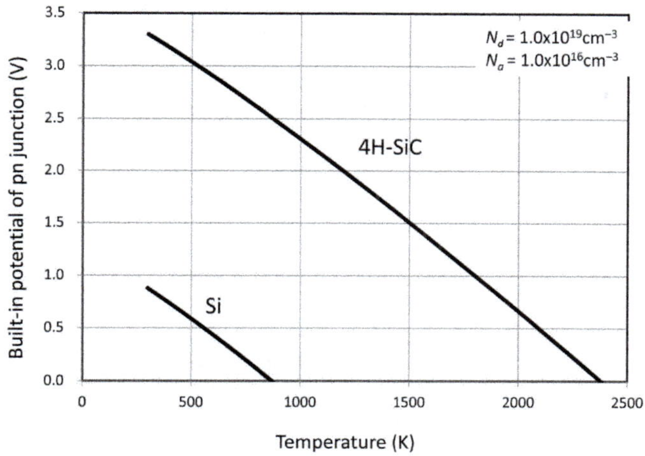

Figure 4.34 Built-in potential for pn junction in SiC and 4H-SiC.

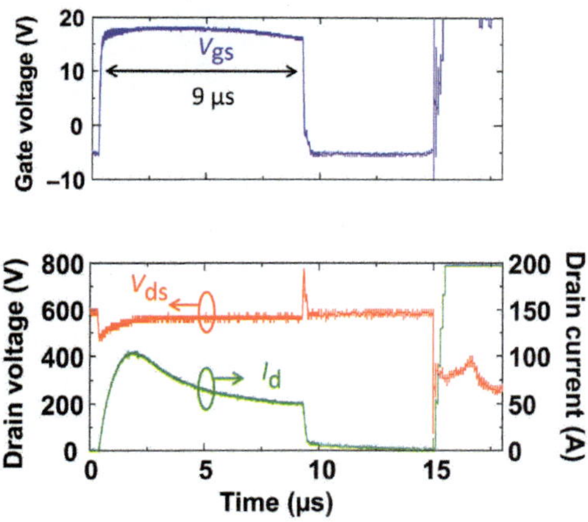

Figure 4.35 An example of measured waveforms of 1200 V SiC DMOSFET in short-circuit test. (@DC bus voltage = 600 V, R.T).

of pn junction. In Si-IGBT, the critical temperature is about 800K; therefore, the IGBT suffers the latch up of parasitic thyristor structure when the internal temperature reaches the critical one of 800K. In SiC-MOSFET, it is as high as ~2300K. This means that peripheral parts of SiC-MOSFET such as source metal and thin gate oxide layer could suffer such the high temperature before the pn junction in SiC-MOSFET reaches the critical temperature. In this situation, there are some possibilities of melting the source metal and the increase of gate oxidation leakage current followed by the gate oxide rapture [94]. Fig. 4.35 exhibits an example of measured drain current/voltage and gate voltage waveforms of the 1200 V SiC-DMOSFET with gate oxide thickness of about 40 nm. This measurement was performed at room temperature with the 600 V DC bus. It is obvious that the drain current I_d increases promptly and the device enters from the linear region to the active region until achieving its saturation current at the initial gate pulse time. After that, the drain current decreases owing to the degradation of the carrier mobility constrained by the elevated lattice temperature. The drain current is no longer under gate driver control and eventually breakdown occurs when short-circuit withstanding time is set at 9 μs. After the short-circuit test, this device was severely damaged as shown in Fig. 4.36 and the impedances of three terminals between the gate, drain, and source were smaller than 0.7 Ω. These results indicate that the SiC-DMOSFET was completely destroyed. Fig. 4.37 reveals the simulation results about

Figure 4.36 SiC-DMOSFET device top view after the short-circuit test shown in Fig. 4.35. There can be seen a severely damaged source metal on the device.

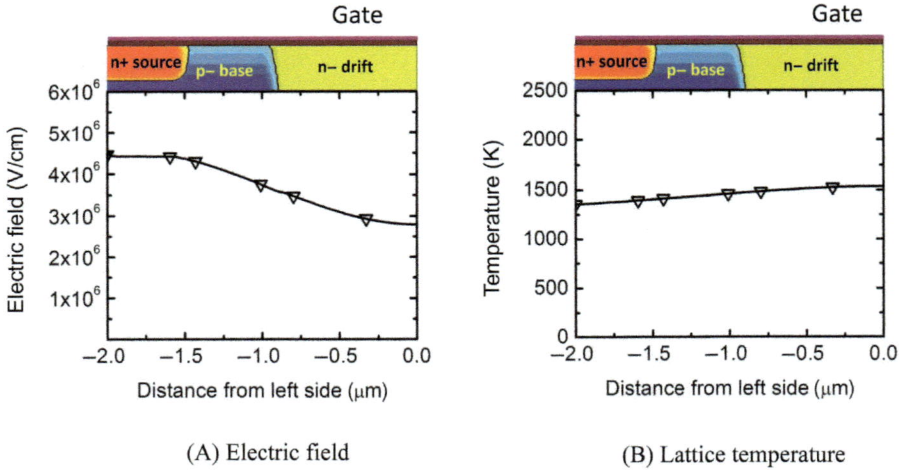

Figure 4.37 Simulation results of electric field and lattice temperature profile for SiC DMOSFET at SiO_2/SiC interface at short-circuit withstand time of 9 μs (@DC bus voltage $V_{cc} = 600$ V, R.T).
This result was calculated based upon the device structure described in Ref. [88] by the author of this chapter.

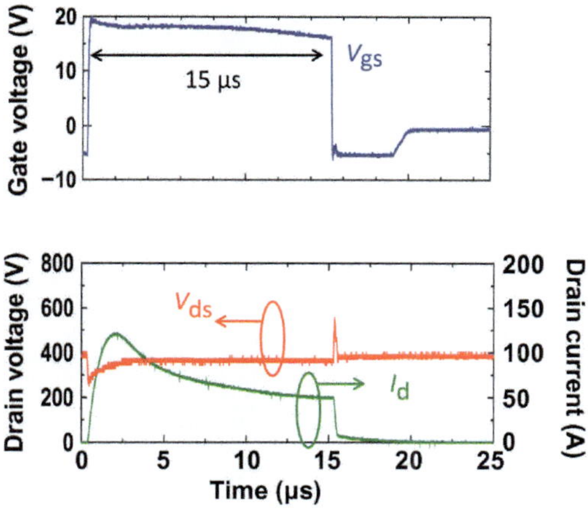

Figure 4.38 An example of measured waveforms of 1200 V SiC DMOSFET in short-circuit test (DC bus voltage = 400 V, @ R.T).

the electric field in gate oxide and lattice temperature profiles at the interface of SiC and oxide ($y = 0$ μm) during the short-circuit test. The lattice temperature rises very high; however, the temperature reached around 1500K, which is below the critical temperature of built-in potential of pn junction shown in Fig. 4.34. It should be pointed out that the critical aluminum melting temperature is around 933K [95]; therefore, gate and/or source metal layers could be damaged during the short-circuit operation [96], followed by triggering the device failure. It can be also seen from Fig. 4.35 that the gate voltage gradually shows a recessive trend when the short-circuit transient is increased. This characteristic was also recognized in the case of lower DC bus voltage of 400 V shown in Fig. 4.38. Since, in the EV/PHEV inverter applications, the DC input voltage could be frequently changed to control an output AC motor speed, it is necessary to investigate the short-circuit capability dependence upon the drain voltage. According to the simulation results about the electric field in gate oxide and lattice temperature profiles at the interface of SiC and oxide layer, a high electric field of 4.3 MV/cm and a high lattice temperature of 844K can be seen [94]. The most prominent oxide degradation mechanism in SiC-MOSFET is the Fowler–Nordheim (FN) tunneling and Poole–Frenkel (PF) emission effect [97] leading to a leakage current, especially at the high gate electric field and ambient temperature. It has also been demonstrated that a substantial increase in FN current occurs as the temperature rises up to 523K, and that the effective barrier height between SiC and oxide decreases to 2.38 eV as the temperature rises up to 573K [98,99]. Therefore, the high temperature will affect the reliability of gate oxide prominently. Fig. 4.39 exhibits the top view of SiC DMOSFET after the short-circuit test of 400 V DC bus. Unlike in the case of 600 V DC, there are no severe damaged on the chip surface. The impedances of the three terminals of the failure device between the gate, drain, and source (R_{gs}, R_{gd}, and R_{ds}) were measured to be 2.6 Ω, 8.6 MΩ, and ∞ Ω, respectively. These results indicate that only the gate–source terminal of the device is damaged and finally shorted.

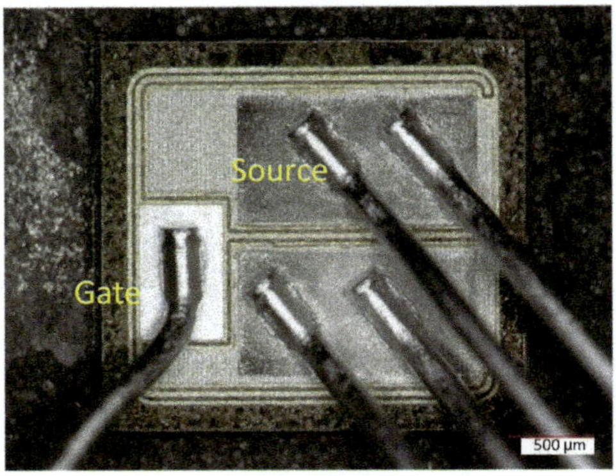

Figure 4.39 SiC DMOSFET device top view after the short-circuit test shown in Fig. 4.36. There cannot be seen any damages on the device.

(A) DC bus voltage: 800 V

(B) DC bus voltage: 400 V

Figure 4.40 SiC-UMOSFET device top view after the short-circuit test.
(A) DC bus voltage = 800 V. There are severe damage in the source metal on the device
(B) DC bus voltage = 400 V. There cannot be seen any damages on the device.

4.3.2.5 Improvement of trade-off characteristic between on-resistance and SCSOA for SiC UMOSFET structure

Due to the smaller cell pitch and the elimination of JFET resistance, SiC UMOSFET structure could exhibit poor SCSOA characteristics compared with the DMOSFET. This is because its drain saturation current $J_{D,\mathrm{SAT}}$ becomes larger resulting in higher power dissipation P_{D} as shown in Eq. (4.21). Key points of the SiC UMOSFET design are to achieve the lower on-resistance with moderately restricted drain saturation current. In addition to this, other key points are the high gate oxide reliability to sustain the harsh condition of gate electric field of about 4 MV/cm with high temperature of over 1000K mentioned earlier. Therefore, it is strongly required to analyze short-circuit failure mechanism deeply as in the case of DMOSFET structure. Fig. 4.40 exhibits the top view of the SiC UMOSFET after the short-circuit test. This result suggests there are two types of failure mechanism

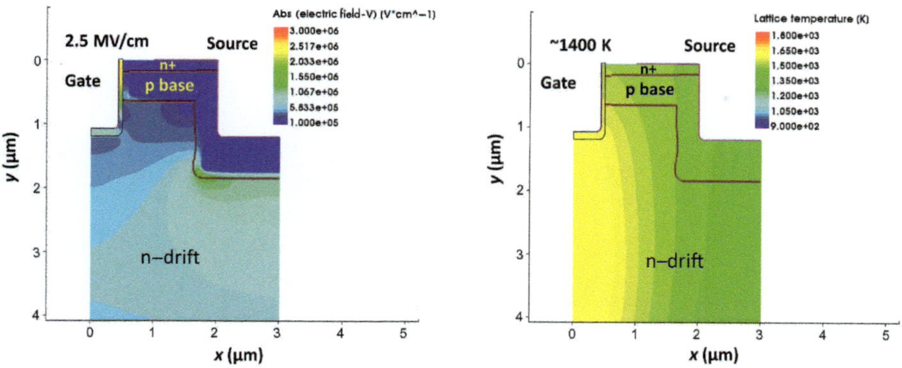

(A) Electric field (B) Lattice temperature

Figure 4.41 Simulation results of electric field and lattice temperature profile for SiC UMOSFET at short-circuit withstand time of 20 μs (DC bus voltage = 400 V, R.T). Gate oxide layer nearby n+ source/p-base suffers from both high electric field and high temperature.
This result was calculated based upon the device structure described in Ref. [94] by the author of this chapter.

in the UMOSFET, leading to the source and/or gate metal damages and the gate oxidation layer rapture [100], which are the same with the SiC-DMOSFETs case. Especially, in the 400 V DC bus case, numerical simulation revealed in Fig. 4.41 that gate oxide layer nearby n+ source/p-base region suffers from both high electric field and high temperature, simultaneously; as a result, gate–source terminal can be easily damaged. This result suggests that this type of failure cannot be protected by the gate shielding structure mentioned earlier because this shield structure only protect at the bottom and corner of trench gate and that thicker gate oxide layer application might be essential for SiC UMOSFET to avoid the oxide layer damage during the short-circuit operation. In practice, the gate oxide thickness of the recent SiC-UMOSFET structures has been set at a thicker value of 80 nm and more [84], resulting in the enhancement of withstanding capability for the gate oxide rapture during the short-circuit test.

It should be noted that, to break through the trade-off characteristic between the on-resistance and the SCSOA, new type UMOSFET structure was revealed [101]. This is the 1200 V UMOSFET structure based upon an asymmetric concept as shown in Fig. 4.42. Features of this device are that only one side of the trench side-wall is used as MOS channel which is exactly aligned to the preferred $\langle 11\bar{2}0 \rangle$ plane. Making use of this favorite crystal plane is the key to achieving a minimum of interface states, resulting in higher channel mobility and lower threshold voltage instability. And, deep p wells are used in order to limit the electric field in gate oxide at the bottom and the corner of the trench. These p-well regions are connected to the source electrode, resulting in low Crss. It should be noted that the JFET region formed by the adjacent deep p wells is not only good at limiting the oxide field in trench corner but also lowers the drain saturation current to the

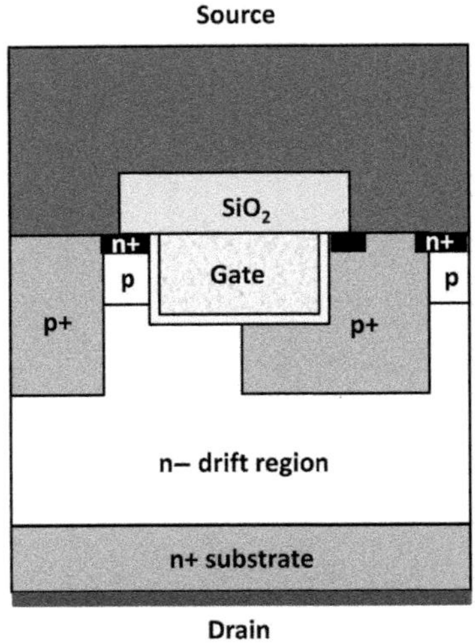

Figure 4.42 Cross section of UMOSFET structure with asymmetric concept.

device. This lowered drain saturation current leads to improve the short-circuit capability [102]. As a result, this 1200 V exhibits an excellent low on-resistance while showing a good SCSOA characteristic of more than 5 μs (@175°C).

4.3.3 Future SiC-MOSFET structure

4.3.3.1 Monolithically integrated SiC-MOSFET and SBD structure

In SiC-MOSFET, its bipolar operation of the inherent body pin diode is undesirable because the bipolar degradation following the expansion of SFs will happen as described in Section 4.2.7. This degradation leads to the increase of on-resistance in the MOSFET operation as well as that of conduction loss of the inherent pin diode. The SBD device, which is unipolar device, can be externally connected in parallel to an SiC-MOSFET as a separate chip in order to accommodate current in the opposite direction. In this case, the inherent body pin diode formed by the p-base and n− drift regions in the MOSFET will not work. Chip size of the external SBD has to be designed larger, especially for higher blocking voltage device, due to its high on- resistance. According to Ref. [103], the active area of the external SBD could be in excess of three times larger than that of coupled MOSFET for 6.5 kV case. When an SiC SBD structure is integrated in an SiC-MOSFET one on a single chip, it is so beneficial because both MOSFET and SBD not only share the forward conduction layer but also they share the edge termination regions, resulting in a significant reduction of SiC wafer area. In addition, this approach will reduce a number of dies which have to be assembled in the SiC-MOSFET module and will save their assembling time bringing down to the module cost. Also, it should be noted that this could improve the power density in power converter/inverter apparatus by increasing the operating switching frequency due to the elimination of parasitic inductance between separate packaged devices, because fewer numbers and smaller size of parasitic components are required in the apparatus. In 2011, Uchida et al. presented the SiC-MOSFET structure with unipolar internal diode named "MOS-channel diode" [104]. After that, a lot of papers have been published and demonstrated the SiC-MOSFET structures with embedded SBD/JBS diodes [105–107]; as a result, the monolithically integrated SiC-MOSFET and SBD/JBS structure becomes one of the candidates for future SiC-MOSFET structures. Kawahara et al. presented the 6.5 kV SBD-embedded SiC-MOSFET structure [103]. By integrating an SBD into each cell of SiC-MOSFET, its specific on-resistance was achieved of 37 mΩ cm^2 (@R. T), 113 mΩ cm^2 (@175°C) while showing high breakdown voltage of 7650 V. Also, this device was attained that the expansion of active area is restricted by 10% or less; therefore, the total active area of a MOSFET and SBD can be greatly reduced by factor of four and more. This paper also explained the effect of SBD integration on the activation current density of the internal pin diode. Fig. 4.43 shows the schematic cross section of a conventional MOSFET with an external SBD and an SBD embedded MOSFET. In order to avoid activation of the internal pin diode, applied voltage at the pn junction (V_{pn}) must be kept lower than the built-in potential of the pn junction. In an SBD embedded MOSFET, source–drain voltage (V_{SDb}) can be higher than V_{pn} owing to voltage drops due to spreading SBD current at the n− drift region and n+ substrate connected in series to the pn junction, whereas anode–cathode voltage in external SBD coupled with a conventional MOSFET (V_{SDa}) should be lower than V_{pn} because the entire voltage on the MOSFET is applied to the pn junction. Therefore, higher SBD current

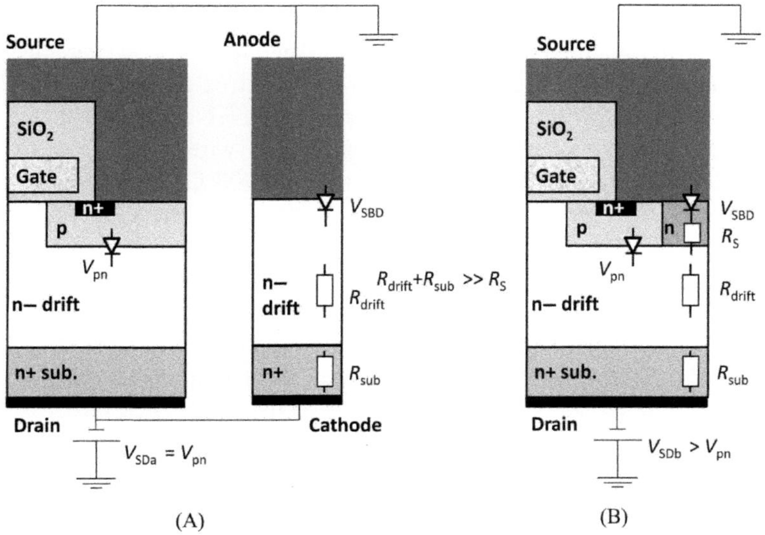

Figure 4.43 Schematic cross section of (A) Conventional MOSFET with an external SBD and (B) SBD-embedded MOSFET.

can be achieved in an SBD-embedded MOSFET without activation of the body diode. The maximum current density without body diode activation J_{Db} in an embedded SBD and J_{Da} in external SBD can be calculated as follows:

$$J_{Da} = \frac{V_{pn} - V_{SBD}}{R_{drift} + R_{sub}} \tag{4.22}$$

$$J_{Db} = \frac{V_{pn} - V_{SBD}}{R_s} \tag{4.23}$$

where V_{SBD} denotes the voltage drop at Schottky junction, R_{drift}, R_{sub}, and R_s signify the resistance of drift region, substrate, and JFET region beneath the Schottky contact, respectively. Therefore, even at high temperature of 175°C, the bipolar current conduction starts at a current density of 120 A/cm^2, which is sufficient high compared with the rated current densities of 6.5 kV devices. The switching characteristics of the SBD embedded in a SiC-MOSFET were also measured and its recovery current was very small and equal to that the calculated results from the output capacitance originating from depletion regions expanding from SiO_2/SiC interfaces and p-base/n− drift layer interface. This indicates that the inherent body pin diode conduction and subsequent carrier storage do not occur during on-state conduction of the diode.

It is very important for fabricating an SBD integrated MOSFET structure to attain a simple top metal process to form ohmic contacts on n$^+$ source, p$^+$ contact regions, and Schottky contact on n− drift layer, simultaneously. Sung and Baliga proposed and reported their original structure named JBSFET using a single ohmic/Schottky process as shown in Fig. 4.44 [107]. Nickel (Ni) is the most commonly used metal for ohmic contact formation on n$^+$ region and is also able to form an ohmic contact on p$^+$ region at the same time. Therefore, in this literature, Ni was chosen and applied and its RTA condition was optimized carefully. In the JBSFET devices, a JBS diode area is surrounded by the MOSFET cell because the gate pad does not interrupt the source pad, which makes the wire bonding easier and the wire to the gate pad shorter. The RTA condition of 900°C/2 min was the optimum one to meet the requirement of low specific on-resistance and blocking characteristic.

Fig. 4.45 exhibits trench-SBD integrated UMOSFET structure [108]. In order to shrink the cell pitch of the SBD integrated MOSFET, SBD-wall integrated UMOSFET, in which the trench SBD is integrated into the IE-UMOSFET (named SWITCH-MOS), was proposed. Unlike the planar SBD, the trench SBD effectively shrinks the cell pitch. However, a high electrical field is applied to the trench bottom and the bottom corner of the integrated trench SBD during the blocking state, and this causes an increase in leakage current and deterioration in breakdown voltage. To avoid this issue, the SBD trench bottom and bottom corner are covered by a wide shielding p+ region, as is used in the gate trench bottom of an IE-UMOSFET. These shielding p+ regions protects the gate and SBD trench bottom and corners each other, the gate and Schottky reliability could be much improved

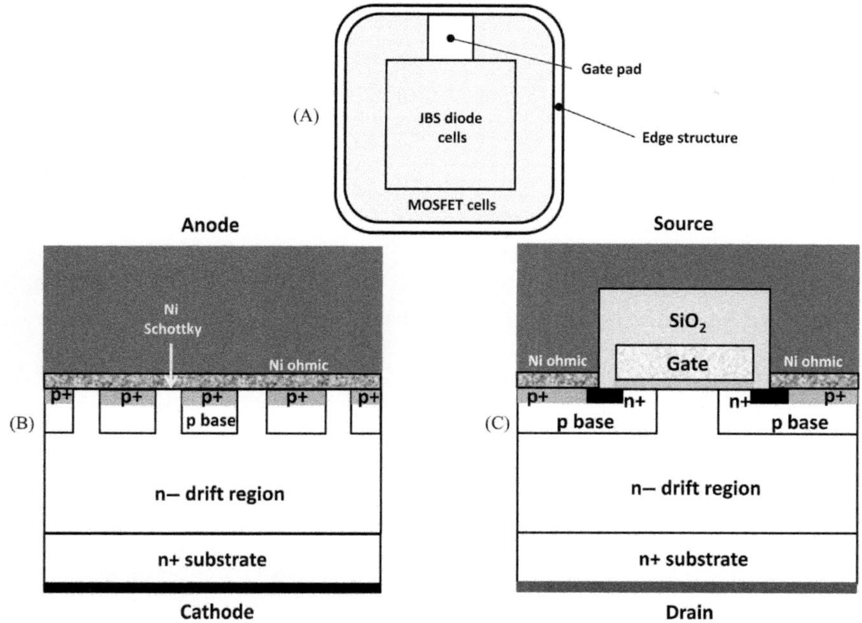

Figure 4.44 Schematic diagram of (A) JBSFET layout, (B) cross section of JBS diode cell structure, and (C) MOSFET cell structure (bottom-right).

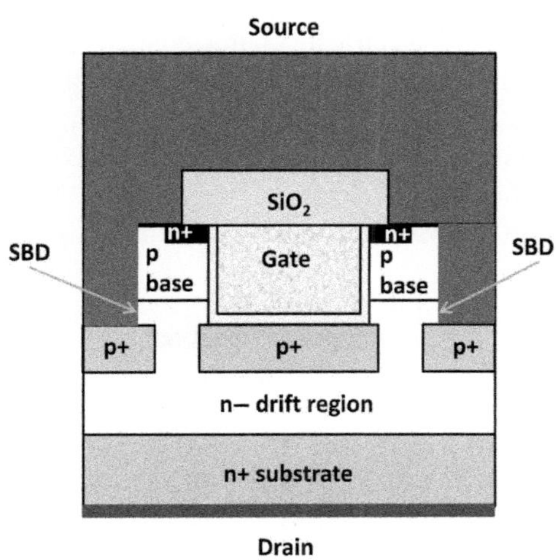

Figure 4.45 Cross section of trench-SBD integrated UMOSFET.

while exhibiting lower on-resistance. Fabrication process is almost same with the IE-UMOSFET; however, Schottky metal layer has to be deposited after the trench etching of the SBD regions. Nickel or titanium is a candidate for the Schottky metal.

4.3.3.2 Superjunction MOSFET device

Si superjunction (SJ) MOSFET structure was proposed and developed in late of 1990s in order to break through the trade-off characteristics between the breakdown voltage and the specific on-resistance [54,55]. This trade-off characteristics have been improved and its specific on-resistance of ~10 mΩ cm^2 with the breakdown voltage of as high as 650 V can be successfully attained. The charge-coupling concept has allowed extending the breakdown voltage of Si power MOSFET and could be applicable to SiC power MOSFET. In the SJ structure, it is very important to set the doping concentration and thickness of the n− drift and p− regions. According to Ref. [16], the optimized doping Q_{opt} and the width are given in the following equation:

$$Q_{\text{opt}} = 2qN_d d = \varepsilon_s E_c \tag{4.24}$$

where q is the charge of electron ($q = 1.6 \times 10^{-19}$ C), N_d is the doping concentration of the n− drift region, and d is half width of the n− drift layer as shown in Fig. 4.46, ε_s is the dielectric constant of the semiconductor and E_c is the critical electric field for breakdown in the semiconductor. For SiC, the optimum charge is found to be 1.71×10^{13} cm^{-2} based upon the E_c of 3.0×10^6 V/cm and the ε_s of 10.32. This optimum charge value is 8.7 times larger than the Si.

There are two types of fabrication process for SJ structure, multiepitaxial growth [109] and trench-filling epitaxial growth methods [110,111]. Kosugi et al. presented the fabrication process of the pn-pillar structure by using multiepitaxial growth method and the measured electrical characteristics [112]. As described in Section 4.3.2, there are major differences between ion implantation technologies for SiC and Si; therefore, some special ion implantation technologies were applied to form the pn-pillar structure. Al^+ implantation was carried out utilizing an MeV-class implanter with a tandem accelerator. This multiple Al^+ implantations at the acceleration energies of 0–7 MeV were used to form a 3 μm-deep Al box profile as

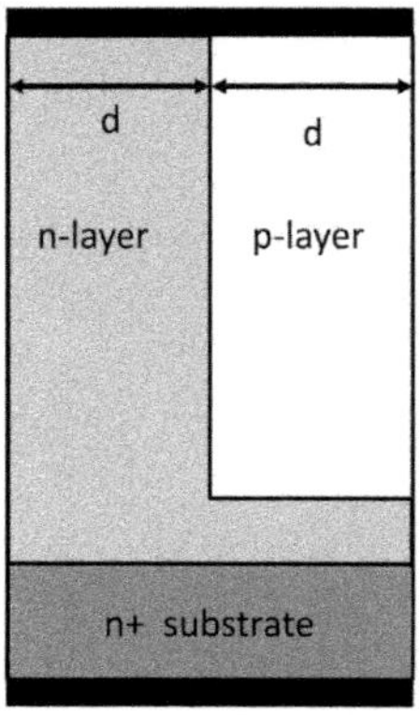

Figure 4.46 Schematic cross section of basic SJ structure.

the first implantation for an SiC wafer with n− epi/n+ substrate, where the pn-pillar width was 2.5 μm at equal intervals. The SiC wafer for this device fabrication were 4 degrees-off (0001) 4H-SiC n+ substrate and doping density and thickness of the n-epitaxial layer were $\sim 3.0 \times 10^{16}\,cm^{-3}$ and 6−10 μm. Following the implantation, the second epitaxial layer was grown on that partially implanted surface. After the second epitaxial growth, the multiple Al implantations at 0−5.5 MeV were conducted to form a 2.5 μm-deep Al box profile. As a result, the total depth of the pn-pillar structure becomes 5.5 μm thick. It should be noted that, in this fabrication process, some unique process techniques were needed such as epitaxial growth on an implanted layer, SiO_2 mask formation for MeV-class implantation, and a novel alignment method to align pn-pillars in the vertical direction during multiepitaxial growth shown as follows:

1. To prevent the implanted layer from being thinner which causes an increase in number of implantation steps, an intentional pretreatment etching in ambient hydrogen was skipped at the beginning of the epitaxial growth. Also, the epitaxial growth was performed at a relatively low temperature.
2. Thick implantation mask of 6.4 μm SiO_2 layer is required to make 3.5 μm depth Al^+ layer.
3. Alignment mark on the wafer backside is required to form a straight-line pn-pillar by multiepitaxial growth. According to this method, alignment accuracy was estimated to be 0.3−0.5 μm.

The breakdown voltage and specific on-resistance of the 5.5 μm-thick pn-pillar structure were 1545 V and 1.06 $m\Omega\ cm^2$, respectively, which are almost the same with the SiC theoretical limits. These results clearly show the SiC SJ-MOSFET structure could improves the trade-off characteristic so much and, when designed as ultrahigh-voltage device such as 6.5 kV and more, the SiC SJ-MOSFET could demonstrate a superior characteristic with much lower on-resistance while showing its fast switching speed.

4.3.3.3 SiC MOSFET with reverse blocking capability

The high-voltage bidirectional switches are utilized for many applications such as matrix converters and multilevel inverters. Conventional bidirectional switches are configured by the two Si-IGBTs and two Si-pin diodes to attain the reverse blocking capability. Since current flows through both an IGBT and a pin diode during on state, the total on-state power loss is determined by the sum of on-state voltage drops for these two devices. To reduce the on-state power loss of bidirectional switches, reverse blocking IGBTs (RB-IGBTs) have been developed [113−115]. Recently, SiC matrix converter has been presented [116]. In these circuits, two SiC MOSFETs were connected in antiseries and constructed a bidirectional switch. Because the on-state power loss of the bidirectional switch is almost twice higher than that of a single SiC-MOSFET as in the case of Si-IGBTs, it is possible to reduce the conduction power loss by using an anti-parallel connection of two SiC MOSFETs with reverse blocking capability. Mori et al. have reported SiC RB-MOSFET by embedding the SBD to the backside of wafers, for the first time [117].

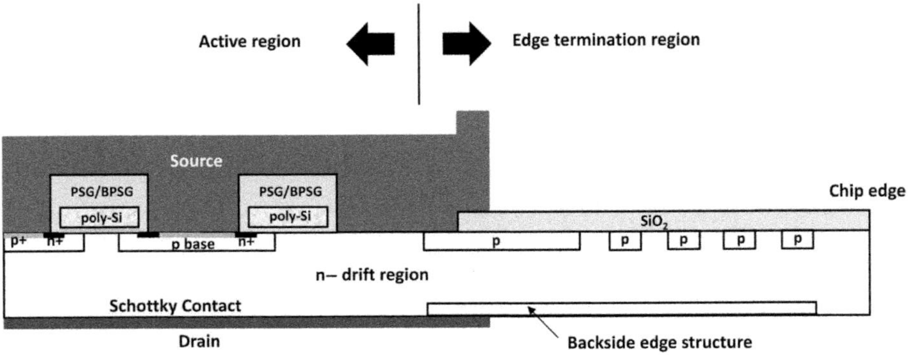

Figure 4.47 Schematic cross section of SiC RB-MOSFET.

Fig. 4.47 exhibits a cross section of the SiC RB-MOSFET. In order to have a reverse blocking capability, SBD was embedded on the backside of the n− drift layer. Also, it should be noted that edge termination structure have to be also formed on the backside.

The fabrication process was a little bit similar to the Si-IGBT with thin wafer technologies [1]. An n− drift epi layer grown on n+ substrate was utilized as a starting material. First, a standard MOSFET without any backside structure was fabricated. Thickness of n− drift layer was set of 50 μm and doping density of this layer was 2×10^{15} cm^{-3}. Then, the n+ substrate was totally removed and chemical mechanical polishing (CMP) was applied. After that, the edge termination structure was fabricated by ion implantation, then, titanium was deposited as backside drain Schottky metal. It should be noted that the thin wafer technology (in this case: 50 μm or less) for the SiC RB-MOSFET is especially required not only to make the wafer thin but also to form the edge termination structure by ion implantation and Schottky metal on the backside of the thin wafer with absolute no wafer breaking. Experimental results of forward and reverse blocking voltages were in excess of 3 kV and the differential specific on-resistance was 20 mΩ cm^2, respectively. Although the reverse leakage current of the backside Schottky was so high, this reverse current can be reduced by introduction of the JBS structure described in Section 4.2. In the on-state characteristics, a built-in potential of about 1.0 V was observed due to the backside of Schottky barrier. It should be noted that the on-state voltage drop of the RB-MOSFET shows the lower values than that of the standard SiC-MOSFETs connected in anti-series at a current density of 70 A/cm^2. Switching characteristics of the RB-MOSFET were almost same with the standard SiC-MOSFETs; therefore, the RB- MOSFETs are suitable for high-voltage bidirectional switches.

4.3.3.4 Complementary p-channel MOSFET device

Most of the analysis and discussion in this chapter has been focused on the n-channel power MOSFET (n-MOSFET) structure because of its predominance in applications. The n-MOSFET structure is favored over the p-channel power MOSFET (p-MOSFET) structure because the larger mobility of electrons than holes results in smaller on-resistance for n-channel device. However, p-MOSFET devices are preferable in many power electronics circuit and its application, especially for medium and high power ranges. In such applications, increasing switching frequency is an important challenge in power converters including an inverter and a DC–DC chopper, which consist of one or more "legs," a pair of semiconductor switches connected to a DC voltage source and has an output terminal in the middle. Higher switching frequency results in the size reduction of inductors and/or capacitors and the harmonics reduction in the output voltage and current. However, a dead time has to be set for the converter using the legs to avoid the short-through in which the upper-side and lower-side arms of the leg conduct, simultaneously. SiC-MOSFET has no tail current in its switching operation; therefore, it allows increasing the switching frequency with keeping low power losses. However, still the dead time corresponding to the period of a signal delay through gate-drive ICs is required, and that is usually longer than 0.1 μs with the state-of-the-art technology. The need for the dead time makes it difficult to increase the switching frequency further. Complementary topologies can reduce the dead time, so the problems caused by the dead time can be directly improved [118]. A leg with the complementary topologies as shown in Fig. 4.48, which consists of the complementary arranged n-MOSFET and p-MOSFET, can minimize the dead time, and allows increasing the switching frequency without possibility of the short-through. The Si p-MOSFET devices have poor characteristics of higher on-resistance compared with the n-MOSFET device. Also, the Si-pIGBT is not available in the market because it has poor operating tolerance, for example, the SCSOA [119]. A vertical SiC-pMOSFET with the blocking voltage of −730 V and the gate threshold voltage of −5.32 V was successfully fabricated and its SCSOA characteristics were first reported [120]. This device was fabricated based upon the design of n-channel

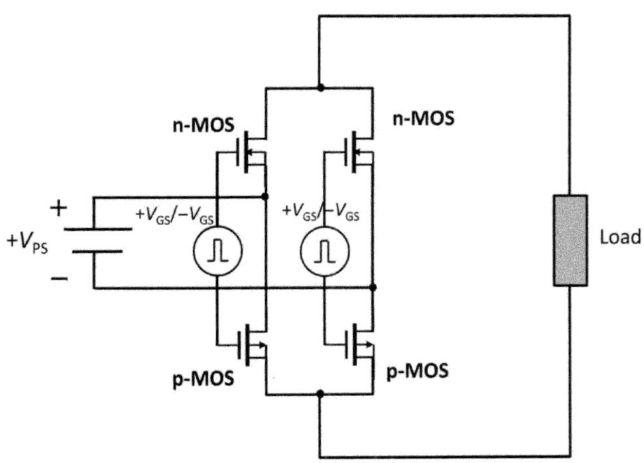

Figure 4.48 Complementary inverter circuit.

IE-MOSFET [78]. Substrate used for p-MOSFET was Si-face p-type 4H-SiC with the thickness of 350 μm and resistivity of 2.0 Ω cm. The fabricated device featured the breakdown voltage of −730 V and could tolerate short-circuit energy of 16.1 J/cm^2 which is 15% higher than that of SiC n-MOSFET. Also, the fabricated device exhibits higher gate oxide reliability and avalanche immunity during the short-circuit test. However, this device showed the higher specific on-resistance of 218 mΩ cm^2, which is about 50 times higher than that of n-MOSFET due to not only the hole lower mobility (channel and bulk) but also fabrication process technologies such as ohmic metal process to get lower p$^+$ contact resistance [121]. These results aim at evaluating actual safety area for the SiC p-MOSFET for complementary inverter circuit in the future. At the same time, it could provide a bench mark to improve the future SiC p-MOSFET device technologies.

4.3.4 Summary

At present, SiC and GaN power switching devices have begun to be available on the market. In terms of structure, "SiC on SiC" is the main stream with the SiC devices while GaN device adopt the "GaN on Si" structure. The reason behind this is that it is difficult to form GaN directly on Si, and it is necessary to sandwich a high-resistance buffer layer between in the "GaN on Si" structure. For this reason, GaN devices must employ the lateral device structure. In contrast, SiC devices can achieve vertical structure devices such as Si power MOSFETs and Si-IGBTs, which make up majority of power device market at present. Hence, SiC devices can be used for high-voltage and high-current application with blocking voltage of 1000 V and higher, and current of 50 A and higher. Thus, SiC-MOSFETs can control large-capacity motors with 50 to 100 kW, which are, for example, incorporated in next-generation vehicles. Hence, SiC-MOSFETs are used for boost converter and inverter elements in power control units (PCUs). As discussed in this section, SiC-MOSFETs feature low on-resistance, low switching loss characteristics, and furthermore enable high-temperature operation. They are expected to realize further miniaturization of cooling components and passive components in PCUs. It is known that air resistance (CD value) is the largest factor that determines the improvement of vehicles' traveling fuel economy. In other words, the improvement of the vehicle structure significantly contributes to the improvement of traveling fuel economy. Hence, the miniaturization of PCUs through the use of SiC power devices have advantages to enhance the degree of design freedom to improve the vehicle structure and the achievement of low air resistance (CD) due to the improved vehicle structure.

This section is mainly focused on the discussion for SiC-MOSFETs device technologies; however, it is strongly required to adopt high reliable packaging technologies for SiC-MOSFET modules that ensure operation at high frequency and high temperature. For this purpose, there are some papers for newly proposed package [122]. For example, copper pins were utilized to ensure a high current flow of the module in the new package. The newly structured package has reduced internal inductance which is approximately a quarter of that of the conventional structure with aluminum wire bonding. The increased switching speed of SiC-MOSFET modules is accompanied by a high surge voltage due to parasitic inductances in the packages; therefore, this small internal inductance is so effective to reduce the surge voltage. In addition, thermal resistance is reduced by the ceramic insulating substrate bonded with thick copper plates. A highly thermal-resistant epoxy resin also suppresses deformation in the bonding portions of the chip and copper pins. By adopting these improvements, the newly developed package realizes higher reliability of ΔTj power cycle capability which is ten times better than that of conventional products. It is very important for not only SiC-MOSFET technologies but also the packaging technologies to utilize the ability that the SiC-MOSFETs have for expansion of SiC-MOSFETs device application. It will not be exaggeration that both technological improvements for the packaging and SiC semiconductor devices control progress of future high efficiency and high add-value power electronics.

4.4 SiC-IGBT

4.4.1 Introduction

IGBT was developed using in Si technology to provide a superior alternative to bipolar power transistors in the 1980s. The IGBT represents a power MOSFET-bipolar transistor integration in the sense of combining the physics of MOSFET operation and bipolar transistor's one. The gate drive signal for the integrated device configuration is applied to the power MOSFET structure, providing the advantage of compact, low cost gate drive circuits made possible high input impedance, voltage controlled operations. In the IGBT, power MOSFET structure is operated to provide the base current to inherent p−n−p bipolar power transistor with bipolar transistor used to modulate the conductivity of the drift region for the MOSFET structure; therefore, low forward voltage drop can be achieved. In recent years, Si-IGBTs are applied to most power electronic applications, especially medium and high power equipment, such as ac drive motion control, UPS, renewables (wind and solar), and others. In the last quarter of centuries, the IGBT has been still playing the leading role for controlling the power electronics equipment through the improvement of IGBT chip and packaging technologies. The optimization of the IGBT structure by the development of thin wafer technologies with low collector injection efficiency, together with the introduction of trench-gate design, has made the device competitive at much higher voltage. The IGBT structure exhibits an excellent current saturation capability, which the thyristor cannot have, is now considered an essential characteristic for many power electronics applications in practice, so that the IGBT has been scaled to the rated voltages up to 6.5 kV and has already replaced the gate turn-off (GTO) thyristor in some applications.

Due to the high breakdown field in the SiC, the drift layer thickness in SiC devices are approximately one-tenth of those of Si devices, so that SiC-IGBT would have an order of magnitude smaller storage charge in the drift layer compared with that of the Si-IGBT providing an improvement in switching behavior. However, as mentioned in Section 4.2.5, since the built-in potential of SiC pn junction is as high as ~2.5 V and more in room temperature, the on-state voltage drop of the SiC-IGBT structure becomes high. Therefore, the development of ultrahigh-voltage SiC-IGBT has been given significant attention for very high voltage range above 10 kV where the Si-IGBT does not have its competitiveness. For example, in the future smart grids and high-voltage power supplies, this ultrahigh-voltage and low loss power devices are key components. In this section, some recent progresses of device and process technologies for the SiC-IGBT are introduced.

4.4.2 Device structure and its fabrication process

4.4.2.1 p-channel IGBT or n-channel IGBT

The basic structure of the IGBT is shown in Fig. 4.49, which is same with the Si-IGBT. However, in recognition of some difficulties of fabrication process shown in Section 4.3, the design concept and process procedure of the state-of-the-are Si IGBT cannot be utilized and have to be totally different from the Si-IGBT counterpart. In 2005, experimental study on 10 kV p-channel SiC-IGBT was reported [123]. As in the case of MOSFETs, this early developed SiC-IGBTs were trench structure (UIGBT) because the base regions can be formed epitaxially without the need for ion implantation and associated high temperature annealing. Furthermore, a disadvantage of the SiC UIGBT structure is the field in the oxide exceeds the semiconductor field by the ratio of the dielectric constants, a factor of 2.5. This places the field in the oxide at ~4 MV/cm, a value dangerously close to oxide breakdown. The geometry of the trench corner further increases the oxide field due to field crowding, resulting in local oxide failure. An obvious way to avoid the problem with oxide breakdown at the trench corners is to utilize the planar gate structures. Fig. 4.50 exhibits the UIGBT structure with its internal resistance

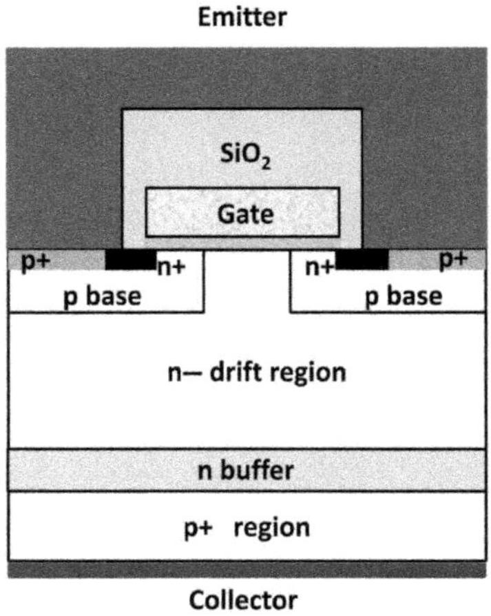

Figure 4.49 Cross section of planar gate IGBT structure.

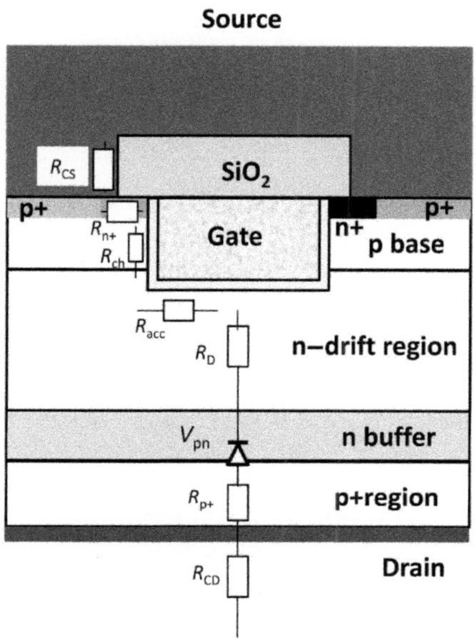

Figure 4.50 UIGBT structure with its internal resistance.

components. The total on-resistance for the UIGBT structure is obtained by the addition of the all resistance because they are considered to be in series the current path between the emitter and the collector electrodes. Since the SiC-IGBT is mainly focused on the ultrahigh-voltage application above 10 kV, its total on-resistance would be occupied mainly by the resistance of thick drift layer even though this drift layer was in conductivity modulation.

The p-channel SiC-IGBT (SiC p-IGBT) can be easily fabricated without the complicated device processing involved in backside grinding and backside ion implantation to eliminate the harmful effects of high resistivity p-type substrate for n-channel IGBT [124,125]. However, the p-MOSFET shows a peak field-effect mobility of as small as 13.5 cm^2/Vs, while the n-MOSFET exhibits a peak field-effect mobility of about 25 cm^2/Vs and more at this time. Hence, the current provided by the p-MOSFET channel is much smaller than that by n-MOSFET channel, which limited the on-state performance of the SiC p-IGBT. It is also suspected that p-type contacts, which have a contact resistance of $\sim 1 \times 10^{-3}\ \Omega\ cm^2$, contributes to the differences between the p- and n-IGBTs. For SiC n-IGBTs, the p-contact forms the wide area contact on the backside, which helps minimizing the impact of the poor ohmic contacts. However, on p-IGBTs, the p-contacts are placed on the source regions of the p-MOSFET channel. This makes the impact of the poor p-contacts much worse because the smaller contact area results in larger values of resistances. In addition, the resistance placed in the source of the MOS channel creates a negative feedback loop, which reduces the gain of the p-MOSFET channel even further. Also, it makes clear that the n-IGBT structure have superior turn-off characteristics to the p-IGBT while their on-state voltage drop is very similar. This is because the reduced gain of the wide base pnp transistor in the n-IGBT coupled with large electron–hole mobility anisotropy in the SiC [125]. This leads to a far more favorable trade-off between on-state voltage drop and switching loss in the n-IGBT structures. The smaller transistor gain also results in a higher dynamic avalanche voltage for n-IGBTs, and thus a larger reverse-bias safe operating area (RBSOA). These results demonstrate that although the fabrication of SiC n-IGBT is currently more difficult due to the lack of and its very high resistivity of p-type substrates, lower switching losses, and better ruggedness make it a more compelling choice for ultrahigh-voltage applications. From these reasons mentioned earlier, there are many planar gate SiC n-IGBT structures reported in many papers [126–131]. Brunt et al. reported that the ultrahigh-voltage SiC n-IGBT devices with their blocking voltage of 27 kV [129]. Vertical design was based upon the punch-through one and the 210 μm thick and low doped n− drift region of $1.0 \times 10^{14}\ cm^{-3}$ and n-type field stop buffer layer of about 1–2 μm were grown on a 4 degrees off-axis 4H-SiC substrate. A 1.5 mm wide edge termination consisting of floating guard rings was used to alleviate high electric fields at the active region edge. Prior to device fabrication, a lifetime enhancement procedure consisting of 15 h of thermal oxidation at 1300°C was performed to increase of the lifetime more than 10 μs. By the application of the lifetime enhancement process, the on-state voltage drop was improved by in excess of 1.0 V [128]. In addition to these static characteristics, inductive load turn-off switching characteristics were measured. The turn-off dV/dt was measured and had a peak value of 120 kV/μs, which is lower than the other n-IGBT device designed with 230 μm thick and doping of $2.5 \times 10^{14}\ cm^{-3}$ (the peak dV/dt:170 kV/μs). The high dV/dt value leads to the increase of electromagnetic compatibility (EMC), so it is very important for the ultrahigh-voltage IGBT to design to satisfy smaller turn-off loss and EMC, simultaneously.

4.4.2.2 Flip-type IEIGBT structure

For the purpose of fabricating an n-channel IGBT, the crystal quality of the p+ SiC substrate is very poor with its high crystal defect density and resistivity compared with those of the n+ substrate from the viewpoint of using it as the collector side. For solving these problems related to the SiC n-IGBTs, new type IGBT structure named flip-type IEIGBT was proposed [130]. Fig. 4.51 exhibits a schematic cross section of the IEIGBT. In the MOSFET region, the IEIGBT employed the IEMOSFET concept, that means the bottom and top of the IEMOSFETs p well are formed by ion implantation and epitaxial growth, respectively. The smooth top surface of the p well affords higher channel mobility than can be achieved using the conventional double implantation method. Fig. 4.52 shows the process flow of a flip-type $(000\overline{1})$ carbon face wafer. First, an n-type drift layer thicker than 150 μm with nitrogen doping ($4 \times 10^{14}\ cm^{-3}$) was grown on the silicon face of an n+ substrate after a buffer layer was formed for converting the BPDs to TEDs. Then, a 1–2 μm thick buffer layer was grown. After the p collector with an aluminum-doped layer was formed, the p+ substrate layer ($2 \times 10^{19}\ cm^{-3}$) was grown to a

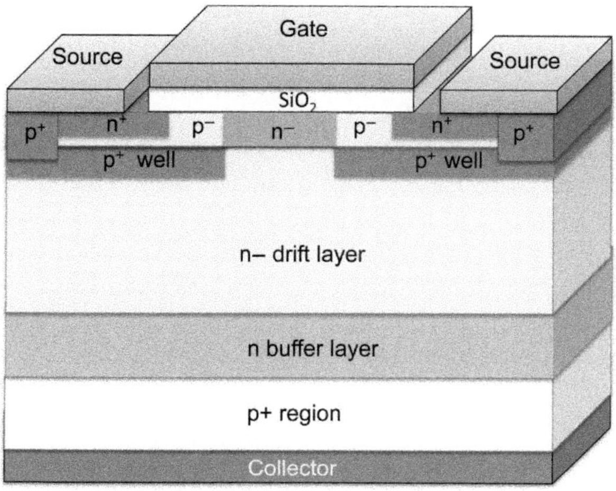

Figure 4.51 Schematic cross section of IEIGBT.

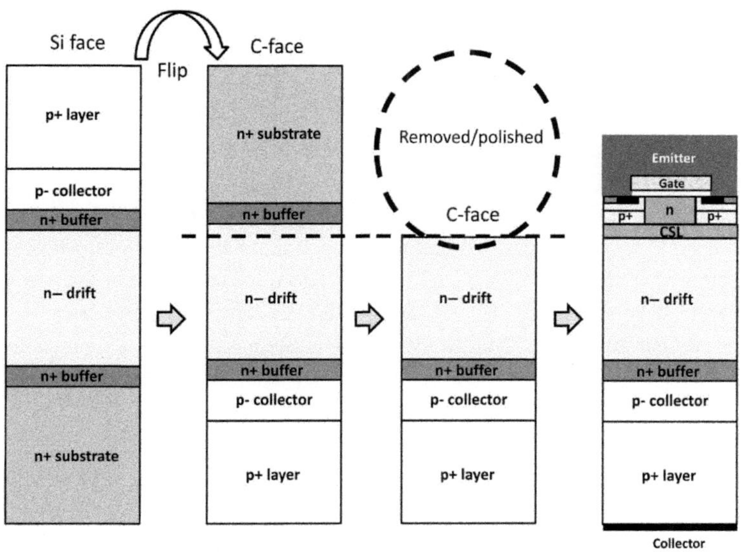

Figure 4.52 Process flow of the flip-type IEIGBT.

thickness of over 100 μm. At this time, turning the substrate over (flip), the n+ substrate was removed and the surface was polished by using the CMP process. Following this flip-type method, a high quality n− drift layer with a p+ substrate layer was obtained. For device fabrication, aluminum ions (Al^{+}) was selectively implanted to form the bottom of p well, then a 0.5 μm thick p-epitaxial layer was grown as the top p-layer. After that, almost the same process with the IEMOSFET was applied for the SiC-IEIGBT surface structure fabrication. It should be noted that stripe cell structure was adopted in the IEIGBT in order to enhance the latch-up immunity whereas the hexagonal cell structure is utilized for the SiC IEMOSFET. The layout of the cell structure on the surface of the IGBT has an impact on the latch-up current density and most of the Si IGBT utilized the stripe cell structure due to its high latch-up current density; therefore, the SiC IEIGBT applied to the stripe cell structure. On the contrary, many MOSFET devices adopted the square or hexagonal cell structures to maximize the packing density of cells because there are no hole-current flows which trigged the latch up in the device. An $8 \times 8\ mm^2$ IEIGBT (active area: $0.37\ cm^2$) was fabricated and its blocking voltage was in excess of 16 kV [131]. The on-state voltage drop is as low as 4.8 V (@54 A/cm^2) and its high turn-off capability of 60 A ($J = 162\ A/cm^2$) was achieved at high temperature of 250°C (@Vce = 6.5 kV).

4.4.3 Summary

The development of ultrahigh-voltage SiC-IGBT has been given significant attention for very high voltage range above 10 kV where the Si-IGBT does not have its competitiveness. The early stage of developed SiC-IGBT structures were p-channel ones because the p-IGBTs can be easily fabricated with n+ substrate. However, the current provided by the p-MOSFET channel is much smaller than that by n-MOSFET one, the on-state performance was so poor. Further, due to reduced gain of the wide base pnp transistor with the large electron–hole mobility anisotropy, the n-IGBTs exhibited better turn-off characteristics and larger RBSOAs than the p-IGBTs; therefore, there were a lot of n-IGBT structures reported as the ultrahigh-voltage devices. Fabrication of the n-IGBTs is currently more difficult due to the lack of good p+ substrate, so that the new type of fabrication process like the flip-type IEIGBTs were developed. Studies on not only the device structure/fabrication process but also on new stable crystal growth methodologies such as p-type substrate using aluminum and nitrogen codoping [132] are strongly required and the achievement of these studies could lead to realization of ultrahigh-voltage SiC n-IGBT devices with very high performance.

References

[1] T. Matsudai, K. Kinoshita, A. Nakagawa, New 600 V trench gate punch-through IGBT concept with very thin wafer and low efficiency p-emitter, having an on-state voltage drop lower than diodes, Proceeding of IPEC Tokyo 2000 April, 2000, pp. 292–296.

[2] A. Nakagawa, Theoretical investigation of silicon limit characteristics of IGBT, in Proc. Int. Symp. Power Semiconductors and ICs, June 2006, pp. 5–8, https://doi.org/10.1109/ISPSD.2006.1666057.

[3] M. Sumitomo, J. Asai, H. Sakane, K. Arakawa, Y. Higuchi, M. Matsui, Low loss IGBT with partially narrow mesa structure (PNM-IGBT), in Proc. Int. Symp. Power Semiconductors and ICs, June 2012, pp. 17–20, https://doi.org/10.1109/ISPSD.2012.6229012.

[4] H. Feng, W. Yang, Y. Onozawa, T. Yoshimura, A. Tamenori, J.K.O. Sin, A 1200 V-class fin P-body IGBT with ultra-narrow-mesas for low conduction loss, in Proc. Int. Symp. Power Semiconductors and ICs, June 2016, pp. 203−206, https://doi.org/10.1109/ISPSD.2016.7520813.
[5] K. Eikyu, A. Sakai, H. Matsuura, Y. Nakazawa, Y. Akiyama, Y. Yamaguchi, et al., On the scaling limit of the Si-IGBTs with very narrow mesa structure, in Proc. Int. Symp. Power Semiconductors and ICs, June 2016, pp. 211−214, https://doi.org/10.1109/ISPSD.2016.7520815.
[6] H. Takahashi, A. Yamamoto, S. Aono, T. Minato, 1200 V reverse conducting IGBT, in Proc. Int. Symp. Power Semiconductors and ICs, May 2004, pp. 133−136, https://doi.org/10.1109/WCT.2004.239844.
[7] O. Hellmund, L. Lorenz, H. Rüthing, 1200 V reverse conducting IGBTs for soft-switching applications, China Power Electron. J. 5 (2005) 20−22.
[8] K. Satoh, T. Iwagami, H. Kawafuji, S. Shirakawa, M. Honsberg, E. Thal, A new 3A/600 V transfer mold IPM with RC (reverse conducting)-IGBT, in Proc. PCIM Europe., May 2006, pp. 73−78.
[9] H. Rüthing, F. Hille, F.J. Niedernostheide, H.J. Schulze, B. Brunner, 600 V reverse conducting (RC-) IGBT for drives applications in ultra-thin wafer technology, in Proc. Int. Symp. Power Semiconductors and ICs, May 2007, pp. 89−92, https://doi.org/10.1109/ISPSD.2007.4294939.
[10] M. Rahimo, U. Schlapbach, A. Kopta, J. Vobecky, D. Schneider, A. Baschnagel, A high current 3300 V module employing reverse conducting IGBTs setting a new benchmark in output power capability, in Proc. Int. Symp. Power Semiconductors and ICs, May 2008, pp. 68−71, https://doi.org/10.1109/ISPSD.2008.4538899.
[11] K. Takahashi, S. Yoshida, S. Noguchi, H. Kuribayashi, N. Nashida, Y. Kobayashi, et al., New reverse-conducting IGBT (1200 V) with revolutionary compact package, in Proc. Int. Symp. Power Semiconductors and ICs, June 2014, pp. 131−134, https://doi.org/10.1109/ISPSD.2014.6855993.
[12] M. Rahimo, M. Andenna, L. Storasta, C. Corvasce, and A. Kopta, Demonstration of an enhanced trench Bimode Insulated Gate Transistor ET-BIGT, in Proc. Int. Symp. Power Semiconductors and ICs, June 2016, pp. 151−154, https://doi.org/10.1109/ISPSD.2016.7520800.
[13] R. Gejo, T. Ogura, S. Misu, Y. Maeda, Y. Matsuoka, N. Yasuhara, and et al., High switching speed trench diode for 1200 V RC-IGBT based on the concept of Schottky controlled injection (SC), in Proc. Int. Symp. Power Semiconductors and ICs, June 2016, pp. 155−158, https://doi.org/10.1109/ISPSD.2016.7520801.
[14] T. Yoshida, T. Takahashi, K. Suzuki, and M. Tarutani, The Second-generation 600 V RC-IGBT with Optimized FWD, in Proc. Int. Symp. Power Semiconductors and ICs, June 2016, pp. 159−162, https://doi.org/10.1109/ISPSD.2016.7520802.
[15] B.J. Baliga, Power semiconductor device figure of merit for high-frequency applications, IEEE Electron Device Lett. 10 (10) (1989) 455−458. Available from: https://doi.org/10.1109/55.43098.
[16] B.J. Baliga, Fundamentals of Power Semiconductor Devices, Springer, 2008.
[17] M. Bhatnager, P.K. McLarty, B.J. Baliga, Silicon-carbide high-voltage (400 V) Schottky barrier diodes, IEEE Electron Device Lett. 13 (10) (1992) 501−503. Available from: https://doi.org/10.1109/55.192814.
[18] F. Dahlquist, A 2.8 kV, forward drop JBS diode with low leakage, Mater. Sci. Forum 338−342 (2000) 1179−1182. Available from: https://doi.org/10.4028/www.scientific.net/MSF.338-342.1179.

[19] Y. Sugawara, K. Asano, R. Saito, 3.6 kV 4H-SiC JBS diode with low Ron, Mater. Sci. Forum 338–342 (2000) 1183–1186. Available from: https://doi.org/10.4028/www.scientific.net/MSF.338-342.1183.
[20] T. Hatakeyama, T. Shinohe, Reverse characteristics of a 4H-SiC Schottky barrier diode, Mater. Sci. Forum 389–393 (2002) 1169–1172. Available from: https://doi.org/10.4028/www.scientific.net/MSF.389-393.1169.
[21] B.J. Baliga, The pinch rectifier: A low-forward-drop high-speed power diode, IEEE Electron Device Lett. 5 (6) (1984) 194–196. Available from: https://doi.org/10.1109/EDL.1984.25884.
[22] T. Kimoto, Material science and device physics in SiC technology for high-voltage power devicesJpn J. Appl. Phys. 54 (4) (2015) 0401031–27 . Available from: https://doi.org/10.7567/JJAP.54.040103.
[23] B.J. Baliga, Analysis of junction-barrier-controlled Schottky (JBS) rectifier characteristics, Solid-State Electron. 28 (11) (1985) 1089–1093. Available from: https://doi.org/10.1016/0038-1101(85)90188-1.
[24] F. Dahlquist, H. Lendenmann, M. Östling, A high performance JBS rectifier-design consideration, Mater. Sci. Forum 353–356 (2001) 683–686. Available from: https://doi.org/10.4028/www.scientific.net/MSF.353-356.683.
[25] X. Jordá, D. Tournier, J. Rebolla, J. Millán, P. Godignon, Temperature impact on high-current 1.2 kV SiC Schottky rectifiers, Mater. Sci. Forum 483–485 (2005) 929–932. Available from: https://doi.org/10.4028/www.scientific.net/MSF.483-485.929.
[26] H. Linchao, S. Huajun, L. Kean, W. Yiyu, T. Yidan, B. Yun, et al., Improved adhesion and interface ohmic contact on n-type 4H-SiC substrate by using Ni/Ti/Ni, J. Semicond. 35 (7) (2014) 072003. Available from: https://doi.org/10.1088/1674-4926/35/7/072003. 1–4.
[27] T. Nakamura, Y. Nakano, M. Aketa, R. Nakamura, S. Mitani, H. Sakairi, et al., High performance SiC trench devices with ultra-low Ron, in IEEE IEDM Tech. Dig., Dec. 2011, pp. 599–601, https://doi.org/10.1109/IEDM.2011.613619.
[28] B.J. Baliga, Closed form analytical solutions for the breakdown voltage of planar junctions terminated with single floating field ring, Solid-State Electron. 33 (5) (1990) 485–488. Available from: https://doi.org/10.1016/0038-1101(90)90231-3.
[29] B.J. Baliga, Silicon Carbide Power Device, World Scientific, 2005.
[30] V. Veliadis, M. McCoy, T. McNutt, H. Hearne, L.-S. Chen, G. deSalvo, et al., Fabrication of a robust high-performance floating guard ring edge termination for power silicon carbide vertical junction field effect transistors, CS MANTECH 2007 Conference, pp. 217–220.
[31] C.-F. Huang, H.-C. Hsu, K.-W. Chu, L.-H. Lee, M.-J. Tsai, K.-Y. Lee, et al., Counter-doped JTE, an edge termination for HV SiC devices with increased tolerance to the surface charge, IEEE Trans. Electron Devices 62 (2) (2015) 354–358. Available from: https://doi.org/10.1109/TED.2014.2361535.
[32] B.J. Baliga, Analysis of a high-voltage merged P-i-N Schottky (MPS) rectifier, IEEE Electron Device Lett. 8 (9) (1987) 407–409. Available from: https://doi.org/10.1109/EDL.1987.26676.
[33] R. Rupp, M. Treu, S. Voss, F. Björk, T. Reimann, 2nd generation SiC Schottky diode: a new benchmark in SiC device ruggedness, in Proc. Int. Symp. Power Semiconductors and ICs, June 2006, pp. 269–272, https://doi.org/10.1109/ISPSD.2006.1666123.
[34] T. Tsuji, A. Kinoshita, N. Iwamuro, K. Fukuda, K. Tezuka, T. Tsuyuki, et al., Experimental demonstration of 1200 V SiC-SBDs with low forward voltage drop at

high temperature, Mater. Sci. Forum 717–720 (2012) 917–920. Available from: https://doi.org/10.4028/www.scientific.net/MSF.717-720.917.

[35] R. Rupp, R. Gerlach, A. Kabakow, R. Schorner, Ch. Hecht, R. Elpelt, et al., Avalanche behavior and its temperature dependence of commercial SiC MPS diode: influence of design and voltage class, in Proc. Int. Symp. Power Semiconductors and ICs, June 2014, pp. 67–70, https://doi.org/10.1109/ISPSD.2014.6855977.

[36] S. Harada, Y. Hoshi, Y. Harada, T. Tsuji, A. Kinoshita, M. Okamoto, et al., High performance SiC IEMOSFET/SBD module, Mater. Sci. Forum 717–720 (2012) 1053–1058. Available from: https://doi.org/10.4028/www.scientific.net/MSF.717-720.1053.

[37] A. Salemi, H. Elahipanah, A. Hallėn, G. Malm, C.-M. Zetterling, M. Östling, Conductivity modulated On-Axis 4H-SiC 10 + kV PiN Diodes, in Proc. Int. Symp. Power Semiconductors and ICs, June 2015, pp. 269–272, https://doi.org/10.1109/ISPSD.2015.7123441.

[38] J. Wang, A.Q. Huang, W. Sung, Y. Liu, B.J. Baliga, Development of 15-kV SiC IGBTs and their impact on utility applications, IEEE Ind. Electron. Mag. 3 (2) (2009) 16–23. Available from: https://doi.org/10.1109/MIE.2009.932583.

[39] Y. Sugawara, D. Takayama, K. Asano, R. Singh, J. Palmour, T. Hayashi, 12–19 kV 4H-SiC PiN diodes with low power loss, in Proc. Int. Symp. Power Semiconductors and ICs, June 2001, pp. 27–30, https://doi.org/10.1109/ISPSD.2001.934552.

[40] R. Singh, J.A. Cooper, M.R. Melloch, T.P. Chow, J.W. Palmour, Large area, ultra-high voltage 4H-SiC p-i-n rectifiers, IEEE Trans. Electron Devices 49 (12) (2002) 2308–2316. Available from: https://doi.org/10.1109/TED.2002.805576.

[41] T. Kimoto, J. Suda, Y. Yonezawa, K. Asano, K. Fukuda, H. Okumura, Progress in ultrahigh-voltage SiC devices for future power infrastructure, in IEEE IEDM Tech., Dig., December 2014, pp. 36–39, https://doi.org/10.1109/IEDM.2014.7046967.

[42] G. Feng, J. Suda, T. Kimoto, Space-modulated junction termination extension for ultrahigh-voltage pin diodes in 4H-SiC, IEEE Trans. Electron Devices 59 (2) (2012) 414–418. Available from: https://doi.org/10.1109/TED.2011.2175486.

[43] T. Kimoto, K. Danno, J. Suda, Lifetime-killing defects in 4H-SiC epilayers and lifetime control by low-energy electron irradiation, Phys.Status Solidi B 245 (7) (2008) 1327–1336. Available from: https://doi.org/10.1002/pssb.200844076.

[44] T. Hiyoshi, T. Kimoto, Reduction of deep levels and improvement of carrier lifetime in n-type 4H-SiC by thermal oxidation, Appl. Phys. Exp 2 (4) (2009) 041101. Available from: https://doi.org/10.1143/APEX.2.041101. 1–3.

[45] T.P. Chow, N. Ramangul, M. Ghezzo, Wide bandgap semiconductor power devices, MRS Symposium Proceedings, Vol. 483, 1998, pp. 89–102.

[46] I.V. Grekhov, V.M. Efanov, A.F. Kardo-Sysoev, S.V. Shenderey, Power drift setp recovery diodes (DSRD), Solid-State Electron. 28 (6) (1985) 597–599. Available from: https://doi.org/10.1016/0038-1101(85)90130-3.

[47] I.V. Grekhov, S.V. Korotkov, A.L. Stepaniants, D.V. Khristyuk, V.B. Voronkov, Y.V. Aristov, High-power semiconductor-based nano and subnanosecond pulse generator with a low delay time, IEEE Trans. Plasma Sci. 33 (4) (2005) 1240–1244. Available from: https://doi.org/10.1109/TPS.2005.852349.

[48] I.V. Grekhov, P.A. Ivanov, D.V. Khrisyuk, A.O. Konstantinov, S.V. Korotkov, T.P. Samsonova, Sub-nanosecond semiconductor opening switches based on 4H-SiC p^+p_o n^+-diodes, Solid-State Electron 47 (10) (2003) 1769–1774. Available from: https://doi.org/10.1016/S0038-1101(03)00157-6.

[49] T. Goto, T. Shirai, A. Tokuchi, N. Naito, K. Fukuda, N. Iwamuro, Experimental demonstration on ultra-high voltage and high speed 4H-SiC DSRD with smaller numbers of

die stacks for pulse power applications, in Abstract of the International Conference on Silicon Carbide and Related Materials, Washington DC, USA, Sept. 2017, MO.D1.2.

[50] J.P. Bergman, H. Lendenmann, P.A. Nilsson, U. Lindefelt, P. Skytt, Crystal defects as source of anomalous forward voltage increase of 4H-SiC diodes, Mater. Sci. Forum 353–356 (2001) 299–302. Available from: https://doi.org/10.4028/www.scientific.net/MSF.353-356.299.

[51] T. Kimoto, J.A. Cooper, Fundamentals of Silicon Carbide Technology: Growth, Characterization, Devices, and Applications, Wiley, Singapore, 2014.

[52] T. Kimoto, A. Iijima, H. Tsuchida, T. Tawara, A. Otsuki, T. Kato, et al., Understanding and reduction of degradation phenomena in SiC power devices, IEEE Reliability Physics Symposium, April 2017, pp. 2A-1.1–1.7, https://doi.org/10.1109/IRPS.2017.7936253.

[53] T. Tawara, T. Miyazawa, M. Ryo, M. Miyazato, T. Fujimoto, K. Takenaka, et al., Suppression of the forward degradation in 4H-SiC PiN diodes by employing a recombination-enhanced buffer layer, Mater. Sci. Forum 897 (2017) 419–422. Available from: https://doi.org/10.4028/www.scientific.net/MSF.897.419.

[54] T. Fujihira, Theory of semiconductor superjunction devices, Jpn J. Appl. Phys. 36 (Part 1 no. 10) (1997) 6254–6262.

[55] G. Deboy, M. März, J.-P. Stengl, H. Strack, J. Tihanyi, H. Weber, A new generation of high voltage MOSFETs breaks the limit line of silicon, in IEEE IEDM Tech. Dig., Dec. 1998, pp. 683–685, https://doi.org/10.1109/IEDM.1998.746448.

[56] J.W. Palmour, H.S. Kong, R.F. Davis, High-temperature depletion-mode metal-oxide semiconductor field-effect transistors in beta-SiC thin films, Appl. Phys. Lett. 51 (24) (1987) 2028–2030. Available from: https://doi.org/10.1063/1.98282.

[57] J.W. Palmour, H.S. Kong, R.F. Davis, Characterization of device parameters in high-temperature metal-oxide-semiconductor field-effect transistors inβ-SiC thin films, J. Appl. Phys. 64 (4) (1988) 2168. Available from: https://doi.org/10.1063/1.341731.

[58] J.W. Palmour, J.A. Edmond, H.S. Kong, C.H. Carter, Jr., Vertical power devices in silicon carbide, in Proc. Silicon Carbide and Related Materials, 1994, pp. 499.

[59] J.W. Palmour, S.T. Allen, R. Singh, L.A. Lipkin, D.F. Waltz, 4H-silicon carbide power switching devices, in Silicon Carbide and Related Materials 1995, ser. Inst. Phys. Conf. Series no. 142, 1996, pp. 813.

[60] J.N. Shenoy, J.A. Cooper, M.R. Melloch, High voltage double-implanted power MOSFETs in 6H-SiC, IEEE Electron Device Lett. 18 (3) (1997) 93–95. Available from: https://doi.org/10.1109/55.556091.

[61] T. Fujiwara, Y. Tanigaki, Y. Furukawa, K. Tonari, A. Otsuki, T. Imai, et al., Low cost ion implantation process with high heat resistant photoresist in silicon carbide device fabrication, Mater. Sci. Forum 778–780 (2014) 677–680. Available from: https://doi.org/10.4028/www.scientific.net/MSF.778-780.677.

[62] M.K. Das, G.Y. Chung, J.R. Williams, N.S. Saks, L.A. Lipkin, J.W. Palmour, High-current NO-annealed Lateral 4H-SiC MOSFETs, Mater. Sci. Forum 389–393 (2002) 981–984.

[63] E.M. Handy, M.V. Rao, K.A. Jones, M.A. Derenge, P.H. Chi, R.D. Vispute, et al., Effectiveness of AlN encapsulant in annealing ion-implanted SiC, J. Appl. Phys. 86 (2) (1999) 746–751. Available from: https://doi.org/10.1063/1.370798.

[64] Y. Negoro, K. Katsumoto, T. Kimoto, H. Matsunami, Electronic behaviors of high-dose phosphorus-ion implanted 4H–SiC (0001), J. Appl. Phys. 96 (1) (2004) 224–228. Available from: https://doi.org/10.1063/1.1756213.

[65] T. Tsuji, Japanese patent #JP 3760688 B, January 20, 2006.

[66] E. Okuno, T. Endo, J. Kawai, T. Sakakibara, S. Onda, (11-20) face channel MOSFET with low on-resistance, Mater. Sci. Forum 600−603 (2008) 1119−1122.
[67] H. Shimizu, A. Shima, Y. Shimamoto, N. Iwamuro, Ohmic contact on n- and p-type ion-implanted 4H-SiC with low-temperature metallization process for SiC MOSFETs, Jpn J. Appl. Phys. 66 (4S) (2017) 04CR15. Available from: https://doi.org/10.7567/JJAP.56.04CR15. 1−4.
[68] G.Y. Chung, C.C. Tin, J.R. Williams, K. McDonald, R.K. Chanana, R.A. Weller, et al., Improved inversion channel mobility for 4H-SiC MOSFETs following high temperature anneals in nitric oxide, IEEE Electron Device Lett 22 (4) (2001) 176−178. Available from: https://doi.org/10.1109/55.915604.
[69] J. Rozen, S. Dhar, M.E. Zvanut, J.R. Williams, L.C. Feldman, Density of interface state, electron traps, and hole traps as a function of the nitrogen density in SiO_2 on SiC, J. Appl. Phys. 105 (12) (2009) 124506. Available from: https://doi.org/10.1063/1.313845. 1−11.
[70] A.J. Lelis, R. Green, D.B. Habersat, M. Le, Basic mechanisms of threshold-voltage instability and implications for reliability testing of SiC MOSFETs, IEEE Trans. Electron Devices 62 (2) (2015) 316−323. Available from: https://doi.org/10.1109/TED.2014.2356172.
[71] H. Kitai, Y. Hozumi, H. Shiomi, K. Fukuda, M. Furumai, Low on-resistance and fast switching of 13-kV SiC MOSFETs with optimized junction field-effect transistor region, in Proc. Int. Symp. Power Semiconductors and ICs, May 2017, pp. 343−346, https://doi.org/10.23919/ISPSD.2017.7988982.
[72] D. Okamoto, H. Yano, K. Hirata, T. Hatayama, T. Fuyuki, Improved inversion channel mobility in 4H-SiC MOSFETs on Si face utilizing phosphorus-doped gate oxide, IEEE Electron Device Lett. 31 (7) (2010) 710−712. Available from: https://doi.org/10.1109/LED.2010.2047239.
[73] D. Okamoto, H. Yano, T. Hatayama, T. Fuyuki, Removal of near-interface traps at SiO_2/4H-SiC (0001) interfaces by phosphorus incorporation, Appl. Phys. Lett. 96 (20) (2010) 203508. Available from: https://doi.org/10.1063/1.3432404. 1−3.
[74] D. Okamoto, M. Sometani, S. Harada, R. Kosugi, Y. Yonezawa, H. Yano, Effect of boron incorporation on slow interface traps in SiO_2/4H-SiC structures, Appl. Phys. A 123 (2) (2017) 133. Available from: https://doi.org/10.1007/s00339-016-0724-1. 1−6.
[75] A. Saha, J.A. Cooper, 1-kV 4H-SiC power DMOSFET optimized for low on-resistance, IEEE Trans. Electron Devices 54 (10) (2007) 2786−2791. Available from: https://doi.org/10.1109/TED.2007.904577.
[76] C. Hu, M.-H. Chi, V.M. Patel, Optimum design of power MOSFETs, IEEE Trans. Electron Devices 31 (12) (1984) 1693−1700. Available from: https://doi.org/10.1109/T-ED.1984.21773.
[77] J.W. Palmour, L. Cheng, V. Pala, E.V. Brunt, D.J. Lichtenwalner, G.-Y. Wang, et al., Silicon carbide power MOSFETs: breakthrough performance from 900 V up to 15 kV, in Proc. Int. Symp. Power Semiconductors and ICs, June 2014, pp. 79−82, https://doi.org/10.1109/ISPSD.2014.6855980.
[78] S. Harada, M. Kato, K. Suzuki, M. Okamoto, T. Yatsuo, K. Fukuda, et al., 1.8 $m\Omega\ cm^2$, 10A Power MOSFET in 4H-SiC, in IEEE IEDM Tech. Dig., Dec. 2006, pp. 1−4, https://doi.org/10.1109/IEDM.2006.346929.
[79] H. Yano, T. Hirao, T. Kimoto, H. Matsunami, K. Asano, Y. Sugawara, High channel mobility in inversion layers of 4H-SiC MOSFETs by utilizing (11-20) face, IEEE Electron Device Lett 20 (12) (1999) 611−613. Available from: https://doi.org/10.1109/55.806101.

[80] J. Senzaki, K. Kojima, S. Harada, R. Kosugi, S. Suzuki, T. Suzuki, et al., Excellent effects of hydrogen postoxidation annealing on inversion channel mobility of 4H-SiC MOSFET fabricated on (11-20) face, IEEE Electron Device Lett. 23 (1) (2002) 13–15. Available from: https://doi.org/10.1109/55.974797.
[81] Y. Nanen, M. Kato, J. Suda, T. Kimoto, Effects of nitridation on 4H-SiC MOSFETs fabricated on various crystal faces, IEEE Trans. Electron Devices 60 (3) (2013) 1260–1262. Available from: https://doi.org/10.1109/TED.2012.2236333.
[82] K. Hara, Vital issues for SiC power devices, Mater. Sci. Forum 264–268 (1998) 901–906.
[83] J. Tan, J.A. Cooper Jr., M.R. Melloch, High-voltage accumulation-layer UMOSFETs in 4H-SiC, IEEE Electron Device Lett. 19 (12) (1998) 487–489. Available from: https://doi.org/10.1109/55.735755.
[84] S. Harada, Y. Kobayashi, K. Ariyoshi, T. Kojima, J. Senzaki, Y. Tanaka, et al., 3.3-kV-class 4H-SiC MeV-implanted UMOSFET with reduced gate oxide field, IEEE Electron Device Lett 37 (3) (2016) 314–316. Available from: https://doi.org/10.1109/LED.2016.2520464.
[85] Y. Kagawa, N. Fujiwara, K. Sugawara, R. Tanaka, Y. Fukui, Y. Yamamoto, et al., 4H-SiC trench MOSFET with bottom oxide protection, Mater. Sci. Forum 778–780 (2014) 919–922. Available from: https://doi.org/10.1108/www.scientific.net/MSF.778-780.919.
[86] J. Wei, M. Zhang, H. Jiang, H. Wang, K.J. Chen, Dynamic degradation in SiC trench MOSFET with a floating p-shield revealed with numerical simulations, IEEE Trans. Electron Devices 64 (6) (2017) 2592–2598. Available from: https://doi.org/10.1109/TED.2017.2697763.
[87] S. Harada, Y. Kobayashi, K. Kinoshita, N. Ohse, T. Kojima, M. Iwaya, et al., 1200 V SiC IE-UMOSFET with low on-resistance and high threshold voltage, Mater. Sci. Forum 897 (2017) 497–500. Available from: https://doi.org/10.4028/www.scientific.net/MSF.897.497.
[88] H. Yano, T. Hirao, T. Kimoto, H. Matsunami, H. Shiomi, Interface properties in metal-oxide-semiconductor structures on n-type 4HSiC(03-38), Appl. Phys. Lett. 81 (25) (2002) 4772–4774. Available from: https://doi.org/10.1063/1.1529313.
[89] T. Hiyoshi, T. Masuda, K. Wada, S. Harada, Y. Namikawa, Improvement of interface state and channel mobility using 4H-SiC (0-33-8), Mater. Sci. Forum 740–742 (2013) 506–509. Available from: https://doi.org/10.4028/www.scientific.net/MSF.740-742.506.
[90] K. Uchida, Y. Saitoh, T. Hiyoshi, T. Masuda, K. Wada, H. Tamaso, et al., The optimized design and characterization of 1200 V/2.0 mΩ cm^2 4H-SiC V-groove trench MOSFETs, in Proc. Int. Symp. Power Semiconductors and ICs, May 2015, pp. 85–88, https://doi.org/10.23919/ISPSD.2015.7123395.
[91] Y. Saitoh, T. Masuda, H. Tamaso, H. Notsu, H. Michikoshi, K. Hiratsuka, et al., Switching performance of V-Groove trench gate SiC MOSFETs with grounded buried p+ regions, Mater. Sci. Forum 897 (2017) 505–508. Available from: https://doi.org/10.4028/www.scientific.net/MSF.897.505.
[92] K. Matocha, Challenges in SiC power MOSFET design, Solod-State Electron 52 (10) (2008) 1631–1635. Available from: https://doi.org/10.1016/j.sse.2008.06.034.
[93] H. Hagino, J. Yamashita, A. Uenishi, H. Haruguchi, An experimental and numerical study on the forward biased SOA of IGBTs, IEEE Trans. Electron Devices 43 (3) (1996) 490–500. Available from: https://doi.org/10.1109/16.485667.

[94] J. An, M. Namai, N. Iwamuro, Experimental and theoretical analyses of gate oxide and junction reliability for 4H-SiC MOSFET under short-circuit operation, Jpn. J. Appl. Phys 55 (12) (2016) 124102. Available from: https://doi.org/10.7567/JJAP.55.124102. 1–4.

[95] Chronological Science Tables, National Institute of Natural Sciences, National Astronomical Observatory of Japan, Maruzen Publishing Co., Ltd, 2015.

[96] A. Castellazzi, A. Fayyaz, Y. Li, Michele Riccio, A. Irace, Short-circuit robustness of SiC Power MOSFETs: experimental analysis, in Proc. Int. Symp. Power Semiconductors and ICs, June 2014, pp. 71–74, https://doi.org/10.1109/ISPSD.2014.6855978.

[97] M. Sometani, D. Okamoto, S. Harada, H. Ishimori, S. Takasu, T. Hatakeyama, et al., Temperature-dependent analysis of conduction mechanism of leakage current in thermally grown oxide on 4H-SiC, J. Appl. Phys 117 (1) (2015) 024505. Available from: https://doi.org/10.1063/1.4905916. 1–6.

[98] R. Singh, S.R. Hefner, Reliability of SiC MOS devices, Solid-State Electron. 48 (10–11) (2004) 1717–1720. Available from: https://doi.org/10.1016/j.sse.2004.05.005.

[99] A.K. Agarwal, S. Seshadri, L.B. Rowland, Temperature dependence of Fowler–Nordheim current in 6H- and 4H-SiC MOS capacitors, IEEE Electron Device Lett 18 (12) (1997) 592–594. Available from: https://doi.org/10.1109/55.644081.

[100] M. Namai, J. An, H. Yano, N. Iwamuro, Experimental and numerical demonstration and optimized methods for SiC trench MOSFET short-circuit capability, in Proc. Int. Symp. Power Semiconductors and ICs, May 2017, pp. 363–366, https://doi.org/10.23919/ISPSD.2017.7988993.

[101] D. Peters, R. Siemieniec, T. Aichinger, T. Basler, R. Esteve, W. Bergner, et al., Performance and ruggedness of 1200 V SiC-trench-MOSFET, in Proc. Int. Symp. Power Semiconductors and ICs, May 2017, pp. 239–242, https://doi.org/10.23919/ISPSD.2017.7988904.

[102] R. Tanaka, Y. Kagawa, N. Fujiwara, K. Sugawara, Y. Fukui, N. Miura, et al., Impact of grounding the bottom oxide protection layer on the short-circuit ruggedness of 4H-SiC trench MOSFETs, in Proc. Int. Symp. Power Semiconductors and ICs, June 2014, pp. 75–78, https://doi.org/10.1109/ISPSD.2014.6855979.

[103] K. Kawahara, S. Hino, K. Sadamatsu, Y. Nakao, Y. Yamashiro, Y. Yamamoto, et al., 6.5 kV Schottky-barrier diode-embedded SiC-MOSFET for compact full-unipolar mode, in Proc. Int. Symp. Power Semiconductors and ICs, May 2017, pp. 41–44, https://doi.org/10.23919/ISPSD.2017.7988888.

[104] M. Uchida, H. Horikawa, K. Taanaka, T. Takahashi, T. Kiyosawa, M. Hayashi, et al., Novel SiC power MOSFET with integrated unipolar internal inverse MOS-channel diode, in IEEE IEDM Tech. Dig., Dec. 2011, pp. 26.6.1–26.6.4, https://doi.org/10.1109/IEDM.2011.6131620.

[105] C.-T. Yen, C.-C. Hung, H.-T. Hung, L.-S. Lee, C.-Y. Lee, T.-M. Yang, et al., 1700 V/30A 4H-SiC MOSFET with low cut-in voltage embedded diode and room temperature boron implanted termination, in Proc. Int. Symp. Power Semiconductors and ICs, May 2015, pp. 265–268, https://doi.org/10.1109/ISPSD.2015.7123440.

[106] S. Hino, T. Hatta, S. Sadamatsu, Y. Nagashita, S. Yamamoto, T. Iwamatsu, et al., Demonstration of SiC-MOSFET embedding Schottky barrier diode for inactivation of parasitic body diode, in Proc. of 11th Eur. Conf., Silicon Carbide Related Materials., Sept. 2016, Tu3.01.

[107] W. Sung, B.J. Baliga, Monolithically integrated 4H-SiC MOSFET and JBS diode (JBSFET) using a single ohmic/Schottky process scheme, IEEE Electron Device Lett. 37 (12) (2016) 1605–1608. Available from: https://doi.org/10.1109/LED.2016.2618720.

[108] Y. Kobayashi, H. Ishimori, A. Kinoshita, T. Kojima, M. Takei, H. Kimura, et al., Evaluation of Schottky barrier height on 4H-SiC m-face $\{1\bar{1}00\}$ for schottky barrier diode wall integrated trench MOSFET, Jpn J. Appl. Phys 56 (4S) (2017) 04CR08. Available from: https://doi.org/10.7567/JJAP.56.04CR08. 1–6.

[109] R. Kosugi, Y. Sakuma, K. Kojima, S. Itoh, A. Nagata, T. Yatsuo, et al., Development of SiC super-junction (SJ) devices by multi-epitaxial growth, Mater. Sci. Forum 740–742 (2013) 793–796.

[110] R. Kosugi, Y. Sakuma, K. Kojima, S. Itoh, A. Nagata, T. Yatsuo, et al., Development of SiC super-junction (SJ) device by deep trench-filling epitaxial growth, Materials Science Forum 740–742 (2013) 785–788. Available from: https://doi.org/10.4028/www.scientific.net/MSF.740-742.785.

[111] K. Kojima, A. Nagata, S. Ito, Y. Sakumad, R. Kosugi, Y. Tanaka, Filling of deep trench by epitaxial SiC growth, Mater. Sci. Forum 740–742 (2013) 793–796. Available from: https://doi.org/10.4028/www.scientific.net/MSF.740-742.793.

[112] R. Kosugi, Y. Sakuma, K. Kojima, S. Itoh, A. Nagata, T. Yatsuo, et al., First experimental demonstration of SiC super-junction (SJ) structure by multi-epitaxial growth method, in Proc. Int. Symp. Power Semiconductors and ICs, June 2014, pp. 346–349, https://doi.org/10.1109/ISPSD.2014.6856047.

[113] M. Takei, Y. Harada, K. Ueno, 600 V-IGBT with reverse blocking capability, in Proc. Int. Symp. Power Semiconductors and ICs, June 2001, pp. 413–416, https://doi.org/10.1109/ISPSD.2001.934641.

[114] N. Tokuda, M. Kaneda, T. Minato, An ultra-small isolation area for 600 V class reverse blocking IGBT with deep trench isolation process (TI-RB-IGBT), in Proc. Int. Symp. Power Semiconductors and ICs, May 2004, pp. 129–132, https://doi.org/10.1109/WCT.2004.239843.

[115] H. Nakazawa, M. Ogino, H. Wakimoto, T. Nakajima, Y. Takahashi, Hybrid isolation process with deep diffusion and V-groove for reverse blocking IGBTs, in Proc. Int. Symp. Power Semiconductors and ICs, May 2011, pp. 116–119, https://doi.org/10.1109/ISPSD.2011.5890804.

[116] S. Safari, A. Castellazzi, P. Wheeler, Experimental and analytical performance evaluation of SiC power devices in the matrix converter, IEEE Trans. Power Electron. 29 (5) (2014) 2584–2596. Available from: https://doi.org/10.1109/TPEL.2013.2289746.

[117] S. Mori, M. Aketa, T. Sakaguchi, T. Kimoto, Demonstration of 3 kV 4H-SiC reverse blocking MOSFET, in Proc. Int. Symp. Power Semiconductors and ICs, June 2016, pp. 271–274, https://doi.org/10.1109/ISPSD.2016.7520830.

[118] K. Okuda, T. Isobe, H. Tadano, N. Iwamuro, A dead-time minimized inverter by using complementary topology and its experimental evaluation of harmonics reduction, in Proc. of Power Electronics and Applications (EPE'16 ECCE Europe), Sept. 2016, pp. 1–10, https://doi.org/10.1109/EPE.2016.7695673.

[119] N. Iwamuro, A. Okamoto, S. Tagami, H. Motoyama, Numerical analysis of short-circuit safe operating area for p-channel and n-channel IGBTs, IEEE Trans. Electron Devices 38 (2) (1991) 303–309. Available from: https://doi.org/10.1109/16.69910.

[120] J. An, M. Namai, M. Tanabe, D. Okamoto, H. Yano, N. Iwamuro, Experimental demonstration of −730 V vertical SiC p-MOSFET with high short circuit withstand

capability for complementary inverter applications, in IEEE IEDM Tech., Dig., Dec. 2016, pp. 272–275, https://doi.org/10.1109/IEDM.2016.7838391.
[121] J. An, M. Namai, H. Yano, N. Iwamuro, Investigation of robustness capability of −730 V p channel vertical SiC power MOSFET for complementary inverter applications, IEEE Trans. Electron Devices 64 (10) (2017) 4219–4225. Available from: https://doi.org/10.1109/TED.2017.2742542.
[122] N. Nashida, Y. Hinata, M. Horio, R. Yamada, Y. Ikeda, All-SiC power module for photovoltaic power conditioner system, in Proc. Int. Symp. Power Semiconductors and ICs, June 2014, pp. 342–345, https://doi.org/10.1109/ISPSD.2014.6856046.
[123] O. Zhang, H.-R. Chang, M. Gomez, C. Bui, E. Hanna, J.A. Higgins, et al., 10 kV trench gate IGBTs on 4H-SiC, in Proc. Int. Symp. Power Semiconductors and ICs, May 2005, pp. 303–306, https://doi.org/10.1109/ISPSD.2005.1488011.
[124] S. Katakami, H. Fujisawa, K. Takenaka, H. Ishimori, S. Takasu, M. Okamoto, et al., Fabrication of a p-channel SiC-IGBT with high channel mobility, Mater. Sci. Forum 740–742 (2013) 938–961. Available from: https://doi.org/10.4028/www.scientific.net/MSF.740-742.958.
[125] S. Chowdhury, T.P. Chow, Performance tradeoffs for ultra-high voltage (15 kV to 25 kV) 4H-SiC n-channel and p-channel IGBTs), in Proc. Proc. Int. Symp. Power Semiconductors and ICs, June 2016, pp. 75–78, https://doi.org/10.1109/ISPSD.2016.7520781.
[126] S.-H. Ryu, L. Cheng, S. Dhar, C. Capell, C. Jonas, J. Clayton, et al., Development of 15 kV 4H-SiC IGBTs, Mater. Sci. Forum 717–720 (2012) 1135–1138. Available from: https://doi.org/10.4028/www.scientific.net/MSF.717-720.1135.
[127] S. Ryu, C. Capell, C. Jonas, M. O'Loughlin, J. Clayton, E.V. Brunt, et al., 20 kV 4H-SiC N-IGBTs, Mater. Sci. Forum 778–780 (2014) 1030–1033. Available from: https://doi.org/10.4028/www.scientific.net/MSF.778-780.1030.
[128] E.V. Brunt, L. Cheng, M. O'Loughlin, C. Capell, C. Jonas, K. Lam, et al., 22 kV, 1 cm^2, 4H-SiC n-IGBTs with improved conductivity modulation, in Proc. Proc. Int. Symp. Power Semiconductors and ICs, June 2014, pp. 358–361, https://doi.org/10.1109/ISPSD.2014.6856050.
[129] E.V. Brunt, L. Cheng, M. O'Loughlin, J. Richmond, V. Pala, J.W. Palmour, et al., 27 kV, 20A, 4H-SiC n-IGBTs, Mater. Sci. Forum 821–823 (2015) 847–850. Available from: https://doi.org/10.4028/www.scientific.net/MSF.821-823.847.
[130] Y. Yonezawa, T. Mizushima, K. Takenaka, H. Fujisawa, T. Kato, S. Harada, et al., Low Vf and highly reliable 16 kV ultrahigh voltage SiC flip-type n-channel implantation and epitaxial IGBT, in IEEE IEDM Tech. Dig., Dec. 2013, pp. 6.6.1–6.6.4, https://doi.org/10.1109/IEDM.2013.6724576.
[131] Y. Yonezawa, T. Mizushima, K. Takenaka, H. Fujisawa, T. Deguchi, T. Kato, et al., Device performance and switching characteristics of 16 kV ultrahigh-voltage SiC flip-type n-channel IE-IGBTs, Mater. Sci. Forum 821–823 (2015) 842–846. Available from: https://doi.org/10.4028/www.scientific.net/MSF.821-823.842.
[132] K. Eto, T. Kato, S. Takagi, T. Miura, Y. Urakami, H. Kondo, et al., Growth study of p-type 4H-SiC with using aluminum and nitrogen co-doping by 2-zone heating sublimation method, Mater. Sci. Forum 821–823 (2015) 47–50. Available from: https://doi.org/10.4028/www.scientific.net/MSF.821-823.47.

GaN smart power devices and integrated circuits

5

T. Paul Chow and Zhibo Guo
Rensselaer Polytechnic Institute, Troy, NY, United States

5.1 Introduction

Silicon is the dominant semiconductor of choice for high-voltage power electronics applications [1,2]. However, recently, due to intriguing material properties, SiC and GaN are under active development and selective commercialization to challenge silicon [3–8]. For gallium nitride, because of its photonic and high-frequency rf applications, it has built up material and processing infrastructures that can be exploited for the implementation of cost-effective, power switching devices in energy efficient power systems.

In this chapter, we will first review the material properties, heterojunction, and metal–oxide–silicon (MOS) technologies, which are the building blocks of power semiconductor devices, for GaN. Then, we will present the lateral and vertical power device structures that have been explored and some of them commercialized. Next, we will highlight the integrated process sequences that have been used to fabricate these devices and compare them to those in commercial Si and GaAs foundries. Further, we will show and discuss the static and dynamic performance of these devices. We will also present some commercial device offerings and their performance. We will then describe some salient examples of monolithic integration of these devices into power electronic and optoelectronic integrated circuits (ICs). Finally, we will discuss the future trend and technology obstacles that need to be addressed and overcome for the success of widespread commercialization of GaN power devices.

5.1.1 Material properties

The bulk material properties of SiC and GaN, together with those of silicon, GaAs, Ga_2O_3, diamond, and AlN are shown in Table 5.1 [8]. It can be noted that the avalanche breakdown field tends to increase with increasing bandgap and that of SiC and GaN is about an order of magnitude higher than that of silicon. In addition, the thermal conductivity of SiC and GaN is 3 and 1.5 times better than that of Si. Both of these properties make them very attractive for power switching devices. Further,

Wide Bandgap Semiconductor Power Devices. DOI: https://doi.org/10.1016/B978-0-08-102306-8.00005-8

Table 5.1 **Physical and electrical properties of various conventional (SiC, GaAs), wide (SiC, GaN, Ga_2O_3), and extreme (Diamond, AlN) bandgap semiconductors [8]**

Material	E_g (eV)	Direct/indirect	n_i (cm^{-3})	ε_r	μ_n (cm^2/V s)	E_c (MV/cm)	v_{sat} (10^7 cm/s)	λ (W/cm K)
Si	**1.12**	**I**	$\mathbf{1.5 \times 10^{10}}$	**11.8**	**1350**	**0.25**	**1.0**	**1.5**
GaAs	**1.42**	**D**	$\mathbf{1.8 \times 10^{6}}$	**13.1**	**8500**	**0.4**	**1.2**	**0.55**
2H-GaN	3.39	D	1.9×10^{-10}	9.9	1000[a]	3.3*	2.5	2.5
					2000**	3.75[a]		4.1*
4H-SiC	3.26	I	8.2×10^{-9}	10	720[a]	2.0[a]	2.0	4.5
					650[c]			
Ga_2O_3	4.5–4.9	D	2.6×10^{-19}	10	300	8	–	0.13–0.21
			-1.2×10^{-22}					
Diamond	5.45	I	1.6×10^{-27}	5.5	2800	10	2.7	22
2H-A1N	6.2	D	$\sim 10^{-34}$	8.5	300	12*	1.7	2.85

Note: a – mobility along a-axis, c – mobility along c-axis.
*Estimated value.
**2DEG.

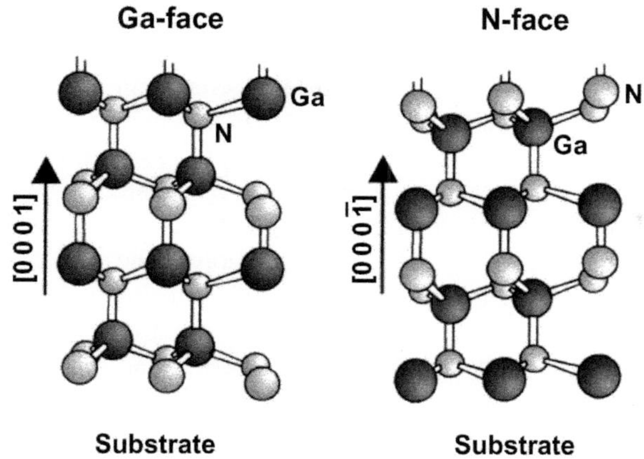

Figure 5.1 Schematic drawings of the Ga or (0001) face and N or (000-1) face of the GaN wurtzite crystal structure [9].

GaN is a direct bandgap semiconductor and can form heterojunction with AlN and InN alloys pseudomorphically. In this regard, GaN and its AlGaInN alloys strongly resemble GaAs and the AlGaInAs alloys. However, GaN has a hexagonal wurtzite structure (Fig. 5.1 [9]) and anisotropic electrical properties while GaAs has a cubic zinc blende crystal structure and hence its electrical properties are isotropic. Nevertheless, as will be seen later in the integrated process section, many of the GaN device fabrication steps are adapted from GaAs processes. Furthermore, its direct bandgap implies only unipolar power device structures, such as Schottky diodes, metal-oxide-field-effect transistor (MOSFET) or high electron mobility transistor (HEMT), but not pindiode or insulated gate bipolar transistor (IGBT), need to be considered, due to a short carrier recombination lifetime and thus short minority carrier diffusion length.

5.1.2 Epitaxy and doping

The late development of bulk GaN substrates led to the development of heteroepitaxy of GaN on many foreign substrates, particularly sapphire, SiC, and silicon substrates. The photonic and rf electronic device commercialization depended on such heteroepitaxial technologies. The economic incentive of using silicon substrates is so overwhelming that much effort has been spent on making viable also for power device applications for over the last decade.

(111) Si and (0001) GaN have a lattice mismatch of 17% and a thermal expansion coefficient difference of more than twice, making the heteroepitaxy extremely challenging, demanding the buffer layers to mitigate these mismatches to minimize and localize the extended defects to the interfacial layers [10]. Generally, there are two approaches used for the buffer layer. One is to use a superlattice of many alternating, thin AlN and AlGaN layers at the interface to accommodate the strain and the other is to use a graded $Al_xGa_{1-x}N$ layer, with the Al-rich region near the silicon

substrate and Ga-rich region near the GaN top layer. In addition, deep levels, such as iron or carbon, are often added to the buffer to ensure its high resistivity and suppress any possible current leakage path. At least some of the undesirable reliability issues discussed later in the chapter can be traced to the buffer layer as the origin.

To compensate for the difference in thermal expansion between the GaN epi and the Si substrate, the GaN epi layer is grown under tensile stress such that upon cooling to room temperature, the epi film is changed to under reasonable compressive strain, resulting in acceptable wafer bow and suppressing film cracking. Such a stress/strain optimization is increasing challenging with increasing wafer diameter. Nevertheless, AlGaN/GaN layers have been successful grown on 200 mm diameter (111) silicon wafers, though the substrate is thicker than usual to act as a stiff mechanical balancer. Shown in Fig. 5.2 is the reciprocal space mapping (RSM) of AlGaN/GaN layers on silicon substrates, clearly indicating the level of tensile or compressive strain in each layer [11].

The most often used shallow donor and acceptor in GaN are silicon and magnesium respectively. In situ doping during epi growth is routinely performed even though magnesium usage leads to the undesirable dopant memory effect in the growth reactor. Doping by ion implantation is only possible for silicon and other donor dopants, but not yet too successful for acceptors. The difficulties in acceptor activation are attributed to the generation of ion-implanted defects, which tend to be donor-like and compensate for activated acceptor atoms. Limited electrical activation of implanted magnesium has been demonstrated using a pulsating annealing technique at elevated temperatures to prevent GaN from dissociation but the activation percentage is still very low (up to only 8%) [12,13].

Iron and carbon are the most popular deep levels introduced to the buffer layer to control its resistivity. Iron has energy level of about 0.6–0.7 eV below the conduction bandgap edge, whereas carbon is actually a deep acceptor in GaN, with an energy level of 0.84 eV above the valence band edge [14,15]. Carbon used to be employed to overcome any buffer defect problem but, at present, it is becoming increasingly clear to the researchers that it is better to solve the buffer epi problem than to do carbon doping to compensate the electrical active defects.

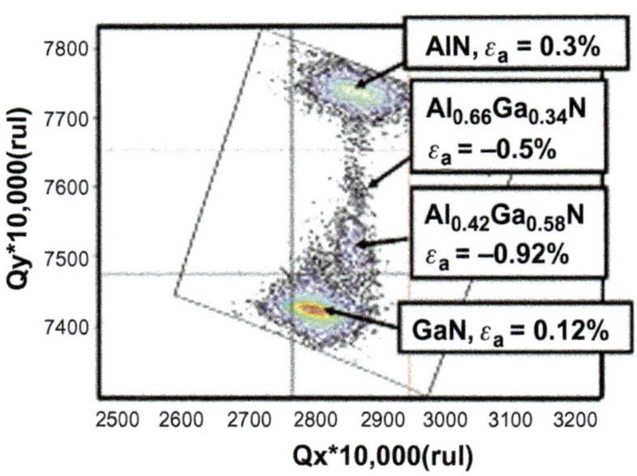

Figure 5.2 RSM of AlGaN and GaN epi films on silicon substrates [11].

5.1.3 Polarization and 2DEG

Group III–V nitrides are the only III–V materials that show spontaneous polarization, P_{SP}, because of an intrinsic asymmetry in the bonding of the wurtzite crystal structure. This asymmetry results from the smaller than ideal c/a ratio, with AlN having the largest deviation, and hence polarization, among them. This P_{SP} has a negative sign and the polarization vector directs toward the substrate for the (0001) orientation case (see Fig. 5.3A [9]). In addition, mechanical strain introduces piezoelectric polarization, P_{PZ}, which is negative for compressive strained and positive for tensile strained layers. Thus, the piezoelectric and spontaneous polarizations aid each other in the former but oppose each other in the latter. Consequently, since pseudomorphically grown AlGaN layers on GaN heterojunctions always have a tensile strain, both polarizations are in the same direction and the total polarization is the sum of these two polarizations. The spontaneous polarization at the AlGaN/GaN interface is usually estimated as a linearly interpolated sum, based on the mole fraction, of the GaN and AlN polarization values [16].

Fig. 5.3B shows the sheet carrier concentration induced by polarization charges present at the AlGaN/GaN heterojunction interface as a function of aluminum mole fraction in AlGaN alloys for varying pseudomorphic AlGaN film thickness [9]. At an Al mole fraction of 0.2 and an AlGaN thickness 20 nm, a polarization charge density and 2DEG concentration of $8 \times 10^{12}/cm^2$ are obtained. It may be worth pointing out that this charge density is close to the optimal value for designing a uniform surface electric field in lateral high-voltage RESURF devices, as will be explained later. The electron mobility of the 2DEG layer at the AlGaN/GaN

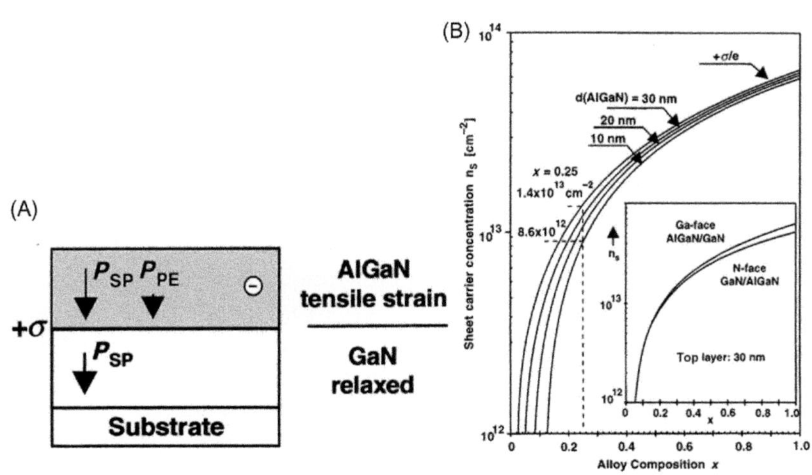

Figure 5.3 (A) Direction of the spontaneous and piezoelectric polarization vectors at the AlGaN/GaN heterojunction interface and the polarization charges. (B) Sheet 2DEG carrier concentration as a function of AlGaN alloy composition for varying AlGaN film thickness [9].

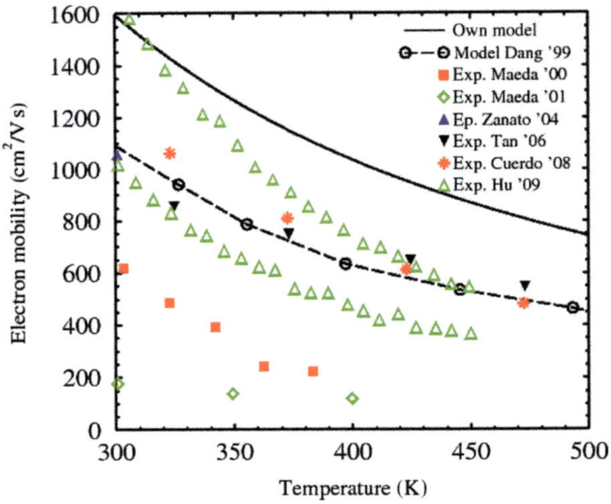

Figure 5.4 Modeled and experimental mobility of the 2DEG layer at the AlGaN/GaN heterojunction interface [17].

heterojunction interface has been modeled and compared to experimental data at room and elevated temperatures (see Fig. 5.4) [17]. Typical values of 1500 cm^2/V s for the electron mobility can be achieved.

5.1.4 MOS

While it is very difficult to form a high-quality MOS interface for GaAs due to surface potential pinning, it is relatively easy to make MOS devices for GaN. The surface potential of GaN appears to be partially pinned but good n-channel MOS devices have been demonstrated with several deposited oxides (SiO_2, Al_2O_3, ZrO_2, HfO_2, and others) or nitrides (Si_3N_4) as gate dielectrics. Besides being used as MOS gate insulators, these dielectrics are often also used as the passivation layers on GaN surface. However, common GaAs surface passivation layers, such as plasma-enhanced chemical vapor deposition (PECVD) silicon nitride, have been proven not to be adequate for GaN. It is worth pointing that the exact conditions for the gate dielectrics (precleaning procedure, precursors used, deposition method, and temperature as well as postdeposition ambient and temperature) are very important in determining the interfacial properties of the MOS devices. In addition, no native thermally grown oxide is available for usage as gate insulator.

The requirements for an acceptable gate dielectric include: high bulk breakdown field, low interface state density, good inversion carrier mobility, and adequate reliability [under positive-bias temperature instability (PBTI) and negative-bias temperature instability (NBTI) stressing conditions]. One way to evaluate the gate dielectric is to examine its bandgap magnitude and offsets with the conduction and valence bands in GaN, as shown in Fig. 5.5 [18]. The bandgap magnitude determines the bulk dielectric breakdown field. Actually, this field is dependent on the

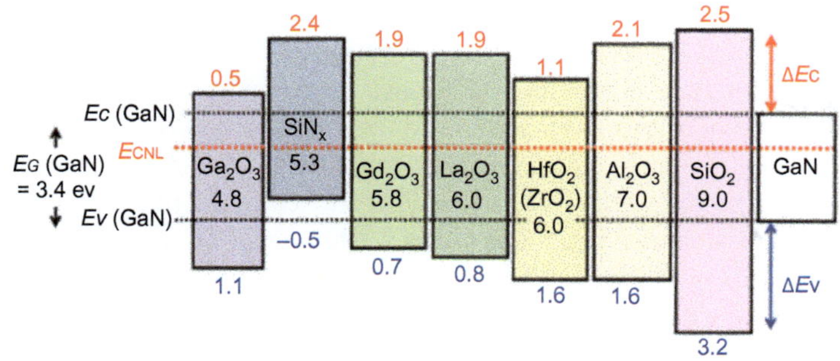

Figure 5.5 Bandgaps and band offsets of various dielectrics with GaN. Adapted from [22].

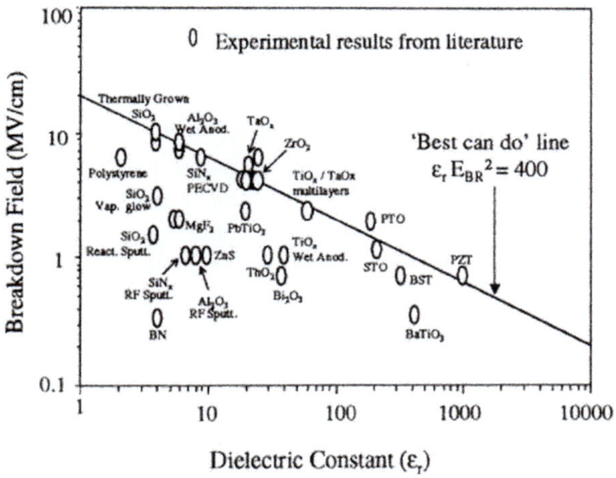

Figure 5.6 Breakdown field versus permittivity for various dielectrics [19].

permittivity (Fig. 5.6 [19]) and varies as $1/\sqrt{\varepsilon}$. For example, when the permittivity increases from 3.9 (SiO_2) to 200 [strontium titanate oxide (STO)], the breakdown field is decreased from 10 to 1 MV/cm. The conduction and valence band offsets for all the gate dielectrics on GaN are smaller than those on Si due to the larger bandgap of GaN. This raises hot carrier reliability of the gate dielectrics, particularly at elevated temperatures. The most popular and widely studied gate insulators examined are Si_3N_4, SiO_2, and Al_2O_3. Among these, SiO_2/GaN has the largest ΔE_C of 2.56 eV, whereas the corresponding values for Si_3N_4 and Al_2O_3 are 1.3 and 2.16 eV, respectively. The valence band offsets for SiO_2 and Si_3N_4 are estimated to be 3.2 and 0.81 eV, respectively [18]. The band offset for the Al_2O_3/GaN case depends on the Al_2O_3 bandgap value assumed [18]. However, photoemission measurements of Si_3N_4 on GaN (0001) surface have shown that the Si_3N_4/GaN has a negative valence band offset, or, equivalently, a Type II band alignment, of a value of −0.4 eV [20]. Also, the conduction band offset for Al_2O_3/GaN (0001) interface has been extracted to be 2.2 eV from photoemission measurements [21].

The interface charge densities, both Q_f and Q_{it}, of Si_3N_4, SiO_2 and Al_2O_3 on (0001) GaN surface has been studied and optimized in terms of film deposition as well as postdeposition annealing conditions. Generally, reasonably low and well-controlled charge densities have been obtained. For PECVD or atomic layer deposition (ALD) SiO_2, a fairly low interface state density has been measured but the fixed oxide charge density ($\sim 10^{12}$ q/cm^2) is still relatively high [22–28]. A cross-sectional transmission electron microscopy (TEM) image of SiO_2/(0001) GaN interface is shown in Fig. 5.7, clearly indicating a sharp and abrupt interface with little roughness and no extraneous layer present. Reactive-ion-etched GaN surfaces have more inferior properties, presumably due to plasma damage, that cannot be fully restored with subsequent processing [25]. Oxide charge densities on nonpolar m plane or (1–100) GaN are higher than those on (0001) GaN [29] but, interestingly, the temperature dependence of its MOS properties lacks the anomalous feature, from the pyroeletric effects, that has been observed on (0001) GaN capacitors [28–30]. In addition, the quality of the GaN epi films has a large influence on the MOS properties, with the films on sapphire superior to those on silicon substrates, attributed to the lower density of extended defects, such as dislocations.

Al_2O_3 is the mostly widely used in GaN devices, both as gate dielectric and surface passivation layers. ALD is the mostly often-employed technique. A fairly high fixed oxide charge density of 4.6×10^{12} q/cm^2 has been extracted [31]. The interfacial charge densities have been optimized with postdeposition and postmetallization annealing [32]. The substrate effect of the electrical properties of aluminum oxide capacitors has clearly been demonstrated with the recent results of superiorly low interface state densities achieved on homoepitaxial GaN layers on free standing substrates [33]. The orientation effect (c-plane vs m-plane), similar to that observed for SiO_2/GaN capacitors, has also been seen for Al_2O_3/GaN capacitors [34].

Silicon nitride films were the first dielectric films explored as gate insulator for GaN metal-insulator-semiconductor (MIS) devices. However, those PECVD SiN_x films generally have inferior bulk electrical properties and the resulting MIS structures are quite poor. More recently, thermally grown or LPCVD silicon nitride films have been found to be much better in quality as gate dielectric and passivation layers [35–37]. The best MIS channel-high electron mobility transistor (HEMT)

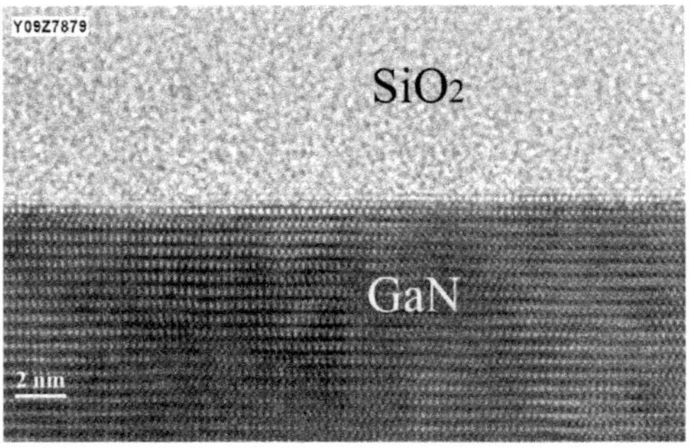

Figure 5.7 Cross-sectional TEM picture of SiO_2/GaN interface.

and metal-insulator-semiconductor field-effect transistor (MISFET) reported to date using these silicon nitride films are those with a bilayer PECVD SiN_x/LPCVD SiN_x gate stack [37]. The insertion of the thin PECVD SiN_x layer between the GaN channel and the LPCVD SiN_x film results in a smooth interface and hence better electrical characteristics.

5.1.5 Power device applications

Power electronics applications span a wide range of power and switching frequency (from mW to GW, 60 Hz to 100 MHz), from portable consumer electronics to large-scale infrastructure utility grid electronics. Present day silicon power devices are specifically designed and optimized for each power range. For the low power and the high frequency applications, silicon power MOSFETs are preferred. In the medium power area, the IGBT, which is an MOS-gated bipolar conductivity-modulated transistor, dominates. For the very large power traction and grid applications, thryistors are still the major device of choice. These choices are a direct consequence of the material properties of silicon. If a wider bandgap semiconductor, such as GaN, is to replace silicon, it offers a different device of choice. For example, if the silicon IGBT is replaced by a GaN power HEMT or MOSFET, the GaN transistors are able to operate at a higher switching frequency and hence a lower switching energy loss, while simultaneously reducing the size of the reactive elements, such as inductors, and the heat sink size and weight. In addition, the unipolar FETs have a better robustness and ruggedness than the bipolar transistor, offering more performance advantages. This situation is illustrated in Fig. 5.8, where the power level versus switching frequency is shown and the silicon device constraint can be shown to be relieved with GaN unipolar or SiC unipolar/bipolar power devices.

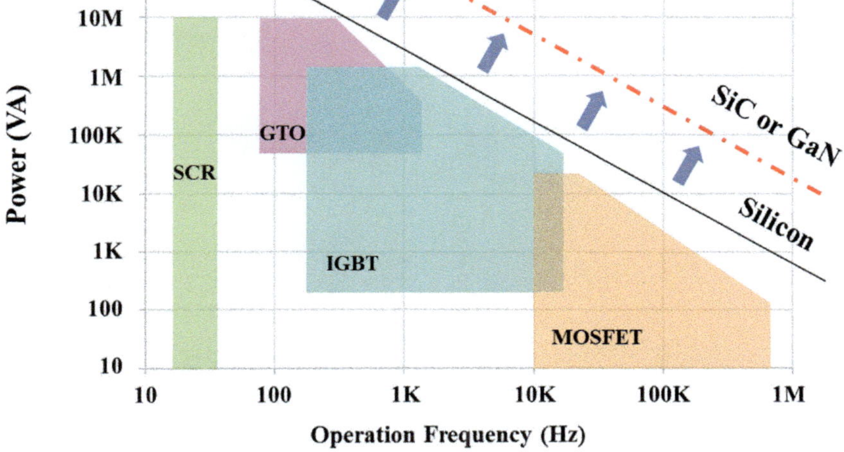

Figure 5.8 Power versus switching frequency for various power electronics applications. Modified from an application note of Powerex.

5.2 Device structures and design

Generally, the high-voltage power device structures can be classified as vertical and lateral types. Vertical device structures tend to be discrete and lateral structures are integrable and can be implemented in power ICs. Because the root of GaN transistors is in high-frequency (rf) amplifying applications, all the power devices commercialized to date are lateral, even when they are sold as discrete power transistors.

5.2.1 Lateral

The schematic cross-sections of several high-voltage lateral GaN transistor structures are shown in Fig. 5.9A–D. Power GaN transistors are interesting in that they can either resemble that of lateral silicon power MOSFETs or AlGaAs/GaAs Schottky-gate HEMTs because high-quality MOS and heterojunction interfaces are possible. Besides an MOS channel, lateral high-voltage GaN MOSFETs have a lightly doped drain region to support high drain voltages and ion-implanted source/drain regions. By contrast, power AlGaN/GaN HEMTs have Schottky-gate controlled heterojunction 2DEG channel, a heterojunction drift region and alloyed source/drain contacts. In either type of transistors, the surface electric field must be

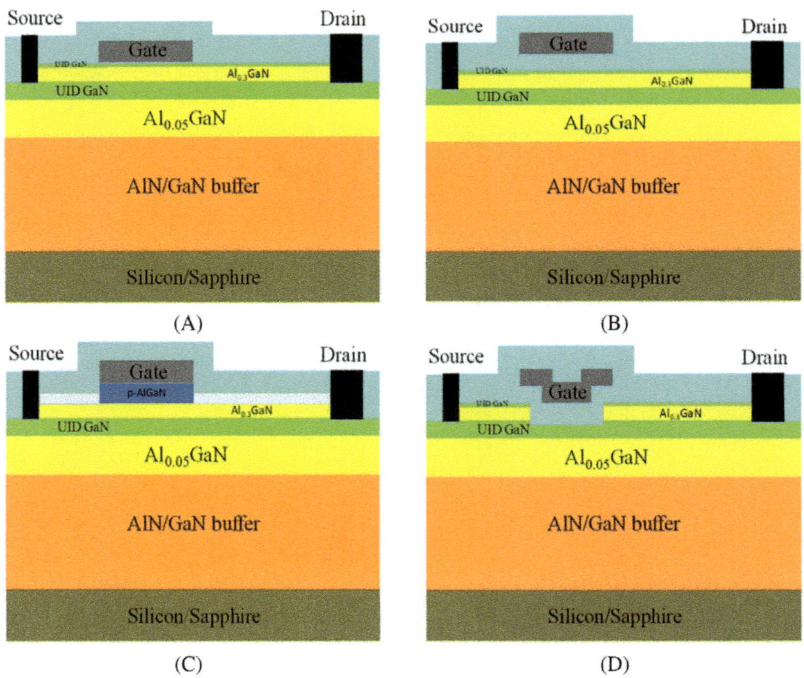

Figure 5.9 Schematic cross-sections of various high voltage lateral GaN power transistor structures [5].

suppressed so as to maximum the blocking voltage. In the MOSHC-HEMT (Fig. 5.9B), the gate insulator in between the Schottky metal and the AlGaN layer reduces the gate leakage current significantly, particularly at elevated temperatures. The MOS channel-HEMT (MOSC-HEMT, previously called hybrid MOS-HEMT [38–40]), shown in Fig. 5.9D, combines the best features of the MOSFET and HEMT, in that it has an MOS channel for enhancement mode operation and heterojunction drift region for minimum resistivity.

The reduced surface field (RESURF) principle [41] is utilized to maximize the breakdown voltage of lateral high-voltage power devices since it can reshape and suppress the surface electric field and force the avalanche process to be initiated in the bulk junction region. The key design parameter is the space charge per unit area (N_{RESURF}) ($=\int N_{epi}\,dx$) in the drift region. Adoption of an optimum N_{RESURF} leads to inherent lateral and vertical electric field shaping due to nonplanar, two- or three-dimensional depletion action originating from interactions of lateral and vertical pn or MIS junctions [42–44], and in GaN, polarization junctions [5].

To determine the optimum RESURF charge density (NRESURF), we use Gauss' law to yield

$$\int \xi \cdot dA \approx \xi \cdot A = -\int \rho \; dV/\varepsilon_s = -Q/\varepsilon_s = -q \; N/\varepsilon_s$$

where ξ is the electric field in the depleted drift region, A is the area, Q is the total amount of space charges enclosed, and ε_s is the semiconductor permittivity. Since $N_{RESURF} = Q_{RESURF}/q\ A \approx \xi_s/q$ and $\xi \approx 1.5 \times 10^5$ V/cm in silicon, the optimum N_{RESURF} is about $10^{12}/cm^2$, but, for GaN and SiC, where the avalanche field is 10 times higher, it is about $10^{13}/cm^2$. Additional features, such as field plates, are needed to further reduce surface field crowding, either near the gate/source or drain side of the device. In addition, the RESURF action can be achieved in many ways: (1) space charge can be introduced with in situ epi doping, ion implantation or polarization charges; and (2) the substrates can be semiconductors (like Si or GaN), insulators (like sapphire) or semiconductor on insulator/semiconductor (SOI) structures (like silicon/oxide/silicon or GaN/AlN/silicon). Lateral high-voltage GaN MOSFETs using implanted or in situ doped epi RESURF layers have been experimentally designed and demonstrated [45–48].

Here, we are focusing on lateral GaN power devices on SOI substrates consisting of AlGaN/GaN or GaN active epitaxial layers/AlN, GaN or AlGaN insulating buffer layers on silicon substrates. Interestingly, the space charges in the AlGaN/GaN case arise from polarization charges and with an aluminum mole fraction of about 22% and an AlGaN/GaN layer thickness of 20 nm, these polarization charges are about $10^{13}/cm^2$ in concentration [9]. Consequently, such a structure can be called a polarization junction or "natural" RESURF structure [49,50], which can be optimized with the thickness of the AlGaN layer as well as the thickness of the GaN cap layer [51]. In the case of infinite number of parallel RESURF layers, a natural superjunction structure [52]. To illustrate the impact of balanced space charges in the drift region, we have simulated two structures, one with uncompensated positive charges at the AlGaN/GaN interface and the other with balanced positive and negative charges on the two AlGaN interfaces. The equipotential line profiles, shown in Fig. 5.10, clearly indicate field crowding at the gate corner in the former but much more uniform electric field in the latter [5].

Besides the drift region resistance, the channel resistance is the next dominant component. Compared to the conventional HEMT or MOSHC-HEMT, the MOSC-HEMT has a higher channel resistance because of the lower (about 10 times) inversion channel electron mobility when compared to the 2DEG channel electron mobility in the HEMT or MOSHC-HEMT (~ 150 vs $\sim 1500\ cm^2/V\ s$). Thus, scaling down of the channel length can substantially improve the on-resistance and on-state performance, particularly for the MOSC-HEMT [40,53]. Also, a systematic channel scaling of all the high-voltage AlGaN/GaN HEMT structures have been performed and its impact on the transistor performance evaluated [53]. This study can quantitatively determine the advantages as well as the limitations of scaling down the channel length. Furthermore, to enhance the channel width and other three-dimensional channel effects, FinFET or trigate channel structures have been applied to both HEMT [54] and MOSC-HEMT [55,56], resulting in lower specific on-resistance and lower off-state leakage.

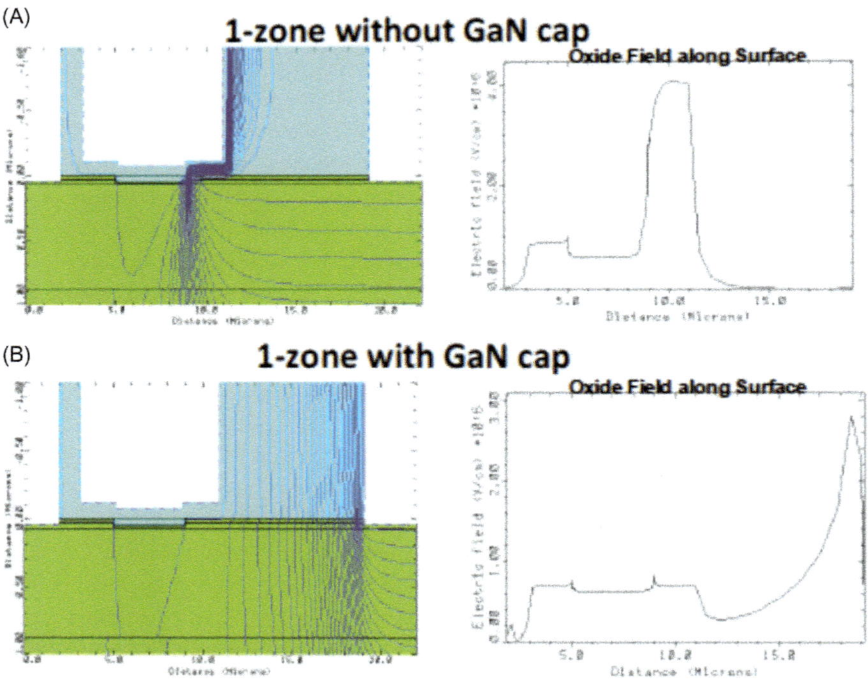

Figure 5.10 Simulated equivalent potential line distribution for 1-zone RESURF AlGaN/GaN MOSC-HEMT (A) without GaN cap at V_{DS} of 40 V and (B) with GaN cap at V_{DS} of 646 V [39].

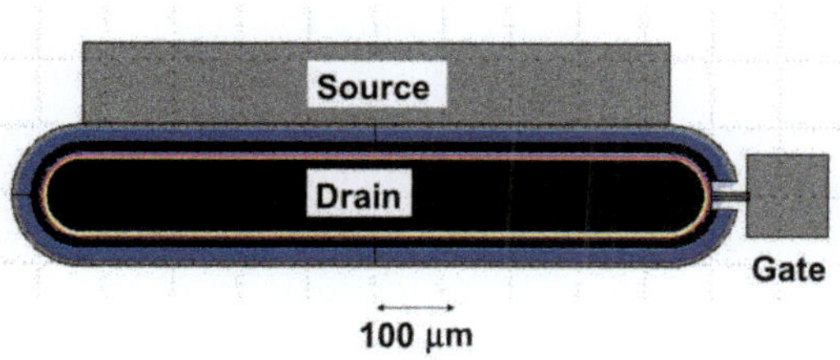

Figure 5.11 Single finger layout of a lateral AlGaN/GaN MOSC-HEMT.

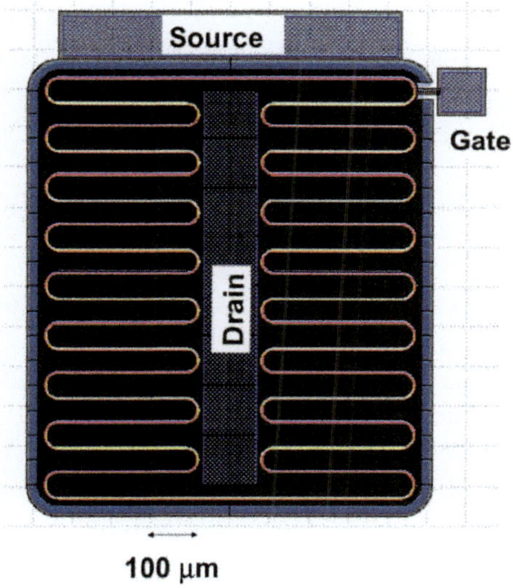

Figure 5.12 Multiple finger layout of a lateral AlGaN/GaN MOSC-HEMT.

Unlike the vertical power transistors, lateral power transistors are actually self-isolating in that the high-voltage drain regions are always surrounded by low-voltage or grounded source regions. Hence, the transistor layout demands the minimization of electric field crowding, as determined by the end-region curvature of transistor fingers (see Fig. 5.11 for a single finger case). Multiple finger structures, needed for high current devices, result in both convex and concave curvature of drain and source junctions (Fig. 5.12). The length and width of the drain/source fingers are determined by the interconnecting metal sheet resistance and the degradation of breakdown voltage due to three-dimensional junction curvature at the finer tips, as well as wire bonding requirements.

Lateral AlGaN/GaN HEMTs tend to be depletion mode or normally on type of transistors due to partial surface potential pinning in the conducting channel. However, for power switching, normally off transistors are needed in case the controlling signal is lost. To implement a normally off transistor, the cascoded pair of a hybrid, copackaged combination of a low voltage, normally off Si MOSFET and a high voltage, normally on AlGaN/GaN HEMT has been proposed and demonstrated (Fig. 5.12) [57,58]. Such a pair would satisfy most requirements of a power transistor switch. However, there are parasitic interconnection and robustness issues that need to be addressed [59], besides the extra chip and packaging costs.

5.2.2 Vertical

Vertical power devices conduct current in the on state and supports blocking voltage in a mostly one-dimensional manner between the top and the bottom terminals. In addition, termination structures are employed to suppress electric field crowding at the device periphery, as well as to spread out and hence significantly lower the peak surface electric field. Furthermore, passivation layers are used to minimize fixed charges on the surface of the semiconductor as well as to prevent exterior from penetrating into the active semiconductor regions and enhance long-term reliability.

Fig. 5.13 shows the schematic cross-sections of two vertical power GaN FETs that have been experimentally demonstrated to date. Since selective p-type doping by implantation followed with high-temperature annealing in GaN has been found to be very difficult, conventional planar Si and SiC power depletion-mode MOSFET (DMOSFET) structures are not used. Instead, for a vertical power MOSFET, the p-body region is epitaxially grown and trench U-shape MOSFETs (UMOSFETs) are the preferred structure [60–68].

A variation of the UMOSFET structure is the elimination of the p-body region altogether. This UMOSFET structure has been previously explored in Si and SiC and called accumulation field-effect transistor (ACCUFET) [69–71]. In GaN, it has been called Fin or nanowire Field-Effect Transistor [72–75]. This ACCUFET is simpler in structure but in its blocking state, its mesa region needs to be completely depleted to prevent the high drain potential from reaching source region and this depletion depends critically on the gate metal function and gate oxide thickness on the mesa sidewall. Also, the mesa pinch-off becomes more difficult with increasing temperature. Thus, mesa widths less than 500 nm are often needed for acceptable off-state leakage current. Because of the hexagonal wurtzite crystal structure of GaN, hexagonal trench cell geometry is desirable over the conventional

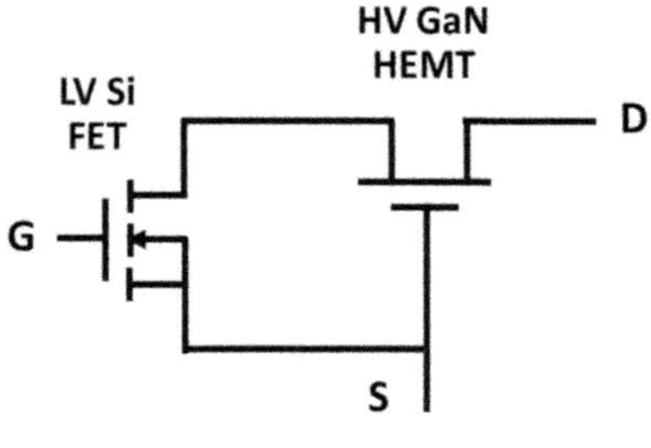

Figure 5.13 Schematic of a cascoded pair of high voltage GaN HEMT and low voltage Si power MOSFET.

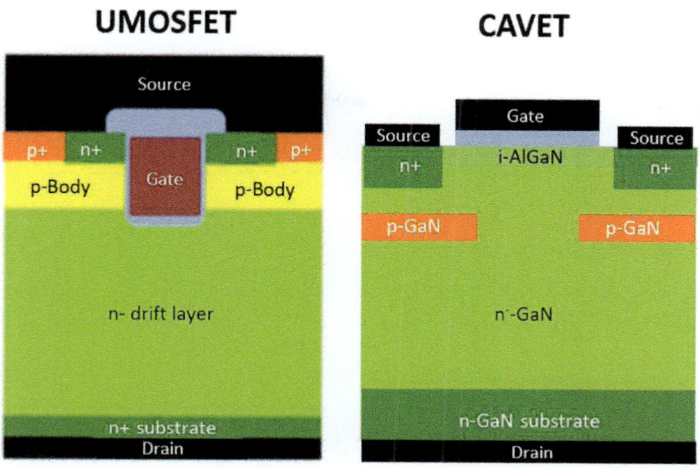

Figure 5.14 Schematic cross-sections of vertical GaN power field-effect transistor structures: (A) UMOSFET, (B) CAVET.

stripe or square cell geometry. Another power transistor structure that has been explored is the adaptation/modification of the lateral AlGaN/GaN HEMT into a vertical transistor, called current aperture vertical electron transistor (CAVET) (Fig. 5.14B) [76–85]. The overall structure resembles that of a vertical ACCUFET [69]. Also, The channel region is lateral and similar to that in the lateral GaN HEMTs by utilizing a heterojunction channel for low channel on-resistance. The high blocking voltage is supported by a vertical JFET region, which electrical isolates the channel from the drain potential.

Regardless of which vertical unipolar transistor structures, the drift region is designed with the triangular electric field profile so that the junction electric field reaches avalanche magnitude when the whole drift region is just completely depleted. This choice of field profile is close to the optimal tradeoff between on-resistance and breakdown voltage [2].

To further improve the $R_{on,\ sp}$ versus BV tradeoff for vertical unipolar transistors, superjunction devices, which utilizes two- and three-dimensional field shaping to achieve an approximate rectangular electric field profile at blocking, will be needed [86]. Schematic cross-sections of vertical GaN superjunction HEMT and MOSFET are shown in Fig. 5.15. Using alternating n- and p-type doped pillars for the drift region, lateral and vertical depletion take place simultaneously with increasing drain bias, leading to a more uniform electric field than the conventional triangular field profile with uniform doped drift region. Actually, the vertical superjunction devices can be considered as the vertical version of the lateral multi-RESURF type device described earlier. The performance of vertical GaN superjunction transistors has been projected [87] and will be discussed later.

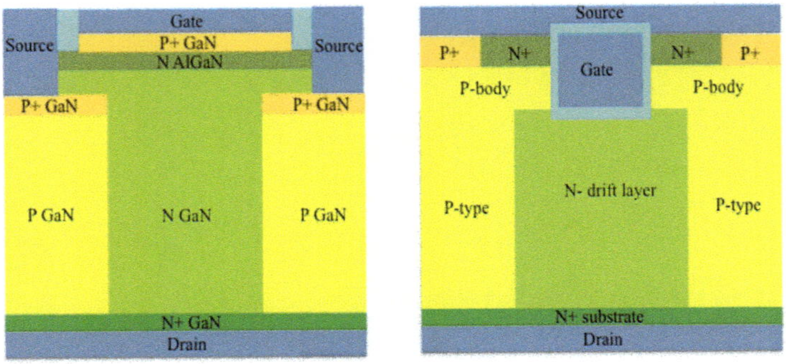

Figure 5.15 Schematic cross-sections of vertical GaN superjunction HEMT and MOSFET.

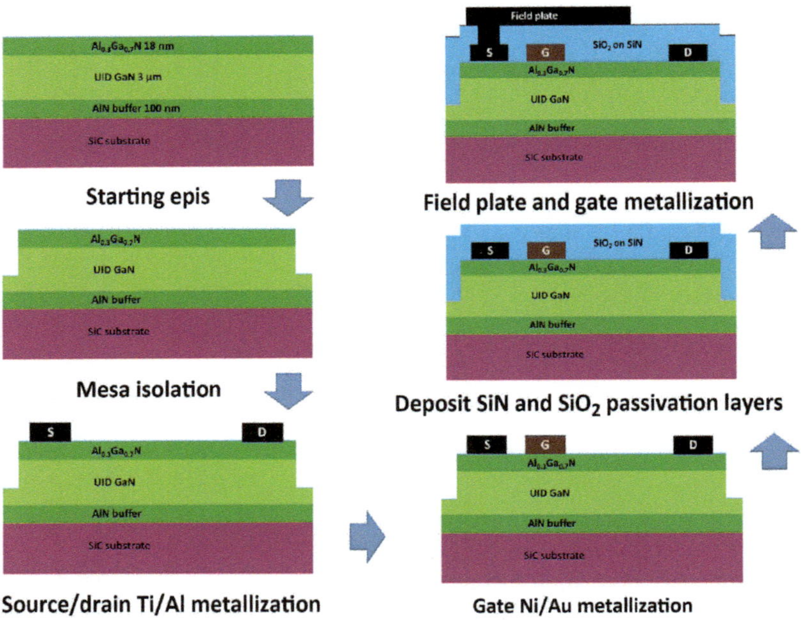

Figure 5.16 Integrated process flow for a lateral AlGaN/GaN HEMT.

5.3 Integrated device processes

5.3.1 Lateral

The salient highlights of a typical integrated process flow for a lateral power AlGaN/GaN HEMT are summarized in Fig. 5.16 [88]. The process architecture strongly resembles that of an rf AlGaAs/GaAs HEMT process, in that it utilizes lift-off metal gate, PECVD silicon nitride passivation, alloyed contacts and gold-based metallization, and sometimes unpassivated GaN or AlGaN surfaces. Such a process sequence is not compatible with standard silicon CMOS IC foundries. The key steps

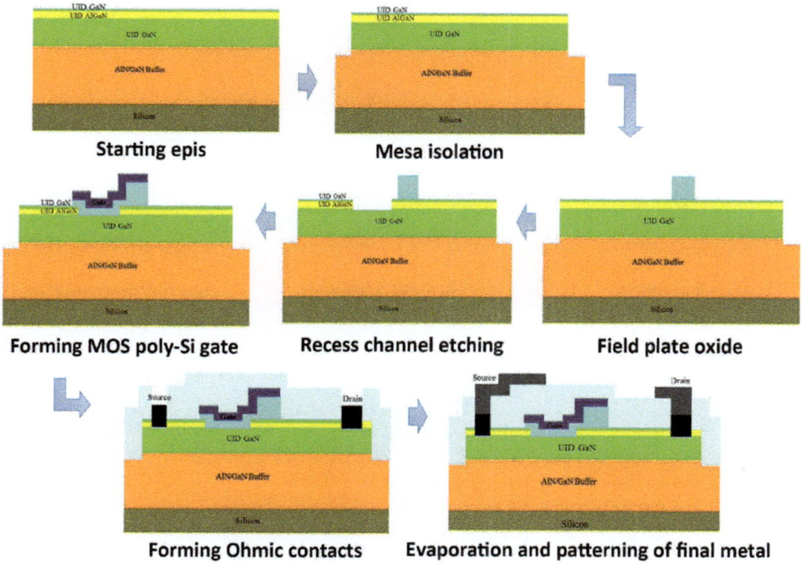

Figure 5.17 Integrated process flow for a lateral AlGaN/GaN MOS channel-HEMT [89].

of the integrated process flow for the MOS channel-HEMT is schematically depicted in Fig. 5.17 [89]. It adopts some of the silicon CMOS process features by using doped poly-Si gate, MOS channel, high temperature densified SiO_2 passivation. However, the temperature in the gate and field oxide annealing must be sufficiently low so as not to degrade the AlGaN/GaN heterojunction. Fig. 5.18 shows the salient features of an integrated process for a lateral GaN MOSFET with implanted source/drain and RESURF regions [46]. It shares many of the commonalities to those employed in the fabrication of silicon lateral RESURF-type MOSFETs. Nevertheless, it differs from the silicon process in having (1) source/drain and RESURF implants followed by activation annealing before gate oxide formation and hence a non-self-aligned gate process, (2) deposited, instead of thermally grown, gate and field oxide layers, and (3) lift-off Ti/Mo/Au contact metallization, instead of etched aluminum-based metallization.

To make the lateral AlGaN/GaN HEMT process more compatible with silicon foundry processes, attempts, such as aluminum-based metallization, have been made [90]. Also, unlike silicon carbide, the maximum process temperature is within the range of standard silicon CMOS foundries and no additional high-temperature annealing or oxidation equipment needs to be purchased and facilitized.

5.3.2 Vertical

As mentioned earlier, selective controlled p-type doping by implantation and annealing has so far very difficult. In addition, the only feasible acceptor dopant in GaN is magnesium. Consequently, the process architecture in vertical GaN power

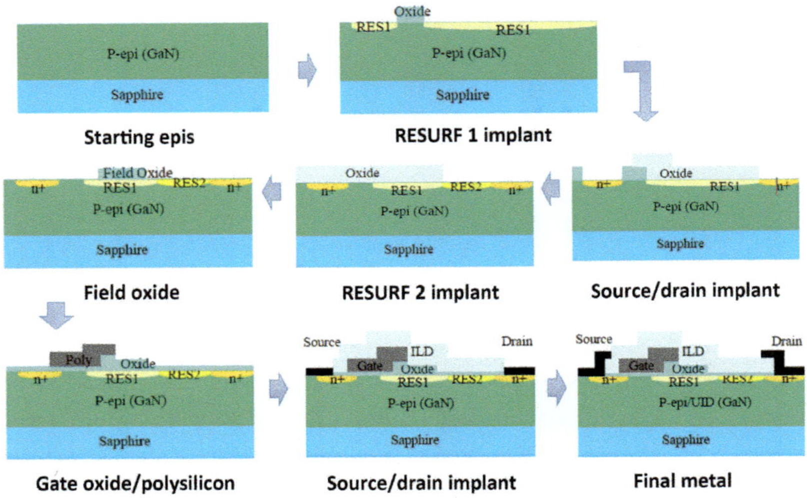

Figure 5.18 Integrated process flow for a lateral GaN MOSFET with an implanted RESURF region.

transistors involves in situ Mg-doped p-type layer. On the other hand, degenerately doped, n-type regions can be accomplished with high dosage silicon implant, followed by a short, 1100°C argon anneal with an SiO_2 cap, resulting in substantially activation [66]. Thus, the n + source regions can be either epi grown or implanted. The anisotropic etching needed for trench formation employs chlorine-based, often containing BCl_3, plasmas, followed by a wet etch by TMAH, which makes smooth the nonuniform sidewall remnants left over from the dry etching. Another complicating factor is the hexagonal nature of the GaN wurtzite structure. Hexagonal trenches can be formed on the m or (1−100) or a (11−20) planes and these planes have different electrical properties, such as interface state densities, from those of the Ga face or (0001) plane of the planar GaN surface [29,30]. Furthermore, for Mg-doped p-layer grown by MOCVD, a 650−700°C annealing in oxygen or air is performed to eliminate the incorporate hydrogen to maximize the electrical activation of magnesium. Also, the top n + source layer, if present during this annealing step, has been found to block the hydrogen out-diffusion, and hence must be partially removed. All these considerations are taken into account to design the integrated flow. Fig. 5.19 depicts schematically the highlights of a typical integrated fabrication process of the GaN UMOSFET that has been used to demonstrate prototype devices [65].

The integrated process flow for a planar CAVET, schematically shown in Fig. 5.20, incorporates additional steps not in the UMOSFET [60]. The buried p-layer, used to pinch off the JFET region, is grown and then patterned and selectively etched away in the JFET region. Then, an n-type epi regrowth step is done to overgrow the nonplanar surface consisting of the p-layer and the etched n-region. The AlGaN/GaN heterojunction is subsequently grown on this mostly planar or

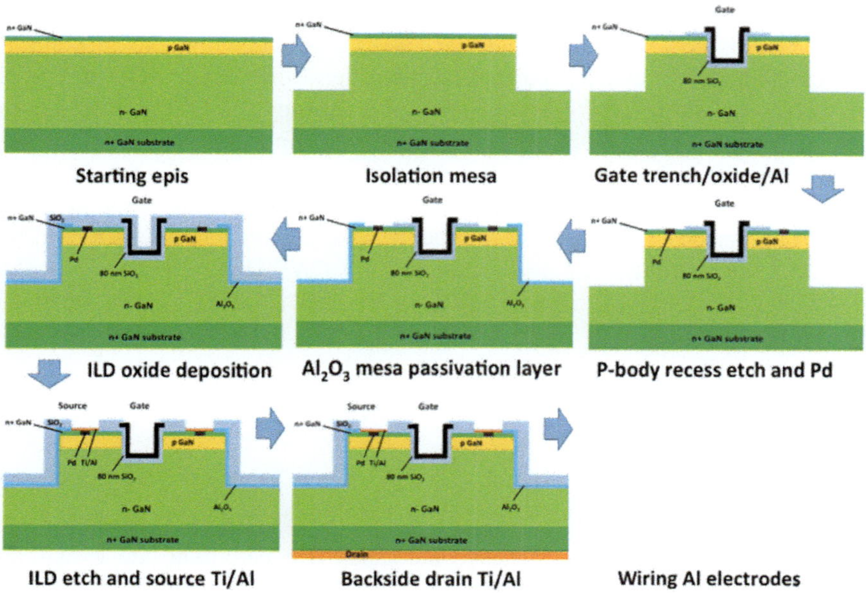

Figure 5.19 Integrated process flow for a vertical GaN UMOSFET.

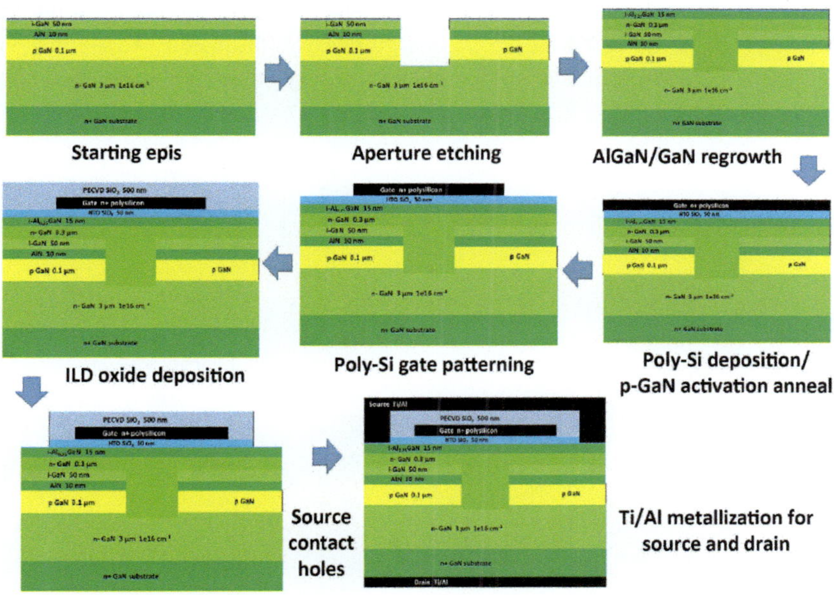

Figure 5.20 Integrated process flow for a planar AlGaN/GaN CAVET.

planarized n-layer. The main problems in this regrown step include autodoping of the n-type JFET region from the adjacent magnesium-doped regions resulting in a compensated, high-resistivity region, and, faceting of the regrown n-layer that may require planarization (like chem-mechanical polishing). In addition, if a trench CAVET structure is desired, an extra anisotropic, controlled sloped GaN etching step is needed. Further, the pseudomorphic growth of AlGaN over the sloped GaN region is different from the growth over planar areas and both needs to be optimized simultaneously. The integrated process flow of a nonplanar CAVET is shown in Fig. 5.21 [82]. Refinement and scale-up of this CAVET process to large-volume device manufacturing will be more difficult than the UMOSFET process described in the Fig. 5.19.

5.4 Device performance

5.4.1 Static

The various dominant on-resistance components that determine the overall total specific on-resistance ($R_{on,\ sp}$) of a 600 V lateral AlGaN/GaN HEMT are estimated in Table 5.2 and the specific on-resistance dependence on channel length is shown in Fig. 5.22. The major ones are from the contact, channel and drift regions, denoted by R_C, R_{ch}, and R_{drift}, respectively. The specific on-resistances of the source and the drain contacts have been optimized and reduced to less 10% of the total $R_{on,\ sp}$. The drift region component is determined by the polarization charge density, which is the RESURF space charge density, discussed earlier, that yields the BV, but it also induces the 2DEG density that specifies the on-state

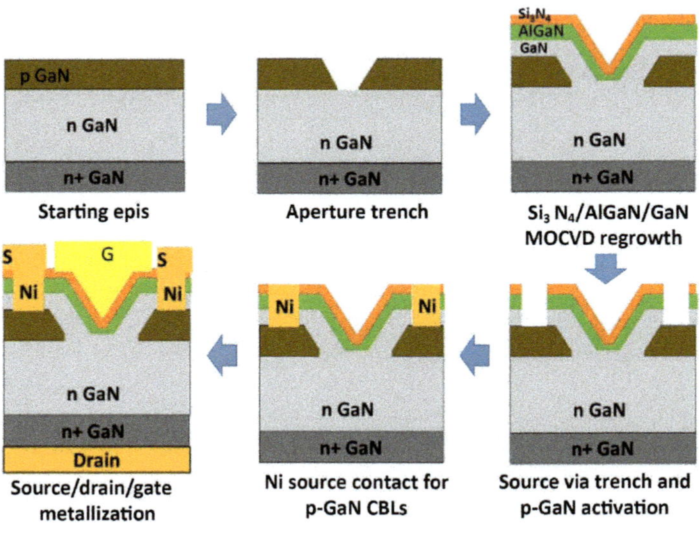

Figure 5.21 Integrated process flow for a nonplanar AlGaN/GaN CAVET [82].

Table 5.2 Various estimated specific on-resistance components in 600 V lateral GaN HEMT for difference channel lengths, together with the assumed structural device parameters

L_{ch} (μm)	$R_{on,\ sp}$ (mΩ/cm²)	$R_{S(\%)}$	$R_{ch(\%)}$	$R_{drift(\%)}$	$f_{m,\ IG}$ (MHz)
0.3	0.44	22	3	75	552
1	0.50	20	10	70	469
3	0.70	17	25	58	313
Design parameters					
Topology					**Linear**
2DEG density					1e13 q/cm²
2DEG channel mobility					1500 cm²/V s
Assumed E_C, lateral					1 MV/cm
Breakdown voltage					700 V

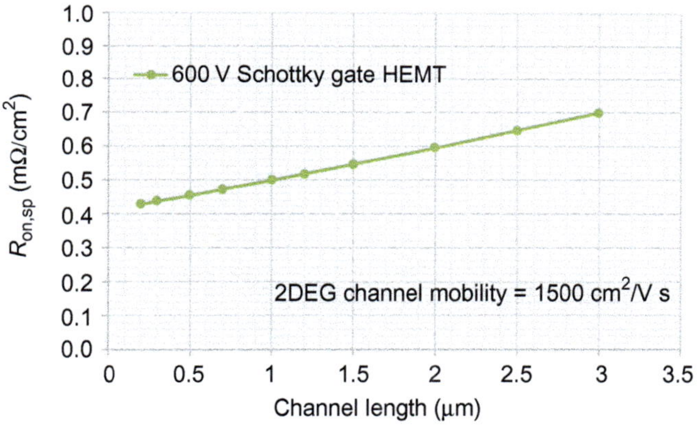

Figure 5.22 Estimated specific on-resistance of 600 V lateral AlGaN/GaN HEMT on channel length.

conductivity. With a space charge density of typically 10^{13} cm^{-2} and a mobility of 1500 cm²/V s, its specific on-resistance is only 0.2 mΩ/cm². The drift region length, specified by the RESURF rule [41–44], depends linearly on the breakdown. With a surface breakdown field of 1 MV/cm [7], a 600 V HEMT only requires a 6 μm drift region length. The channel on-resistance is proportional to the channel length, till the short-channel effects become significant. The electrons in the heterojunction channel have a higher mobility than those in an MOS inversion channel, usually by 5 to 10 times.

However, the MOS channel-HEMTs can be made competitive with the heterojunction HEMTs by scaling down the channel length to deep submicron dimensions (<0.5 μm) [39,40,53]. The various dominant on-resistance components of a 600 V

Table 5.3 Various estimated specific on-resistance components in 600 V lateral GaN MOS channel- HEMT for difference channel lengths, together with the assumed structural device parameters

L_{ch} (μm)	$R_{on,\ sp}$ (mΩ/cm²)	$R_{S(\%)}$	$R_{ch(\%)}$	$R_{drift(\%)}$	$f_{m,\ IG}$ (MHz)
0.3	0.76	12	44	43	322
1	1.7	6	73	21	149
3	4.7	2	89	9	52
Design parameters					
Topology					**Linear**
Gate oxide					50 nm
MOS channel mobility					100 cm²/V s
2DEG density					1e13 q/cm²
2DEG channel mobility					1500 cm²/V s
Gate voltage					15 V
Threshold voltage					0.5 V
Assumed E_C, lateral					1 MV/cm
Breakdown voltage					700 V

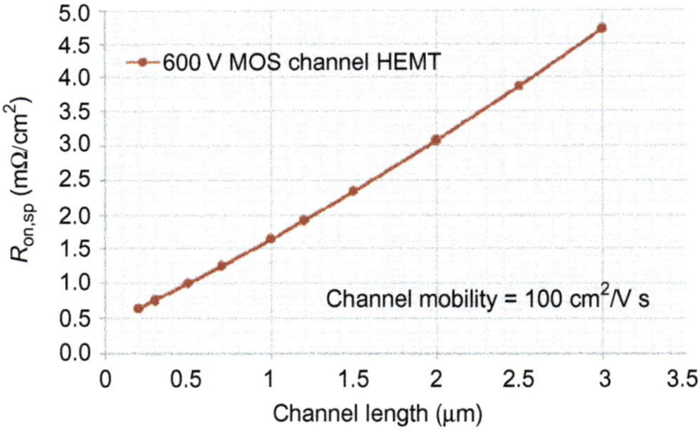

Figure 5.23 Estimated specific on-resistance of 600 V lateral AlGaN/GaN MOS channel-HEMT on channel length.

lateral AlGaN/GaN MOS channel-HEMT are estimated in Table 5.3 and the specific on-resistance dependence on channel length is illustrated in Fig. 5.23. In addition, three-dimensional effects can be exploited with the trigate or FinFET type structures to further reduce the channel on-resistance by about a factor of 3 and suppress short channel effects [55,56]. (The trigate concept was first demonstrated in GaN with a conventional AlGaN/GaN HEMT [54].) In addition, lateral GaN devices with multiple RESURF layers in the drift region, which reduce the drift-layer specific on-resistance by a factor equal to the number of RESURF layers, have been implemented on sapphire substrates [52].

The on-resistance of a cascoded Si MOSFET/AlGaN/GaN HEMT pair is the serial combination of the on-resistances of the two constituent transistors. Since the Si MOSFET is low voltage (typically BV of 30 V), its on-resistance is a fraction (<10%) of that for the GaN HEMT. Also, its size can be made larger to reduce the R_{on}, but at the expense of a larger input capacitance.

Interestingly, the specific on-state resistance dependence on the breakdown voltage is different for lateral RESURF power devices when compared to that of vertical power devices. For lateral RESURF devices, it can be shown that in the ideal case, when the device drift layer resistance dominates, $R_{on,\ sp} \propto (BV)^n$, where n ranges from 1 to 1.3 or higher, dependent on the doping profile, carrier mobility, critical field dependence on doping, and substrate character (junction vs SOI) for silicon [91,92] and GaN [93].

This dependence is substantially different from the n value of 2.4 to 2.6 for one-dimensional vertical devices and the commonly derived value of 2 for lateral RESURF devices deduced from the linear dependence of both BV and on-resistance on drift length. However, for lateral GaN HEMTs on silicon, which can be viewed as functionally equivalent to SOI RESURF structures, the 2DEG mobility and the breakdown field do not depend on the drift layer doping, and the ultimate n value can approach unity when the lateral space charge (polarization or dopant charge) concentration profile is nonuniform [92]. We have shown the polarization charge concentration at the AlGaN/GaN heterojunction can be modified by varying the GaN cap thickness [51]. Experimentally, the specific on-resistance of lateral GaN HEMTs versus breakdown voltage is shown in Fig. 5.24. An empirical n value of ~1.3 has been extracted from the reported data for BV ranges from 30 to 1 kV.

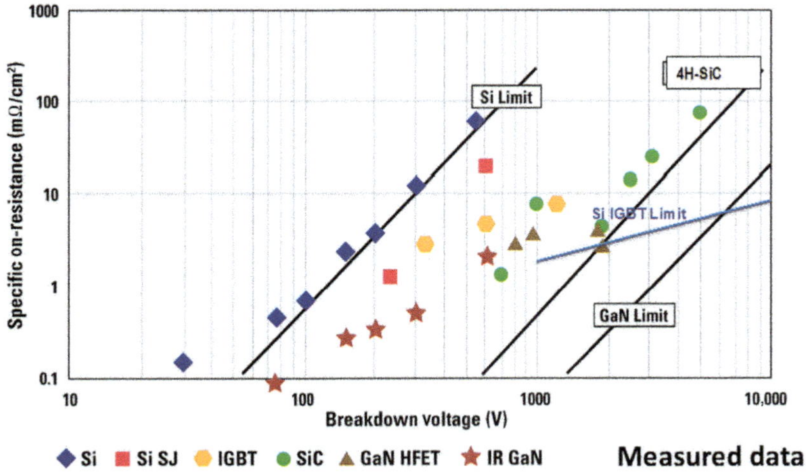

Figure 5.24 Specific on-resistance of lateral AlGaN/GaN HEMTs versus breakdown voltage. It be seen that the experimental data yields a n value of ~1.3. The ideal specific on-resistance of Si and GaN vertical devices are also shown for comparison.
Adapted from [94].

The schematic cross-section and top-view of a unit cell of a vertical 600 V GaN UMOSFET are depicted in Fig. 5.25. While the channel and drift on-resistances are also dominant here, the current spreading in the drift layer, determined by the mesa and trench widths, is different than that in the lateral HEMT or MOSFET. Also, typical device design parameters used or assumed are given. In Fig. 5.26, the specific on-resistance of these UMOSFETs as a function of channel length and mobility is shown. In Table 5.4, these various resistance components, together with the maximum switching frequency, are shown for 600 V GaN UMOSFET as a function of channel length. The channel on-resistance is clearly one of the most important

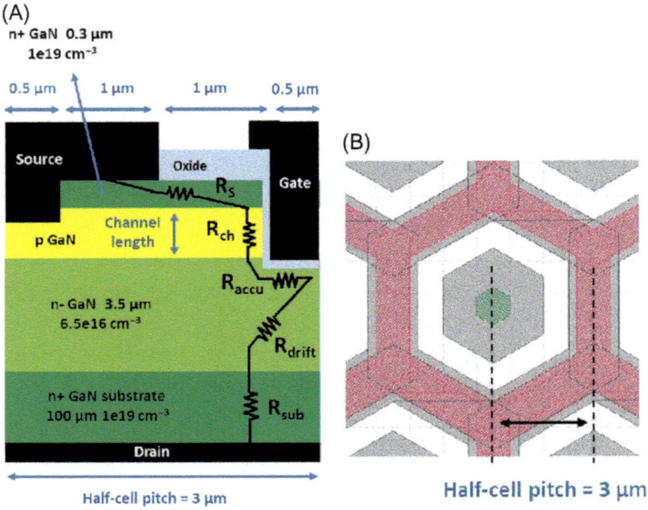

Figure 5.25 (A) Schematic cross-section of a half unit cell of a 600 V vertical GaN UMOSFET and (B) top view of the hexagonal unit cell.

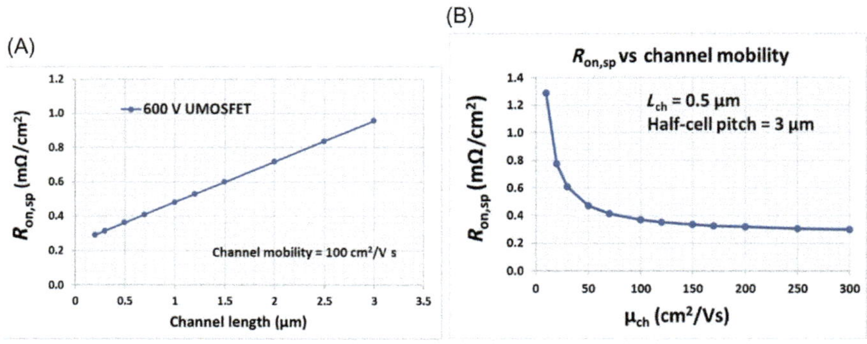

Figure 5.26 Estimated specific on-resistance of 600 V vertical GaN UMOSFET as a function of (A) channel length for a channel mobility of 100 cm^2/V s, and (B) channel mobility for a 0.5 μm channel length.

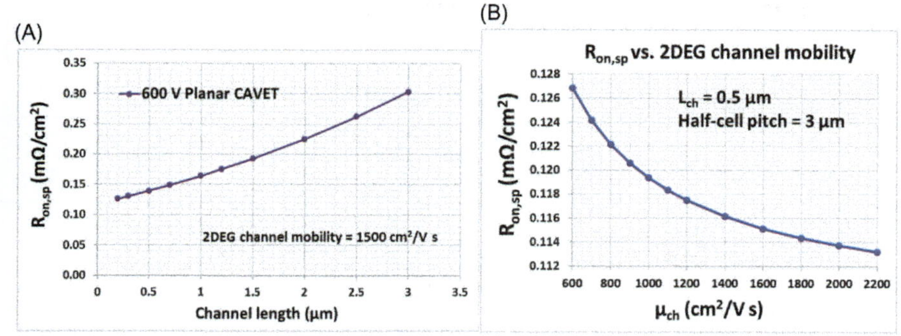

Figure 5.27 Estimated specific on-resistance of 600 V vertical GaN CAVET as a function of (A) channel length for a channel mobility of 1500 cm^2/V s, and (B) channel mobility for a 0.5 μm channel length.

Table 5.4 Various estimated specific on-resistance components in 600 V vertical GaN UMOSFET for difference channel lengths, together with the assumed structural device parameters

L_{ch} (μm)	$R_{on,\ sp}$ (mΩ/cm^2)	$R_{S(\%)}$	$R_{ch(\%)}$	$R_{accu(\%)}$	R_{drift} (%)	R_{sub} (%)	$f_{m,\ IG}$ (MHz)
0.3	0.30	6	24	22	20	28	532
1	0.46	4	51	14	13	18	288
3	0.94	2	76	7	6	9	98

Design parameters	
Topology	**Hexagonal**
Gate oxide	50 nm
Channel mobility	100 cm^2/V s
Gate voltage	15 V
Threshold voltage	4 V
Accumulation layer spreading factor	0.6
Breakdown voltage	700 V

components and can be as high as 50% or more of the total. Table 5.5 summarizes the on-resistance components and the maximum switching frequency for a 1200 V GaN UMOSFET. It can be seen that the drift region has become more dominant and the total specific on-resistance is less dependent on the channel resistance and hence channel length (see Fig. 5.28). Another similar vertical GaN UMOSFET is the ACCUFET or FINFET. It also has similar on-resistance components as those shown for the conventional GaN UMOSFET. However, the mesa design for the ACCUFET is more stringent because of a lack of a blocking pn junction and the

Table 5.5 Various estimated specific on-resistance components in 600 V vertical planar GaN CAVET for difference channel lengths, together with the assumed structural device parameters

L_{ch} (μm)	$R_{on,\ sp}$ (mΩ/cm^2)	$R_{S(\%)}$	$R_{ch(\%)}$	$R_{accu(\%)}$	$R_{JFET(\%)}$	$R_{drift(\%)}$	$R_{sub(\%)}$	$f_{m,\ IG}$ (MHz)
0.3	0.20	11	2	4	18	23	42	321
1	0.4	14	7	4	18	21	36	294
3	0.37	20	20	3	16	18	23	217

Design parameters	
Topology	**Hexagonal**
2DEG channel density	1e13 q/cm^2
2DEG channel mobility	1500 cm^2/V s
JFET region doping	1e16 cm^{-3}
Accumulation layer spreading factor	0.6
Breakdown voltage	700 V

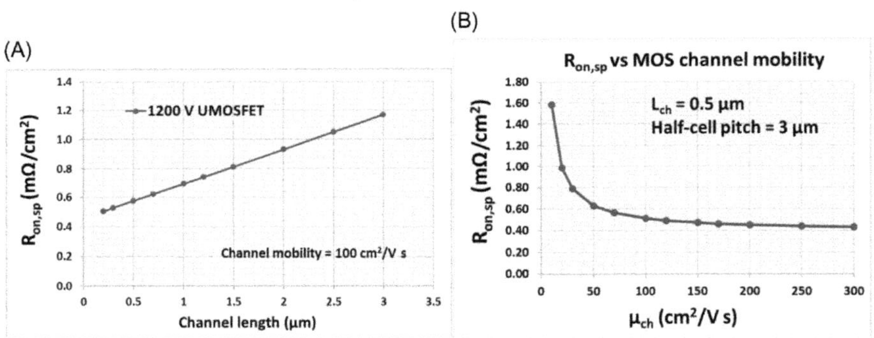

Figure 5.28 Estimated specific on-resistance of 1.2 kV vertical GaN UMOSFET as a function of (A) channel length for a channel mobility of 100 cm^2/V s, and (B) channel mobility for a 0.5 μm channel length.

mesa depletion is entirely dependent on the MOS gate parameters and the mesa width and doping.

While the device structure of the CAVET resembles that of a vertical DMOSFET, its heterojunction channel has a 10 times higher electron mobility than the MOS inversion channel. Consequently, on the other hand, the drift region on-resistance component is almost very similar to the vertical UMOSFETs. In addition, there is an extra JFET on-resistance component, R_{JFET}, determined by the spacing between adjacent p-regions. This on-resistance component rises sharply with decreasing JFET spacing from down scaling the unit cell pitch. Typical unit cell designs for 600 and 1200 V GaN CAVETs are given and the on-resistance

Table 5.6 Various estimated specific on-resistance components in 1200 V vertical GaN UMOSFET for difference channel lengths, together with the assumed structural device parameters

L_{ch} (μm)	$R_{on, sp}$ (mΩ/cm^2)	$R_{S(\%)}$	$R_{ch(\%)}$	$R_{accu(\%)}$	$R_{drift(\%)}$	$R_{sub(\%)}$	$f_{m, IG}$ (MHz)
0.3	0.54	3	13	12	56	16	321
1	0.71	3	34	9	43	12	204
3	1.2	2	60	5	26	7	82
Design parameters							
Topology							**Hexagonal**
Gate oxide							**50 nm**
Channel mobility							100 cm^2/V s
Gate voltage							15 V
Threshold voltage							4 V
Accumulation layer spreading factor							0.6
Breakdown voltage							1400 V

Table 5.7 Various estimated specific on-resistance components in 1.2 kV vertical planar GaN CAVET for difference channel lengths, together with the assumed structural device parameters

L_{ch} (μm)	$R_{on, sp}$ (mΩ/cm^2)	$R_{S(\%)}$	$R_{ch(\%)}$	$R_{accu(\%)}$	$R_{JFET(\%)}$	$R_{drift(\%)}$	$R_{sub(\%)}$	$f_{m, IG}$ (MHz)
0.3	0.42	5	1	2	9	63	20	167
1	0.46	7	4	2	9	60	18	162
3	0.62	12	12	2	10	50	14	139
Design parameters								
Topology								**Hexagonal**
2DEG channel density								**1e13 q/cm^2**
2DEG channel mobility								1500 cm^2/V s
JFET region doping								1e16 cm^{-3}
Accumulation layer spreading factor								0.6
Breakdown voltage								1400 V

components are depicted. In Tables 5.6 and 5.7, the various the on-resistance components and the maximum switching frequency for 600 and 1200 V GaN CAVETs are given. The specific on-resistance of these transistors versus channel length and mobility is shown in Figs. 5.27 and 5.29. Due to the JFET component, this

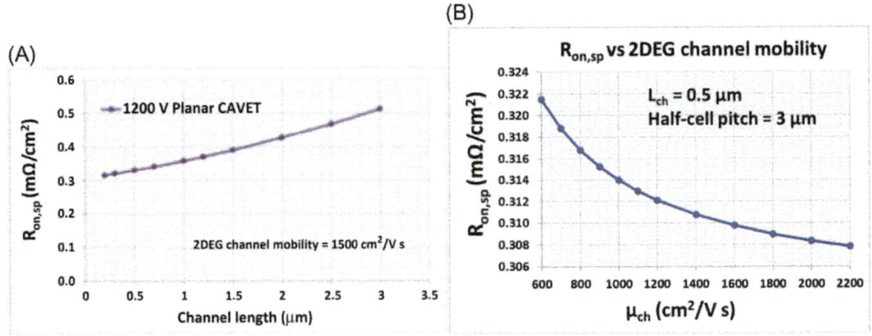

Figure 5.29 Estimated specific on-resistance of 1.2 kV vertical GaN CAVET as a function of (A) channel length for a channel mobility of 1500 cm^2/V s, and (B) channel mobility for a 0.5 μm channel length.

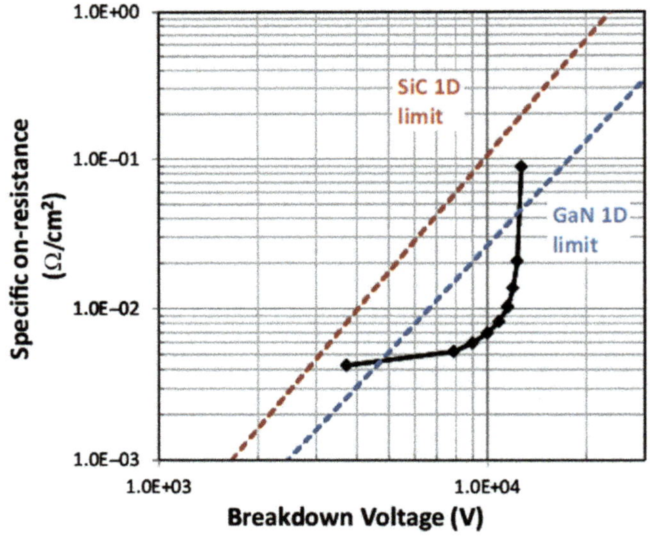

Figure 5.30 Drift-layer specific on-resistance versus breakdown voltage for GaN superjunction devices with 60 μm drift layer thickness, half pillar width of 8 μm and varying pillar doping [87].

transistor structure is more difficult to achieve a small cell pitch than the conventional UMOSFETs.

The design and optimization of 5–20 kV GaN vertical superjunction HEMTs have been systematically performed [87]. The breakdown voltage and specific on-resistance can be optimized by varying the doping and width of the p and n pillars with a fixed drift layer thickness, as shown in Fig. 5.30 [87]. The sensitivity of the charge balance in the pillar on both BV and $R_{on,\ sp}$ is shown in Fig. 5.31 [87]. It can be seen that only 10% deviation in doping is allowed so as to keep the BV

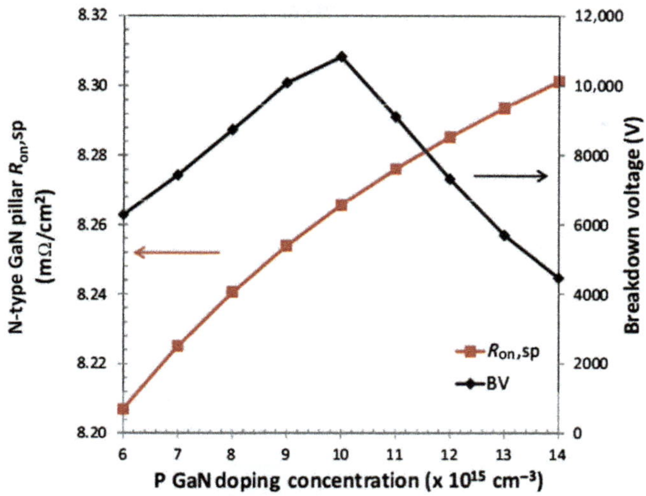

Figure 5.31 Dependence of breakdown voltage and drift-layer specific on-resistance on the pillar doping concentration in GaN superjunction devices [87].

degradation to within 10%. Fig. 5.31 shows the drift-layer and total specific on-resistances of vertical GaN superjunction HEMTs with different pillar widths. It is clear that for BV over 5 kV, the vertical superjunction HEMTs are superior to conventional vertical HEMTs in GaN [87].

While the $R_{on,\ sp}$ versus BV relationship generally has an n exponent of about 2.5, it can deviate from that value for two reasons. First, the drift layer doping for GaN is about 2 orders of magnitude higher for the same breakdown voltage when compared to silicon (e.g., 1 kV GaN transistor has a doping of $\sim 10^{16}\ cm^{-3}$ vs $\sim 10^{14}\ cm^{-3}$ for silicon counterpart). The bulk carrier mobility start to degrade with increasing doping density due to impurity scattering, leading to a smaller n value. Second, if the superjunction structure, in which the uniformly doped drift region is replaced by alternating n and p regions of higher and equal doping concentration, is adopted, the specific on-resistance is linearly dependent on the breakdown voltage (Fig. 5.32).

Table 5.8 summarizes selected experimentally demonstrated vertical high-voltage GaN FETs [60,62,64,65,67,68,73–76,81,84,85]. In Fig. 5.33, the specific on-resistance versus breakdown voltage of vertical GaN power transistor for conventional and superjunction vertical FETs is shown.

5.4.2 Dynamic switching

Switching of power transistors depends on the load and the gate driving circuitry and their switching characteristics are determined by the switching energy and times. Since all the GaN power transistors are unipolar, the intrinsic on-state and gate resistance, together with the extrinsic inductances and resistances, dominate

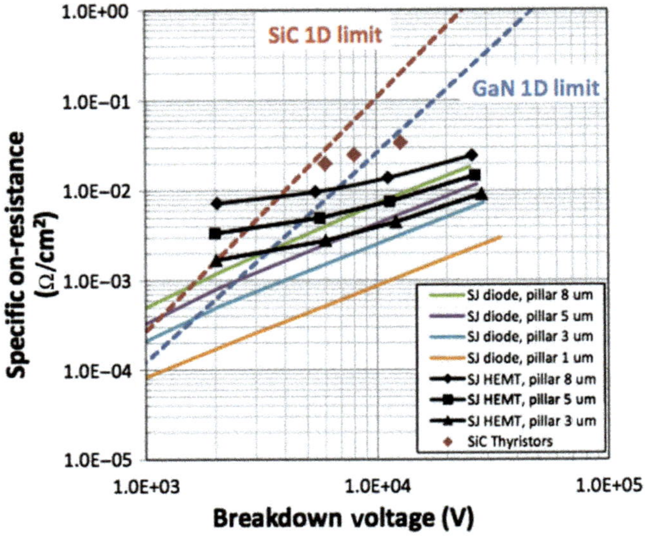

Figure 5.32 Drift-layer and total specific on-resistances versus BV for vertical GaN superjunction HEMT with varying pillar width and doping for a fixed pillar space charge dosage of $8 \times 10^{12}\ cm^{-2}$ [87].

their switching behavior, which is similar to those of silicon power MOSFETs [2,95−98].

Figs. 5.34 and 5.35 show the numerical simulation results of comparative switching performance of 600 and 1200 V, 10 A, lateral HEMT and MOSC-HEMT and vertical UMOSFET and CAVET under typical load conditions for transistors [98]. Due to the high channel mobility and low gate capacitance of lateral HEMT and vertical CAVET, these have a lower switching energy than their MOS channel counterparts. In addition, the vertical CAVET is superior over lateral HEMT due to a better a smaller active area. These trends hold for both 600 and 1200 V.

Alternatively, but intuitively, we can estimate the main switching energy by using the gate charge (Q_g) and output drain charge (Q_{oss}) [99]. The gate charge is usually larger than the output drain charge.

Here, we have used the gate charge (Q_g) and the derived figure of merit (FOM), the maximum operating frequency ($f_{m,\ IG}$), to comparatively evaluate the switching characteristics of the various lateral and vertical GaN power FETs [53]. The results are shown in each GaN field-effect transistor structure in Tables 5.2−5.7. It can be seen that the lateral transistors have a higher $f_{m,\ IG}$ than the vertical transistors because of a smaller C_{gd}. Also, this FOM increases with decreasing channel length, emphasizing the importance of device downscaling for improved performance. A maximum operating frequency of higher than 5 MHz has been projected for both lateral and vertical GaN power FETs and can be further enhanced with refined device structures to reduce the gate-to-drain capacitance. Furthermore, by

Table 5.8 Experimentally demonstrated vertical GaN high-voltage FETs

Device	BV (V)	V_T (V)	$R_{on,sp}$ (mΩ/cm^2)	$I_{DS,\ max}$ (A/cm^2)	Features	Group	Year	Ref.
CAVET	–	−4	13Ω-mm	430 mA/m m	AlGaN/GaN channel $I_{DS,\ max}$ = 430 mA/mm	UCSB	2002	[76]
CAVET	–	−16	2.6	400	MOS-HEMT Channel	Toyota	2007	[60]
CAVET	250	−6	2.2	4000	Mg-implanted current blocking layer	UCSB	2012	[79]
CAVET	1500	+0.5	2.2	1500	P + GaN gate AlGaN/GaN channel	Avogy	2014	[80]
VHFET	672	−1.1	7.6	267	Regrown AlGaN/GaN channel	Sumitomo Electric	2010	[83]
VHEMT	1700	+2.5	1.0	4000	p-GaN gate Truncated V shaped AlGaN/GaN channel	Panasonic	2016	[84]
UMISFET	180	+10	–	0.5 mA/mm	m-face channel SiN gate dielectric	Toyota	2008	[62]
UMOSFET	1200 1300	+3.5 +3.5	1.8 3.4	2100 1000	SiO_2 gate dielectric Inversion MOS channel Hexagonal cells	Toyoda Gosei	2015 2016	[64] [65]
UMISFET	600	+4.8	8.5	420	SiN gate dielectric	HRL	2016	[67]

(*Continued*)

Table 5.8 (Continued)

Device	BV (V)	V_T (V)	$R_{on,sp}$ (mΩ/cm^2)	$I_{DS, max}$ (A/cm^2)	Features	Group	Year	Ref.
					Hexagonal cells			
UMOSFET	990	+ 3	2.6	700	Al_2O_3 gate dielectric Inversion MOS channel	UCSB	2017	[68]
FinFET	800	+ 1.5	4.4* (0.36)	1390 (17,000)	Molybdenum gate Al_2O_3 gate dielectric Accumulation MOS channel	MIT	2017	[73]
Nanowire FET	140	+ 1.2	2.2	314 mA/m m	Wrap-around gate PEALD SiO_2 gate dielectric	TU Brauns-ch-weig	2016	[74]
FinFET/ nanowire FET	513 @$V_{GS} = -15$V	< 0.5	0.4 (normalized to fin area, not device area)	12,000 (normalized to fin area, not device area)	ALD Al_2O_3 gate dielectric	Cornell	2017	[75]

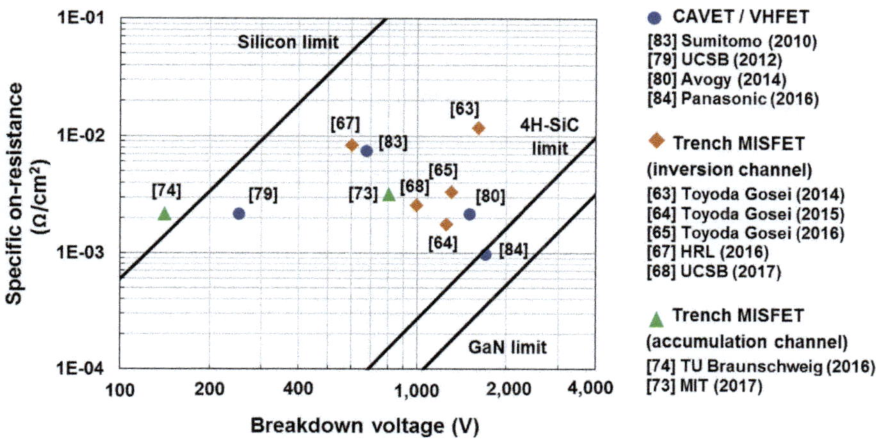

Figure 5.33 Specific on-resistance of experimentally demonstrated vertical GaN UMOSFETs and CAVETs.

optimizing the gate resistance, we have projected the switching conditions for possible lossless turn-off energy in the usage of these GaN transistors [99].

Summarizing Tables 5.9 and 5.10 summarizes the static or on-state and dynamic or switching performance results for the 600 and 1200 V lateral HEMT and MOSC-HEMT and vertical GaN UMOSFET and CAVET in GaN, respectively. It can be readily seen that the vertical GaN CAVET has the best performance due to its higher 2DEG channel mobility and low junction gate capacitance as well as small active area from vertical voltage blocking. However, there are other design factors, such as robustness, as well as competing silicon and SiC power transistors, that must also be considered. These will be discussed in detail in the Sections 5.4.3 and 5.4.4.

5.4.3 Robustness

One unique feature concerning GaN transistors that incorporate a heterojunction AlGaN/GaN channel is the current collapse phenomenon. This current collapse or dynamic R_{on} degradation can be defined as a reduction of conduction current, after application of a pulse of high drain bias, resulting in an increase in the on-resistance of the transistor. The possible mechanisms and trapping sites of the current collapse phenomenon is schematically depicted in Fig. 5.36 [100]. It is generally agreed upon by now that this degradation is caused by a transient, reversible channel electron trapping. Possible trapping areas include the heterojunction interface at the drift region as well as the buffer layers between the substrate and the active channel. This phenomenon has been observed in both GaN on Si and on sapphire substrates but the actual degradation rate is highly dependent on the exact buffer structure, substrate kind, pulsed drain voltage duration and magnitude and passivation layer used [101–106], and thus, vary widely among different suppliers.

(A)

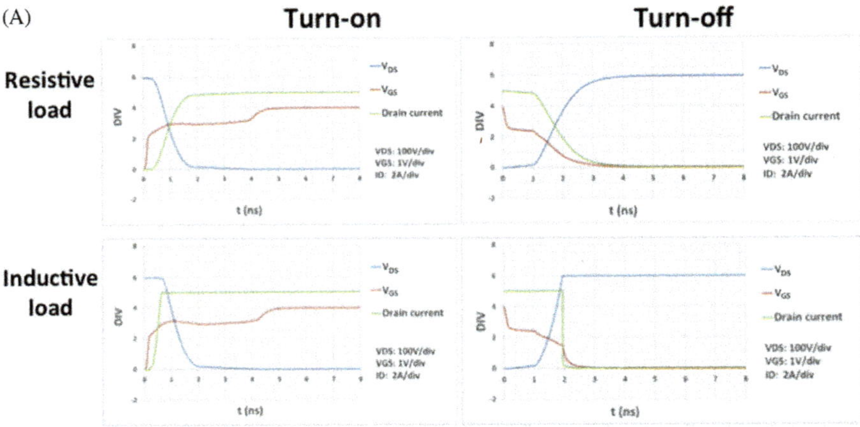

	Load	t_{on} (ns)	t_{off} (ns)	E_{on} (μJ)	E_{off} (μJ)	E_{total} (μJ)
600 V p-GaN gate HEMT	Resistive	1.5	2.8	1.5	2.4	3.9
	Inductive	0.6	2.0	4.6	2.5	7.1

(B)

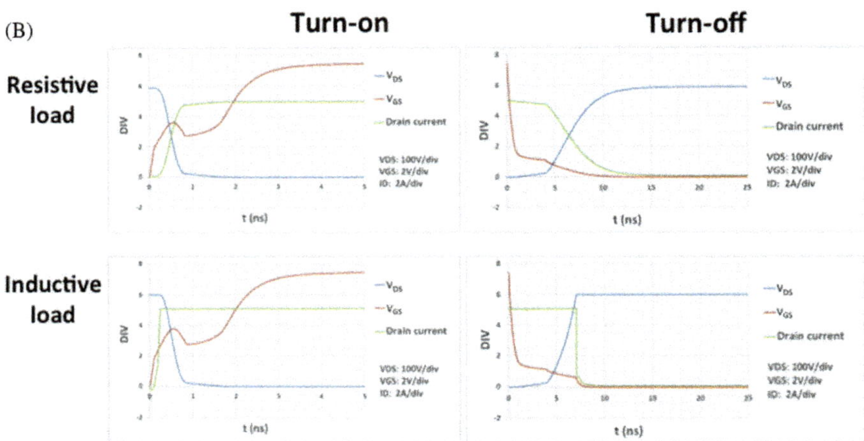

	Load	t_{on} (ns)	t_{off} (ns)	E_{on} (μJ)	E_{off} (μJ)	E_{total} (μJ)
600 V MOSC HEMT	Resistive	0.7	10.5	0.7	8.5	9.2
	Inductive	0.2	7.5	2.2	9.5	11.7

Figure 5.34 Switching waveforms of 600 V, 10 A GaN (A) lateral HEMT and (B) lateral MOSC-HEMT, and (C) vertical UMOSFET and (D) vertical CAVET under resistive and inductive (L = 120 μH) load conditions and a gate resistance of 0.5 Ω. [98]

(C)

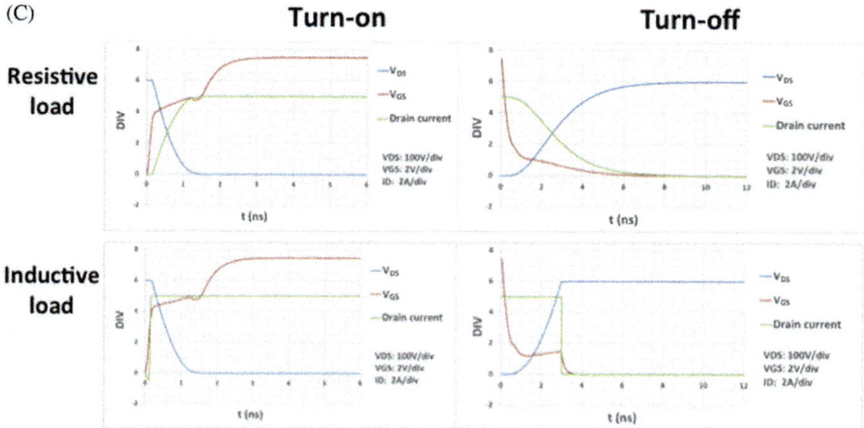

	Load	t_{on} (ns)	t_{off} (ns)	E_{on} (μJ)	E_{off} (μJ)	E_{total} (μJ)
600 V UMOSFET	Resistive	1.0	5.1	1.0	5.2	6.2
	Inductive	0.2	3.0	2.9	5.2	8.1

(D)

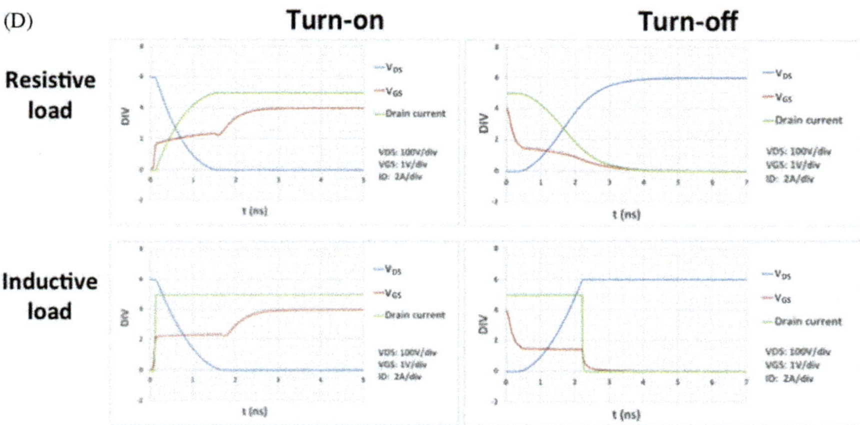

	Load	t_{on} (ns)	t_{off} (ns)	E_{on} (μJ)	E_{off} (μJ)	E_{total} (μJ)
600 V p-GaN gate CAVET	Resistive	1.2	2.9	1.2	2.7	3.9
	Inductive	0.1	2.2	3.5	4.1	7.6

Figure 5.34 (Continued)

Fig. 5.37 illustrates the varying degree of current collapse with applied pulsed drain bias for two suppliers [100].

To quantitatively measure this phenomenon, a substrate bias sweeping technique, which monitors the drain current with a linear ramping of the substrate voltage, has been found to be very useful for GaN-on-Si HEMTs [100]. The substrate bias can

(A)

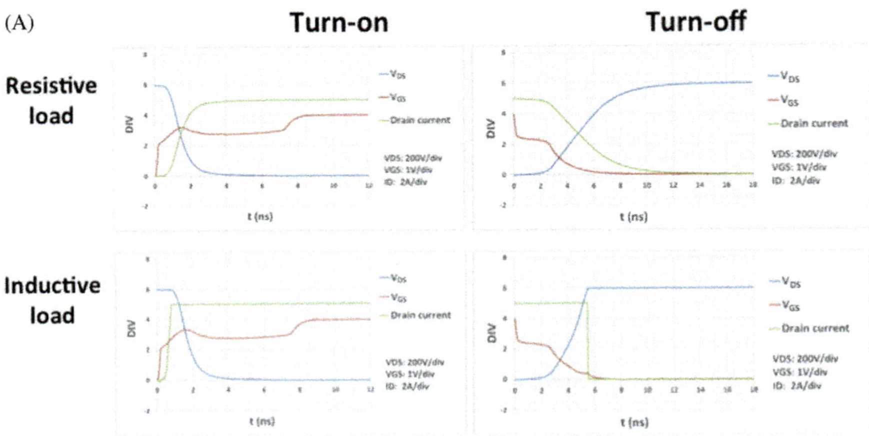

	Load	t_{on} (ns)	t_{off} (ns)	E_{on} (μJ)	E_{off} (μJ)	E_{total} (μJ)
1200 V p-GaN gate HEMT	Resistive	2.4	8.5	4.4	16	20
	Inductive	0.8	5.5	14	16	30

(B)

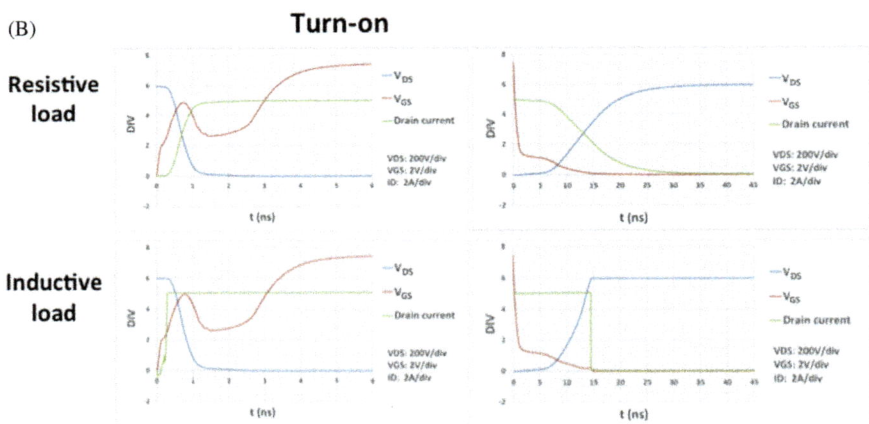

	Load	t_{on} (ns)	t_{off} (ns)	E_{on} (μJ)	E_{off} (μJ)	E_{total} (μJ)
1200 V MOSC HEMT	Resistive	1.0	23	1.9	40	42
	Inductive	0.3	15	6.2	40	46

Figure 5.35 Switching waveforms of 1.2 kV, 10 A GaN (A) lateral HEMT and (B) lateral MOSC-HEMT, and (C) vertical UMOSFET and (D) vertical CAVET under resistive and inductive (L = 120 μH) load conditions and a gate resistance of 0.5 Ω. [98].

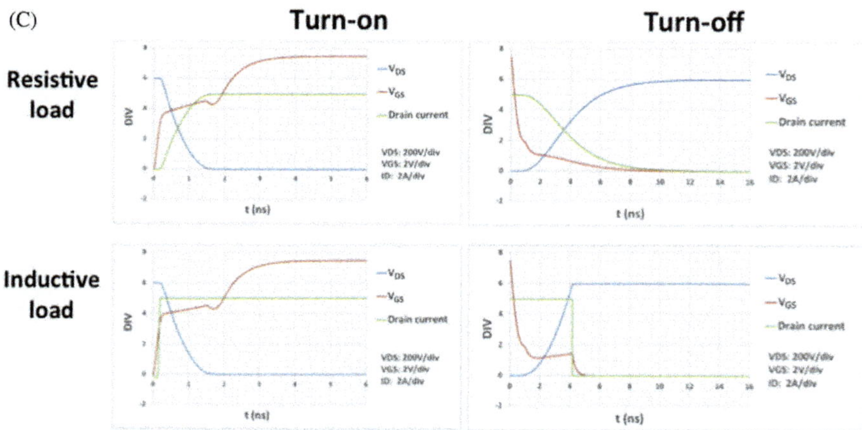

	Load	t_{on} (ns)	t_{off} (ns)	E_{on} (μJ)	E_{off} (μJ)	E_{total} (μJ)
1200 V UMOSFET	Resistive	1.2	7.0	2.3	14	16
	Inductive	0.2	4.2	6.7	14	21

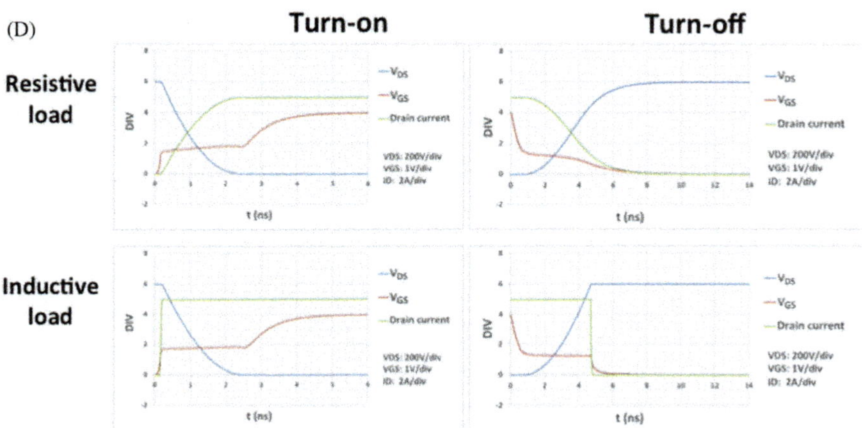

	Load	t_{on} (ns)	t_{off} (ns)	E_{on} (μJ)	E_{off} (μJ)	E_{total} (μJ)
1200 V p-GaN gate CAVET	Resistive	1.7	5.8	3.6	10	14
	Inductive	0.2	4.8	9.9	17	27

Figure 5.35 (Continued)

directly influence the charging/discharging of the carrier trapping centers in the buffer. These charging/discharging rates of these electron and/or hole trap centers are usually slow when compared to the sweep rate of the substrate bias. Consequently, a hysteresis is present in the drain current versus substrate bias plot.

Table 5.9 Comparative static and dynamic performance simulation results for 600 V lateral GaN HEMT and MOSC-HEMT, and vertical GaN UMOSFET and CAVET [98]

	Load	t_{on}	t_{off}	E_{total}	$R_{on}{}^{*}Q_g$	V_{GS} swing	$R_{on}{}^{*}C_{gd}$	$R_{on,\,sp}$	$Q_{g,\,sp}$	$C_{gd,\,sp}$
		(ns)	(ns)	(μJ)	(Ω/nC)	(V)	(Ω/pF)	(mΩ/cm^2)	(μC/cm^2)	(nF/cm^2)
600 V MOSC HEMT	Resistive	0.7	10.5	9.2	1.9	15	30	1.9	1.0	16
	Inductive	0.2	7.5	11.7						
600 V UMOSFET	Resistive	1.0	5.1	6.2	1.1	15	10	0.4	2.7	25
	Inductive	0.2	3.0	8.1						
600 V p-GaN gate HEMT	Resisitve	1.5	2.8	3.9	0.4	4	12	0.6	0.7	20
	Inductive	0.6	2.0	7.1						
600 V p-GaN gate CAVET	Resistive	1.2	2.9	3.9	0.4	4	4.6	0.2	1.9	23
	Inductive	0.1	2.2	7.6						

Table 5.10 Comparative static and dynamic performance simulation results for 1200 V lateral GaN HEMT and MOSC-HEMT, and vertical GaN UMOSFET and CAVET [98]

	Load	t_{on}	t_{off}	E_{total}	R_{on}* Q_g	V_{GS} swing	R_{on} *C_{gd}	$R_{on, sp}$	$Q_{g, sp}$	$C_{gd, sp}$
		(ns)	(ns)	(μJ)	(Ω/nC)	(V)	(Ω/pF)	(mΩ/ cm^2)	(μC/ cm^2)	(nF/ cm^2)
1200 V MOSC HEMT	Resistive	1.0	23	42	2.7	15	41	3.4	0.8	12
	Inductive	0.3	15	46						
1200 V UMOSFET	Resistive	1.2	7.0	16	1.5	15	14	0.6	2.5	24
	Inductive	0.2	4.2	21						
1200 V p-GaN gate HEMT	Resisitve	2.4	8.5	20	0.9	4	21	1.4	0.6	15
	Inductive	0.8	5.5	30						
1200 V p-GaN gate CAVET	Resistive	1.7	5.8	14	0.7	4	6	0.4	1.8	15
	Inductive	0.2	4.8	27						

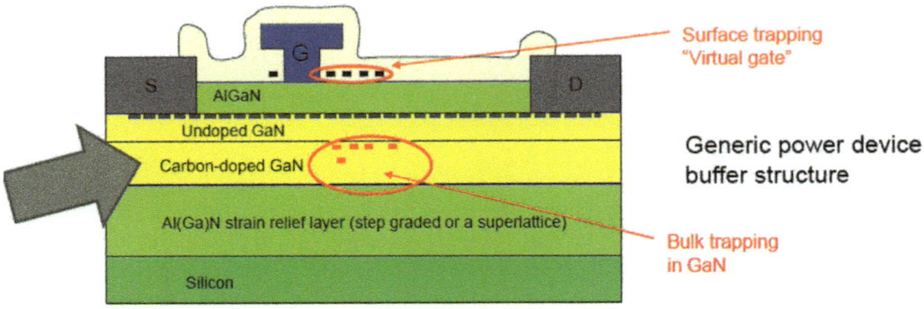

Figure 5.36 Current collapse phenomenon observed in lateral AlGaN/GaN HEMT [100].

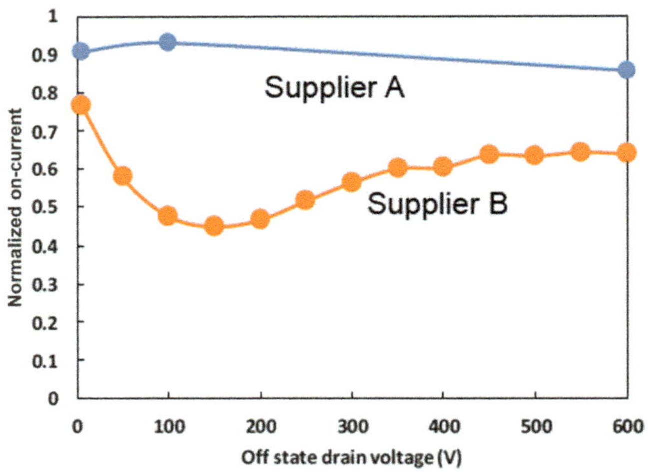

Figure 5.37 Normalized on-current current after the application of a 100 s-long pulsed drain bias, showing different degree of current collapse for two suppliers [100].

This hysteresis can be used to extract the polarity type (positive or negative) and concentration of trapping centers responding to the substrate bias and to comparatively evaluate various buffer structures for optimization. Fig. 5.38 shows RC-equivalent model developed for the buffer and active channel layers and schematic normalized drain conductivity versus substrate voltage plot [100].

While a complete physical understanding of this phenomenon is not quite done at present [100], we can tentatively attribute the underlying reason for it taking place is the attraction and subsequent trapping of electrons from the channel by the bulk positive charges, often holes in the heterojunctions underneath in the buffer layer, when the drain is pulsed to sufficiently high magnitude.

Since the buffer layer is designed and optimized to minimize the mechanical strain experienced by the active channel layer due to the lattice mismatch and thermal conductivity difference between GaN and Si or sapphire substrates, the

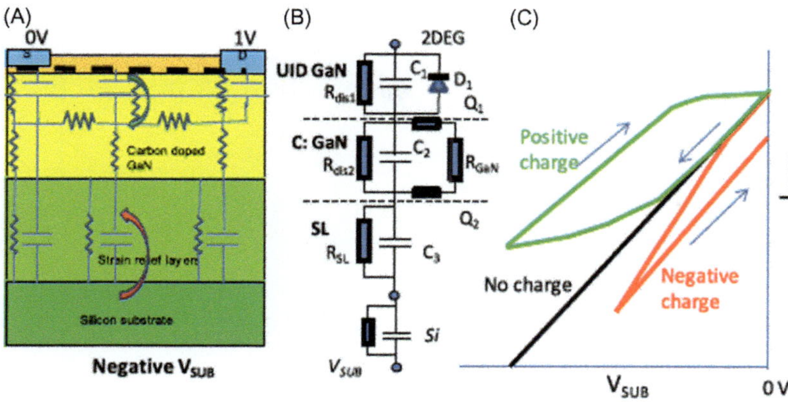

Figure 5.38 (A) Substrate bias ramp measurement configuration, together with (B) one-dimensional lumped element representation. (C) Schematic current versus voltage ramp curves. The *green line* shows leakage current through the UID GaN and the *red line* through the SRL [100].

concurrent minimization of the electrical effects of these charge centers is probably difficult. Instead, device designs, such as implementing a hole injection, a buried p+ layer or light illumination from an integrated light-emitting diode (LED), have been proposed or found to be effective to suppress the current collapse [105,106].

Often in motor inverter drives, the power transistors are expected to survive a short period of time when the load is shorted and before protective accessories can be activated. Hence, we need to determine the robustness and ruggedness of GaN power transistors. The short-circuit survival time (t_{SC}) and the energy dissipated (E_{SC}) are key performance parameters for transistor robustness evaluation. Lateral transistors tend to have a shorter t_{SC} than vertical ones because of current nonuniformities and surface electric field localization. The short-circuit survival performance for 600 V AlGaN/GaN HEMTs with p-GaN gate and MISHEMTs as well as cascoded GaN HEMTs have been experimentally tested [107–112]. The cascoded GaN HEMTs have the shortest t_{SC} at a lower drain bias due to gate loop ringing but even the other GaN HEMTs have t_{SC} less than 10 μs at drain bias of 400 V or less. The dissipated and critical energies for these GaN HEMTs are compared with Si and SiC power transistors as a function of drain voltage in Fig. 5.39 [111]. It can be readily seen that the lateral hybrid and monolithic GaN power HEMTs are inferior to both Si IGBTs and SiC power MOSFETs in terms of short-circuit ruggedness.

5.4.4 Device choices in applications

To comparatively evaluate the lateral and vertical GaN power transistors in applications, we have compared the on-state performance of various lateral and vertical GaN and SiC power transistors in power electronics application requiring devices with blocking voltages from 600 V to 15 kV [113,114]. Figs. 5.40 and 5.41 indicate that lateral GaN HEMTs are competitive at 600 and 1.2 kV and vertical GaN

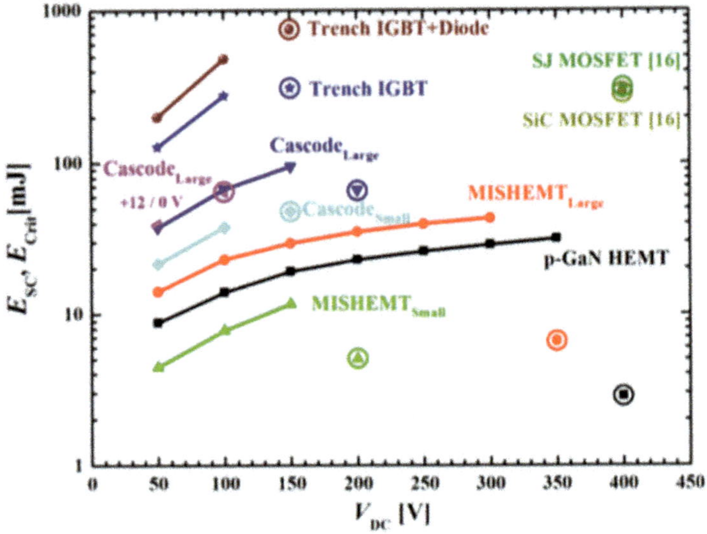

Figure 5.39 Comparison of measured dissipated and critical energies of various lateral hybrid and monolithic GaN power HEMTs, together with those for Si IGBTs and SiC MOSFETs [111].

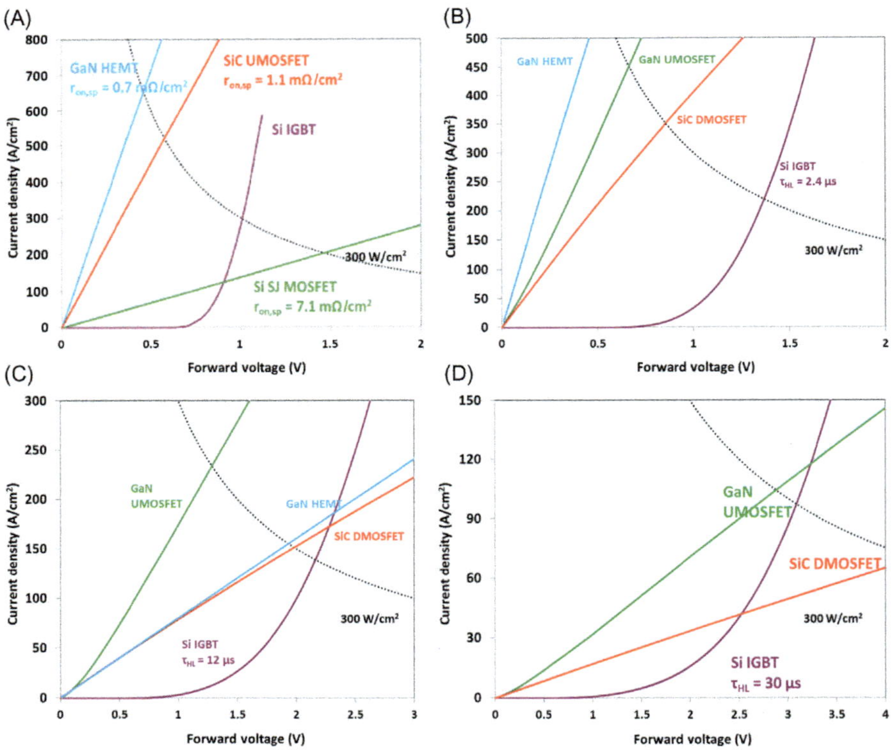

Figure 5.40 Comparison of on-state performance of (A) 600 V, (B) 1.2 kV, (C) 3.3 kV, and (D) 6.5 kV Si, GaN, and SiC power transistors [113,114].

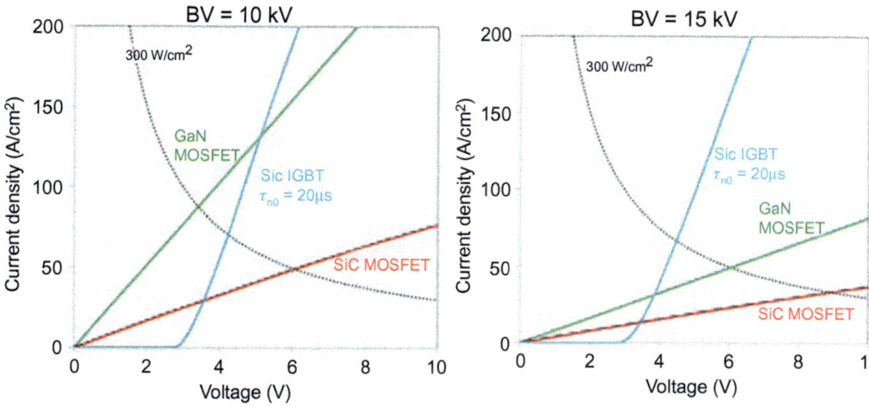

Figure 5.41 Comparison of on-state performance of 10 and 15 kV GaN and SiC power transistors [114].

UMOSFETs are up to at least 6.5 kV. At 10 kV or above, vertical GaN superjunction FETs (not included in the figure) need to be used to compete against vertical SiC IGBTs.

5.5 Commercial device examples

Several commercial GaN power transistors have been commercialized over the last few years [115–124]. However, the state of commercialization is at a state of flux because of the difference in device structures, packages and applications targeted. We can categorize the commercial devices into three classes: discrete transistors, hybrid transistors, and integrated transistors. The discrete transistors are single-chip GaN power transistors whereas the hybrid transistors consist of a high-voltage GaN power transistor copackaged with a low-voltage silicon power transistor, in different cascoded configurations. The integrated transistors are high-voltage GaN transistor monolithically integrated with the gate drive circuitry so as to facilitate gate drive requirements and provide protective features to the power transistor.

5.5.1 Discrete transistors

5.5.1.1 Efficient power conversion

Efficient power conversion (EPC) has been marketing discrete GaN power transistors, initially up to 200 V, but more recently, up to 450 V [115]. It utilizes a p-GaN gate layer over the AlGaN/GaN heterojunction channel so as to shift the threshold voltage sufficiently positive (typically 1.6 V) to provide enhancement mode operation (called JHC-HEMT here, see Fig. 5.9C). The junction gate is not sufficiently forward biased at $V_{GS} = 5$ V so that the maximum gate current is only 0.1 mA

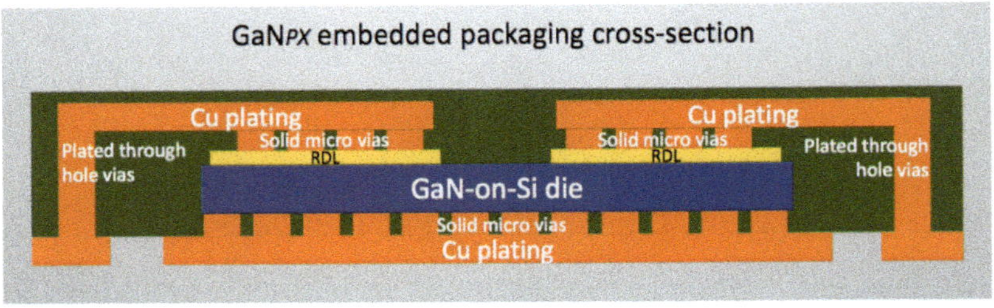

Figure 5.42 Schematic cross-section of a GaN power transistor package [119].

[116] and the transistor essentially operates exclusively in the field-effect transistor mode. The biggest drawback of the EPC transistor is its lack of conventional discrete package since it is only sold in chip form with solder bumps on the front side. Often, when it is assembled onto PCBs, its thermal management during operations is a challenge.

5.5.1.2 Panasonic/Infineon

Panasonic has marketed a normally-off GaN power transistor called the gate injection transistor (GIT) with V_T of 1.2 V [117]. Its structure is very similar to the EPC device but with a p-AlGaN gate. However, in its on state, minority carriers (holes) are injected into the channel from the forward-biased gate. While this local conductivity modulation [118] improves the channel conductivity, it necessitates a substantial gate current. Hence, strictly speaking, this transistor is not a field-effect transistor. Nevertheless, the amount of gate current is small when compared to the drain current [117].

5.5.1.3 GaN systems

In many aspects, the GaN Systems transistor is similar to the EPC device in both structure and operation. At a gate voltage of 6 V, its gate leakage current is only 80 μA. However, it has a higher blocking voltage (650 V) than the EPC transistors and is packaged in its novel proprietary chip-embedding package (Fig. 5.42) [119,120]. Electroplated copper is used as interconnections are used to minimize parasitic inductances as well as backside heat sink [119].

5.5.2 Hybrid transistors

A very popular hybrid approach is to implement a cascoded pair consisting of a high voltage, normally on GaN MOSHC-HEMT and low voltage, normally off silicon power MOSFET, as shown in Fig. 5.14 or Fig. 5.43A [122,123]. The drain of the Si power MOSFET is connected to the source of the GaN HEMT with the gate of GaN HEMT tied to ground. Another cascode approach is to drive the gate of the GaN HEMT independently from the gate of the Si power MOSFET (Fig. 5.43B).

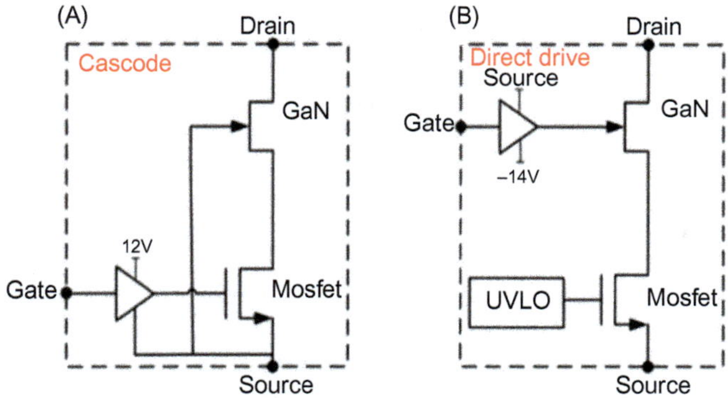

Figure 5.43 Comparison of different cascoded configurations of GaN power transistors [124].

Even though it requires a more complex gate drive circuitry, this configuration allows independent control of both the GaN and Si transistors and may provide more stability than the prior cascoded configuration. This has been called as "direct-drive" configuration by one manufacturer [124,125].

5.5.2.1 Cascoded (International Rectifier (IR)/Infineon)

IR was the first company which commercialized the GaN power transistors and device ratings of 30–600 V were available. The GaN power transistor is a depletion-mode MOSHC-HEMT (Fig. 5.9B) connected in series with a low-voltage Si MOSFET (Fig. 5.14) [121]. Infineon has apparently discontinued the high-voltage version of this product.

5.5.2.2 Cascoded (Transphorm)

Among the GaN power transistor producers, Transphorm has been the most aggressive in manufacturing and commercializing their products in high performance energy systems. The rated blocking voltage is 600 V but it has plenty of voltage margin (up to at least 750 V). To enhance the control between the two stages, a resistor is inserted between the drain end of the silicon MOSFET and the source end of the GaN HEMT. Also, at present, they are apparently the only manufacturer that offers a GaN power transistor half-bridge module of 600 V and 70 A, 30 mΩ (TPD3215M [122]), attempting to challenge SiC MOSFET modules at the lower power end

5.5.2.3 Direct-drive transistors (Texas Instruments)

Texas Instruments recently started to market a direct-drive hybrid GaN power transistor, which stacks a high-voltage GaN HEMT on a low-voltage Si power MOSFET and drives both directly and separately using an Si gate driver IC (see Fig. 5.43 [124]). The main gate drive is for the GaN power HEMT and the auxiliary gate drive for the Si power MOSFET has the undervoltage lockout (UVLO) feature that keeps the Si transistor turned off till it is ready to be deployed. 600 V, 70 mΩ, 12 A GaN power stage is now commercially available [125].

5.5.3 Integrated transistors

5.5.3.1 Navitas

Due to the difficulties of optimizing the threshold voltage, Navitas has decided to integrate all the gate control circuitry and implement it monolithically in an all GaN technology [126]. We can consider this product either as an integrated GaN power transistor or, more accurately, a GaN power IC with a single high-voltage output transistor. High-frequency (27 and 40 MHz) DC/AC converters using Navitas' GaN power IC and air-core inductors with 93%–96% efficiency have been demonstrated [126].

5.6 Monolithic integration

Because all their terminals are on the top surface, lateral power devices, hence the lateral AlGaN/GaN power HEMTs, lend itself to monolithic integration for low to medium power applications. In addition, monolithic integration of GaN for optoelectronics applications is also appealing since it offers cost effectiveness by assembling photonic and electronic devices on a single multifunctional chip. We will discuss recent experimental demonstrations of GaN power and optoelectronic ICs.

5.6.1 Power ICs

Several monolithic power converter ICs have been recently demonstrated [127–131]. Six 700 V lateral GaN GITs were monolithic integrated on Si substrates to drive a 100 W, three-phase motor inverter circuit with peak 93% efficiency [127]. These normally off, p-AlGaN gate GaN power transistors, isolated from each other with iron implantation, also operate in the third quadrant or reverse mode, hence eliminating the need of antiparallel, free-wheeling diodes. More recently, a novel three-phase AC-to-AC matrix converter IC was also demonstrated using GaN GITs operating as bi-directional transistors [128]. A microphotograph of this power converter IC chip is shown in Fig. 5.44, with the schematic cross-section of the bidirectional GaN GIT transistor in Fig. 5.45 [128]. In addition, a remotely controlled gate circuitry powered with microwave was used. The experimental bidirectional, output conduction and the blocking I–V characteristics of this GaN switch are shown in Fig. 5.46 [128]. Also, building blocks of PWM converters in GaN have been demonstrated using HEMTs [129]. Very high frequency (50–100 MHz), monolithic switch-mode power converters have also been implemented using depletion-mode GaN HEMTs on SiC substrates

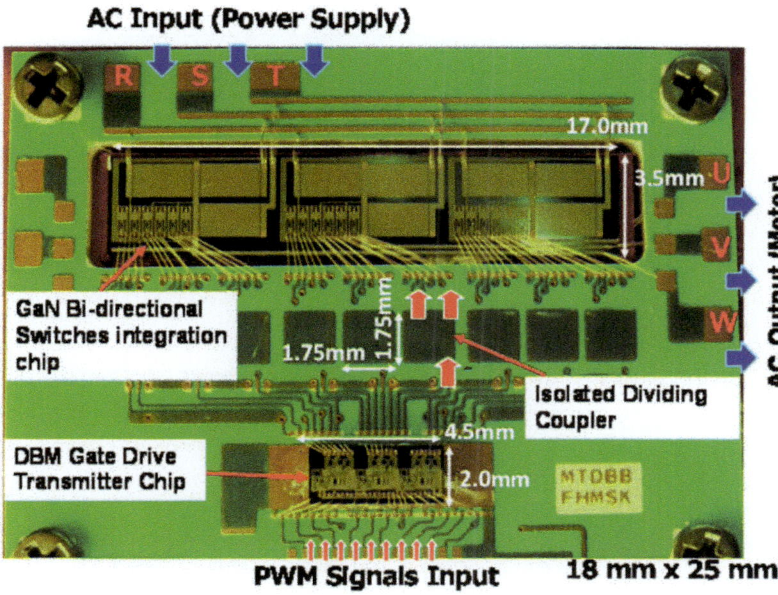

Figure 5.44 Microphotograph of a 3 × 3 matrix converter chip for a 4 kW motor drive implemented with bidirectional GaN GITs and Drive-by-Microwave technology [128].

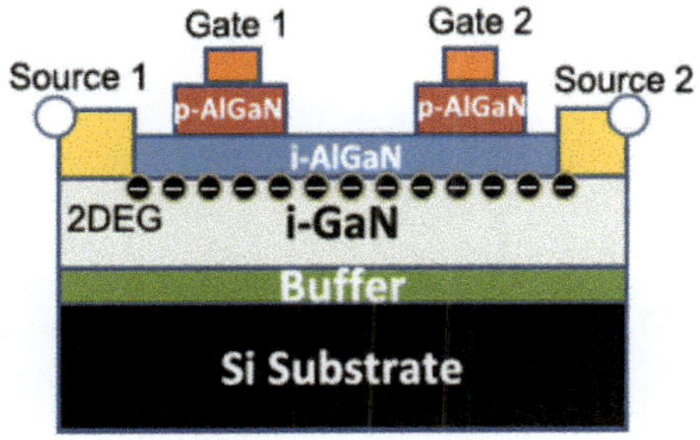

Figure 5.45 Schematic cross-section of a dual-gate, bidirectional power transistor switch implemented with AlGaN/GaN GITs [128].

with 0.15 and 0.25 μm channel lengths [130]. The 0.15 μm channel devices were used to demonstrate a single-phase buck converter with DC input voltage of 25 V and over 90% efficiency at 100 MHz with output power of 7 W. The 0.25 μm process was employed to yield a two-phase buck converter with also input voltage of 25 V and 93% efficiency at 50 MHz with output power of 12.5 W [130]. Three different gate driver topologies were considered to construct 20 V, 5 W, 100 MHz synchronous buck converter prototypes, with the modified active pull-up driver being the best performing [131].

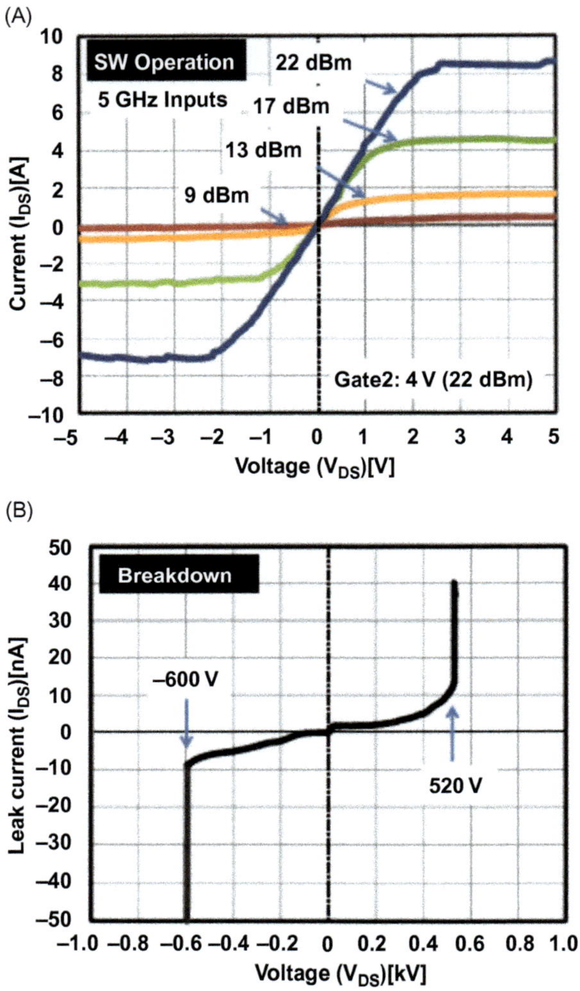

Figure 5.46 (A) Output conduction and (B) blocking *I–V* characteristics of an experimental RF-triggered, bidirectional GaN-GIT power transistor switch [128].

5.6.2 Optoelectronic ICs

There are at least three ways of implementing a GaN optoelectronic IC [132]. These are: selective epi removal, selective epi growth, and 3D integration with wafer bonding. The first GaN monolithic optoelectronic integration was demonstrated by fabrication of a vertical InGaN/GaN MQW LED in series with a lateral, enhancement and depletion mode AlGaN/GaN power MOSC-HEMT using the selective epi removal approach (see Fig. 5.47) [133,134]. Operation up to 225°C was also demonstrated, as shown in Fig. 5.48 [134], showing the superior potential at high switching frequencies and elevated temperature of integrated GaN FETs

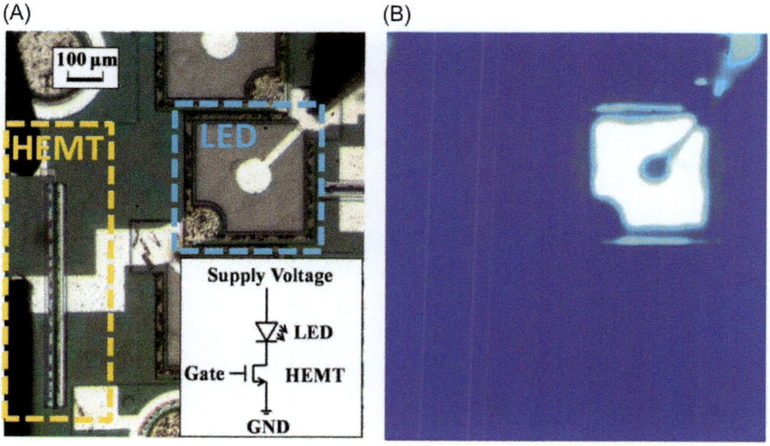

Figure 5.47 Optical images of a monolithically integrated fabricated GaN LED/HEMT pair (A) in off-state, and (B) with the LED lighted up. Inset in (A): Schematic of the circuit configuration [133].

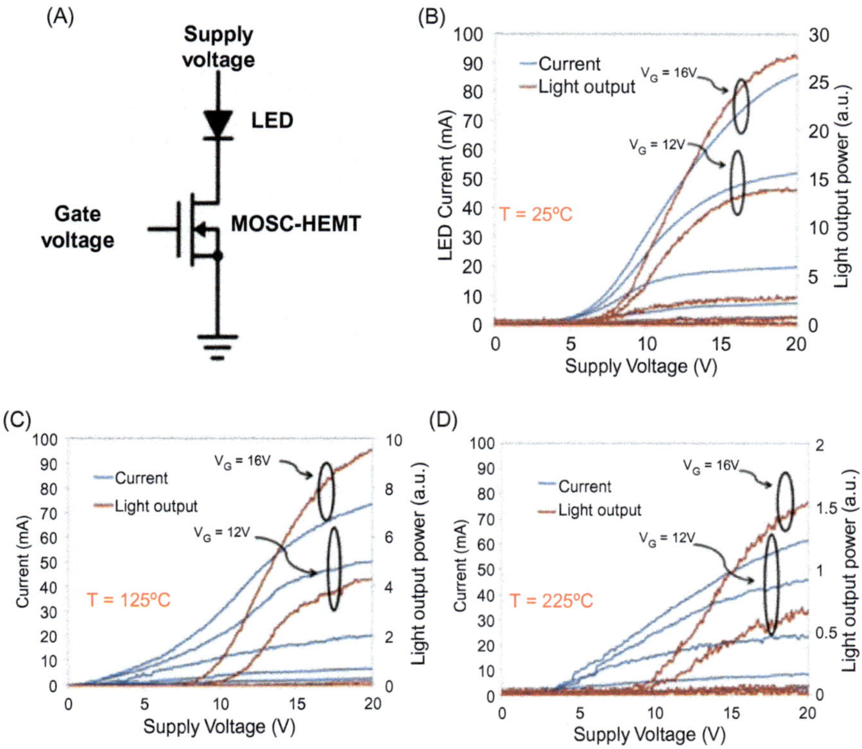

Figure 5.48 (A) Testing circuit configuration, and current and light output power of an integrated LED/MOSC-HEMT as a function of supply voltage and gate voltage at (B) 25°C, (C) 125°C, and (D) 225°C. The gate voltage is varied from 0 to 16 V in steps of 4 V [134].

over conventional silicon LED driver circuits. The first successful selective epi growth efforts for GaN optoelectronic integration [135,136] were done to grow the HEMT over the LED epi using 500 nm−1 μm thick SiO_2 masking layer. However, no on-chip interconnecting metallization between the LED and the HEMT was implemented. Later, the same group demonstrated a metal-interconnection free integration by connecting the cathode of the LED with the 2DEG layer of the adjacent HEMT [137]. Another GaN optoelectronic integration effort [138] was implemented using a lateral normally-on GaN MOSFET driving a GaN LED. The lateral MOSFET was directly fabricated on the n-type GaN epi layer of the LED. More recently, monolithic integration of a vertical, enhancement mode MOSFET with a vertical LED was also demonstrated [139,140]. The vertical GaN MOSFET was made on selectively grown p- and n-type epi over the LED epi using SiO_2 as mask.

5.7 Future trend, possibilities, and challenges

At present, the GaN power device technology is still at a stage of flux. Despite lateral GaN power HEMTs have been commercialized, large-scale deployment or utilization has not taken place because of the concerns for long-term reliability and robustness under harsh conditions as well as cost-effectiveness. In addition, the performance of these commercial GaN power HEMTs are not direct replacement of existing silicon power devices. For example, to replace silicon IGBT, a threshold voltage larger than 4 V is needed and monolithic GaN HEMTs have V_T. Cascoded GaN transistors can match the threshold voltage requirements but its robustness, such as short-circuit survival time, is inferior to the silicon IGBT at this time. Furthermore, being a unipolar transistor, GaN power HEMTs have more limited surge current capabilities than the bipolar silicon IGBT. Nevertheless, lateral GaN power HEMTs can switch at higher frequencies and are easier to integrate and have less interconnecting parasitics. These advantages tend to favor them toward lower power applications.

Vertical GaN power transistors on native GaN or silicon substrates are under active development to address the higher power level capabilities. However, in this area, GaN transistors are competing with the emerging but rapidly maturing SiC power devices, particularly power MOSFETs. The present epi and foundry infrastructures in SiC power are building up very fast. 4H-SiC substrates up to 200 mm diameter have already been demonstrated by two commercial companies while GaN substrates are still at most 100 mm in diameter. From the critical electric field comparisons, the performance advantages of vertical GaN power devices are only apparent for blocking voltages over 3 kV. For these reasons, it would be difficult for vertical GaN power devices to directly challenge the SiC counterparts in the near future. Monolithic integration in gallium nitride (GaN) has several advantages over SiC. Among these are the lateral HEMT structures, insulating buffer layer providing dc electrical isolation from the silicon substrate and possibility of integrating high-frequency electronic and photonic devices/elements. Hence, it is expected that power electronic and optoelectronic-integrated circuits will be continually explored and possibly commercialized to address a large variety of energy efficient, low to medium power electronics as well as photonic systems.

Acknowledgments

This work was supported in part by the Engineering Research Centers Program (ERC) of the National Science Foundation under NSF Cooperative Agreement No. EEC- 0812056 and by New York State under Contract No. C160145.

References

[1] S.K. Ghandhi, Semiconductor Power Devices, Wiley, 1977.
[2] B.J. Baliga, Fundamentals of Power Semiconductor Devices, Springer-Verlag, 2008.
[3] T.P. Chow, R. Tyagi, Wide bandgap compound semiconductors for superior high-voltage unipolar power devices, IEEE Trans. Electron Devices 41 (8) (1994) 1481–1483.
[4] U.K. Mishra, P. Parikh, Y.-F. Wu, AlGaN/GaN HEMTs-an overview of device operation and applications, Proc. IEEE 90 (6) (2002) 1022–1031.
[5] T.P. Chow, Z. Li, Recent advances in high-voltage GaN MOS-gated transistors for power electronics applications. Chap. 8 in: S. Pearton (Ed.), GaN and ZnO-based Materials and Devices, Springer, 2012, pp. 239–250.
[6] B.J. Baliga, Gallium nitride devices for power electronic applications, Semicond. Sci. Technol. 28 (7) (2013) 074011.
[7] D. Ueda, Renovation of power devices by GaN-based materials, IEEE IEDM, Tech. Dig (2015) 422–425.
[8] T.P. Chow, I. Omura, M. Higashiwaki, H. Kawarada, V. Pala, Smart power devices and ICs using GaAs, wide and extreme bandgap semiconductors, IEEE Trans. Electron Devices 64 (3) (2017) 856–873.
[9] O. Ambacher, J. Smart, J.R. Shealy, N.G. Weimann, K. Chu, M. Murphy, et al., Two-dimensional electron gases induced by spontaneous and piezoelectric polarization charges in N- and Ga-face AlGaN heterostructures, J. Appl. Phys. 85 (6) (1999) 3222–3233.
[10] A. Dadger, F. Schulze, M. Wienecke, A. Gadanecz, J. Blasing, P. Veit, et al., Epitaxy of GaN on silicon – Impact of symmetry and surface reconstruction, New J. Phys. 9 (2007) 389–398.
[11] K.-L. Lin, E.Y. Chang, Y.-L. Hsiao, W.-C. Huang, Growth of GaN films on 150 mm Si (111) using multi-layer AlN/AlGaN buffer by metal-organic vapor phase epitaxy method, Appl. Phys. Lett. 91 (2007) 222111.
[12] B.N. Feigelson, T.J. Anderson, M. Abraham, J.A. Freitas, J.K. Hite, C.R. Eddy, et al., Multicycle rapid thermal annealing technique and its application for the electrical activation of Mg implanted in GaN, J. Crystal Growth 350 (2012) 21–26.
[13] T.J. Anderson, B.N. Feigelson, F.J. Kub, M.J. Tadjer, K.D. Hobart, M.A. Mastro, et al., Activation of Mg implanted in GaN by multicycle rapid thermal annealing, Electron. Lett. 50 (3) (2014) 197–198.

[14] M. Silvestri, M.J. Uren, D. Marcon, M. Kuball, GaN buffer design: Electrical characterization and prediction of the effect of deep level centers in GaN/AlGaN HEMTs, Proceedings of the CS Mantech Conference, pp. 195–197, 2013.

[15] M. Meneghini, D. Bisi, I. Rossetto, C. De Santi, A. Stocco, O. Hilt, et al., Trapping processes related to iron and carbon doping in AlGaN/GaN power HEMTs, in: J.-I. Chyi, H. Fujioka, H. Morkoç (Eds.) (Eds.), Proceedings of SPIE 9363, Gallium Nitride Materials and Devices X, 2015, p. 936314.

[16] http://www.iue.tuwien.ac.at/phd/vitanov/node65.html.

[17] S. Vitanov, V. Palankovski, S. Maroldt, R. Quay, High-temperature modeling of AlGaN/GaN HEMTs, Solid-State Electron. 54 (2010) 1105–1112.

[18] J. Robertson, B. Falabretti, Band offsets of high K gate oxides on III–V semiconductors, J. Appl. Phys. 100 (1) (2006) 014111.

[19] P. Jain, E.J. Rymazewski, Embedded thin film capacitors—Theoretical limits, IEEE Trans. Adv. Packag. 25 (3) (2002) 454–458.

[20] T.E. Cook, C.C. Fulton, W.J. Mecouch, R.F. Davis, G. Lucovsky, R.J. Nemanich, Band offset measurements of the Si_3N_4/GaN (0001) interface, J. Appl. Phys. 94 (2003) 3949–3954.

[21] Z. Zhang, C.M. Jackson, A.R. Arehart, B. McSkimming, J.S. Speck, S.A. Ringel, Direct determination of energy band alignments of Ni/Al_2O_3/GaN MOS structures using internal photoemission spectroscopy, J. Electron. Mater. 43 (4) (2014) 828–832.

[22] Z. Yatabe, J.T. Asubar, T. Hashizume, Insulated gate and surface passivation structures for GaN-based power transistors, J. Phys. D. Appl. Phys. 49 (2016) 393001.

[23] W. Huang, T. Khan, T.P. Chow, Comparison of MOS Capacitors on n- and p-type GaN, J. Electron. Mater 35 (4) (2006) 726–732.

[24] W. Huang, T. Khan, T.P. Chow, Enhancement-mode n-channel GaN MOSFETs on p and n-GaN/sapphire substrates, IEEE Electron Device Lett. 27 (10) (2006) 796–798.

[25] K. Tang, T.P. Chow, Performance of MOSFETs on reactive-ion-etched GaN surfaces, Int. J. High Speed Electron. Syst. 19 (1) (2009) 121–127.

[26] S. Takashima, Z. Li, T.P. Chow, Metal-oxide-semiconductor interface and dielectric property of atomic layer deposited SiO_2 on GaN, Jpn. J. Appl. Phys. 52 (8S) (2013).

[27] Z. Guo, K. Tang, T.P. Chow, Temperature dependence of GaN MOS capacitor characteristics, Phys. Status Solidi C 13 (2016) 336–340. 5–6.

[28] K. Matocha, T.P. Chow, R.J. Gutmann, Positive flatband voltage shift in MOS capacitors on n-type GaN, IEEE Electron Device Lett. 23 (2) (2002) 79–81.

[29] K. Matocha, V. Tilak, G. Dunne, Comparison of metal-oxide-semiconductor capacitors on c- and m-plane gallium nitride, Appl. Phys. Lett. 90 (12) (2007) 123511.

[30] V. Tilak, K. Matocha, G. Dunne, Fabrication and characterization of m-plane (1-100) GaN based metal-oxide-semiconductor capacitors, Phys. Status Solidi C 5 (6) (2008) 2019–2021.

[31] M. Esposto, S. Krishnamoorthy, D.N. Nath, S. Bajaj, T.-H. Hung, S. Rajan, Electrical properties of atomic layer deposited aluminum oxide on gallium nitride, Appl. Phys. Lett. 99 (2011) 133503.

[32] T.-H. Hung, S. Krishnamoorthy, M. Esposto, D.N. Nath, P.S. Park, S. Rajan, Interface charge engineering at atomic layer deposited dielectric/III-nitride interface, Appl. Phys. Lett. 102 (7) (2013) 072105.

[33] S. Kaneki, J. Ohira, S. Toiya, Z. Yatabe, J.T. Asubar, T. Hashizume, Highly-stable and low-state density Al_2O_3/GaN interfaces using epitaxial n-GaN layers on free-standing GaN substrates, Appl. Phys. Lett. 109 (2016) 162104.

[34] D. Wei, T. Hossain, N. Nepal, N.Y. Garces, J.K. Hite, H.K. Meyer III, et al., Comparison of the physical, chemical and electrical properties of ALD Al_2O_3 on *c*- and *m*-plane GaN, Phys. Status Solidi C 11 (2014) 898−901. 3−4.
[35] K. Cheng, M. Leys, J. Derluyn, S. Degroote, D.P. Xiao, A. Lorenz, et al., AlGaN/GaN HEMT grown on large size silicon substrates by MOVPE capped with in-situ deposited Si_3N_4, J. Cryst. Growth 298 (2007) 822−825.
[36] Z. Tang, Q. Jiang, Y. Lu, S. Huang, S. Yang, X. Tang, et al., 600-V normally off SiN_x/AlGaN/GaN MIS-HEMT with large gate swing and low current collapse, IEEE Electron Device Lett. 34 (11) (2013) 1373−1375.
[37] J. He, M. Hua, G. Tang, Z. Zhang, K.J. Chen, Comparison of E-mode fully-recessed GaN MIS-FETs and partially-recessed MIS-HEMTs with PECVD-SiNx/LPCVD-SiNx Gate Stack, Proceedimgs of the International Conference on Nitride Semiconductors (ICNS), July 2017.
[38] W. Huang, Z. Li, T.P. Chow, Y. Niiyama, T. Nomura, S. Yoshida, Enhancement-mode GaN hybrid MOS-HEMTs with $R_{on,\ sp}$ of 20 mΩ/cm^2, Proceedings of the International Symp. Power Semicond. and ICs, pp. 295−298, 2008.
[39] Z. Li, J. Waldron, R. Dayal, L. Parsa, M. Hella, T.P. Chow, High voltage normally-off GaN MOSC-HEMTs on silicon substrates for power switching applications, Proceedings of the International Symp. Power Semicond. and ICs, pp. 45−48, 2012.
[40] Z. Li, T.P. Chow, Channel scaling of hybrid GaN MOS-HEMTs, Solid-State Electron. 56 (1) (2011) 111−115.
[41] J.A. Appels, H.M.J. Vaes, High voltage thin layer devices (RESURF devices), IEEE IEDM, Tech. Dig. (1979) 238−241.
[42] H. Vaes, J. Appels, High voltage, high current lateral devices, IEEE IEDM, Tech. Dig (1980) 87−90.
[43] E.J. Wildi, P.V. Gray, T.P. Chow, H.R. Chang, M. Cornell, Modeling and process implementation of implanted RESURF type devices, IEEE IEDM, Tech. Dig (1982) 268−271.
[44] S. Merchant, E. Arnold, H. Baumgart, S. Mukherjee, H. Pein, R. Pinker, Dependence of breakdown voltage on drift length and buried oxide thickness in SOI RESURF LDMOS transistors, Proceedings of the International Symposium on Power Semiconductor Devices and ICs, pp. 31−35, 1991.
[45] K. Matocha, T.P. Chow, R.J. Gutmann, High-voltage normally off MOSFETs on sapphire substrates, IEEE Trans. Electron. Devices 52 (1) (2005) 1−6.
[46] W. Huang, T.P. Chow, Y. Niiyama, T. Nomura, S. Yoshida, Lateral implanted RESURF GaN MOSFETs with BV up to 2.5 kV, Proceedings of the International Symposium on Power Semiconductor Devices and ICs, pp. 291−294, 2008.
[47] W. Huang, T.P. Chow, Y. Niiyama, T. Nomura, S. Yoshida, Experimental demonstration of novel high-voltage epilayer RESURF GaN MOSFET, IEEE Electron Device Lett. 30 (10) (2009) 1018−1020.
[48] Y. Niiyama, S. Ootomo, J. Li, T. Nomura, S. Kato, T.P. Chow, Normally off operation GaN-based MOSFETs for power electronics applications, Semicond. Sci. Technol. 25 (12) (2010) 1125006.
[49] N.-Q. Zhang, B. Moran, S.P. DenBaars, U.K. Mishra, X.W. Wang, T.P. Ma, Effects of surface traps on breakdown voltage and switching speed of GaN power switching HEMTs, IEEE IEDM, Tech. Dig (2001) 589−592.
[50] A. Nakajima, K. Adachi, M. Shimizu, H. Okumura, Improvement of unipolar power device performance using a polarization junction, Appl. Phys. Lett. 89 (19) (2006) 193501.

[51] Z. Li, T.P. Chow, Drift region optimization in high-voltage GaN MOS-Gated HEMTs, Phys. Status Solidi C 8 (7–8) (2011) 2436–2438.
[52] H. Ishida, S. Shibata, H. Matsuo, M. Yanagihara, Y. Uemoto, T. Ueda, et al., GaN-based super junction diodes with multi-channel structure, IEEE IEDM, Tech. Dig (2009) 1–4.
[53] Z. Guo, T.P. Chow, Performance evaluation of channel length downscaling of various high voltage AlGaN/GaN HEMTs, Phys. Status Solidi A 212 (5) (2015) 1137–1144.
[54] K. Ohi, T. Hashizume, Drain current stability and controllability of threshold voltage and subthreshold current in multi-mesa-channel AlGaN/GaN high electron mobility transistor, Jpn. J. Appl. Phys. 48 (8) (2009) 081002.
[55] B. Lu, E. Matioli, T. Palacios, Tri-gate normally-off GaN MISFET, IEEE Electron Device Lett. 33 (3) (2012) 360–362.
[56] S. Takashima, Z. Li, T.P. Chow, Sidewall dominated characteristics on Fin-Gate AlGaN/GaN MOS-Channel-HEMTs, IEEE Trans. Electron Devices 60 (10) (2017) 3025–3031.
[57] Y. Wu, M. Jacob-Mitos, M.L. Moore, S. Heikman, A 97.8% efficient GaN HEMT boost converter with 300-W output power at 1 MHz, IEEE Electron Device Lett. 29 (8) (2008) 824–1826.
[58] P. Parikh, Cascode gallium nitride HEMTs on silicon: Structure, performance, manufacturing, and reliability, in: M. Meneghini, G. Meneghesso, E. Zanoni (Eds.), Power GaN Devices. Power Electronics and Power Systems, Springer, 2017.
[59] R. Siemieniec, G. Nobauer, D. Domes, Stability and performance analysis of a SiC-based cascode switch and an alternative solution, Microelectron. Reliab. 52 (2012) 509–518.
[60] M. Kanechika, M. Sugimoto, N. Soejima, H. Ueda, O. Ishiguro, A vertical insulated gate AlGaN/GaN heterojunction field-effect transistor, Jpn. J. Appl. Phys. 46 (2007) L503–L505. part 2, 20–24.
[61] H. Otake, K. Chimakatsu, A. Yamaguchi, T. Fujishima, H. Ohta, Vertical GaN-based trench gate metal oxide semiconductor field-effect transistors on GaN bulk substrates, Appl. Phys. Exp. 1 (1) (2008) 011105.
[62] M. Kodama, M. Sugimoto, E. Hayashi, N. Soejima, O. Ishiguro, M. Kanechika, et al., GaN-based trench gate metal oxide semiconductor field-effect transistor fabricated with novel wet etching, Appl. Phys. Express 1 (2008) 21104.
[63] T. Oka, Y. Ueno, T. Ina, K. Hasegawa, Vertical GaN-based trench metal oxide semiconductor field-effect transistors on a free-standing GaN substrate with blocking voltage of 1.6 kV, Appl. Phys. Exp. 7 (2014) 021002.
[64] T. Oka, T. Ina, Y. Ueno, J. Nishii, 1.8 mΩ/cm^2 vertical GaN-based trench metal–oxide–semiconductor field-effect transistors on a free-standing GaN substrate for 1.2-kV-class operation, Appl. Phys. Exp 8 (5) (2015) 054101.
[65] T. Oka, T. Ina, Y. Ueno, J. Nishii, Over 10 A operation with switching characeristics of 1.2-kV-class vertical GaN MOSFETs on a bulk GaN substrate, Proceedings of the International Symposium on Power Semiconductor Devices and ICs, pp. 459–462, 2016.
[66] C. Gupta, S.H. Chan, C. Lund, A. Agarwal, O.S. Koksaldi, J. Liu, et al., Comparing electrical performance of GaN trench-gate MOSFETs with a-plane (11-20) and m-plane (1100) sidewall channels, Appl. Phys. Exp 9 (2016) 121001.
[67] R. Li, Y. Cao, M. Chen, R. Chu, 600V/1.7 Ω normally-off GaN vertical trench metal-oxide-semiconductor field-effect transistor, IEEE Electron Device Lett. 37 (11) (2016) 1466–1469.

[68] C. Gupta, C. Lund, S.H. Chan, A. Agarwal, J. Liu, Y. Enatsu, et al., In situ oxide, GaN interlayer-based vertical trench MOSFET (OG-FET) on bulk GaN substrates, IEEE Electron Device Lett. 38 (3) (2017) 353–355.

[69] T. Syau, P. Venkatraman, B.J. Baliga, Comparison of ultralow specific on-resistance UMOSFET structures: The ACCUFET, EXTFET, INVFET, and conventional UMOSFETs, IEEE Trans. Electron Devices 41 (5) (1994) 800–808.

[70] R. Singh, D.C. Capell, M.K. Das, L.A. Lipkin, J.W. Palmour, Development of high-current 4H–SiC ACCUFET, IEEE Trans. Electron Devices 50 (2) (2003) 471–478.

[71] J.-S. Lee, D.-H. Chun, J.-H. Park, Y.-K. Jung, E.G. Kang, M.Y. Sung, Design of a novel SiC MOSFET structure for EV inverter efficiency improvement, Proceedings of the International Symposium on Power Semiconductor and ICs, pp. 281–284, 2014.

[72] M. Sun, M. Pan, X. Gao, T. Palacios, Vertical GaN power FET on bulk GaN substrate, Proceedings of the 74th Annual Device Research Conference—Conference Digest, pp. 1–2, 2006.

[73] M. Sun, Y. Zhang, X. Gao, T. Palacios, High-performance GaN vertical fin power transistors on bulk GaN substrates, IEEE Electron Device Lett. 38 (4) (2017) 509–512.

[74] F. Yu, D. Rümmler, J. Hartmann, L. Caccamo, T. Schimpke, M. Strassburg, et al., Vertical architecture for enhancement mode power transistors based on GaN nanowires, Appl. Phys. Lett. 108 (2016) 213503.

[75] Z. Hu, W. Li, K. Nomoto, M. Zhu, X. Gao, M. Pilla, et al., GaN vertical nanowire and fin power MISFETs, Device Research Conference—Conference Digest DRC, vol. 35, no. 2014, pp. 2016–2017, 2017.

[76] I. Ben-Yaacov, Y.K. Seck, S. Heikman, S.P. DenBaars, U.K. Mishra, AlGaN/GaN current aperture vertical electron transistors, Device Research Conference, Conference Digest, pp. 31–32, 2002.

[77] I. Ben-Yaacov, Y.-K. Seck, U.K. Mishra, S.P. DenBaars, AlGaN/GaN current aperture vertical electron transistors with regrown channels, J. Appl. Phys. 95 (4) (2004) 2073–2078.

[78] S. Chowdhury, B.L. Swenson, U.K. Mishra, Enhancement and depletion mode AlGaN/GaN CAVET with Mg-ion-implanted GaN as current blocking layer, IEEE Electron Device Lett. 29 (6) (2008) 543–545.

[79] S. Chowdhury, M.H. Wong, B.L. Swenson, U.K. Mishra, CAVET on bulk GaN substrates achieved with MBE-regrown AlGaN/GaN layers to suppress dispersion, IEEE Electron Device Lett. 33 (1) (2012) 41–43.

[80] S. Chowdhury, U.K. Mishra, Lateral and vertical transistors using AlGaN/GaN heterostructure, IEEE Trans. Electron Devices 60 (10) (2013) 3060–3066.

[81] H. Nie, Q. Diduck, B. Alvarez, A.P. Edwards, B.M. Kayes, M. Zhang, et al., 1.5-kV and 2.2-mΩ/cm^2 vertical GaN transistors on bulk-GaN substrates, IEEE Electron Device Lett. 35 (9) (2014) 939–941.

[82] D. Ji, M.A. Laurent, A. Agarwal, W. Li, S. Mandal, S. Keller, et al., Normally-off trench CAVET with active Mg-doped GaN as current blocking layer, IEEE Trans. Electron Devices 64 (3) (2017) 805–808.

[83] Z. Li, K. Tang, T.P. Chow, M. Sugimoto, T. Uesugi, T. Kachi, Design and simulations of novel enhancement-mode high-voltage GaN vertical hybrid MOS-HEMTs, Phys. Status Solidi C 7 (7–8) (2010) 1944–1948.

[84] M. Okada, Y. Saitoh, M. Yokoyama, K. Nakata, S. Yaegassi, K. Katayama, et al., Novel vertical heterojunction field- effect transistors with re-grown AlGaN/GaN two-dimensional electron gas channels on GaN substrates, Appl. Phys. Exp 3 (5) (2010) 054201.

[85] D. Shibata, R. Kajitani, M. Ogawa, K. Tanaka, S. Tamura, T. Hatsuda, et al., 1.7kV/1.0 mΩ/cm^2 normally-off vertical GaN transistor on GaN substrate with regrown p-GaN/AlGaN/GaN semipolar gate structure, IEEE IEDM, Tech. Dig (2016) 248–251.

[86] T. Fujihira, Theory of semiconductor superjunction devices, Jpn. J. Appl. Phys. 36 (10) (1997) 6254–6262. part 1.

[87] Z. Li, T.P. Chow, Design and simulation of 5–20 kV GaN enhancement-mode vertical superjunction HEMT, IEEE Trans. Electron Devices 60 (10) (2013) 3230–3237.

[88] W. Saito, Y. Takada, M. Kuraguchi, K. Tsuda, I. Omura, T. Ogura, Design and demonstration of high breakdown voltage GaN high electron mobility transistor (HEMT) using field plate structure for power electronics applications, Jpn. J. Appl. Phys. 43 (4B) (2004) 2239–2242.

[89] Z. Li, High-Voltage Gallium Nitride MOS-Channel HEMTs, PhD Thesis, Rensselaer Polytechnic Institute, 2013.

[90] M. Van Hove, X. Kang, S. Stoffels, D. Wellekens, N. Ronchi, R. Venegas, et al., Fabrication and performance of Au-free AlGaN/GaN-on-silicon power devices with Al_2O_3 and Si_3N_4/Al_2O_3 gate dielectrics, IEEE Trans. Electron Devices 60, 10, pp. 3071–3078.

[91] R. Zingg, On the specific on-resistance of high-voltage and power devices, IEEE Trans. Electron Devices 51 (3) (2004) 492–499.

[92] P. Wessels, Smart power technology on SOI, 6th International Electrostatic Discharge Workshop (IEW), Oud-Turnhout, Belgium, 2012.

[93] W. Saito, I. Omura, T. Ogura, H. Ohashi, Theoretical limit estimation of lateral wide band-gap semiconductor power-switching device, Solid-State Electron. 48 (2004) 1555–1562.

[94] N. Ikeda, S. Kaya, J. Li, Y. Sato, S. Kato, S. Yoshida, High power AlGaN/GaN HFET with a high breakdown voltage of over 1.8 kV on 4 inch Si substrates and the suppression of current collapse, Proceedings on the 20th International Symposium on Power Semiconductor Devices and Ics, pp. 287–290, 2008.

[95] D.A. Grant, J. Gowar, Power MOSFETs: Theory and Applications, Wiley-Interscience, 1989.

[96] G. Lakkas, MOSFET power losses and how they affect power supply efficiency, Analog Appl. J. 1Q (2016) 22–26.

[97] https://toshiba.semicon-storage.com/info/docget.jsp?did = 13415.

[98] Z. Guo and T.P. Chow, International Workshop on Nitride Semiconductors (IWN2018), 2018.

[99] Z. Guo, C. Hitchcock, T.P. Chow, Lossless switching projection of GaN power field-effect transistors, Phys. Status Solidi A 13 (8) (2017) 1600820.

[100] M.J. Uren, S. Karboyan, I. Chatterjee, A. Pooth, P. Moens, A. Banerjee, et al., Leaky dielectric model for the suppression of dynamic R_{ON} in AlGaN/GaN HEMTs, IEEE Trans. Electron Devices 64 (7) (2017) 2826–2834.

[101] Y. Uemoto, D. Shibata, M. Yanagihara, H. Ishida, H. Matsuo, S. Nagai, et al., 8300V blocking voltage AlGaN/GaN power HFET with thick poly-AlN passivation, IEEE IEDM, Tech. Dig. (2007) 861–864.

[102] N. Tsurumi, H. Ueno, T. Murata, H. Ishida, Y. Uemoto, T. Ueda, et al., AlN passivation over AlGaN/GaN HFETs for surface heat spreading, IEEE Trans. Electron Devices 57 (5) (2010) 980–984.

[103] S. Huang, Q. Jiang, S. Yang, C. Zhou, K.J. Chen, Effective passivation of AlGaN/GaN HEMTs with ALD-grown AlN thin film, IEEE Electron Device Lett. 33 (4) (2012) 516–518.

[104] Z. Li, T. Marron, H. Naik, W. Huang, T.P. Chow, Experimental study on current collapse of GaN MOSFETs, HEMTs and MOS-HEMTs, Proceedings on the International Symposium on Power Semiconductor Devices and ICs, pp. 225−228, 2010.

[105] S. Kaneko, M. Kuroda, M. Yanagihara, A. Ikoshi, H. Okita, T. Morita, et al., Current-collapse-free operations up to 850V by GaN-GIT utilizing hole injection from drain, Proceedings on the International Symposium on Power Semiconductor Devices and ICs, pp. 41−44, 2015.

[106] X. Tang, B. Li, H. Wang, J. Wei, G. Tang, Z. Zhang, et al., Impact of integrated photonic-Ohmic drain on static and dynamic characteristics of GaN-on-Si heterojunction power transistors, Proceedings of the International Symposium on Power Semiconductor Devices and ICs, pp. 31−34, 2016.

[107] X. Huang, D.Y. Lee, V. Bondarenko, A. Baker, D.C. Sheridan, A.Q. Huang, et al., Experimental study of 650V AlGaN/GaN HEMT short-circuit safe operating area (SCSOA), Proceedings of the International Symposium on Power Semiconductor Devices and ICs, pp. 273−276, 2014.

[108] T. Nagahisa, H. Ichijoh, T. Suzuki, A. Yudin, A.O. Adan, M. Kubo, Robust 600 V GaN high electron mobility transistor technology on GaN-on-Si with 400 V, 5 μs load-short-circuit withstand capability, Jpn. J. Appl. Phys. 55 (4) (2016), pp. 04EG01-1−04EG01-11.

[109] M. Fernandez, X. Perpina, J. Roig, M. Vellvehi, F. Bauwens, X. Jorda, et al., P-GaN HEMTs drain and gate current analysis under short circuit, IEEE Electron Device Lett 38 (4) (2017) 505−508.

[110] M. Fernandez, X. Perpina, M. Vellvehi, X. Jorda, J. Roig, F. Bauwens, et al., Short-circuit capability in p-GaN HEMTs and GaN MISHEMTs, Proceedings of the International Symposium on Power Semiconductor Devices and ICs, pp. 455−458, 2017.

[111] M. Fernandez, X. Perpina, J. Roig, M. Vellvehi, F. Bauwens, M. Tack, X. Jorda, Short-circuit study in medium voltage GaN cascodes, p-GaN HEMTs and MISHEMTs, IEEE Trans. Industrial Electronics 64 (11) (2017) 9012−9022.

[112] Z. Guo, C. Hitchcock, T.P. Chow, Second breakdown and robustness of vertical and lateral GaN power field-effect transistors, Phys. Status Solidi A 13 (8) (2017). 1600822-1-7.

[113] S. Chowdhury, Z. Stum, Z. Li, K. Ueno, T.P. Chow, Comparison of 600V Si, SiC and GaN power devices, Mater. Sci. Forum 778 (2014) 771−774.

[114] S. Chowdhury, Z. Guo, X. Liu, T.P. Chow, Comparison of silicon, SiC and GaN power transistor technologies with breakdown voltage rating from 1.2 kV to 15 kV, Phys. Status Solidi C 13 (2016) 354−359. 5−6.

[115] https://www.pntpower.com/epc-corp-450v-gan-fet-is-now-available-normally-off/.

[116] http://epc-co.com/epc/Portals/0/epc/documents/datasheets/EPC2046_preliminary.pdf.

[117] https://www.mouser.jp/pdfdocs/PreliminaryTO220GANDatasheetPGA26C09DV.pdf.

[118] https://b2bsol.panasonic.biz/semi-spt/apl/cn/news/contents/2013/apec/panel/APEC2013_GaN_FPD_WEB.pdf.

[119] http://www.apec-conf.org/Portals/0/Industry%20Session%20Presentations/2015/APEC_2015%5B1%5D.pdf.

[120] http://www.gansystems.com/datasheets/GS66504B%20DS%20Rev%20170321.pdf.

[121] https://www.infineon.com/cms/en/product/promopages/gallium-nitride/.

[122] http://www.transphormusa.com/document/600v-cascode-gan-fet-tph3206ps/.

[123] http://www.transphormusa.com/product/product-profile-tpd3215m/.

[124] http://www.ti.com/lit/wp/slpy008/slpy008.pdf.

[125] http://www.ti.com/lit/ds/symlink/lmg3410.pdf.

[126] http://www.apec-conf.org/Portals/0/APEC%202016%20Plenary%20-%203rd%20Speaker%20-%20Dan%20Kinzer.pdf.

[127] Y. Uemoto, T. Morita, A. Ikoshi, H. Umeda, H. Matsuo, J. Shimizu, et al., GaN monolithic inverter IC using normally-off gate injection transistors with planar isolation on Si substrate, IEEE IEDM, Tech. Dig (2009) 165–168.

[128] S. Nagai, Y. Yamada, N. Negoro, H. Handa, M. Hiraiwa, N. Otsuka, et al., A 3-phase ac–ac matrix converter GaN chipset with drive-by-microwave technology, J. Electron Devices Soc. 3 (1) (2015) 7–14.

[129] H. Wang, A.M.K. Ho, Q. Jiang, K.J. Chen, A GaN power width modulation integrated circuit, Proceedings of the International Symposium on Power Semiconductor and ICs, pp. 430–433, 2014.

[130] D. Maksimovic, Y. Zhang, M. Rodriguez, Monolithic very high frequency GaN swich-mode power converters, IEEE Custom Integrated Circuits Conference, 2015.

[131] Y. Zhang, M. Rodriguez, D. Maksimovic, Very high frequency PWM buck converters using monolithic GaN half-bridge power stages with integrated gate drivers, IEEE Trans. Power Electron 31 (11) (2016) 7926–7942.

[132] J. Waldron, R. Karlicek, T.P. Chow, Monolithic optolectronic integration of GaN high-voltage power FETs and LEDs, IEEE Lester Eastman Conference, August 7–9, 2012.

[133] Z. Li, J. Waldron, T. Detchprohm, C. Wetzel, R.F. Karlicek, T.P. Chow, Monolithic integration of light-emitting diodes and power metal-oxide-semiconductor channel high-electron-mobility transistors for light-emitting power integrated circuits in GaN on sapphire substrate, Appl. Phys. Lett. 102 (1–3) (2013) 192107.

[134] Z. Li, J. Waldron, S. Chowdhury, L. Zhao, T. Detchprohm, C. Wetzel, et al., High temperature characteristics of monolithically Integrated LED and MOS-channel HEMT in GaN using selective epi removal, Phys. Status Solidi A 212 (5) (2015) 1110–1115.

[135] Z.J. Liu, T. Huang, J. Ma, C. Liu, K.M. Lau, Monolithic integration of AlGaN/GaN HEMT on LED by MOCVD, IEEE Electron Device Lett. 35 (3) (2014) 330–332.

[136] Z. Liu, J. Ma, T. Huang, C. Liu, K.M. Lau, Selective epitaxial growth of monolithically integrated GaN-based light emitting diodes with AlGaN/GaN driving transistors, Appl. Phys. Lett. 104 (2014), pp. 091103-1-3.

[137] C. Liu, Y. Cai, Z. Liu, J. Ma, K.M. Lau, Metal-interconnection-free integration of InGaN/GaN light emitting diodes with AlGaN/GaN high electron mobility transistors, Appl. Phys. Lett. 105 (2015), pp. 181110-1-3.

[138] Y.-J. Lee, et al., Monolithic integration of GaN-based light-emitting diodes with metal-oxide-semiconductor field-effect transistors, Opt. Express 22 (S6) (2014) A1589–A1595.

[139] X. Lu, C. Liu, H. Jiang, X. Zou, A. Zhang, K.M. Lau, Monolithic integration of enhancement-mode vertical driving transistors on a standard InGaN/GaN light emitting diode structure, Appl. Phys. Lett. 109 (2016), pp. 053504-1-3.

[140] X. Lu, C. Liu, H. Jiang, X. Zou, A. Zhang, K.M. Lau, High performance monolithically integrated GaN driving VMOSFET on LED, IEEE Electron Device Lett. 38 (6) (2017) 752–755.

GaN-on-GaN power device design and fabrication

6

Srabanti Chowdhury
Electrical and Computer Engineering, UC Davis, Davis, California, United States

6.1 Introduction

GaN technology is an ever-expanding topic of research and development, proving its potential to solve several challenges in power conversion that cannot be addressed by Si. For instance, medium voltage (650−900 V) devices using the high electron mobility transistor (HEMT) configuration [1,2] have been able to reduce form factor at the system level by driving circuits at higher frequencies (100 kHz−1 MHz) [3] and eliminating heat sinks or reducing cooling requirements. This alone sparked the interest in GaN research to save space, energy, and ultimately cost of power conversion. However, in power conversion the demand of high current from a single chip for a rated voltage is a standard need. Particularly, when the market is favorable toward electrification of cars and other means of transportations, GaN must expand its scope to provide high power solutions with higher power density compared to Si, and even SiC. Vertical devices have been the choice of power device engineers for economic use of the material and maximum use of its physical properties (which allow more uniform electric field distribution and the highest possible blocking electric field, field mobility, etc.). GaN vertical devices carrying all those advantages offered by vertical geometry are being explored increasingly with emphasis on material and device needs. Fig. 6.1 shows the potential power electronic applications ranging from watts to gigawatts of power. Wide bandgap semiconductors will play a very critical role in all of these applications over time, with SiC and GaN at the helm.

The purpose of this chapter is to take the reader through the various developments that took place over the last decade or so creating the ground for vertical GaN device technology.

In this chapter, we briefly go over the requirements from a high voltage power switch, materials requirement, followed by the design and fabrication techniques for two types of vertical GaN devices: current aperture vertical electron transistor (CAVET), and vertical metal oxide semiconductor field-effect transistor (v-MOSFET or simply MOSFET). We discuss the relevant variants with each type of transistors. Last but not the least, we dedicate a section on diodes.

Wide Bandgap Semiconductor Power Devices. DOI: https://doi.org/10.1016/B978-0-08-102306-8.00006-X

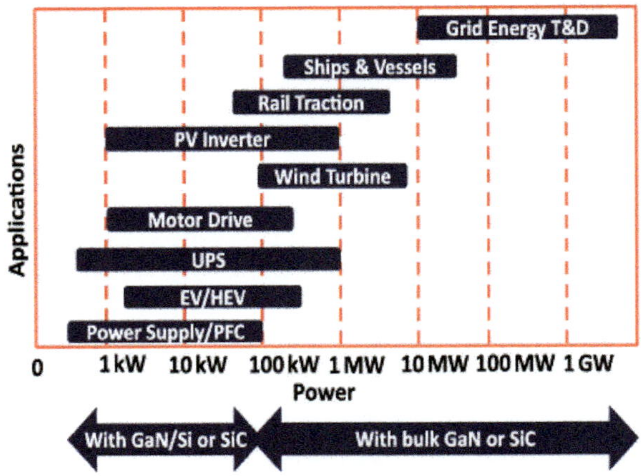

Figure 6.1 Power requirements by various applications. GaN vertical devices will play a major role in power conversions above 30 KW (chart is based on Yole Development report presented in 2012 at CS-Europe).
Source: Adapted from [1].

6.2 Requirements from a power switch

To realize the next generation power electronic switches, the following performance will be required.

6.2.1 Normally-off operation

Most efforts on GaN-based devices have been directed toward normally-on ones, and only a few reliable normally-off ones have been reported to date. In normally-off devices, the drain current does not flow at the gate voltage of 0 V. Si-IGBTs [4] used in the present power converters are normally-off devices. Likewise, a normally-off device is required for automobiles (electric and hybrid electric vehicles) in order to simplify the inverter circuit and make effective use of design techniques and mounting technologies for the inverters. For GaN devices to make an entry into the conventional power conversion architecture, normally-off operation is critical. AlGaN/GaN HEMTs are inherently normally-on devices and thereby necessitates techniques at the circuit or at the device level to make it normally-off to suit power electronics (PE) applications. Some of the methods practiced in HEMTS, also applicable to vertical GaN devices, to achieve normally-off operation at the device level that are

(1) Junction-gated devices to pinch-off the channel under the high work function gate whereas maintaining the channel in the access regions that do not have the p region have been implemented by EPC and Panasonic, the latter as the GIT or gate injection transistor [5].

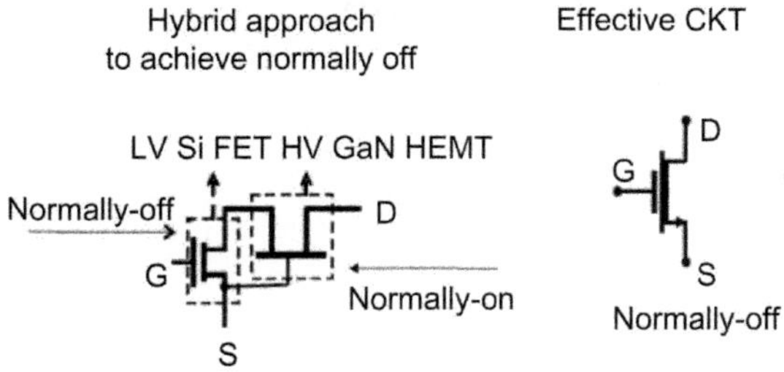

Figure 6.2 A schematic of a cascode normally-off GaN HEMT used by industry to allow for standard silicon-based drive circuits while delivering the high voltage performance of GaN devices.
Source: Adapted from [1].

(2) A multichannel transistor with two AlGaN/GaN interfaces have been developed by Toshiba, Fujitsu, NTT, and Transphorm [6]. Here a conductive access region is defined by a 2DEG formed at the upper AlGaN/GaN interface (as in a normal HEMT). A second channel below is accessed via gate recess etching and MIS gate formation within the recess. The second channel is formed between a thin AlGaN layer and GaN such that there is no charge in the channel at zero bias and charge is induced under forward bias (normally-off).

(3) Normally-off operation by introducing F under the gate structure to shift the threshold voltage positive under the gate while maintaining a conductive channel in the access regions has been developed by HKUST [7].

(4) Finally, the channel etch through (CET) structure MOSFET where the 2DEG is etched through and an MOS gate deposited has been developed by RPI [8].

All these structures, listed above, are designed to ensure single chip solution. However, in MIS gated devices the gate insulator on GaN and AlGaN is not reliable as yet and the reproducibility of the interface properties and stability under operation needs to be significantly improved. The junction-gated devices have low threshold voltage and limited gate-bias drive range because of the turn-on of the p–n junction gate.

At the circuit level, so far the simplest embodiment of normally-off GaN transistors incorporates a cascode approach where a normally-off low-voltage Si FET is connected to a normally-on high-voltage GaN HEMT in series, while the gate of the GaN HEMT is connected to the source of the Si FET as shown in Fig. 6.2 [1]. This hybrid configuration delivers a normally-off solution compatible with existing Si drivers, as well as the freedom to optimize the GaN HEMTs without the complication of special gate drive circuits. Although attractive for the simplicity of design, the single chip solution will be necessary to reduce losses due to bond wire inductances and PCB traces.

6.2.2 High breakdown voltage

Higher power density is the overarching need for all power conversion applications that shapes the need of high power density switches. An example best shown with

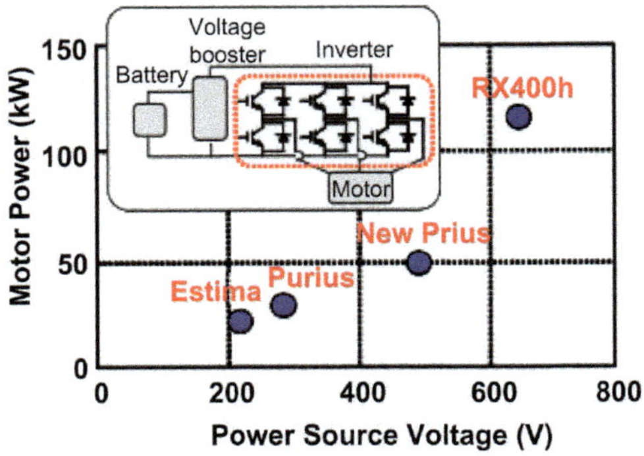

Figure 6.3 Power source voltage versus motor power in Toyota HVs. In the inset is shown the simplified inverter where the transistor is a key element.
Source: Adapted from [40].

Toyota HVs and EVs paints a comprehensive picture, which can be extended qualitatively, to many other high power applications.

Fig. 6.3 shows the relationship between the motor power of Toyota's HVs and power source voltages of these systems. In an HV system, the battery voltage is once raised to a power source voltage by a voltage booster and then supplied to the motor through the inverter. The raised voltage can take values from 202 V of the battery up to a maximum of 500 V. The new HVs need high motor power with high power source voltage as projected in Fig. 6.3. The breakdown voltage of devices used in these inverters is about 1.2 kV. The breakdown voltage of devices used in the inverters will probably become higher in the future due to protection against surge voltage and higher efficiency due to wiring losses, etc.

6.2.3 Low on-resistance and high current density

In order to increase the energy conversion efficiency of any converter systems, it is necessary to decrease the on-resistance of the devices. Moreover, it is demanded for miniaturization of converters/inverters to increase the current capacity up to several hundred A/cm^2.

6.2.4 High temperature operation

Power Si devices are not used over 150 °C, because power loss increases due to an increase of leakage current in the off-state under high temperature environment, thereby decreasing their reliability. GaN devices are expected to work over 200 °C and possibly higher, because of its wider bandgap. An example best shared with the present hybrid vehicles having two cooling systems, one is for the engine and the other is for the inverter, and the temperature of coolant water for the inverter is lower than that for the engine. If GaN devices work over 200 °C, the cooling system of hybrid vehicles will be simplified and its cost will be reduced. The same example can be extended toward other applications like PV inverters, where heat sink forms the majority of the volume. Higher temperature operation will reduce cooling requirement and cooling cost, effectively shrinking the size and overall system cost.

GaN vertical devices have the potential to meet the entire requirement stated in 1–4. Moreover, packaging technologies developed for Si IGBT can be utilized effectively with minimum changes in modules.

6.3 Substrates and epitaxial layers

Gallium nitride when grown on foreign substrates as normally done for HEMTs allows high density of dislocations that form during the nucleation process. Dislocation densities can vary between high $10^9\ cm^{-2}$ (on sapphire with lattice mismatch of 16%) and low $10^7\ cm^{-2}$ (on SiC with lattice mismatch of 3.5%). Table 6.1 captures the parameters of the substrates that influences the growth of the GaN epitaxially.

The mismatch between coefficient of thermal expansion between epilayers and substrate leads to wafer bowing, a drawback that can hinder fabrication processes.

A transmission electron microscopy (TEM) cross-sectional view of GaN on Sapphire shown in Fig. 6.4 illustrates an example of the extended dislocations, which can run through the entire thickness of the material grown on the substrates.

6.4 Availability of GaN substrate

Perhaps one of the landmark achievements in the vertical GaN device technology was development of bulk GaN substrate. The availability of GaN substrate opened up the possibility of homoepitaxial growth in GaN, which has been also extended to other III-nitrides, such as AlN and $Al_xGa_yN_{1-x-y}$. Gallium nitride when grown on GaN substrate does not suffer from lattice mismatch or thermal expansion mismatch, thereby allowing the least amount of defect concentration (between 10^6 and $10^4\ cm^{-2}$ or even as low as $10^2\ cm^{-2}$) in the material grown on the substrate.

The challenges in cost and scalability of the bulk substrates are currently being addressed by various growth methods such as hydride vapor pressure epitaxy (HVPE), ammonothermal growth, Na-flux method, and their various combinations.

Table 6.1 **An overview of the advantages and disadvantages of various substrates for growing epitaxial GaN**

Substrate	*a* (Å)	Thermal conductivity (W/cm/K)	Coefficient of thermal expansion, in-plane (10^{-6} K^{-1})	Lattice mismatch (%)	Thermal mismatch (%)	Advantages	Disadvantages
GaN	3.189	1.3	5.59	–	–	– Homoepitaxy	– Expensive – Lacks scalability – Only available up to 2″
Sapphire	4.758	0.5	7.5	16	−34	– Inexpensive – Simple growth process	– Extreme wafer bowing – Large compressive film stress
6H-SiC	3.080	3.0–3.8	4.2	3.5	25	– High thermal conductivity – Better CTE match	– Expensive – Limited wafer size scalability
Si (111)	3.840	1–1.5	2.59	−16.9	54	– Inexpensive – Highly scalable – Si industry support	– High tensile strain and bowing during growth – Complex growth process

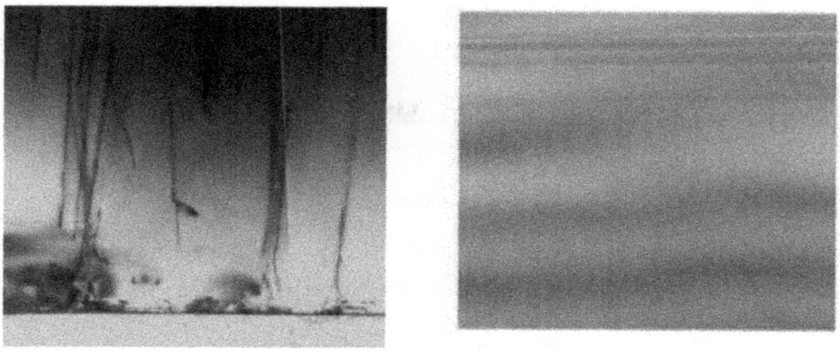

Figure 6.4 (A) TEM cross-section of the nucleation layer and partial buffer layer of GaN on sapphire. Dislocation annihilation takes place in the buffer layer, which is doped with C or Fe to attain high resistivity in the layer. (B) TEM cross-section of GaN on GaN substrate, showing visibly lower dislocation density (could be 10^3-10^5x lower on GaN).
Source: S.-Y. Huang and J.-R. Yang, "A Transmission Electron Microscopy Observation of Dislocations in GaN Grown on (0001) Sapphire by Metal Organic Chemical Vapor Deposition," Jpn. J. Appl. Phys., vol. 47, no. 10R, p. 7998, 2008

6.5 Vertical devices: Current aperture vertical electron transistor

A CAVET is an effective combination of a lateral and a vertical topology, designed to bring the best of both worlds. CAVETs resemble DMOSFETs in many design aspects, although unlike a DMOSFET, it is a normally-on device. The channel formed laterally at the AlGaN/GaN interface source electrons (unipolar), which then are flown vertically into an aperture to the drain as shown in Fig. 6.5. Current blocking layers (CBLs) are placed around the aperture to force the current through it. Since the 2D-electron gas channel region in a CAVET is similar to that in an HEMT, the normally-on behavior is to be expected along with many other HEMT-like characteristics. For example, the gating action occurs under the gate metal between the source-side edge of the metal gate and the source-side edge of the aperture region as shown by the shaded region with a dimension of L_{go} in Fig. 6.5 causes the depletion of the lateral channel determining the threshold voltage of the device.

The major advantage in a vertical device like CAVET and other vertical transistors is the buried high electric field region under the gate into the bulk material. In a lateral HEMT the drain-side edge of the gate carries high field region, which peaks to a very high value and may cause surface arcing, dielectric breakdown, channel breakdown, and encapsulation breakdown leading to premature device failure. Moreover, this high field in an HEMT favors charging of surface states leading to dispersion in an HEMT. Field plate (FP) technology with surface passivation helps reduce dispersion and field shaping leads to higher breakdown voltage. Compared to an HEMT (lateral device) where the electric field peaks at the edge of the metal gate and FPs, the electric field is more uniformly distributed in a vertical device set up by the p + /n − junction, which form an integral part in most of the high power vertical device structures. The buried nature of the peak electric field in

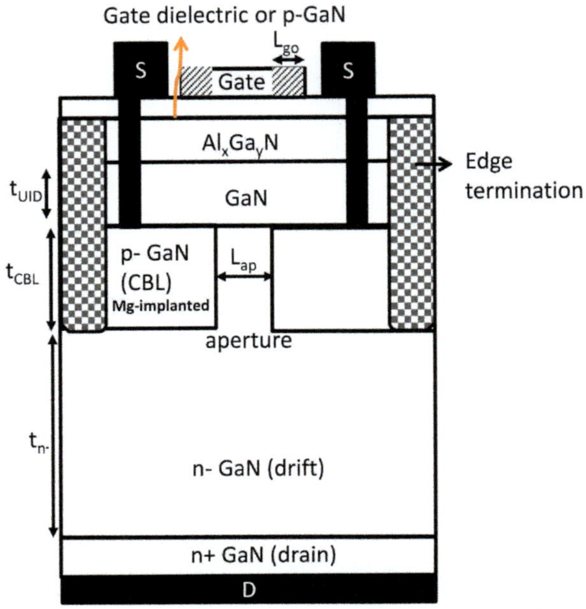

Figure 6.5 Cross-sectional picture of a CAVET. The CBL is Mg-implanted GaN. The effective gate of the device is indicated by the shaded region. Sources are connected to the CBL to offer a discharge path to the traps in the CBL.

a vertical device allows higher breakdown electric field (closer to the critical electric field in GaN), when properly designed, compared to a lateral HEMT.

The CAVET, therefore utilizes the whole area between the source electrodes to block voltage, making it a more economical solution than the HEMT for the same rated voltage. Or in other words, one can say that owing to the buried electric field and with appropriate edge termination scheme, a CAVET should offer higher "volts per micrometer." Fig. 6.6 sketches the argument how a CAVET offers electric field management without extensive FP requirements. The CBL plays a vital role in mitigating the electric field generated under the off-state. For higher voltage designs appropriate edge and field terminations would be necessary to distribute the electric field uniformly in the drift region CBL and the aperture.

6.6 A brief history of GaN vertical devices

Ben-Yaacov et al. designed first generation CAVETs for RF power performance, which was successfully demonstrated in 2000 [9] where the structures were grown on sapphire and with the drift region grown no thicker than 0.5 μm. The CBL was achieved by doping the GaN layer with Mg. Any CAVET fabrication requires a regrowth of the AlGaN/GaN layers which creates the 2DEG. In this design, where Mg-doped GaN formed the CBL, the aperture region needed to be regrown as well

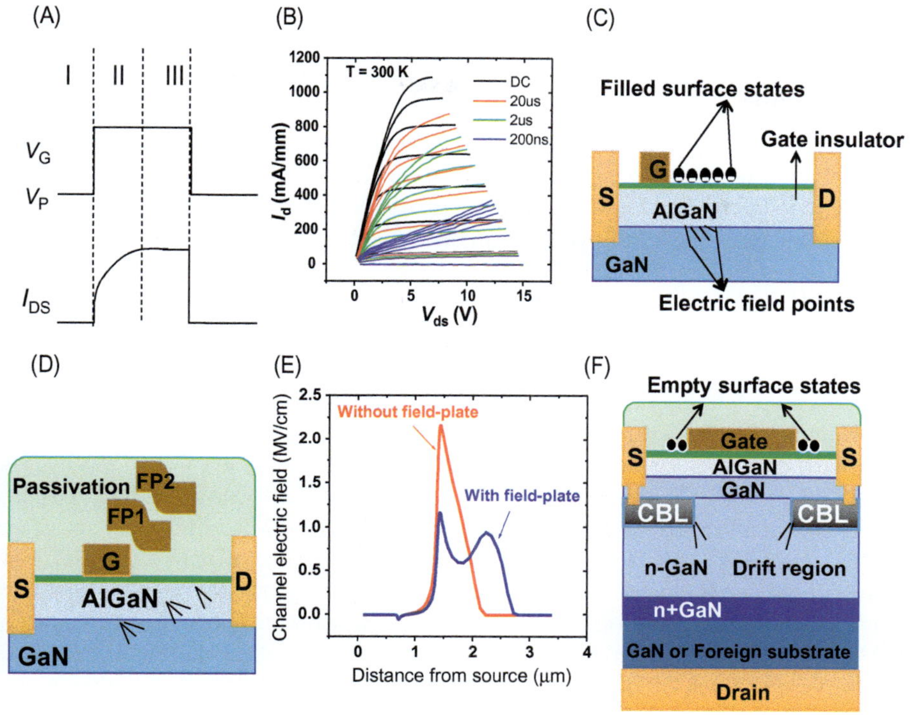

Figure 6.6 (A) $I-V$ characteristics of dispersive device. Dispersion can lead to switching losses. (B) Lateral devices without FP causes high electric field at the drain edge of the gate leading to occupied surface states causing dispersion. (C) FPs are used to mitigate and mange peak electric field to minimize dispersion and enhance breakdown voltages. (D) Two FPs smoothens out the single peak that occurs at the drain edge of the gate without any FP, into two peaks with lower peak values. (E) Vertical transistors have the peak electric field region buried inside the bulk material. This leaves the surface states unoccupied causing no dispersion.

along with the channel and the AlGaN cap. Since the design was directed toward realizing RF power, the gate and the aperture were kept as small as possible for a starting point to minimize capacitances. The devices successfully showed a CAVET operation validating the concept and design, but suffered from high leakage. A major contribution to the leakage came due to the technique of regrown apertures. The next design reported by Gao et al. [10] employed photoelectrochemical (PEC) etching techniques to etch InGaN aperture layer to form air gaps as current blocking layers. This method did not require any regrowth and the whole structure was grown with an InGaN aperture layer, which was then etched away using the bandgap selectivity to form the CBL. The devices fabricated by this technique suffered from a turn-on voltage drop as the polarization charges between the InGaN and the GaN create a barrier to the electron flow. With a lot of design modifications using

Si delta doping, the barrier could be reduced but a voltage drop of 0.7 V still remained. Although it was a significant improvement and the method circumvented the problem of regrowth, the air gap as a current blocking layer is not appropriate at higher voltages since air is a poor dielectric.

Later in 2004, CBLs created by ion-implantation was incorporated into CAVETs, which delivered much improved results. The technique involved regrowth of the AlGaN/GaN layers, but not the aperture. Since the aperture was not regrown many of the earlier problems that existed in the first generation CAVETs were alleviated.

In 2006 CAVETs were revisited foreseeing a different end application. The ultimate benefit of a DMOSFET-like structure represented in CAVET would be best realized in power switching applications. Targeting a goal toward achieving a low-loss power switch, the first research work on high voltage CAVETs was pursued by Chowdhury et al. between 2006 and 2010 [11,12].

Toyota Motor Corporation, Japan focused their attention on GaN CAVET and CAVET-like devices. In their report of a vertical insulated gate AlGaN/GaN HFET Kanechika et al. reported vertical insulated gate HFETs [13] on bulk GaN substrates in 2007 with a R_{on} of 2.6 mΩ/cm^2. The device structure, similar to a CAVET, was achieved with Mg-doped GaN as the CBL enclosing an n-type GaN aperture. Toyota Motor Corporation, Japan funded and supported the development of CAVET for power electronics at UCSB, which resulted in the development of CAVETs and CAVET-like vertical devices targeting 60 KW and higher power switching applications.

In 2007 Otake et al. [14] also reported on bulk GaN their successful demonstration of "vertical GaN-based trench gate metal oxide semiconductor FET" with a threshold voltage of 3.7 V and an R_{on} of 9.3 mΩ/cm^2.

In 2014 Nei et al. from Avogy Inc. reported vertical GaN transistors exhibiting larger than 2.3 A saturation current, breakdown voltages of 1.5 kV, area differential specific R_{on} of 2.2 mΩ/cm^2 [15].

In 2015 Toyota Gosei reported MOSFET-based vertical GaN devices with an MOS gate structure blocking over 1.2 kV and an R_{on} of 2 mΩ/cm^2 [16]. Fig. 6.7 captures the development of CAVET and other vertical devices over the past decade.

The timeline depicted in Fig. 6.7 can be categorized into three periods that mark the historical development of the devices. First 6 years showed the demonstration of various types of vertical devices in GaN, which was enabled by the availability of the crystalline substrates. Following the demonstration phase came the high voltage devices presenting over 1 kV blocking voltages. Over the last 5 years high voltage devices are showing successful scaling to offer high current, often accompanied with switching performances.

The suitable CAVET design for holding high voltages while offering low R_{on} was developed through extensive simulation where every components of the device were analyzed. Some of the key components required for the working of a CAVET have been discussed in Section 6.7.

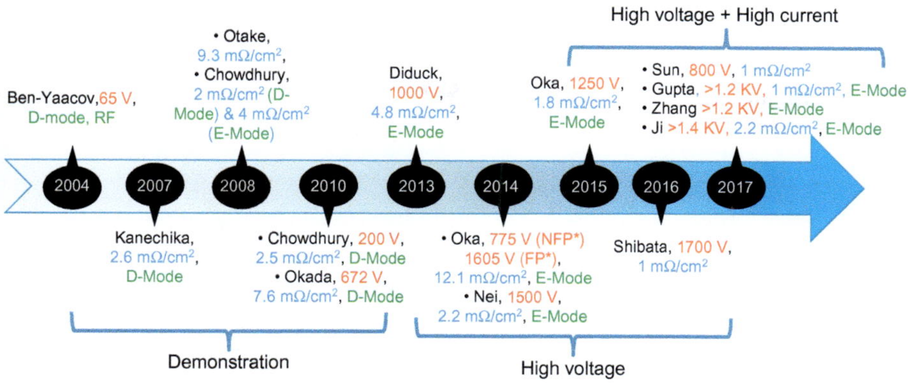

Figure 6.7 A trajectory showing the major developments reported in the vertical GaN technology.

6.7 Design of a current aperture vertical electron transistor and its key components

CAVET differs from conventional vertical devices like vertical JFET or MOSFET in its fusion of an HEMT-like channel to a thick drift region. A CAVET can be formed with simply n-GaN channel instead of AlGaN/GaN, just like an MESFET. The first CAVET fabricated on bulk GaN [11,17] substrates with Mg-implanted CBL showed dispersion-less drain characteristics under 80 μs pulses applied to the gate, as depicted in Fig. 6.8. The breakdown voltages in these CAVETs were around 200 V, limited by the gate dielectric breakdown. These CAVETs were fabricated on a 3 μm thick drift region. A relationship of the on-resistance (R_{on}) on aperture length (L_{ap}) was experimentally determined from these early devices, which later would be verified by its drift-diffusion-based model.

In this section, first we will discuss the design space with respect to a CAVET and bring other device platforms into the discussion as relevant to the design parameters [17–20].

In a CAVET (Fig. 6.5) the intrinsic current flow occurs in two dimensions; electrons first flow horizontally through the 2DEG and then move vertically through the aperture region. This is quite different from the HEMT or the bipolar transistor, where current flow is confined to one dimension. It is therefore critical to develop an accurate model in order to identify which parameters primarily determine the device characteristics.

Electrons (current flows opposite to the flow of electrons) first travel horizontally through the 2DEG, until it reaches the gate. The gate only modulates the electrons in the 2DEG, so the pinch-off occurs in the horizontal direction inside the 2DEG underneath the gate, just like in a standard FET.

Electrons, which pass the pinch-off point in the channel, continue to travel horizontally at their saturated velocity v_{sat} until they arrive at the aperture, travel

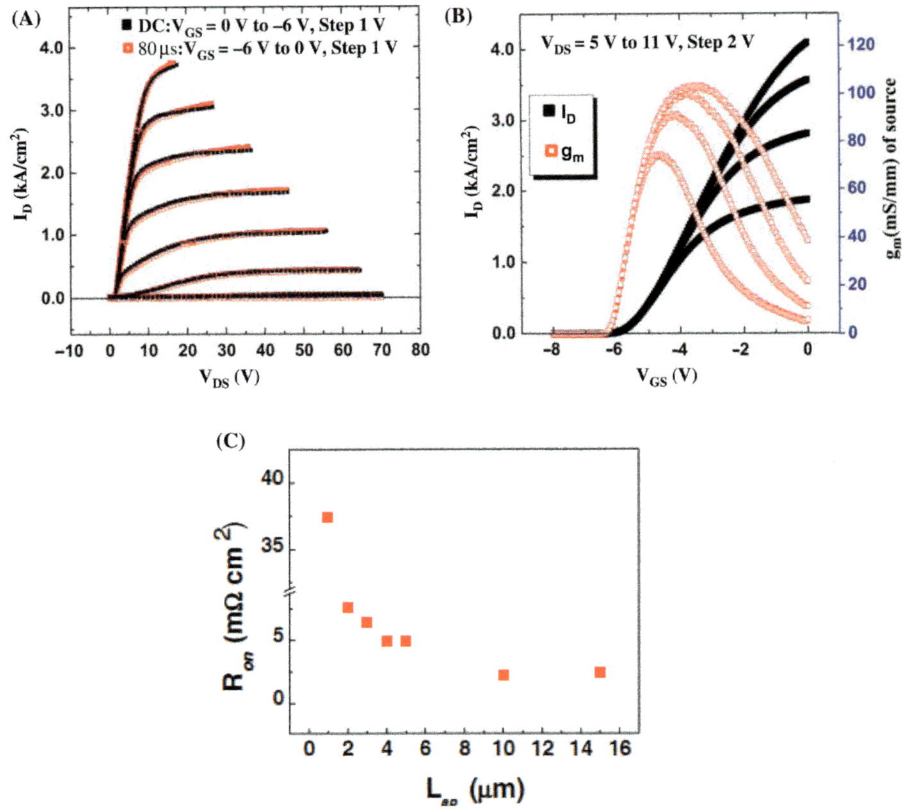

Figure 6.8 (A) DC and pulsed I_D-V_{DS} measurement of a depletion-mode CAVET for $L_{ap} = 10$ μm and $L_{go} = 3$ μm with an active area of 22 Å ~ 75 μm^2. (B) Transfer characteristics of the device in (A). The dependence of R_{on} on the aperture length was experimentally verified in these early CAVETs and plotted in (C).
Source: Adapted from [12].

downward through the aperture, and are collected at the drain. It is critical that the conductivity of the material inside the aperture as well as in the drain region be much larger than that of the 2DEG so that the entire voltage drop between the source and the drain occurs in the 2DEG. This condition ensures that the total current passing through the device is entirely determined by the conductivity of the 2DEG. If such condition is not met, then a significant amount of the applied source−drain voltage is supported across the aperture. In this case, until V_{DS} is very large, the 2DEG does not pinch-off and the current does not reach its saturation value. This is analogous to quasisaturation in a bipolar transistor, which can occur at large injection currents when the ohmic drop I_c. R_c across the collector drift region becomes comparable to the total base-collector voltage V_{CB}. In addition, the conductivity of the 2DEG must be much higher than that of the adjacent bulk GaN directly below the 2DEG to ensure current flow through the 2DEG rather than through the bulk GaN.

Extensive simulation-based studies were conducted during the course of development of a CAVET to optimize the device dimensions for making it suitable for high power application. CAVET has a complex geometry and hence the design rules are not simple. The important parameters also labeled in Fig. 6.5 are detailed out in this section.

6.8 Doping in the aperture (N_{ap}) and length of the aperture (L_{ap})

It is essential to have the aperture region more conductive than the channel region to avoid any downstream choke of the current in the on-state of the transistor. In order to ensure this, the doping of the aperture region (N_{ap}) is chosen such that the conductance of the aperture region ($L_{ap}.q.\mu.N_{ap}.\ W_g/t_{CBL}$) is higher than the channel resistance (where q is the electron charge, L_{ap} is the length of the aperture, t_{CBL} is the thickness of the CBL region, W_g is the gate width of the device, and μ is the mobility of electrons in the drift region).

Increasing the aperture doping is not very desirable because that increases the chance of the breakdown occurring in the aperture. Therefore, lowering R_{on} primarily means increasing L_{ap}. To give an idea of the impact of aperture doping on $I-V$ characteristics, two CAVETs were modeled, one with a higher aperture resistance than the other. The aperture resistance was changed using a combination of L_{ap} and N_{ap} (Fig. 6.9).

From the simulation (Silvaco ATLAS) result shown in Fig. 6.9A and B, it can be seen that the device with higher aperture resistance does not show saturation of the current. Or in other words, since a major part of the applied V_{DS} is now absorbed in the aperture, much higher voltage needs to be applied to gain saturation. Device shown in Fig. 6.9B with lower aperture resistance easily saturates.

The results indicate that for lower aperture doping, which is necessary for high breakdown voltage design, longer apertures (wider opening) are needed. The simulation was done with a 4 μm aperture and as high as $8 \times 10^{17}\ cm^{-3}$ doping was needed in the aperture to saturate the current without the quasisaturation region. This suggests that the aperture needs to be wider than 4 μm for lower N_{ap}. The simulations were done for qualitative understanding of the device. Later experimental results showed the validity of the model. The doping in the aperture was maintained as low as the drift region to avoid premature breakdown in the aperture region, which necessitated an aperture region longer than 4 μm and was experimentally verified.

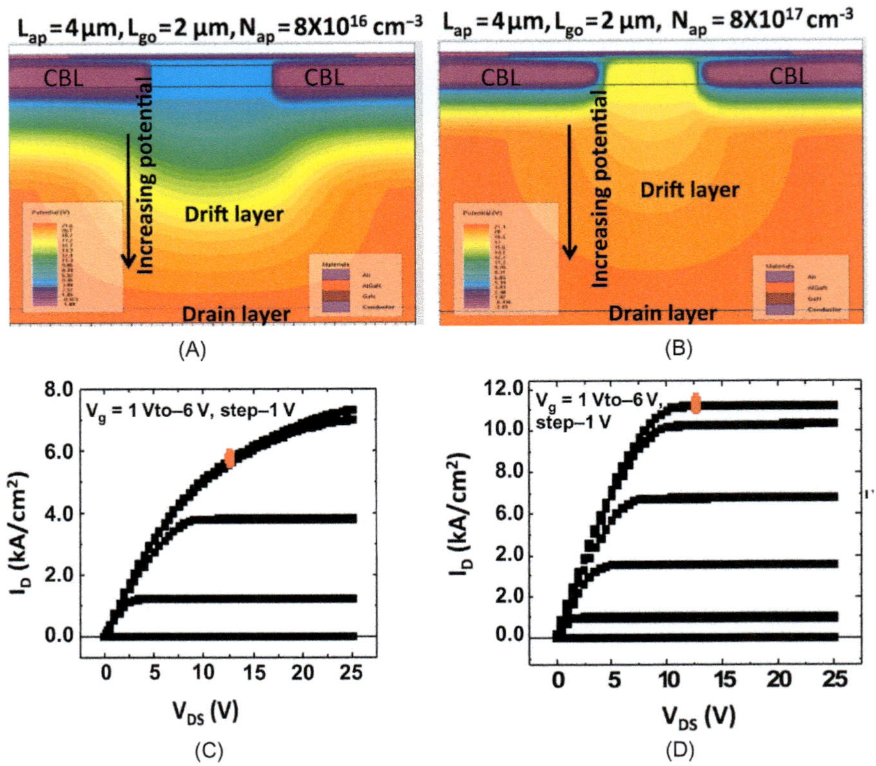

Figure 6.9 (A) A CAVET simulated with Silvaco ATLAS with a resistive aperture. (B) CAVET in (A) with increased conductivity of the aperture. CAVET with resistive aperture leads to slow saturation as seen from the *I–V* curve in (C). As the resistance of the aperture region is decreased, the current saturates as shown in (D). The voltage distribution is shown in the potential contour, which is shown in the simulated snap shots above. The red dots in the plots indicate the bias condition at which the above pictures were taken. *Source*: Adapted from [17].

6.9 Drift region thickness (t_{n-})

The CAVET is designed to support most of the applied voltage in the drift or lightly doped (n −) layer. This is the very principle based on which the vertical devices work. While the channel architecture can vary to form different types of vertical transistors, viz., JFET or MOSFET, the expected properties of the drift region and therefore its design is common to all. Majority of the voltage drop under the off-state operation occurs in the drift region. Ideally, most vertical devices including CAVET use p–n junction formed by the p-GaN base region and the n-GaN drift region to block the high voltage.

In the off-state, a positive voltage is applied onto the cathode, and a triangular-shaped electric field distribution along the depletion region is obtained as shown in Fig. 6.10. According to Poisson's equation, the maximum electric field E_{max} can be written as

$$E_{max} = \frac{qN_D}{\varepsilon_r \varepsilon_0} W_D \tag{6.1}$$

where N_D is the doping concentration in the n- drift region, W_D is the depletion width in the drift region, q is the charge of an electron, ε_r is the relative permittivity and ε_o is the permittivity of free space. If the E_{max} reaches the value of the critical electric field, E_C, ideally the device should avalanche, provided it has a high quality p–n junction. The breakdown voltage, V_{BR}, can be written as

$$V_{BR} = \frac{1}{2} W_D E_C \tag{6.2}$$

Therefore the drift region has to be lightly doped to deplete under the off-state drain bias. However increasing thickness and lowering the doping of the region increases the R_{on} under the on-state. The effect is more pronounced for higher voltage classes of vertical devices, where drift region is made thicker to sustain the high voltage switching. From the potential contours, one can see that most of the applied voltage is supported by the drift region. As the applied voltage is increased, the depletion region extends and the potential contours are pushed into the aperture as seen in the simulated potential contour in Fig. 6.11. The peak field region is buried in the bulk material and occurs at the lower edge of the CBL or the p–n junction (see Fig. 6.10). A plot of the drift region doping and corresponding breakdown voltages reported by various group is shown in Fig. 6.12.

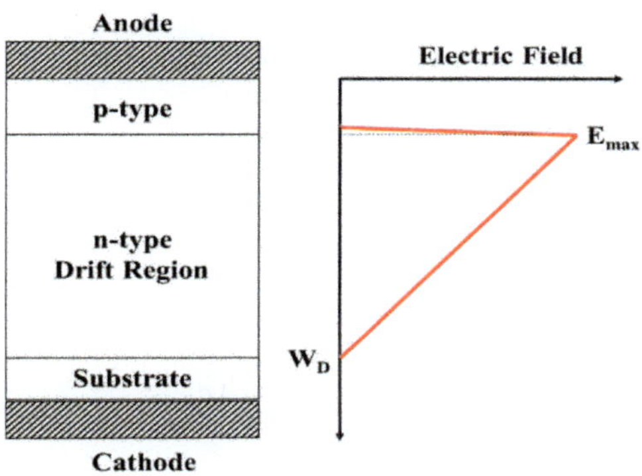

Figure 6.10 In an ideal drift region design the triangular-shaped electric field distribution extends almost all the way to the cathode (here the substrate is n$^+$ GaN).
Source: Figure adopted from [19].

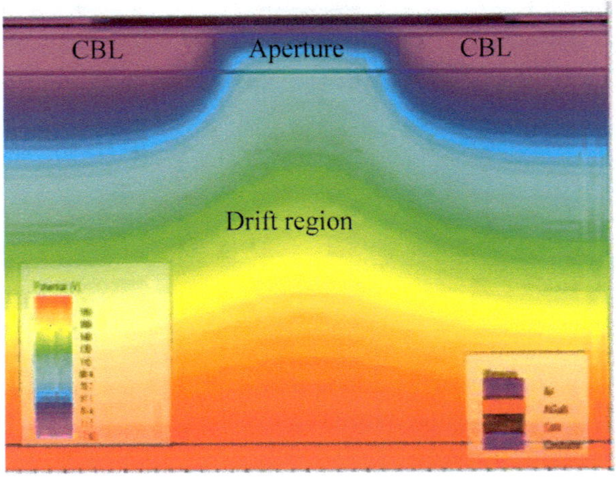

Figure 6.11 Potential contour in a CAVET under the off-state operation showing the majority of the voltage dropping across the drift region.

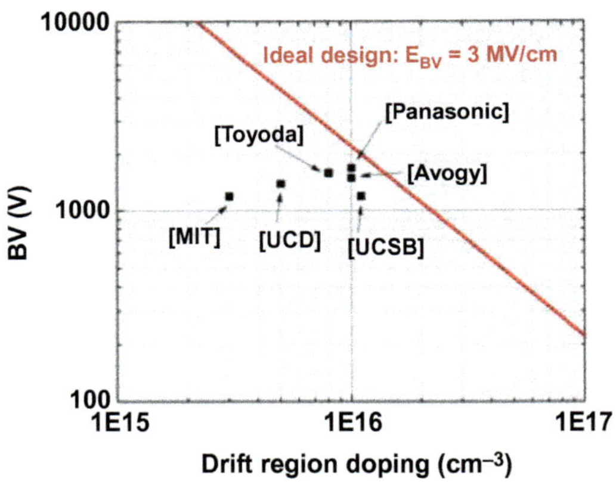

Figure 6.12 Breakdown voltage of vertical GaN transistors as a function of drift region doping concentration.
Source: Adapted from [19].

The electron mobility in the drift region plays an important role in vertical transistor performances. The improvement in the quality of the bulk GaN material over the past decade leads us to predict that the differentiator of GaN technology with other competing technologies like SiC (offering similar advantages due to its comparable bandgap energy and critical electric field) is indeed the mobility of the material. Increase in mobility of the bulk GaN that forms the drift region will lower the R_{on} without any significant penalty paid. The drift region mobility is therefore a key component and impacts the performance and the roadmap just like the critical electric field. The ideal $R_{on,sp}$ of the GaN CAVET can be written as

$$R_{on,sp} = \rho_{2DEG} L_G p + \frac{W_D}{q\mu_n N_D} \tag{6.3}$$

From Fig. 6.13, for low breakdown voltage device design (BV < 2000 V), the on-state resistance is limited by the channel electron mobility; while for BV > 2000 V, the on-state resistance is limited by the bulk GaN mobility. A simplified model of the device simulated in Silvaco ATLAS to demonstrate the sensitivity of R_{on} toward bulk mobility in the drift region is shown in Fig. 6.14, which shows the total on-state resistance ($R_{ON,TOT}$), drift region resistance (R_{DR}), and channel resistance (R_{CH}) versus aperture length L_{ap}. The lowest $R_{ON,TOT}$ is obtained between an L_{ap} of 4 and 10 μm, with a minimum of ~1.4 mΩ/cm^2 [19,21].

Dependence of R_{on} on the L_{ap} for the same doping density and the same aperture to gate overlap (L_{go}), optimized for minimal parasitic leakage, is also illustrated in Fig. 6.10. R_{on} decreases with increase of L_{ap} since the conductance in the aperture increases proportional to L_{ap}. Saturation in R_{on} is observed when the channel and drift region offers the dominant part of the resistance.

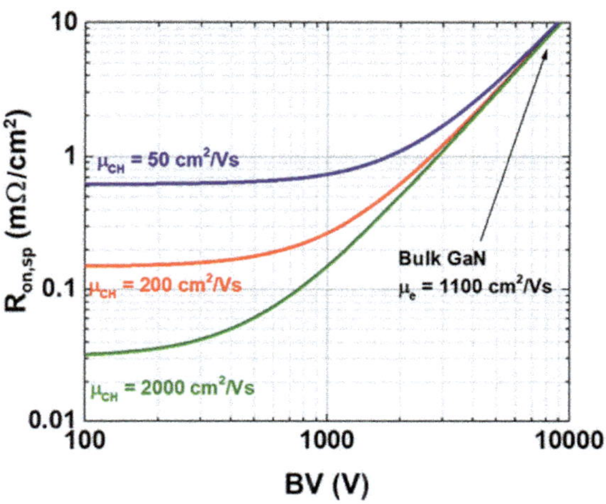

Figure 6.13 Device figure-of-merit of vertical GaN transistors.
Source: Adapted from [19].

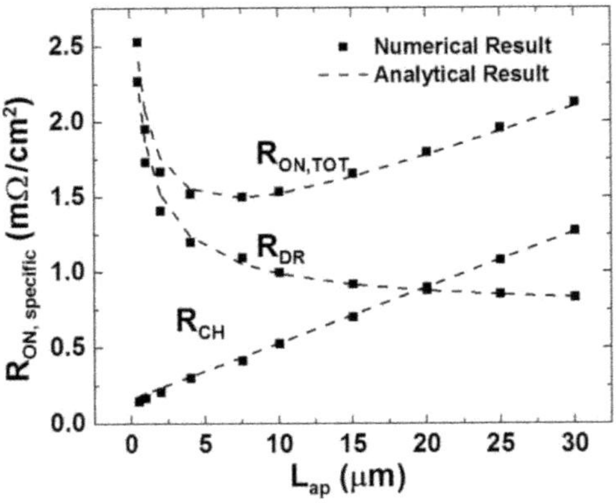

Mobility (drift region) in cm^2/Vs	V_{br} (V)	R_{on} $m\Omega/cm^1$ (Lateral channel JFET)
1100	1240	1.4
900	1260	1.7
700	1280	2.2
500	1300	3.1

Figure 6.14 $R_{ON,TOT}$, R_{DR}, and R_{CH} as a function of L_{ap}. The *solid squares* indicate the results obtained numerically, while the *dashed curves* show the analytical result (channel electron mobility: 1500 cm^2/Vs; bulk GaN electron mobility: 900 cm^2/Vs). Total R_{on} as a function of the electron mobility in the drift region is listed. Bulk GaN mobility will play a key role in distinguishing GaN vertical devices from SiC counterparts.
Source: Adapted from [41].

6.10 The channel thickness t_{UID} and effective gate length (L_{go})

Besides ensuring proper gate control and maintaining a good transconductance (g_m) in the device, channel thickness and effective gate length play another important role in these devices best understood from the leakage analysis of the device. Three main leakage paths exist and are shown in Fig. 6.15.

6.10.1 *Through the CBL*

If the CBL does not provide a high enough barrier to the current, current can flow from source to drain through it. This is an undesirable path for the current and can be of large significance in the device performance.

6.10.2 *Unmodulated electrons*

Under normal functioning, all the electrons from the source should flow through the 2DEG to the gate horizontally, and then vertically through the aperture to the drain.

However if the confinement of the electrons in the channel is not sufficient, then some electrons can find an easy path from the source to the drain via the conductive aperture, bypassing the gate. Such electrons add to the leakage current and should

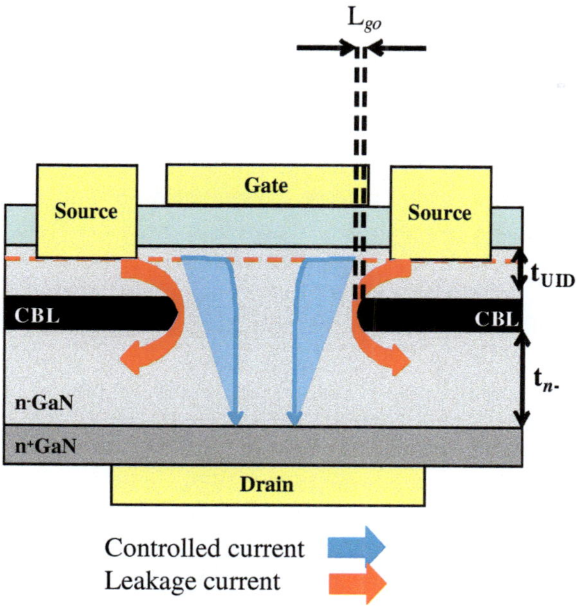

Figure 6.15 Three critical leakage paths in a CAVET: (1) through CBL, (2) unmodulated electrons, and (3) gate leakage.
Source: Adapted from [17].

be cut off from flowing. L_{go}, the length of the gate that overlaps the aperture (which is analogous to a std. HEMT gate length) and t_{UID} are the two knobs that control this path.

If L_{go} is small and or t_{UID} is large, the effective distance from the source to the aperture is decreased and the chance of electrons flowing to the drain, without being modulated by the gate, increases. So L_{go} has to be large enough to enhance gate control injected by the source. On the other hand, t_{UID} needs to be of minimum thickness.

However if t_{UID} is too small, the electrons in the 2DEG would be close to the CBL. This increases the chance of the electrons getting captured (especially if the CBL has traps) causing dispersion when the gate signal is pulsed.

This sets a lower limit in t_{UID} determined by the dispersion criterion.

6.10.3 Through the gate

The other leakage takes place through the gate and can be addressed by suitable gate dielectric or a p-GaN layer under the gate.

6.11 Current blocking layers

6.11.1 A discussion on doped versus implanted current blocking layer

The design of a CAVET involves designing a robust CBL, an aperture conductive enough to not choke the current but not so conductive to limit the breakdown voltage of the device and a low doped drift region to hold the blocking voltage. The CBL is designed to provide a barrier to the electron flowing from the source to the drain through any other path except the aperture. A p-type GaN layer therefore would be a very effective current block providing a barrier of over 3 eV. Although from the band diagram as high as 3 eV barrier could be achieved by a doping (Mg) level of $10^{19}/cm^3$ (1% active at RT) there are other fabrication related issues that make it less attractive. To complete the device fabrication, AlGaN/GaN (25 nm/140 nm) layers need to be regrown on top of the CBL. This implies that the p-doped layer underneath has to be activated. While buried p-layer activation is difficult to achieve, this scheme also necessitates regrowth++ on etched trenches. It was seen in previous studies, that etching of the aperture region exposed non-c planes and regrowth over such inclined facets resulted in a depression that propagated to the surface. Other studies [18,22] have found evidences that GaN, which is grown on facets other than ⟨0001⟩ plane, tends to incorporate large concentrations of n-type impurities. Since the peak electric field in a CAVET is located directly beneath the gate, such high n-type doping in that region causes an increase in the peak field limiting the device breakdown. Such highly doped region if under the gate metal would increase the gate leakage. Use of a good gate insulator could prove beneficial in terms of gate leakage, but premature breakdown because of increase in peak electric field, which is as a result of the unintentional high n-type impurities, still cannot be ruled out. One of the best alternative paths to circumvent these problems was identified to be the ion-implantation techniques to achieve current blocking functionality. The advantage of using an ion-implanted CBL comes from the fact that aperture layer need not be regrown. Fig. 6.16 illustrates the process flow of fabricating a CAVET with ion-implanted CBL. In this approach, the aperture layer is grown along with the base structure. Then the aperture is masked and implanted to form the CBL. Since this method involves no etching of the aperture layer, the regrowth of the AlGaN/GaN layers take place on the c-plane and remains planar. Both implanted and doped p-GaN CBL presents their distinctive realm of advantages and disadvantages. CBL obtained by [Mg] ion-implantation technique has proved to be advantageous over a partially activated Mg-doped CBL, particularly if the implanted Mg is activated using a very high temperature process. Although such high temperature activation process is a common practice in SiC, it required extensive modification and development to adapt to GaN. Recent reports demonstrated excellent p-type property of Mg-implanted GaN when activated at temperatures above 1400 °C [23]

Avogy, working on the commercialization of GaN vertical devices reported their devices based on doped CBL [15]. The critical most part of a doped p-GaN CBL

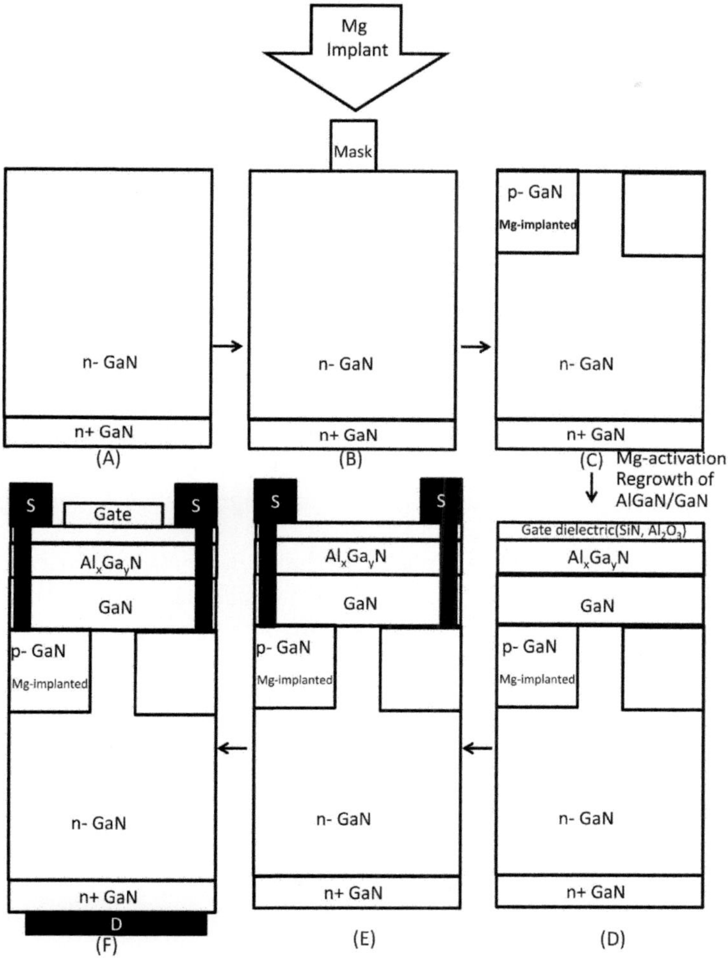

Figure 6.16 Fabrication steps of a CAVET. (A) Thick drift region (n − GaN) is grown on the drain (n + GaN) region. (B) Implantation of Mg is carried out protecting the aperture region of the device. (C) p-GaN implantation is achieved which should be activated (not shown) using high temperature multiple cycle rapid thermal annealing (RTA) scheme to activate the Mg ion. (D) Regrowth of AlGaN and GaN is done in MBE or MOCVD reactor to form the channel. An in situ gate dielectric is often grown if MOCVD regrowth technique is used. (E) Sources connected to the p-GaN region through via-holes are fabricated. (F) Gate and drain metal are deposited in separate steps to complete the device.

vertical FET is the realization of an excellent p–n junction, buried into the bulk GaN material. Other associated challenges related to regrowth on nonplanar surface, etc. have been well tackled with extensive etch and regrowth studies. The p–n junction controls the breakdown electric field (or more appropriately the electric field at the designed blocking voltage) as well as dispersion characteristics under switching operation. The predicted electric field in GaN is around 3 MV/cm and the operation close to the predicted limits relies on a nearly ideal p–n junction. Mg-doped CBL approach is discussed in the following section.

6.12 Trench-current aperture vertical electron transistor

A CAVET with the gate fabricated on the sidewalls of a trench, as shown in Fig. 6.17 [24,25] has been successfully demonstrated using Mg-doped p-GaN as the CBL. The sidewall gate geometry works well with the Mg-doped scheme. The trench CAVET relies on regrown AlGaN/GaN layers on the trench sidewall as the channel. The trench sidewall angle determines the polarization scale of the channel where a 90° angle indicates a nonpolar plane, and a 45° angle indicates a semipolar plane. The threshold voltage in a trench CAVET strongly depends on the trench sidewall angle, and can be used as a design parameter for normally-off devices.

The functioning of a trench CAVET is similar to a CAVET, except now the AlGaN/GaN channel continues at an angle on the sidewall. The high mobility 2D-electron gas, therefore sets the electron mobility of the channel, although it must be noted that the AlGaN/GaN channel is regrown on an etched trench. However, channel mobility in a trench CAVET as high as 1690 cm^2/Vs was reported [8].

In GaN device technology, where a robust oxide technology can still be a challenge, in situ grown Si_3N_4 on AlGaN/GaN structure has proven to be an excellent dielectric with low interface traps allowing JEDEC-standard reliability in HEMTs.

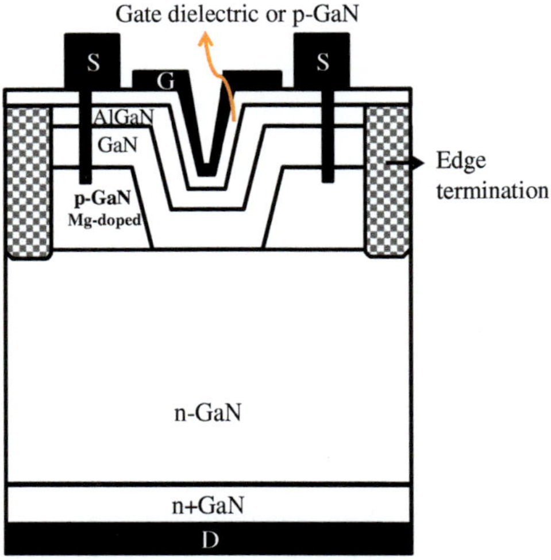

Figure 6.17 Schematic representation of a Trench CAVET with Mg-doped CBL. An MIS gate with SiN_x or Al_2O_3 or a junction gate with optimized p-type GaN are available options to allow the device a normally-off behavior.

In our work we took advantage of an in situ grown Si_3N_4 to cap the AlGaN/GaN to suppress gate leakage.

The first MIS gate trench CAVET was reported by Ji et al. in 2016 [24] on bulk GaN substrates. From Fig. 6.18 one can identify the process flow of a trench CAVET. The activation of the p-GaN layer occurs at two stages. First, after the base structure is grown ending in around 400 nm (for a 1.2 kV design) of p-GaN activation of the p-GaN is carried out in a baking over at 700 °C to break the Mg-H complexes and release hydrogen out of the crystal. However, when the AlGaN/GaN layers are regrown, the hydrogen can enter once again into the crystal lattice thereby deactivating the Mg in the CBL. Therefore, before the sources are

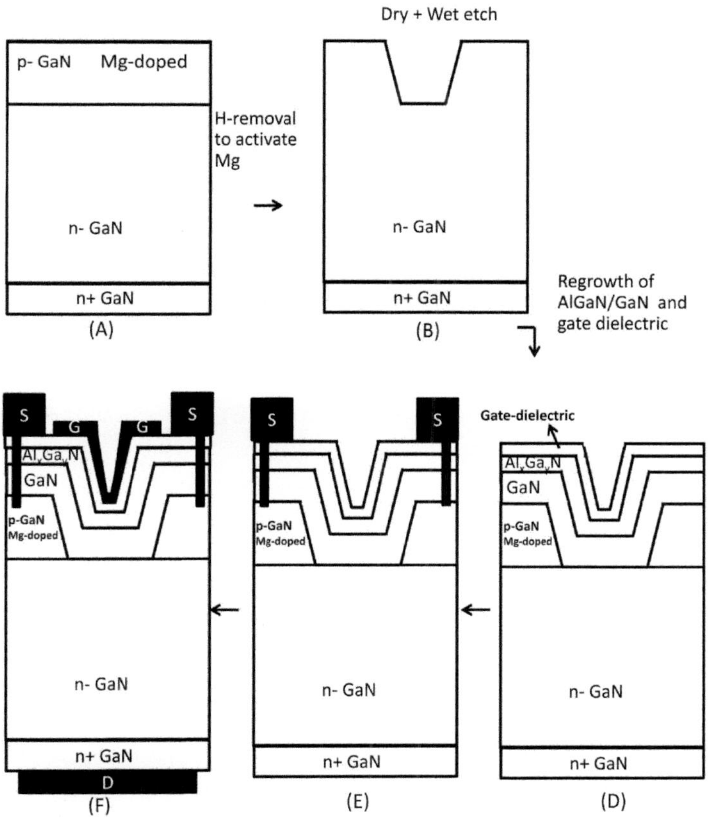

Figure 6.18 Fabrication steps of a trench CAVET. (A) Current blocking layer (p-GaN) and thick drift region (n − GaN) is grown on the drain (n + GaN) region; p-GaN is activated using 700 °C baking for 15 min, (B) the p-GaN is then etched using dry etch and sidewall is smoothened by wet TMAH etch. (C) Regrowth of AlGaN and GaN is done in MOCVD reactor to form the channel. An in situ gate dielectric is often grown in the same MOCVD or alternatively ALD oxide is deposited. (D) Sources connected to the p-GaN region through via-holes are fabricated. (E) Gate and drain metal are deposited in separate steps to complete the device.

fabricated, another 700 °C baking is done to diffuse out any trapped hydrogen from the crystal through the via-holes. Fig. 6.19 shows a scanning electron microscopy (SEM) cross-section of the p–n junction in the trench CAVET where p-GaN was successfully activated by dehydrogenation baking.

Although the first generation devices reported by Ji et al. had a breakdown voltage of 225 V [24] limited by the gate dielectric breakdown, subsequent improvements led to realization of over 880 V devices with an R_{on} less than 2.7 mΩ/cm^2 as shown in Fig. 6.20 [25]. It must be noted that controlling the threshold voltage in these devices could be difficult since it implies an accurate control of the slope and a very reproducible regrowth technology. In the 880 V device, regrown GaN layer was significantly n-type leading to a negative threshold voltage device, whereas the

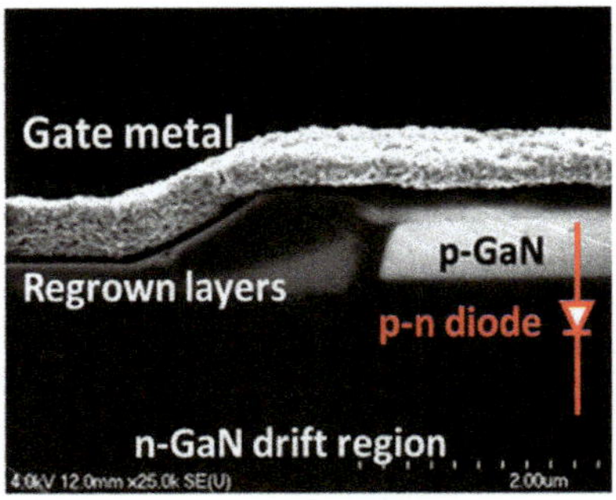

Figure 6.19 Cross-sectional SEM image of a trench CAVET after activation of Mg by releasing the hydrogen.
Source: Adapted from [24].

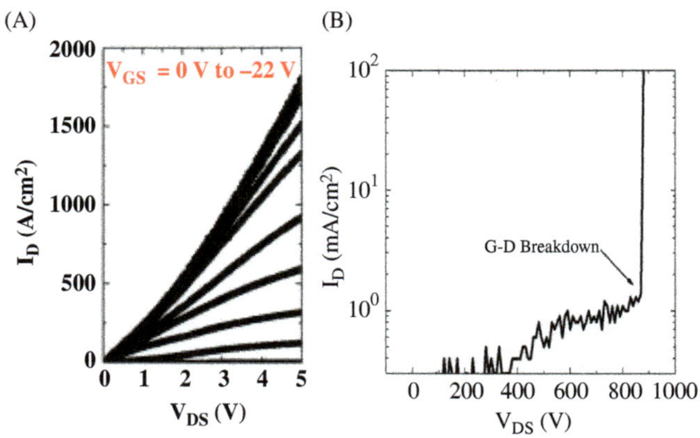

Figure 6.20 (A) Drain and (B) breakdown characteristics of a trench CAVET.
Source: Adapted from [25].

225 V device showed a very positive threshold voltage. The p-GaN layer as a gating layer can be an alternative to the aforesaid dielectric approach. The p-GaN gate structure is widely used in normally-off lateral GaN HEMTs. Due to the high electron density of 2DEG induced by the polarization charge, the threshold voltage of p-GaN gated HEMT is typical less than 2 V. However, a more positive threshold voltage can be realized in a p-GaN gated trench CAVET with the channel formed at the semipolar plane as demonstrated by Shibata et al. in 2016 [26]. On a p−n epitaxial structure grown on bulk GaN substrate, a "V"-shaped trench was formed using inductively coupled plasma etching, the p-GaN/AlGaN/GaN triple layers were regrown over the trench by MOCVD. Because the channel is located at the semipolar plane instead of c-plane, the threshold voltage shifted toward the positive side by 1.5 V. A high positive threshold voltage of 2.5 V was demonstrated with blocking voltages as high as 1700 V using a drift region thickness of 13 μm. Remarkably low specific on-resistance of 1 $m\Omega/cm^2$ was reported in this device. An important feature of the vertical GaN transistor reported by Shibata et al. [26] is the insertion of carbon-doped layer on top of the p-GaN well/n-GaN drift epitaxial layer on GaN substrate. Here carbon-doped GaN layer served as a semi-insulating layer, because carbon introduces deep acceptor levels. The high voltage definitely underlines the role of a robust blocking layer offered by the combination of C-doped GaN and Mg-doped GaN layers.

A variant of a trench CAVET with a p-GaN gating, therefore called Vertical JFET (V-JFET), is seen in the work reported by Nie et al., [15] where the regrowth of AlGaN/GaN layers was designed to fill in the trench completely. The devices demonstrated excellent DC characteristics blocking over 1.5 kV and 2.2 $m\Omega/cm^2$. Fig. 6.21 shows Avogy's device schematic adopted from their reported work [15].

6.13 Metal oxide semiconductor field-effect transistor

GaN MOSFETs are the other branch of devices that are showing promising performance offering a normally-off solution which is a significant drawback of any GaN HEMT-based design.

There are two types of MOSFET reported so far: (1) non-regrowth-based MOSFET and (2) regrowth-based MOSFET. The latter one involves regrowth of a GaN interlayer and therefore call **O**xide, **G**aN interlayer FET (OGFET).

6.13.1 Non-regrowth-based metal oxide semiconductor field-effect transistors

Since the developments of dry etch technology in the early 1990s, the trench MOSFET has been a dominant device structure for power electronics [4]. To date, both Si- and SiC-based trench MOSFET have been commercialized and shown excellent performance. However, because of the absence of bulk GaN substrate, the GaN-based MOSFET development has been sporadic until recently. The first GaN

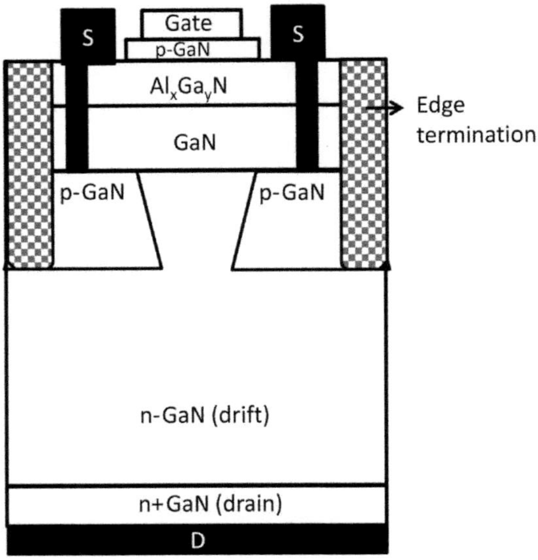

Figure 6.21 A schematic diagram of a V-JFET using Mg-doped CBL and a regrown aperture and channel region. In this type of devices, the trench is filled in and planarized before the AlGaN/GaN layers are regrown.
Source: Figure adapted from [15].

MOSFET was reported by Otake et al. [27] in 2007; the device offering a remarkably high threshold voltage of 5.1 V. After 7 years Oka et al. reported a 1.6 kV device in 2014 [28].

Fig. 6.22 shows the structure of a vertical GaN MOSFET. P-base and n-drift regions form the main p–n junction between source and drain. The device breakdown voltage is determined by the reverse characteristics of this p–n junction. An n+ source region is formed partially on top of the p-base region, the junction between the n+ source region and the p-base region is shorted by source contact to improve the breakdown voltage by eliminating the n–p–n open base effect. The channel is formed at the etched sidewall as inversion layer of the MOS structure.

Compared to the CAVET structure, there are two basic advantages of the MOSFET: (1) the MOSFET is a readily normally-off device with a high threshold voltage over 2 V and (2) the absence of the regrowth makes the process significantly simple, reducing the cost. These advantages of the MOSFET make it an attractive design for vertical GaN transistors.

However, for GaN MOSFET, the biggest challenge lies in the poor electron mobility in the channel of the device. During on-state of the device, the electrons flow through the inversion layer of the sidewall MOS structure, the channel electron mobility is limited by the surface roughness and impurity scattering. Another issue along with the poor channel property is the dielectric (oxide) reliability. The GaN MOSFET cannot be extensively recognized without a strong reliability track.

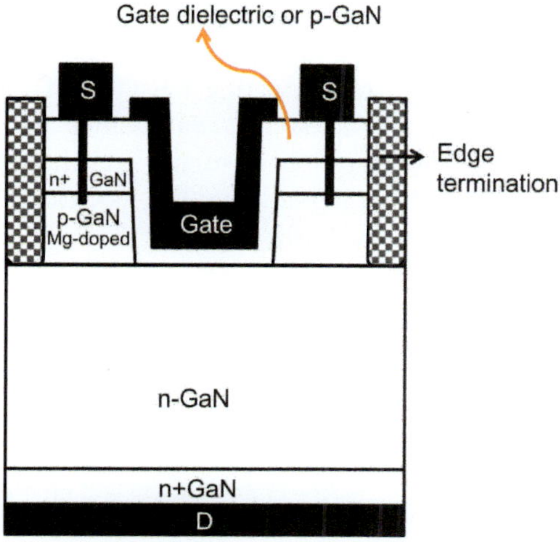

Figure 6.22 Structure of conventional GaN MOSFET without any regrowth involved.

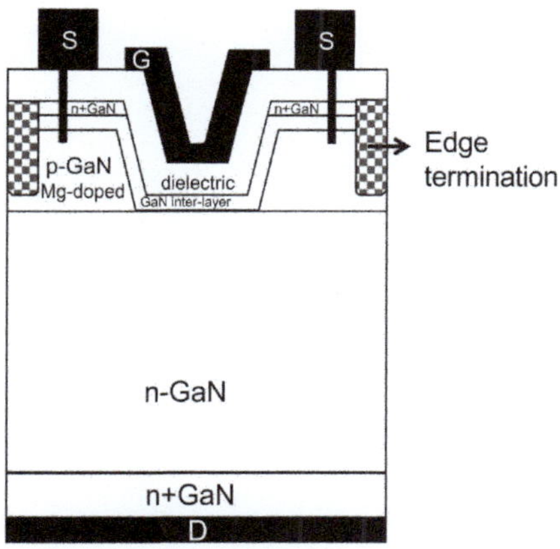

Figure 6.23 Structure of GaN OGFET.

6.13.2 Regrowth-based metal oxide semiconductor field-effect transistor (OGFET)

GaN OGFET is a modified structure based on the conventional trench MOSFET as shown in Fig. 6.23. Compared to the conventional trench MOSFET, the OGFET has two features: (1) an unintentional-doped (UID) GaN interlayer is used as the channel region, which enhance the channel electron mobility to reduce the coulomb scattering by the dopants; (2) the oxide was in situ grown by MOCVD, which reduces the interface states, and improve the gate oxide reliability. The novelty of the OGFET lies in enhancing the channel electron mobility without scarifying the normally-off behavior. The process flow is shown in Fig. 6.24.

The working principle of the OGFET is similar to the MOSFET. The OGFET is designed such that under a zero V_{GS}, the electrons in the regrown-thin-GaN layer are depleted by the p-GaN base region taking the OGFET into its off-state. The energy band diagram and the simulated electron contour are shown in Fig. 6.25. The p–n diode formed by the p-GaN base region and the n-GaN drift region is used to hold the high off-state blocking voltage. The electric field distribution along the p-GaN base region and the n-GaN drift region is shown in Fig. 6.26.

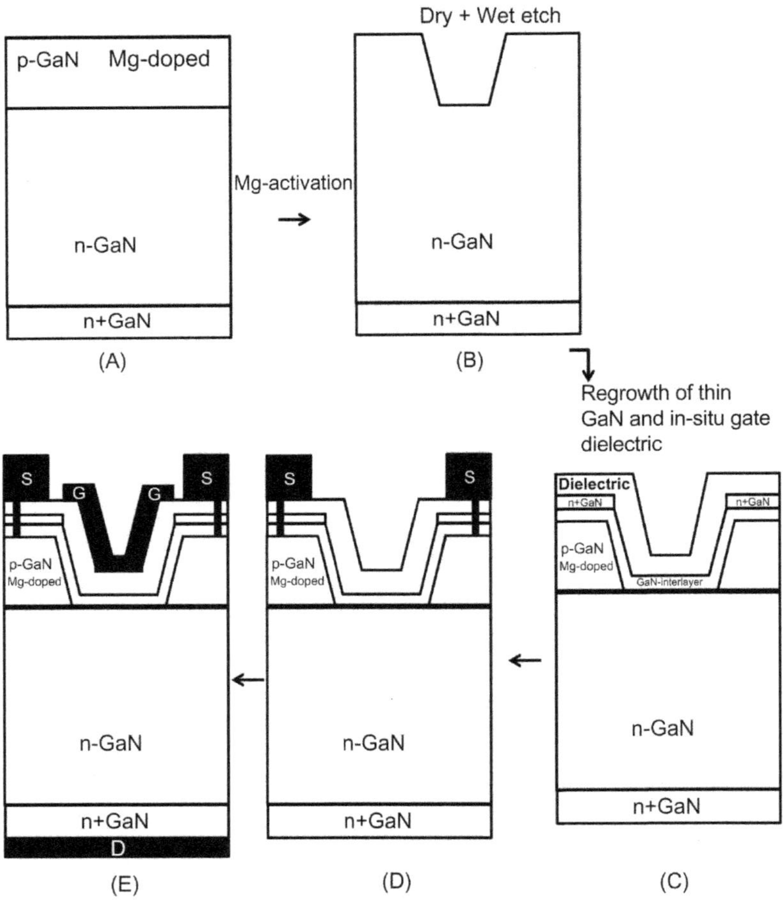

Figure 6.24 Fabrication steps of an OGFET. (A) Current blocking layer (p-GaN) and thick drift region (n − GaN) are grown on the drain (n + GaN) region; p-GaN is activated using 700 °C baking for 15 min, (B) the p-GaN is then etched using dry etch and sidewall is smoothened by wet Tetramethylammonium hydroxide (TMAH) etch. (C) Regrowth of a thin layer of GaN is done in MOCVD reactor to form the channel. This regrowth of thin GaN distinguishes OGFET from an MOSFET. An in situ gate dielectric is grown in the same MOCVD or alternatively ALD oxide is deposited. (D) Sources connected to the p-GaN region through via-holes are fabricated. (E) Gate and drain metal are deposited in separate steps to complete the device.

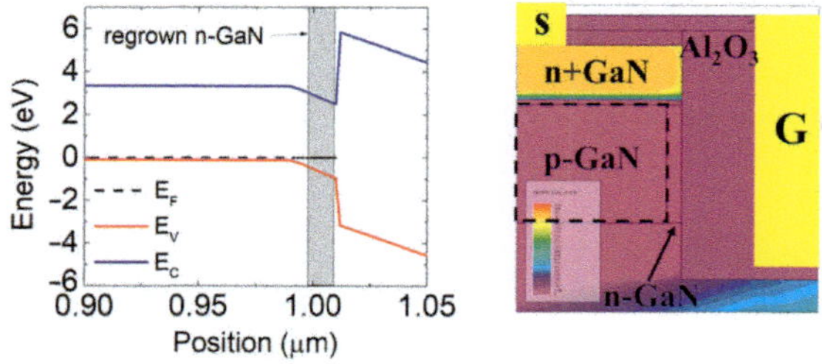

Figure 6.25 Energy band diagram and the electron distribution in an its OGFET during its off-state [19].

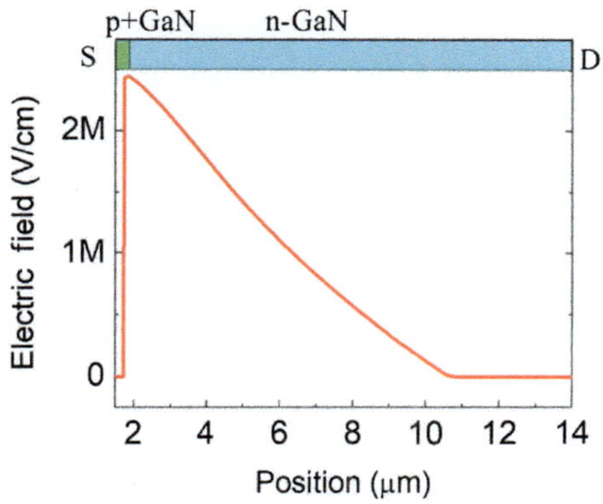

Figure 6.26 Electric field distribution along the drift region.

Under a positive V_{GS} (15 V), the electrons are accumulated in the UID GaN insert layer, and the transistor is turned into its on-state. The energy band diagram and the electron concentration are shown in Fig. 6.27. Because of the enhanced channel electron mobility, the OGFET has a lower $R_{on,sp}$ compared to the conventional trench MOSFET.

In 2016, Gupta et al., reported the first OGFET results based on the sapphire substrates where the device showed a 60% $R_{on,\ sp}$ reduction compared to a conventional MOSFET while main the threshold voltage >2 V [29]. In 2017, they further demonstrated an OGFET on bulk GaN substrates with an breakdown voltage of 990 V and a low $R_{on,\ sp}$ of 2.6 mΩ/cm^2 [30]. In the same year, Ji et al., demonstrated a high performance OGFET with an breakdown voltage over 1.43 kV and a low $R_{on,sp}$ of 2.2 mΩ/cm^2 [31].

Using a 10 nm thick, unintentionally doped, regrown GaN interlayer as the channel, a low $R_{on,sp}$ of 2.2 mΩ/cm^2 was achieved, establishing a superior on-state performance [19,31]. The drain characteristics of the fabricated OGFET are shown in Fig. 6.28. Fig. 6.29 shows the transfer I_D-V_{GS} characteristics and the gate leakage. The threshold voltage, V_{TH}, defined at a current level of 10^{-4} A/cm^2 ($I_{on}/I_{off} = 10^6$), obtained was 4.7 V (when V_{GS} sweeps up). A clockwise hysteresis of

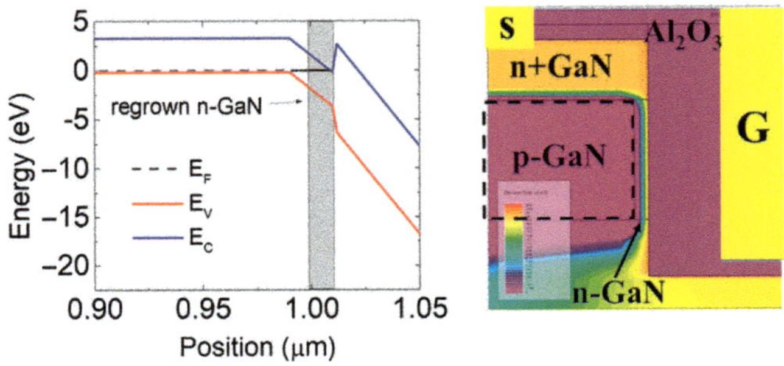

Figure 6.27 Energy band diagram and the electron distribution in an OGFET under its on-state [19].

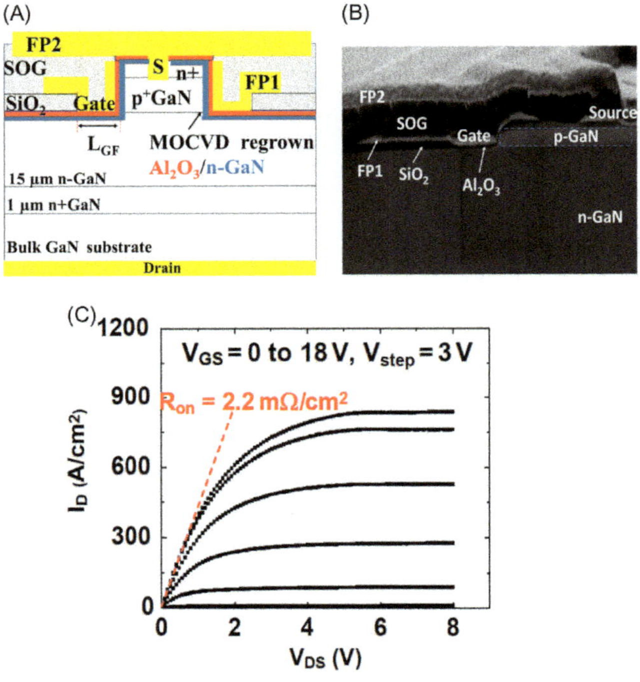

Figure 6.28 (A) A two-field plated OGFET reported by Ji et al. at IEDM 2017 with its cross-sectional SEM view in (B). (C) $I-V$ characteristics of fabricated single unit cell OGFET with saturation current density of 850 A/cm^2 and $R_{on,sp}$ of 2.2 mΩ/cm^2.
Source: Adapted from [31].

ΔV_{TH} of 0.3 V was observed. Low subthreshold slope of 283 mV/decade was measured from $I_D = 10^{-5}$to 10^{-2} A/cm^2. Fig. 6.30 shows the off-state measurement of a unit cell device with an L_{GF} of 0.5 μm under a V_{GS} of −10 V. It is shown that a breakdown voltage of 1435 V was obtained at a current level of 50 mA/cm^2. Scaling in current implies arranging unit OGFETs in a hexagonal-closed-packing layout resembling a honeycomb as shown in Fig. 6.31

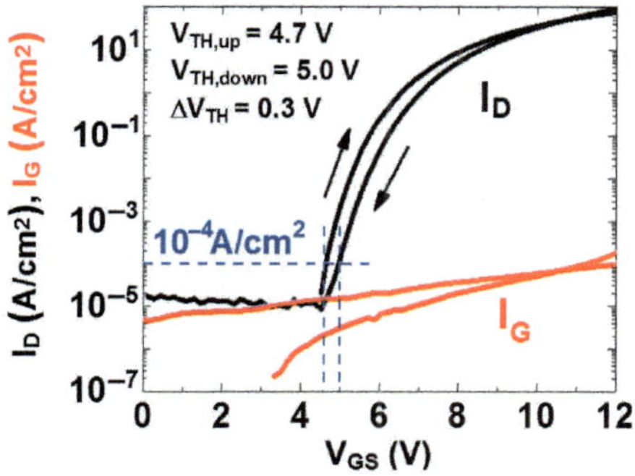

Figure 6.29 Transfer characteristics of the fabricated single unit cell OG-FET (*black curves*) and gate leakage (*red curves*). The threshold voltages of sweep up and down were 4.7 and 5 V (defined at a current level of 10^{-4} A/cm^2). The subthreshold slope was 283 mV/decade measured from 10^{-5} to 10^{-2} A/cm^2.
Source: Adapted from [31].

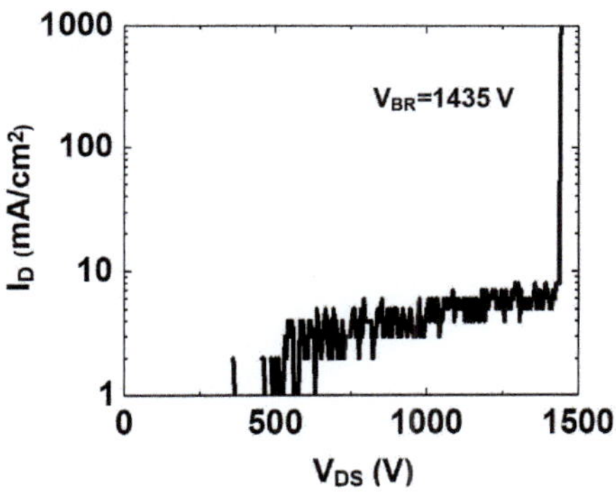

Figure 6.30 Off-state characteristics of the fabricated OG-FET. Breakdown voltage (V_{BR}) was 1435 V defined at off-state leakage of 50 mA/cm^2.
Source: Adapted from [31].

6.13.3 OGFET switching performance

Based on the reported 1.4 kV OGFET fabricated on the bulk GaN substrate [19], a physics-based device model was developed, and then integrated it with a circuit model to study the dynamic characteristics and power losses. The switching waveforms of the transistor during the turn-off and turn-on transients in a double-pulse test circuit are shown in Figs. 6.32 and 6.33 The total turn-off time is ~30 ns, while the turn-on time is 47 ns. The total gate charge is ~89 nC, which includes a Q_{GD} of ~17 nC.

Table 6.2 shows the comparison of CREE SiC MOSFET [32], simulated SiC MOSFET, simulated GaN CAVET, and simulated GaN OGFET with the same voltage and current rating [19]. Owing to the high channel electron mobility as well as the bulk electron mobility, GaN vertical devices show faster switching speed and lower energy loss.

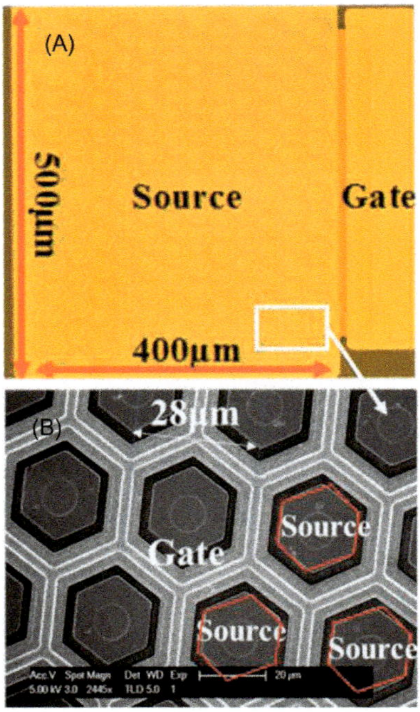

Figure 6.31 (A) A cell layout of large area OGFET along with the (B) SEM view of a unit cell. *Source:* Adapted from [31].

6.14 GaN-high voltage diodes

SiC high voltage diodes in the form of Schottky barrier (SBD), junction bipolar Schottky (JBS), and p–i–n diodes have remarkably filled in the space beyond Si in power applications with a growing market in discrete parts.

GaN diodes present similar potential to serve the future power electronics market, particularly in integrated power modules. The Schottky barrier in GaN is consistently reported to be between 0.9 and 1 V due to Fermi-level pinning, which allows Schottky diodes with turn-on voltage between 1 and 2 V based on particular design. p–i–n diodes in GaN are of high interest due to their well-behaved turn-on voltage (around 3 V) and good forward current density.

In 2013, Kizilyalli et al. reported vertical p–n diodes (Fig. 6.34) fabricated on pseudobulk gallium nitride (GaN) substrates [33,34].

The blocking voltage in a p–n diode is determined and designed by the n-type drift-layer doping and the thickness. Typical drift-layer thicknesses for 600–1.7 kV diodes are between 6 and 20 μm with targeted doping densities of $N_D \approx (1-3)\times 10^{16}\ cm^{-3}$. The p+ region is achieved by in situ growth of Mg-doped $((5-10)\times 10^{19})\ cm^{-3}$ GaN epitaxial layer on top of the n-type GaN epitaxial drift region.

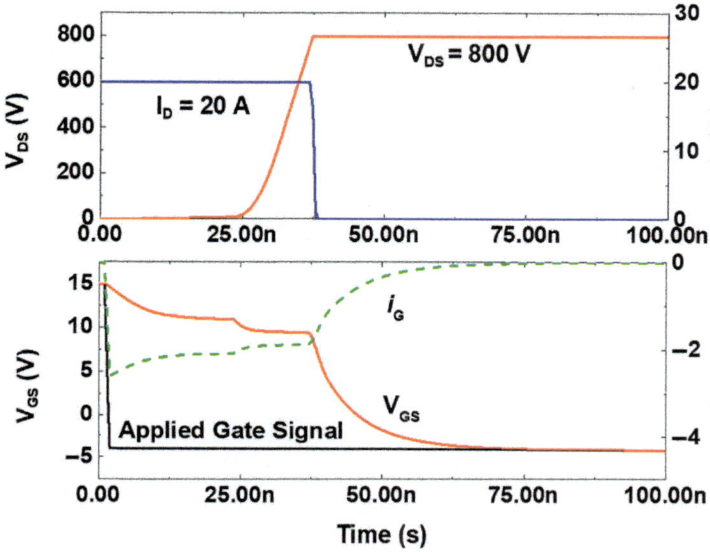

Figure 6.32 Waveforms in an OGFET during the turn-off transient. The $t_{d(off)}$ is 21.3 ns, t_f is 9 ns, and the $\Delta v/\Delta t$ is 71 kV/μs. The gate charge is 89 nC. The gate resistance is 7.5 Ω, and the frequency is 100 kHz.
Source: Adapted from [19].

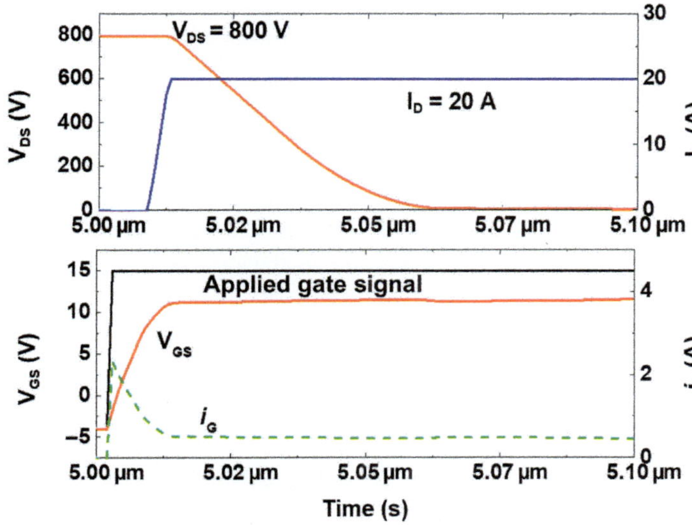

Figure 6.33 Waveforms in an OGFET during the turn-on transient. The $t_{d(on)}$ is 14 ns, t_r is 33 ns, and the $\Delta v/\Delta t$ is 19.4 kV/μs.
Source: Adapted from [19].

Table 6.2 Comparison of simulated cascoded CAVET, simulated GaN OGFET, simulated SiC MOSFET, and SiC MOSFET from CREE [19]

Parameters	CREE SiC MOSFET	Simulated SiC MOSFET	Simulated GaN CAVET	Simulated GaN OGFET
V_{BR} (kV)	1.2	1.2	1.3	1.4
R_{on} (mΩ)	80	80	80	75
$T_{off\text{-}delay}$ (ns)	40	55	29.5	21.3
T_f (ns)	38	25	18	9
$T_{on\text{-}delay}$ (ns)	13	10	2	14
T_r (ns)	24	27	16	33
Q_G (nC)	90.8	157	88	89
E_{on} (μJ)	305	312	57	348
E_{off} (μJ)	305	304	152	92
E_{ts} (μJ)	610	616	209	440

Hole concentration of $(5-10) \times 10^{17}$ cm^{-3} and a hole mobility of between 10 and 50 cm^2/Vs at 25 °C are normally expected from Hall effect measurements. Appropriate edge termination techniques are required for high voltage performance.

In their work, Kiziyalli et al. measured devices demonstrating breakdown voltages of 2600 V with a differential specific on-resistance of 2 mΩ/cm^2. This performance placed GaN diodes beyond the SiC theoretical limit on the power device figure-of-merit chart as shown in Fig. 6.35 adopted from the work of Kiziyalli et al. Moreover, these diodes demonstrated avalanche capability contrary to common belief. That GaN devices do not avalanche was a common belief developed largely by the lack of avalanching in unipolar GaN HEMTs. The temperature coefficient of the breakdown voltage was reported positive, showing that the breakdown is indeed because of impact ionization and avalanche. In their work, the GaN diodes were driven into avalanche breakdown using 30 mS and 15 mA current pulses demonstrating avalanche energy capability of 1000 mJ. The reverse recovery time of the vertical GaN p−n diode was not noticeable, as the authors suggest, "because it was limited by capacitance rather than minority carrier storage, and because of this its switching performance exceeds the highest speed silicon diode." A snap shot of the GaN diode characteristics reported in by the same team is shown in Fig. 6.36.

6.15 Edge termination, leakage, and active area of the device

In this chapter I have discussed devices in terms of their design and fabrication techniques. While edge termination structures have been included in the schematics, I have not described it in detail. Edge-termination structures and techniques, often a trade secret or proprietary to the companies, are of unique importance and often

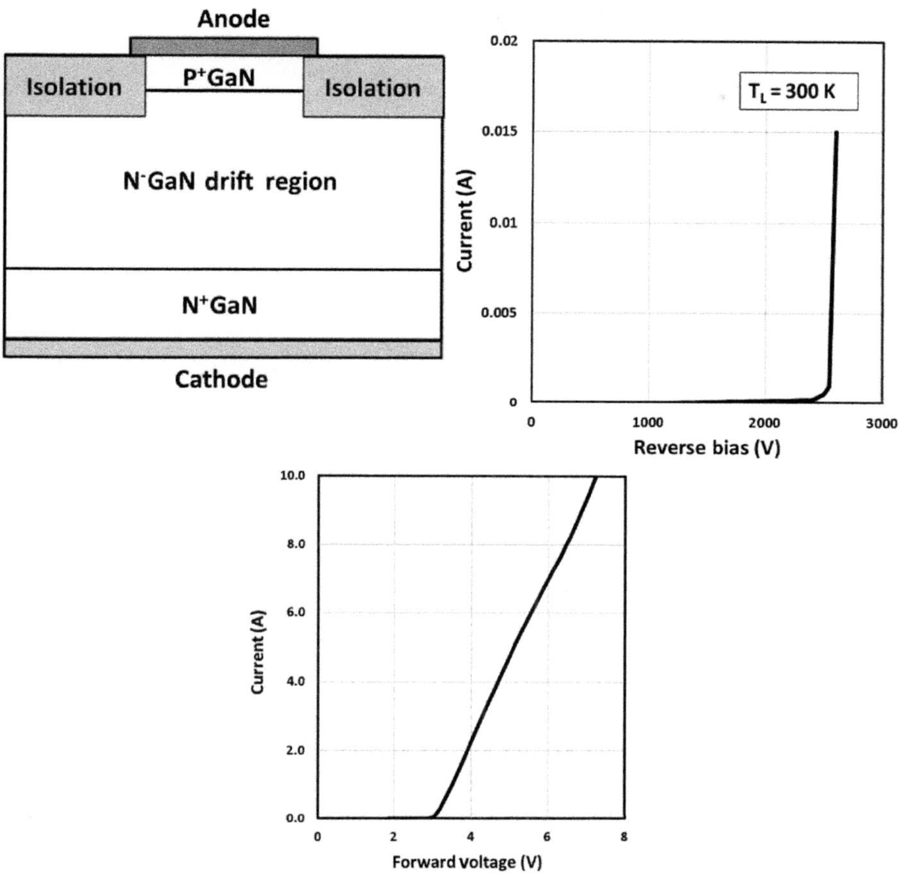

Figure 6.34 A schematic cross-sectional view of the vertical GaN p–n diode on bulk GaN [34].

very little discussed in the literature. Junction termination based on ion-implantation with Mg (often nitrogen, aluminum, and others) is carried out along the edge of the devices to supply enough surface area with negative charge to image the positive charges in a depleted drift region. FP termination techniques have the benefit of being simple with fabrication process and that it avoids defects induced reverse leakage current associated with implantation-based JTE terminations. However, severe electric field crowding in the oxide associated with the FP technique can cause reliability concern and limit device breakdown voltage. Fig. 6.37 depicts the role of FPs in reducing the peak electric field in an OGFET allowing the breakdown voltage to exceed 1400 V [19,31].

In this section I would like to emphasize that leakage current in GaN vertical devices, a basis of reporting blocking voltage, doesn't often follow a universal standard like in Si or even SiC. This is due to the early stage of development that GaN devices are currently undergoing. Among the cited work in this chapter the blocking voltage reported by various groups are often for leakage currents between 1 mA/cm^2

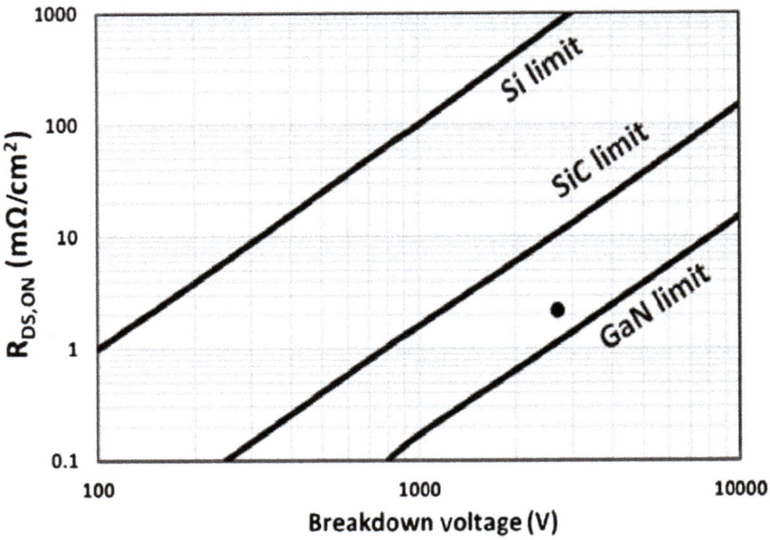

Figure 6.35 Power device figure-of-merit comparison of p–i–n diodes reported by Kiziyalli et al. compared to Si, SiC, and GaN theoretical values.
Source: Adapted from [34].

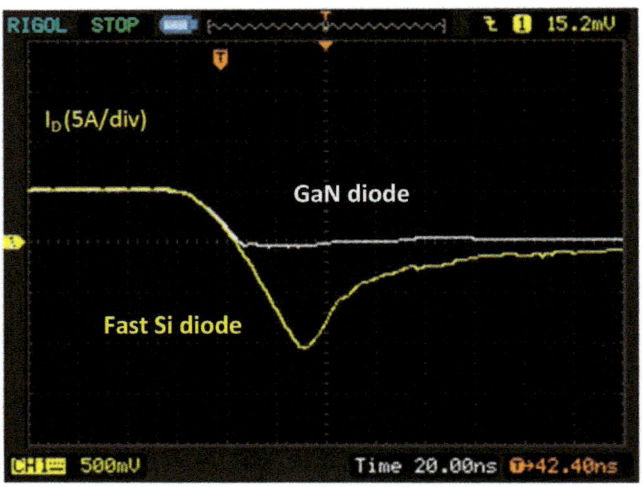

Figure 6.36 Double pulse testing of vertical GaN p–n diode and a high speed 1200 V rated Si diode.
Source: Adapted from [34].

and 1 μA/cm^2. The range is large, however, the acceptability of leakage current in power devices is strictly application based. An acceptable rule of thumb for leakage current is often regarded as 1 μA/cm^2, when normalized by the device area. The device area in unit devices are often measured from source-to-source edges. However, in this

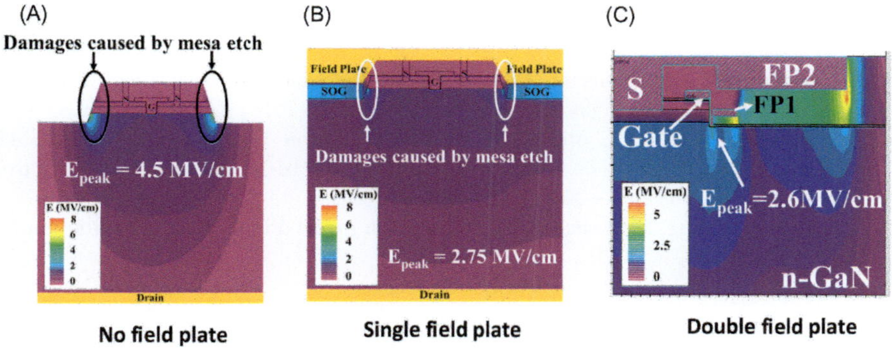

Figure 6.37 An OGFET simulated with 0, 1, and 2 FPs to manage the peak electric field occurring p–n junction shows the role of FP in reducing the peak electric field.
Source: Adapted from [31].

approximation, the current spreading is ignored and therefore current densities are overestimated and therefore normalized R_{on}s are underestimated. The best practice is to take the whole die area as an active device area, which includes the pads and edge terminations often. For large area device, the active area is typically the die area.

6.16 Conclusion

The chapter accounts the history and on-going development of vertical GaN devices, while discussing the design and fabrication techniques of the devices. The availability of single crystalline GaN substrates unlocked the opportunity to explore GaN-on-GaN devices. The research in this field is growing in both quality and quantity [26,31,35–37] opening up unprecedented possibilities to serve the future power electronics market.

Acknowledgment

First of all, I would like to thank my former student Dr. Dong Ji for his contribution to the devices discussed in this chapter. A significant amount of the work on OGFET is discussed in this chapter and was carried out by Dong during his PhD work with me.

My sincere acknowledgment goes to Prof. Umesh Mishra, my PhD advisor, with whom the early work on CAVETs took place between 2006 and 2010 during my PhD years, and later we collaborated on OGFETs.

I sincerely thank Prof. Baliga for creating this opportunity to share our understanding of GaN vertical devices with a broader audience. His books on power devices have been an invaluable source of enrichment for me.

Last but not the least, I am very thankful toToyota Motor Corporation and ARPA-E for funding and supporting the vertical GaN device work.

References

[1] S. Chowdhury, U.K. Mishra, Lateral and vertical transistors using the AlGaN/GaN heterostructure, IEEE Trans. Electron Devices 60 (10) (2013) 3060–3066.

[2] Y.-F. Wu, J. Gritters, L. Shen, R.P. Smith, B. Swenson, kV-class GaN-on-Si HEMTs enabling 99% efficiency converter at 800 V and 100 kHz, IEEE Trans. Power Electron. 29 (6) (June 2014) 2634–2637.

[3] J. Honea, J. Kang, High-speed GaN switches for motor drives, Power Electron. Europe 3 (2012) 38–41.

[4] B. Jayant Baliga, Fundamentals of Power Semiconductor Devices, Springer, New York, 2008.

[5] T. Ueda, T. Tanaka, D. Ueda, Gate injection transistor (GIT)-A normally-off AlGaN/GaN power transistor using conductivity modulation, IEEE Trans. Electron Devices 54 (12) (2007) 3393–3399.

[6] M. Kanamura, T. Ohki, T. Kikkawa, K. Imanishi, T. Imada, A. Yamada, et al., Enhancement-mode GaN MIS-HEMTs with n-GaN/i-AlN/n-GaN triple cap layer and high-k gate dielectrics, IEEE Electron Device Lett. 31 (3) (2010) 189–191.

[7] Y. Cai, Y. Zhou, K.J. Chen, K.M. Lau, High-performance enhancement-mode AlGaN/GaN HEMTs using fluoride-based plasma treatment, IEEE Electron Device Lett 26 (7) (2005) 435–437.

[8] Y. Niiyama, H. Kambayashi, S. Ootomo, S.T. Nomura, S. Yoshida, T.P. Chow, Over 2 A operation at 250°C of GaN metal-oxide semiconductor field effect transistors on sapphire substrates, Jpn. J. Appl. Phys. 47 (9) (2008) 7128–7130.

[9] I. Ben-Yaacov, Y.-K. Seck, U.K. Mishra, S.P. Denbaars, AlGaN/GaN current aperture vertical electron transistors with regrown channels, J. Appl. Phys. 95 (4) (2004) 2073–2078.

[10] Y. Gao, A. Stonas, I. Ben-Yaacov, U. Mishra, S. Denbaars, E. Hu, AlGaN/GaN current aperture vertical electron transistors fabricated by photoelectrochemical wet etching, Electron. Lett. 39 (1) (2003) 148.

[11] S. Chowdhury, B.L. Swenson, U.K. Mishra, Enhancement and depletion mode AlGaN/GaN CAVET with Mg-ion-implanted GaN as current blocking layer, IEEE Electron Device Lett. 29 (6) (2008) 543–545.

[12] S. Chowdhury, M.H. Wong, B.L. Swenson, U.K. Mishra, CAVET on bulk GaN substrates achieved with MBE-regrown AlGaN/GaN layers to suppress dispersion, IEEE Electron Device Lett. 33 (1) (2012) 41–43.

[13] M. Kanechika, M. Sugimoto, N. Soeima, H. Ueda, O. Ishiguro, M. Kodama, et al., A vertical insulated gate AlGaN/GaN hetrojunction field effect transisitor, Jpn. J. Appl. Phys. 46 (21) (2007) L503–L505.

[14] H. Otake, K. Chikamatsu, A. Yamaguchi, T. Fujishima, H. Ohta, Vertical GaN-based trench gate metal oxide semiconductor field-effect transistors on GaN bulk substrates, Appl. Phys. Express 1 (2008) 011105–1–011105-3.

[15] H. Nie, Q. Diduck, B. Alvarez, A. Edwards, B. Kayes, M. Zhang, et al., 1.5 kV and 2.2 mΩ/cm^2 vertical GaN transistors on bulk-GaN substrates, IEEE Electron Device Lett 35 (9) (2014) 939–941.

[16] T. Oka, T. Ina, Y. Ueno, J. Nishii, 1.8 mΩ cm^2 vertical GaN-based trench metal–oxide–semiconductor field-effect transistors on a free-standing GaN substrate for 1.2-kV-class operation, Appl. Phys. Express 8 (5) (2015) 054101.

[17] S. Chowdhury, PhD thesis, AlGaN/GaN CAVETs for high power switching application, 2010, ProQuest Dissertations Publishing, University of California, Davis.
[18] I. Ben Yaacov, PhD thesis, AlGaN/GaN current aperture vertical electron transistor, 2004, ProQuest Dissertations Publishing, University of California, Davis.
[19] D. Ji, Electrical and Computer Engineering Department, UC Davis, 2017, Design and Development of GaN-Based Vertical Transistors for Increased Power Density in Power, ProQuest Dissertations Publishing, University of California, Davis.
[20] S. Mandal, Electrical and Computer Engineering Department, UC Davis, 2017, Gallium Nitride Vertical Devices for Power Electronics Applications, ProQuest Dissertations Publishing, University of California, Davis.
[21] D. Ji, S. Chowdhury, Design of 1.2 kV power switches with low R_{ON} using GaN-based vertical JFET, IEEE Trans. Electron Devices 62 (8) (2015) 2571–2578.
[22] I. Ben-Yaacov, Y.-K. Seck, U.K. Mishra, S.P. DenBaars, AlGaN/GaN current aperture vertical electron transistors with regrown channels, J. Appl. Phys. 95 (4) (2004) 2073. Available from: https://doi.org/10.1063/1.1641520.
[23] T. Anderson, F. Kub, C. Eddy, J. Hite, B. Feigelson, M. Mastro, et al., Activation of Mg implanted in GaN by multicycle rapid thermal annealing, Electron. Lett. 50 (3) (2014) 197–198.
[24] D. Ji, M.A. Laurent, A. Agarwal, W. Li, S. Mandal, S. Keller, et al., Normally OFF trench CAVET with active Mg-doped GaN as current blocking layer, IEEE Trans. Electron Device 64 (3) (2016) 805–808. Available from: https://doi.org/10.1109/TED.2016.2632150.
[25] D. Ji, A. Agarwal, H. Li, W. Li, S. Keller, S. Chowdhury, 880 V/2.7 mΩ cm^2 MIS gate trench CAVET on bulk GaN substrates, *IEEE Electron Device Lett.* 39 (6), 863–865.
[26] D. Shibata, R. Kajitani, M. Ogawa, K. Tanaka, S. Tamura, T. Hatsuda, et al., 1.7 kV/1.0 mΩ/cm^2 normally-off vertical GaN transistor on GaN substrate with regrown p-GaN/AlGaN/GaN semipolar gate structure, Proceedings of IEEE Electron Devices Meeting (IEDM), 2016, pp. 248–251. https://doi.org/10.1109/IEDM.2016.7838385.
[27] H. Otake, et al., GaN-based trench gate metal oxide semiconductor field effect transistors with over 100 cm^2/(Vs) channel mobility, Jpn. J. Appl. Phys. 46 (7L) (2007) L599–L601.
[28] T. Oka, Y. Ueno, T. Ina, K. Hasegawa, Vertical GaN-based trench metal oxide semiconductor field-effect transistors on a free-standing GaN substrate with blocking voltage over 1.6 kV, Appl. Phys. Express 7 (2) (2014) 021002. Available from: https://doi.org/10.7567/APEX.7.021002.
[29] C. Gupta, S.H. Chan, Y. Enatsu, A. Agarwal, S. Keller, U.K. Mishra, OG-FET: An in-situ oxide, GaN interlayer based vertical trench MOSFET, IEEE Electron Device Lett. 37 (12) (2016) 1601–1604. Available from: https://doi.org/10.1109/LED.2016.2616508.
[30] C. Gupta, C. Lund, S.H. Chan, A. Agarwal, J. Liu, Y. Enatsu, et al., In-situ oxide, GaN interlayer based vertical trench MOSFET (OG-FET) on bulk GaN substrates, IEEE Electron Device Lett. 38 (3) (2017) 353–355. Available from: https://doi.org/10.1109/LED.2017.2649599.
[31] D. Ji, C. Gupta, S.H. Chan, A. Agarwal, W. Li, S. Keller, et al., Demonstrating $>$ 1.4 kV OG-FET performance with a novel double field-plated geometry and the successful scaling of large-area devices, IEDM Tech. Dig. pp. 9.4.1-9.4.4 Dec. 2017
[32] CMF 20120D datasheet, Available: http://www.cree.com/~/media/Files/Cree/Power/Data%20Sheets/CMF20120D.pdf.
[33] D. Ji, W. Li, S. Chowdhury, Switching Performance Analysis of GaN OG-FET Using TCAD Device-Circuit-Integrated Model, ISPSD, Chicago, 2018.

[34] I.C. Kizilyalli, A. Edwards, H. Nie, et al., High voltage vertical GaN p-n diodes with avalanche capability, IEEE Trans. Electron Devices 60 (10) (2013) 3067–3070.
[35] R. Li, Y. Cao, M. Chen, R. Chu, 600V/1.7 Ω normally-off GaN vertical trench metal-oxide-semiconductor field-effect transistor, IEEE Electron Device Lett. 37 (11) (2016) 1466–1469. Available from: https://doi.org/10.1109/LED.2016.2614515.
[36] M. Sun, Y. Zhang, X. Gao, T. Palacios, High-performance GaN vertical fin power transistors on bulk GaN substrates, IEEE Electron Device Lett. 38 (4) (2017) 509–512. Available from: https://doi.org/10.1109/LED.2017.2670925.
[37] S. Mandal, A. Agarwal, E. Ahmadi, K.M. Bhat, D. Ji, M.A. Laurent, et al., Dispersion-free 450V p GaN-gated CAVETs with Mg-ion implanted blocking layer, IEEE Electron Device Lett. 38 (7) (2017) 933–936. Available from: https://doi.org/10.1109/LED.2017.2709940.
[38] A.M. Ozbek, B.J. Baliga, Planar Nearly Ideal Edge-Termination Technique for GaN Devices, in IEEE Electron Device Letters 32 (3) (2011) 300–302. Available from: https://doi.org/10.1109/LED.2010.2095825.
[39] A.M. Ozbek, B.J. Baliga, Finite-Zone Argon Implant Edge Termination for High-Voltage GaN Schottky Rectifiers, in IEEE Electron Device Letters 32 (10) (2011) 1361–1363. Available from: https://doi.org/10.1109/LED.2011.2162221.
[40] H. Ueda et al., Wide-Bandgap semiconductor devices for automobile applications, Paper presented at CS MANTECH Conference, Vancouver, BC, Canada, April 2006, Available from: http://csmantech.pairserver.com/Digests/2006/2006Digests/3A.pdf.
[41] D. Ji, J. Gao, Y. Yue, S. Chowdhury, Dynamic modeling and power loss analysis of high frequency power switches based on GaN CAVET, IEEE Trans. Electron Devices 63 (10) (2016) 4011–4017.

Further Reading

I.C. Kizilyalli, A. Edwards, H. Nie, et al., 3.7 kV vertical GaN pn diodes, IEEE Electron Dev Lett 35 (2) (2014) 247–249.
M. Okada, Y. Saitoh, M. Yokoyama, K. Nakata, S. Yaegassi, K. Katayama, et al., Novel vertical heterojunction field-effect transistors with regrown AlGaN/GaN two-dimensional electron gas channels on GaN substrates, Appl. Phys. Express 3 (5) (2010) 054201.

Gate drivers for wide bandgap power devices

7

Subhashish Bhattacharya
North Carolina State University, Raleigh, NC, United States

7.1 Introduction

The gate-drive design for a power Metal-oxide Semiconductor Field-Effect Transistor (MOSFET) is essential to ensure proper switching and conduction characteristics. It enables control of the switching rate by selection of the gate resistance and ensures reduced on-state resistance based on the gate turn-on voltage. The design requirements for the gate-drive circuits for Si Power MOSFETs have been thoroughly enumerated in literature [1]. With the increased use of wide bandgap (WBG) semiconductors like SiC and GaN, for enabling power devices with improved switching and conduction characteristics, there is a need to study the gate-drive requirements and challenges associated with such devices to ensure optimal performance. SiC power devices have turn-on and turn-off voltage requirements similar to Si power devices. However, the lower threshold voltage and the ability to operate at higher switching speeds introduces new design challenges like miller turn-on and signal ground-bouncing due to increased common-mode currents. Moreover, due to the reduced switching transition-times, the dead-time duration can be reduced, compared with Si power devices. Hence, shoot-through protection must be included in the gate-drive design. In addition to the abovementioned challenges, gate-drive design for GaN devices includes additional considerations. The turn-on and turn-off voltage requirements for GaN devices are much lower than Si power devices. This limits the availability of commercially available gate-drive integrated circuits (ICs). Also, the absolute maximum ratings for the device gate are very close to the optimum drive voltages, mandating a layout approach to ensure reduced gate-loop inductance. This chapter explains in detail the aforementioned design challenges along with possible solutions to ensure reliable device operation. It is divided into five sections, treating the gate-drive design for low voltage (LV) SiC devices in section II, high voltage (HV) SiC devices in section III, and GaN devices in section IV. Finally, section V deals with the qualification of gate-drives.

Wide bandgap Semiconductor Power Devices. DOI: https://doi.org/10.1016/B978-0-08-102306-8.00007-1

7.2 Gate drivers for LV SiC devices (1200 and 1700 V SiC MOSFETs and JFETs)

7.2.1 Introduction

SiC MOSFET design and developments can be mainly divided into two categories as medium voltage (MV) and HV, and LV. The LV category include the MOSFETs with voltage blocking capacity up to 3.3 kV. The MV/HV device category includes devices with voltage blocking capabilities of 10 kV and higher [2,3].

The efficiency of a power supply is one of the critical criteria for modern converter systems. The losses in the devices acts as the main yardstick in determining the thermal solution and, consequently, affects the cost and power density of the entire system. Due to considerably lower switching losses of SiC MOSFETs (as compared to Si insulated-gate bipolar transistor (IGBTs)), SiC MOSFETs are slowly replacing Si IGBTs in the LV applications [4,5]. The short switching transient (as experienced by the SiC MOSFETs) enables high frequency converter operation. This is however associated with a large change in the voltage across the switch in a very short time (associated *dv/dt*). This necessitates a different approach for the gate driver (GD) design for SiC MOSFETs and junction field-effect transistor (JFETs).

7.2.2 Basic structure of a gate driver

The basic structure of a GD circuit is provided in Fig. 7.1. The GD circuit can be divided basically into two parts: signal and power. In addition, an isolation is provided in the GD circuit since the outputs of the GD might be at a floating potential (as in the case of half-bridge/half-bridge-based circuits).

7.2.2.1 PWM signal channel

The gate drive signal goes through additional circuitry before it can be used to switch a device. An isolation is required between the signal from the controller and the signal provided to the GD. This isolation is provided by either a digital isolator or an optocoupler. In some cases, optical fibers are used to provide this isolation.

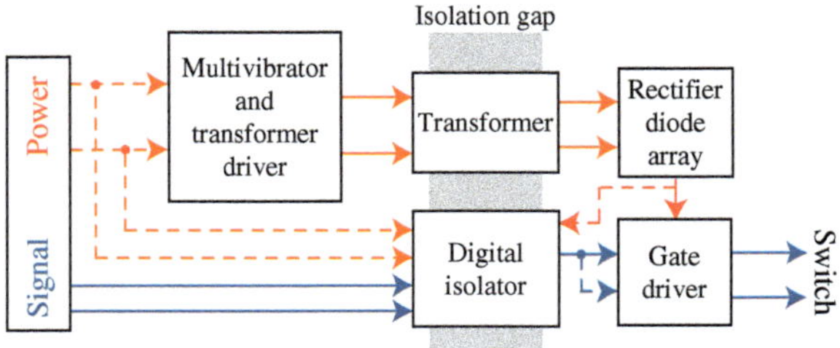

Figure 7.1 Schematic of a generic gate drive circuit board.

7.2.2.2 Power supply

The circuitry which handles the power for all the components can be classified into this category. Usually, power supplies ranging from +5 to +20 V are used to power up the GD. Since a dc power supply is used, a transformer driver is used to converter dc into a high frequency ac signal for the transformer. The multivibrator aids the transformer driver and it is optional. In other cases, an isolated dc–dc converter can be used instead of a multivibrator and a transformer. The secondary side of the transformer/dc–dc converter powers the digital isolator (or the optocoupler) and the GD.

7.2.3 Design considerations for LV SiC MOSFETs

Due to different gate voltage levels and high switching *dv/dt*, the off-the-shelf GDs for Si IGBTs is not suitable for driving SiC devices. Design considerations for such a GD are provided here and an example design is also provided.

7.2.3.1 Gate driver schematic

A designed GD is shown in Fig. 7.2. The GD is built with eight isolated channels on the same printed circuit board (PCB), which can drive eight devices independently. Each of the channels are isolated from the control ground at two nodes. The gate power is isolated through galvanic isolation of a transformer and the gate signal is isolated through optical isolators. The GD can accept gate signal from the control board either through optical cables or through flat ribbon cables (FRC)

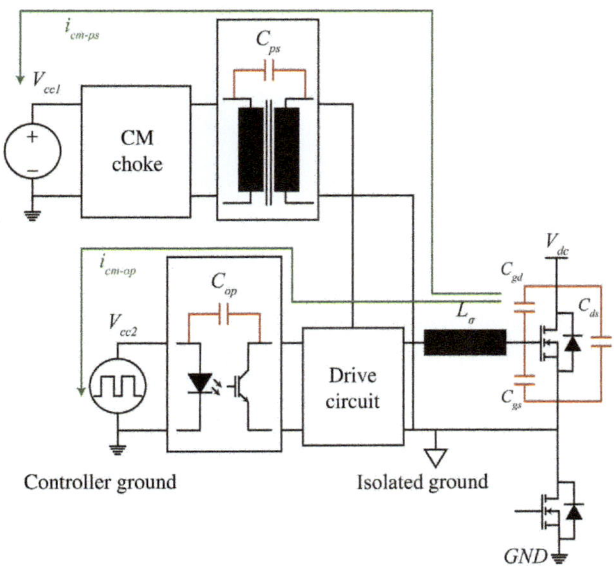

Figure 7.2 Top side of the eight channel gate driver board.

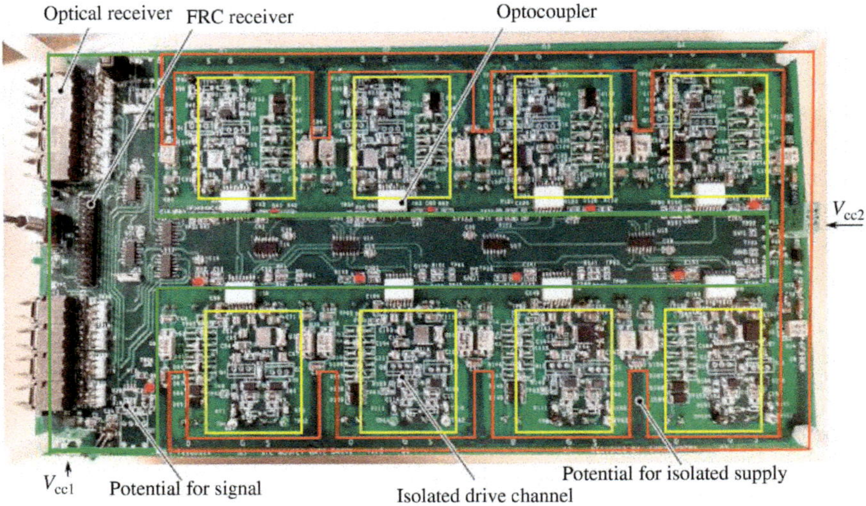

Figure 7.3 Schematic of the gate drive circuit board and common mode current paths.

Figure 7.4 Bottom side of the eight channel gate driver board.

cables. In the case of FRC cables (which is more economical), the common mode switching noise from the power converter can be conducted into the control circuit. Two isolated power supply blocks per channel are mounted at the back of the board generating +20/5 V gate supply. The two different +5 V power supply nodes, V_{cc1} and V_{cc2} are marked in Fig. 7.2. The supply V_{cc1} is routed through the control board of the converter which generates the gate signals for the GD. Both V_{cc1} and V_{cc2} are connected at the auxiliary power supply of the converter. However, they are routed differently between the control board and GD. This enables a separate common-mode current path as shown in Fig. 7.4. The common mode current through the isolated power supply bypasses the control board and directly reaches the auxiliary power supply. A GD design for a eight channel gate driver board is shown in Figs. 7.3 and 7.4.

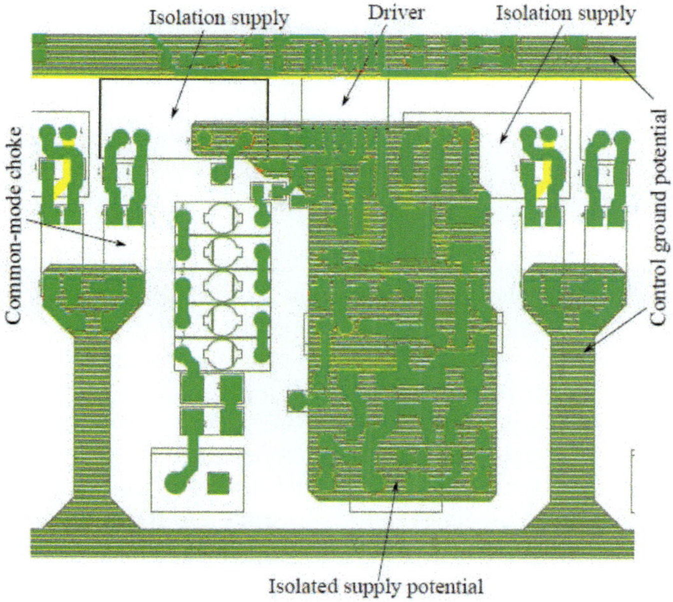

Figure 7.5 Layout of one gate driver channel.

7.2.3.2 Layout design considerations

In order to reduce the switching loss, the SiC MOSFET switching transient time is minimized. However, the presence of stray inductance, L_{σ} in gate current path delays the gate capacitor charging. Also, it forms a resonant circuit with C_{gs} and C_{gd} (as shown in Fig. 7.2) and creates ringing in the gate supply. Care should be taken while laying out the PCB. The gate-source path should be designed as a plane one above the other (thus, minimizing the inductance and maximizing the capacitance). The layout of the GD channel is shown in Fig. 7.5. In the present design, the stray inductance of the gate loop is estimated to be around 7.5 nH.

7.2.3.3 Separate power supply

Common mode current is generated in the switch mode power converter due to the presence of parasitic capacitance between switching potential point to the ground potential. Due to the faster switching transient of the SiC device, the *dv/dt* of the converter switching voltage is higher compared to silicon devices. Therefore, the presence of coupling capacitance either in isolated power supply or in optoisolated driver, causes significant common mode current flow to the control circuit. The high frequency common mode current injection into the control circuit affects the signal integrity, resulting in improper functioning of the circuit. As shown in Fig. 7.3, the common-mode current can flow into the control circuit through two possible paths. Generally, the coupling capacitance in the optodriver is less compared to an isolated power supply. In this design, the coupling capacitor of optodriver (C_{op}) and isolated supply (C_{ps}) are 1.3 and 10 pF, respectively. Therefore, the common-mode current through the optoisolator (i_{cmop}) is expected to be lower than that through the isolated power supplies (i_{cmps}). The total common-mode current which flows into the auxiliary power supply is given by

$$i_{cm} = i_{cmps} + i_{cmop} \tag{7.1}$$

In the layout of the circuit shown in Fig. 7.5, two grounds of the control supply, supplying to the signal and to the input of the isolated power supply are kept separate. Two separate supplies V_{cc1} and V_{cc2} are marked in Fig. 7.2.

Therefore, the power supply to the GD board does not come entirely from the control board. The part of the power supply, associated with pulse width modulation (PWM) signals, comes from control board, and the power supply to the input of the isolated power supply modules comes from outside of the control board. This separation of power sources provide two distinct paths for common mode current to flow to ground potential as shown in Fig. 7.2. Since the coupling capacitance in the isolated power supply is more than that of the optocoupler, the amount of common-mode current flows through the optocoupler is less compared to the isolated power supply line. Since, the only current through the optocoupler flows into the control board from where the PWM signal is supplied, the integrity of the signal in the control broad is preserved better than the case of allowing the total current from the control board. To verify the common mode current flow through two different paths, a boost converter is operated with output voltage of 300 V. The designed GD is used to drive the active device of the boost converter, and common mode current at the control voltage supply to the GD is measured. The common mode current is measured individually on two different control supplies and also as the total current in different cases. When the supply to signal section and at the input of the isolated power supply are supplied from a single source, the total common mode current can be seen from Figs. 7.6 and 7.7 during turn-on and turn-off, respectively. The peak of common mode current during turn-on is 0.28 A, and the peak of common mode current during turn-off is 0.2 A. The turn-on and off switching *dv/dt* are 3.75 and 3.71 kV/s, respectively. Due to higher turn-on *dv/dt*, the peak of the turn-on common mode current is higher. Also, it can be noted that due to higher *dv/dt* which includes higher frequency signals, the frequency of oscillation of common mode current is higher during turn-on.

By separating the power supply of the signal and input supply of gate power section, the converter is operated at the same operating conditions. The common mode current, through the isolated power supply line and optocoupler, can be seen from Figs. 7.8 and 7.9 during turn-on and turn-off, respectively. It can be seen that the current through the isolated power supply is more than that through the optocoupler. During turn-on, the common mode current through isolated power supply has a peak value of 0.32 A, and the peak value of current through optocoupler is 0.18 A. During turn-off, the corresponding peak values are 0.21 and 0.09 A, respectively,

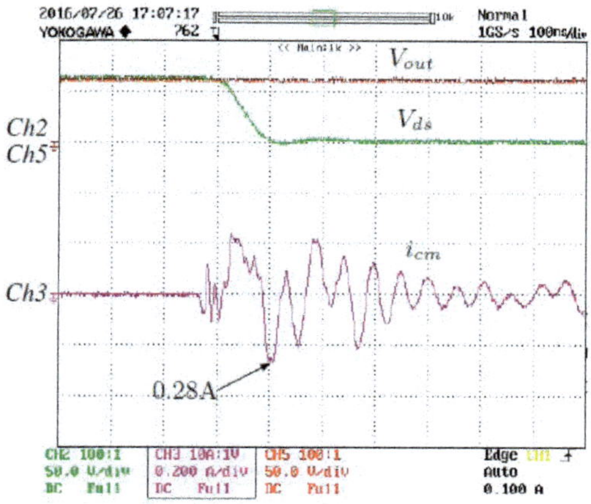

Figure 7.6 Common mode current from single source (V_{cc1} and V_{cc2} tied together) during turn-on of the device. Scale: Ch2, device voltage, V_{ds}: 250 V/div, Ch3, total common mode current, i_{cm}: 0.2 A/div, Ch5, dc-link voltage, V_{out}: 250 V/div, time 100 ns/div.

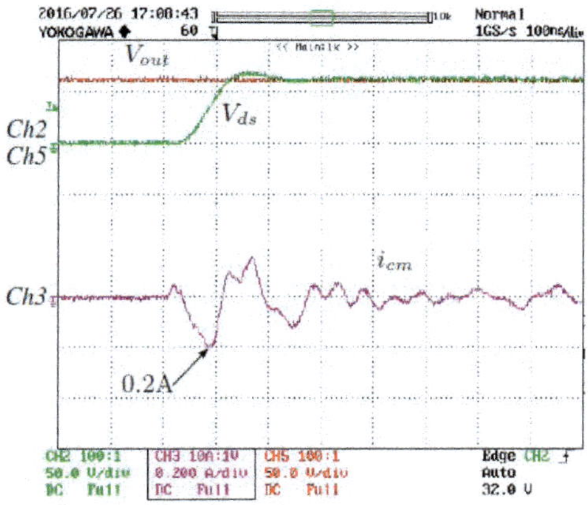

Figure 7.7 Common mode current from single source (V_{cc1} and V_{cc2} tied together) during turn-off of the device. Scale: Ch2, device voltage, V_{ds}: 250 V/div, Ch3, total common mode current, i_{cm}: 0.2 A/div, Ch5, dc-link voltage, V_{out}: 250 V/div, time 100 ns/div.

Also, the evidence of the higher coupling capacitance in the isolated power supply line can be seen from the low frequency of the current (i_{cmps}) through the isolated supply line compared to the optocoupler (i_{cmop}). To minimize the common mode current flow into the control circuit, a common-mode choke (CMC) (CM04RC07T-RC04) is used at the input side of the isolated power supply. High impedance of CMC reduces the common-mode current peak substantially.

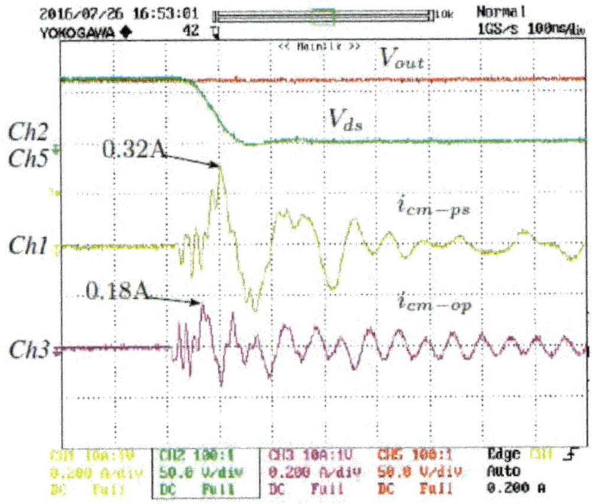

Figure 7.8 Separate paths for common mode current during turn-on of the device. Scale: Ch1, common mode current through isolated power supply, i_{cmps}: 0.2 A/div, Ch2, device voltage, V_{ds}: 250 V/div, Ch3, common mode current through optocoupler, i_{cmop}: 0.2 A/div, Ch5, dc link voltage, V_{out}: 250 V/div, time 100 ns/div.

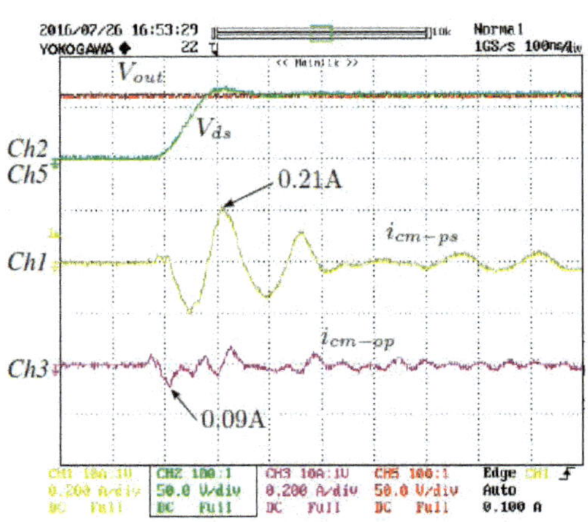

Figure 7.9 Separate paths for common mode current during turn-off of the device. Scale: Ch1, common mode current through isolated power supply, i_{cmps}: 0.2 A/div, Ch2, device voltage, V_{ds}: 250 V/div, Ch3, common mode current through optocoupler, i_{cmop}: 0.2 A/div, Ch5, dc link voltage, V_{out}: 250 V/div, time 100 ns/div.

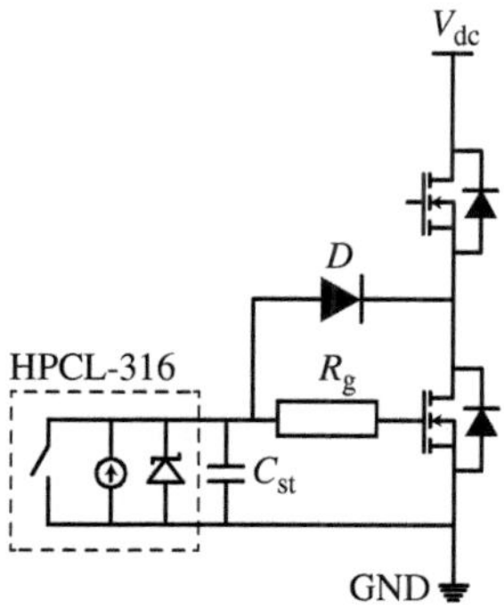

Figure 7.10 Equivalent circuit for the shoot through protection.

7.2.3.4 Shoot through protection

The HCPL-316J driver chip has the capability to detect fault current through the device based on the measurement of the conduction drop of the device. The shoot-through protection feature of the HCPL-316J is adapted for the SiC device by suitable selection of a charging capacitor, C_{st} as shown in representative circuit responsible for shoot-through protection in Fig. 7.10. The switch, S_w, closes and short-circuits (SCs) the equivalent current source, i_{st} at PWM pulse in OFF condition. At ON pulse the switch opens and the current of the source, i_{st} flows through the diode D and the corresponding device. Therefore, the potential across the capacitor is the drop across the device, V_{ds}, and the diode drop. At the event of a shoot-though fault, the device voltage rises, thereby rising the voltage at the capacitor C_{st}. The fault by the chip HCPL-316J is initiated once the voltage rises to 7 V. The value of the C_{st} determines the delay in detecting the shoot-through and thereby, smaller value is desirable. However, to provide blanking time at the start of the switching ON period, to allow the device voltage to fall below 7 V, C_{st} cannot be arbitrarily small. The value of the C_{st} for SiC MOSFET can be selected lower than the value for Insulated-gate bipolar transistor (IGBT) due to shorter turn-on transient. Smaller capacitance reduces the delay time and the fault current. The result of protecting the device in a dead short condition, where the device switches to the shorted load, is shown in Fig. 7.11. The fault is detected and cleared within 2 μs.

7.2.3.5 High current drive

The peak drive of HCPL-316J is 2.5 A. However, for fast switching of the SiC devices and due to high gate capacitance, specially for high current modules, the drive current requirement of the GD is higher. To meet the requirement of higher drive current than the rating of the HCPL-316J, an emitter follower circuit at the terminal of the gate is designed to enhance the required current drive. The peak drive current can reach 12 A with drive current from the optocoupler limited to 1.75 A. The actual gate current during switch-on and off are shown in Figs. 7.12 and 7.13, respectively, while driving a SiC MOSFET of gate capacitance of 4 nF.

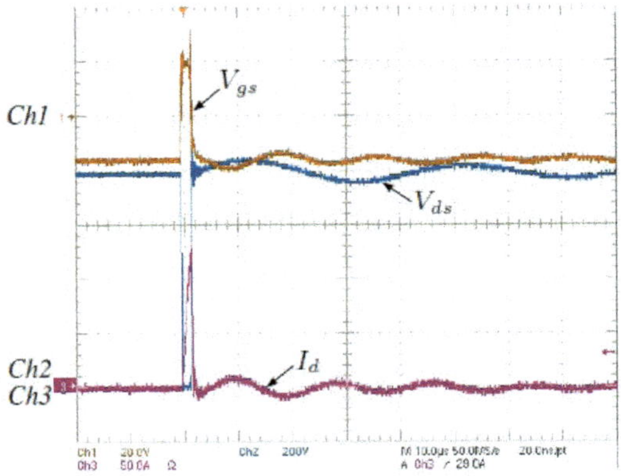

Figure 7.11 Shoot-through protection of the gate driver with the device operating with 800 V dc-link voltage. Scale: Ch1 V_{gs}: 20 V/div, Ch2, V_{ds}: 200 V/div, Ch3, device current, i_d: 50 A/div, time: 10 μs/div.

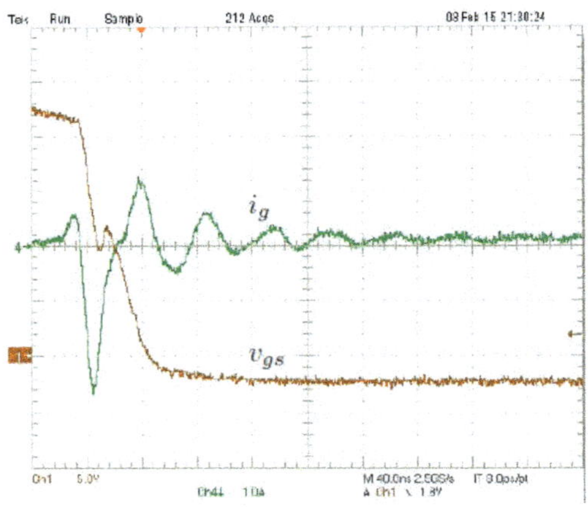

Figure 7.12 Gate current i_g and voltage v_{gs} during turn-on of the device. Scale: gate-source voltage, v_{gs}: 5 V/div, gate current, i_g: 1 A/div, time: 10 ns/div.

The gate resistance during turn-on and off are selected to be 10 and 15 Ω, respectively. Peak gate current during turn-on reaches 3.2 A. The turn-off peak current is around 2.6 A. The oscillation in the gate current supply is around 18.5 MHz. The stray inductance in the gate supply path then can be computed to be around 18.5 nH. As mentioned, the gate path inductance in the board layout is 7.5 nH. So, the additional 11 nH inductance is introduced due the gate wiring.

7.2.4 Active gating

Active gate driving for SiC MOSFETs is desirable to mitigate cross-talk associated with high *dv*/*dt* switching of the device. In phase-leg configuration as shown in

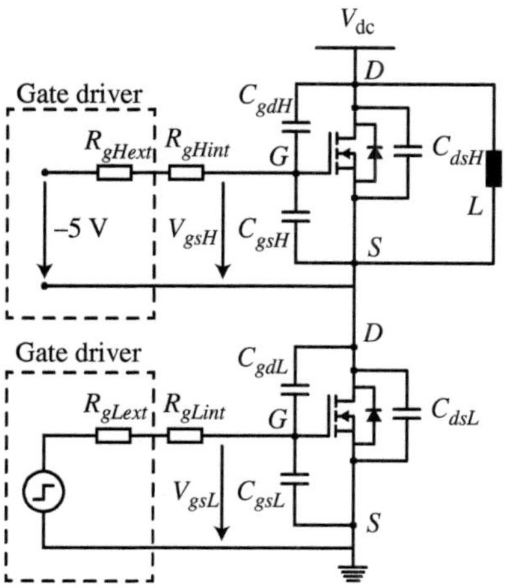

Figure 7.13 Double pulse test setup.

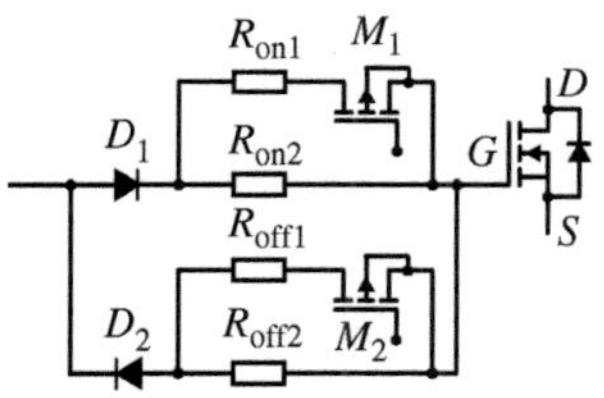

Figure 7.14 Active gate driving logic circuit (PWM input for MOSFETs *M*1 and *M*2 are *P*1 and *P*2, respectively). *PWM*, pulse width modulation.

Fig. 7.13, due to high turn-on *dv/dt* of SiC MOSFET, a positive spurious voltage is induced on the gate to source voltage (V_{gs}) of the complementary MOSFET. This reduces signal to noise margin on the V_{gs} of the complementary MOSFET, making it susceptible to spurious turn-on in the phase-leg. Similarly, during turn-off, a negative voltage is induced on V_{gs} of the complementary MOSFET which reduces the safety margin on the maximum allowed negative V_{gs}. The peak value of the spurious gate voltage [6] can be written as

$$V_{gs_H(\max)} = \frac{dv}{dt} R_{gH} C_{gdH} \left(1 - e^{-V_{dc}/\left((dv/dt) C_{issH} R_{gH}\right)}\right) \tag{7.2}$$

where R_{gH} and C_{gdH} are the associated resistance and capacitance as shown in Fig. 7.1, V_{dc} is the input dc-link voltage. C_{issH} is the sum of C_{gsH} and C_{gdH}.

Fig. 7.14 shows the logic circuit for active gate driving for both turn-on and turn-off. A low turn-on gate resistance leads to lower switching loss but causes higher *dv/dt*. Active gate driving can be used to provide lower turn-on gate resistance during first stage to control *di/dt* and higher turn-on gate resistance during the

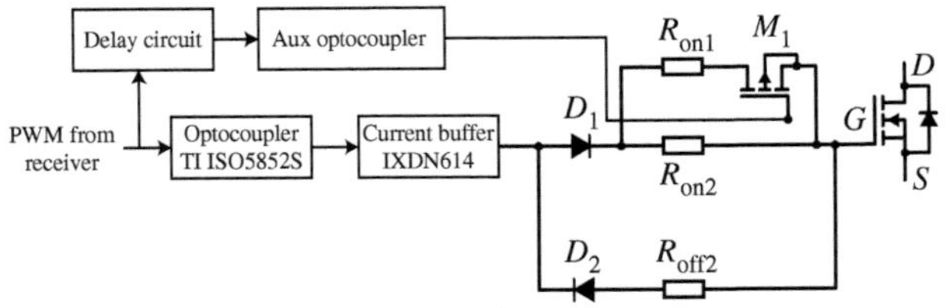

Figure 7.15 Block diagram for turn-on active gate driving.

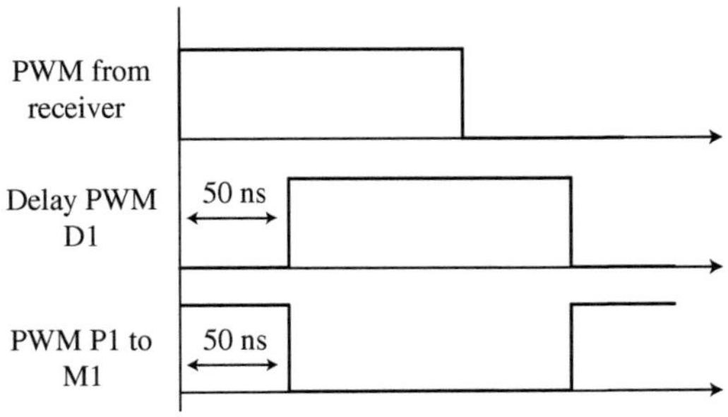

Figure 7.16 Timing Diagram for main and auxiliary MOSFET.

second stage to control *dv/dt*. In Fig. 7.14, MOSFET $M1$ is switched on during first stage leading to effective gate resistance of R_{on1} paralleled R_{on2}, and during the second stage, $M1$ is switched off to get effective gate resistance of R_{on2}.

7.2.4.1 Block diagram

Fig. 7.15 shows the block diagram for active gate driving with delay circuit for generating the PWM signal for driving auxiliary MOSFET $M1$. The delay circuit decouples the *dv/dt* and the *di/dt* control. The timing diagram for the main and the auxiliary circuit is shown in Fig. 7.16. The delay is associated with the time taken by the current to rise to its peak value. $M1$ is turned off after that delay to ensure a higher gate resistance for the second stage. The same optocoupler has been used for the delay circuit to match propagation delays. However, the delay introduced by current buffer and auxiliary MOSFETs needs to be compensated to achieve desired delays.

The triggering time for delay circuit should be tuned according to following timing

$$t_{\text{trigger}} = t_{\text{on-delay}} + t_{i\text{-slow-rise}} + t_{i\text{-buffer-prop-delay}} - t_{\text{aux-MOS-turn-off}} \tag{7.3}$$

which depends on the turn-on delay for the gate resistance, the time period for slow-current rise, propagation delay of current buffer, and the time required to turn-off the auxiliary MOSFET.

Figure 7.17 Active gate driver PCB. *PCB*, printed circuit board.

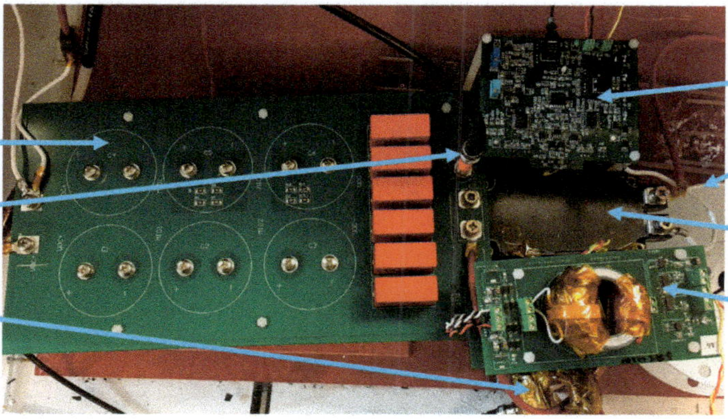

Figure 7.18 Experimental setup for the double pulse test.

The active GD PCB is shown in Fig. 7.17 with different circuits marked on the PCB. For delay circuit, op-amps with high bandwidth and slew rate has been used. Trimmers have been used to fine-tune the delays to achieve desired operating point for lower *dv*/*dt* with reduced switching loss. Gate loop inductance has been optimized by inductance cancellation by overlapping of forward and return path. A jumper has been provided to bypass active gating, if required. A standard double pulse test (DPT) setup is used to compare active gate driving with nonactive gate driving. ROHM 1200 V, 300 A device BSM300D12P2E001, has been characterized using test setup shown in Fig. 7.18. The power loop inductance for DPT is of order of 30 nH to minimize voltage overshoots. An air-core inductor of 113 μH has been used to avoid core saturation issues.

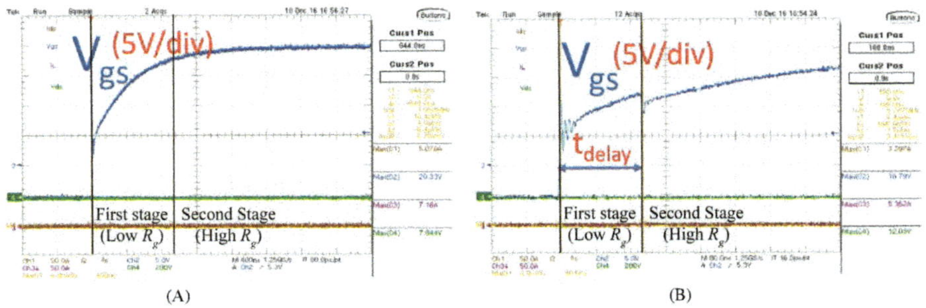

Figure 7.19 Gate-source voltage (A) without active gating and (B) with active gating (V_{gs} = 5 V/div).

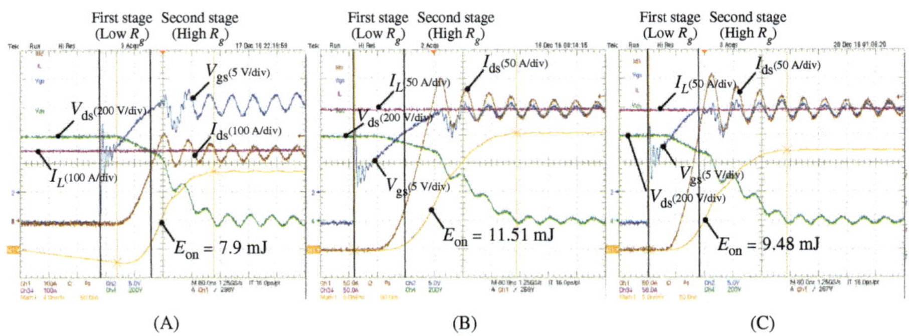

Figure 7.20 Experimental waveforms at turn-on for 600 V at 150 A with a conventional gate driver with gate resistance of (A) $R_g = 2\ \Omega$, (B) $R_g = 3.9\ \Omega$, (C) with active gate driver with a delay of 142 ns and $R_g = 2\ \Omega$ in the first stage and $R_g = 3.9\ \Omega$ in the second stage.

The testing is done at 150, 200, and 250 A at 600 V, and switching loss and *dv*/*dt* is compared for gate resistance of 2 and 3.9 Ω. Active gating is then implemented by using R_g of 3.9 Ω in first stage and 3.9 Ω in second stage. The delay timings are tuned using trimmers for achieving *dv*/*dt* due to higher resistance. Different delays are used for finding optimum operating point to achieve *dv*/*dt* due to higher resistance but lower losses as compared to R_g of 3.9 Ω. The PWM pulses for conventional gate signal and the active GD signal are shown in Fig. 7.19. It can be clearly seen that on applying active gating, the V_{gs} curve has two slopes. After active gating is actuated at t_{delay}, the rate of V_{gs} rise decreases due to switching from lower gate resistance to higher gate resistance. Fig. 7.20 gives the comparison between the conventional and the active GD in terms of switching energy and *dv*/*dt* associated with turn-on. Figs. 7.21 and 7.22 show the comparison of turn-on losses and *dv*/*dt* for with and without active gating. Turn-on *dv*/*dt* is minimum for active gating, whereas turn-on losses are minimized to large extent with active gating. It

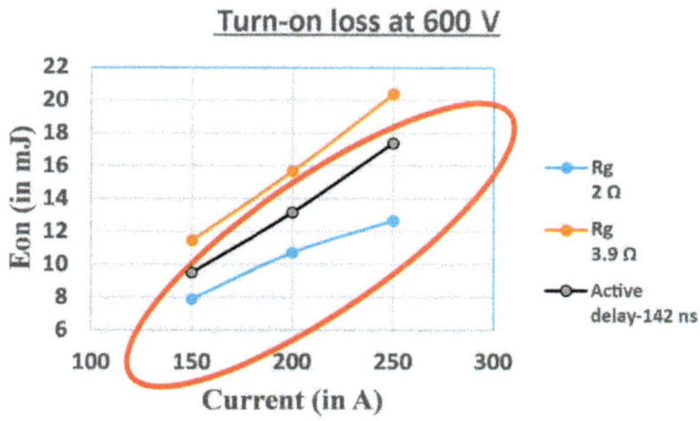

Figure 7.21 Turn-on switching loss at 600 V.

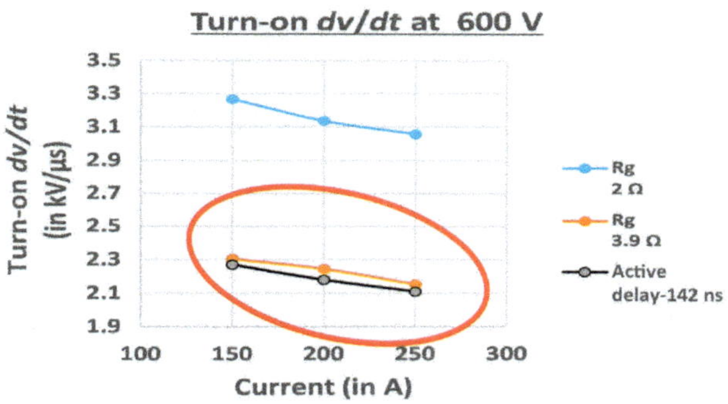

Figure 7.22 Turn-on *dv/dt* at 600 V.

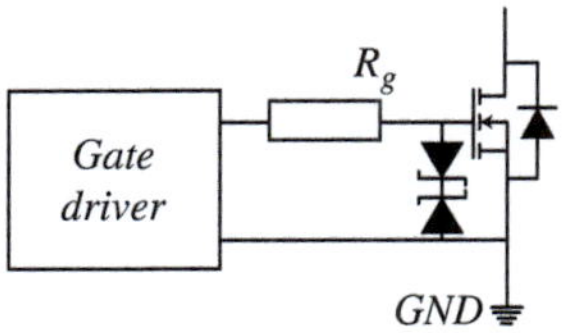

Figure 7.23 Schematic of a voltage clamping circuit.

can be observed from Fig. 7.21 that the turn-on losses can be minimized more without active gating, if gate resistance with smaller values are used for switching.

7.2.4.2 Voltage clamping to reduce voltage overshoot

It is not an uncommon practice to use zener diodes at the output of the gate-source in order to clamp the voltages to nominal levels during static conditions as well as switching transitions. Fig. 7.23 shows the schematic of a generic voltage clamping circuit. Two zeners are placed back-to-back in order to clamp both the positive as well as the negative voltages across the gate-source terminals.

Figure 7.24 1200 V, 100 A SiC MOSFET module.

7.2.5 Evaluation of gate drivers for 1200/1700 V devices

The clamped inductive (or fully hard-switched) characteristics are most widely provided data for power semiconductor devices to cover majority of applications. The circuit schematic of the double-pulse test is shown in Fig. 7.13. While testing a single device, a free-wheeling diode (FWD) is used. Whereas, to evaluate a module, a separate FWD is not required since the antiparallel diode of the top device can enable the free-wheeling function.

7.2.6 Characterization of 1200 V, 100 A SiC MOSFET [1]

A 1200 V, 100 A SiC MOSFET module is shown in Fig. 7.24. The turn-on and turn-off characteristics at 800 V and 100 A with gate resistance (R_g) of 15 Ω are shown in Figs. 7.25 and 7.26. From observation, turn-off loss is more than turn-on losses because of the over voltage appearing in the turn-off conditions due to parasitic and package inductance.

7.2.7 Characterization of 1700 V SiC MOSFET and comparison with 1700 V Si IGBT and 1700 V Si BIMOSFET [7]

In this section, a 1700 V SiC MOSFET is characterized and compared with the similar 1700V Si device results [8]. A 1700 V Si IGBT and Si bidirectional MOSFET (BIMOSFET) results are compared first. Test setup for the discrete device in TO-247 package is shown in Fig. 7.27. The 1700 V, 50 A SiC MOSFET is tested with the dc bus voltage of 1200 V and maximum switch current of 50 A. The gate voltage for SiC MOSFET is +20 V during turn-on condition and −5 V during turn-off condition. In Fig. 7.28A, the turn-on characteristics is shown using 20 Ω gate resistance at junction temperature of 400K. Switching energy is measured by integrating the product of device voltage (V_{ds}) and device current (I_d). Measured energy loss (E_{on}) during turn-on transient is 3.38 mJ. Turn-off behavior of the device is shown in Fig. 7.28B and measured turn-off loss (E_{off}) is 2.16 mJ. The turn-on and off *dv*/*dt*

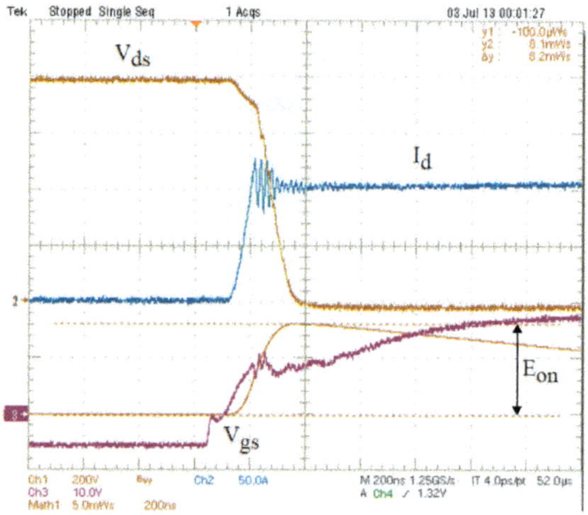

Figure 7.25 Turn-on characteristics with $R_g = 15\ \Omega$ and $T_j = 400$K, measured $E_{on} = 8.2$ mJ, scale: V_{ds}: 200 V/div, I_d: 50 A/div, V_{gs}: 10 V/div, energy: 5 mJ/div, time: 200 ns/div.

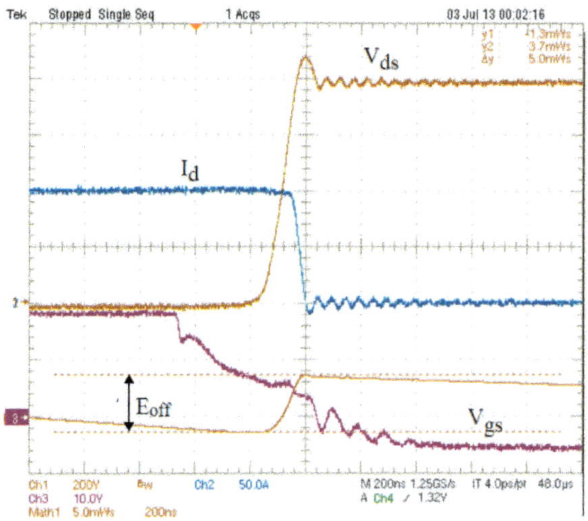

Figure 7.26 Turn-off characteristics with $R_g = 15\ \Omega$ and $T_j = 400$K, measured $E_{off} = 12$ mJ, scale: V_{ds}: 200 V/div, I_d: 50 A/div, V_{gs}: 10 V/div, energy: 5 mJ/div, time: 200 ns/div.

are found to be 23.7 and 24.5 kV/μs, respectively. The turn-on and off *di/dt* are 0.75 and 2.25 kA/μs, respectively. A characterization of the 1700 V, 32 A Si IGBT is performed in the same test setup with the gate voltage of −15 to +15 V. The turn-on and off behaviors are shown in Fig. 7.29A using gate resistance of 5 Ω at junction temperature of 400K. The energy loss during ON and OFF transients is 6.6 and 19.8 mJ, respectively. Turn-off loss of IGBT is much higher due to tailing effect of the current. BiMOSFET, a silicon device from IXYS, is constructed by

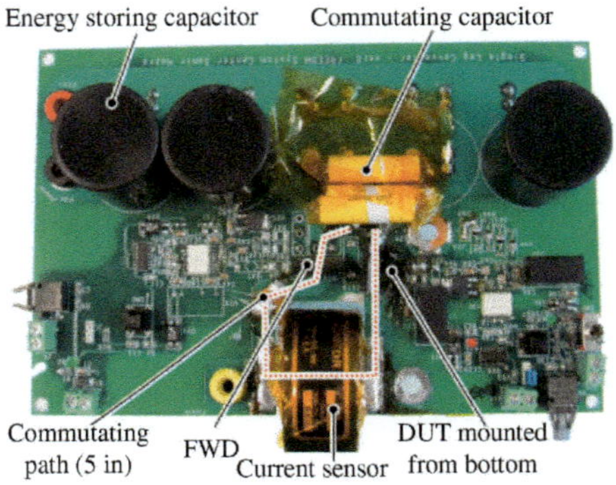

Figure 7.27 Test circuit for characterization of 1700 V SiC MOSFET, 1700 V Si IGBT, and 1700 V Si BIMOSFET.

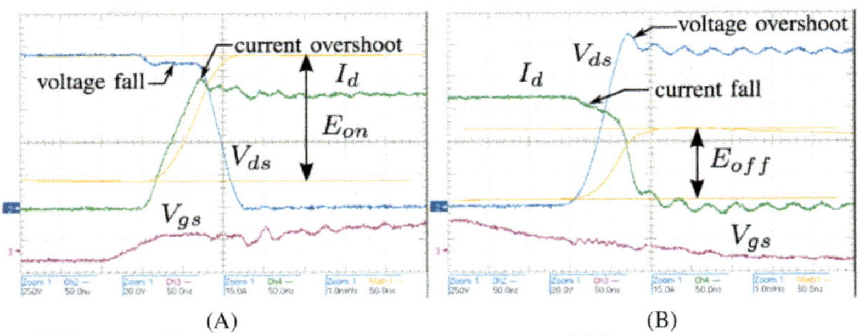

Figure 7.28 Switching characteristics of SiC MOSFET with $R_g = 20\ \Omega$ and $T_j = 400K$, measured $E_{on} = 3.88$ mJ and $E_{off} = 2.16$ mJ, scale: V_{ds}: 250 V/div, I_d: 15 A/div, V_{gs}: 20 V/div, energy: 5 mJ/div, time: 20 ns/div. (A) Turn-on and (B) turn-off.

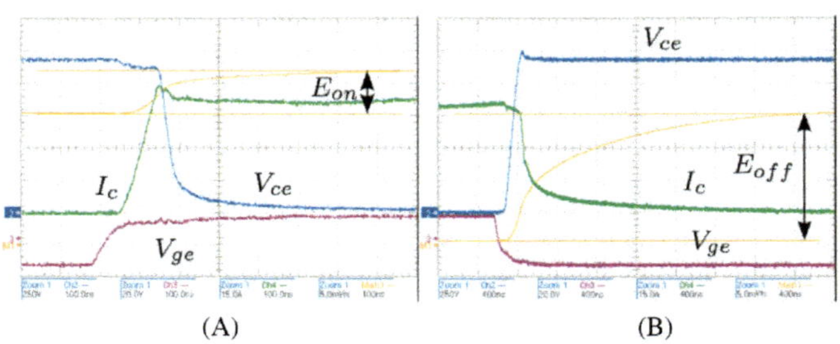

Figure 7.29 Switching characteristics of IGBT with $R_g = 5\ \Omega$ and $T_j = 400K$, measured $E_{on} = 6.6$ mJ and $E_{off} = 19.8$ mJ, scale: V_{ds}: 250 V/div, I_d: 15 A/div, V_{gs}: 20 V/div, energy: 5 mJ/div, time: 20 ns/div. (A) Turn-on and (B) turn-off.

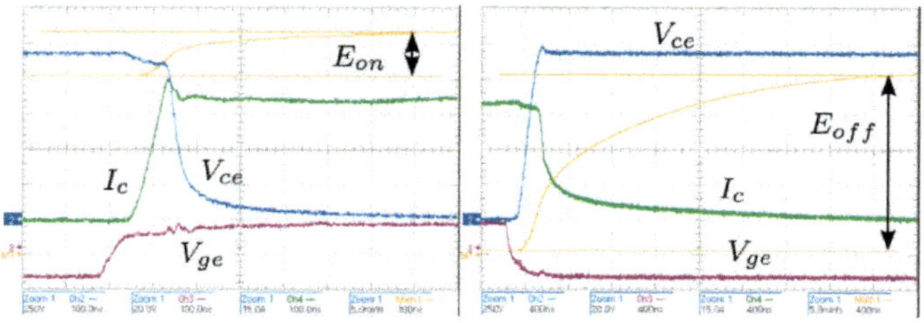

Figure 7.30 Switching characteristics of BiMOSFET with $R_g = 5\ \Omega$ and $T_j = 400$K, measured $E_{on} = 6.1$ mJ and $E_{off} = 25.2$ mJ, scale: V_{ds}: 250 V/div, I_d: 15 A/div, V_{gs}: 20 V/div, energy: 5 mJ/div, time: 20 ns/div. (A) Turn-on and (B) turn-off.

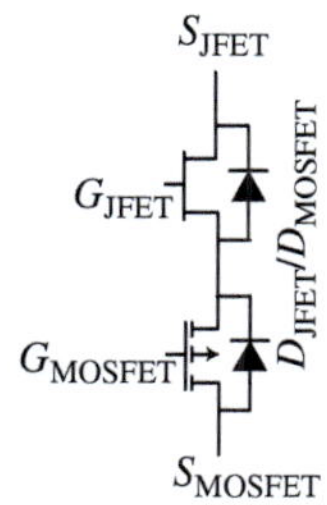

Figure 7.31 Novel cascode topology with p-MOSFET and antiparallel diodes.

keeping MOSFET and IGBT in its structure. The switching losses of the 1700 V, 42 A BiMOSFET is evaluated using the same circuit setup with gate voltage of −15 and +15 V for OFF and ON condition. Fig. 7.30 shows the BiMOSFET turn-on and off characteristics using a gate resistance of 5 Ω at junction temperature of 400K. Turn-on and turn-off losses are 6.1 and 25.2 mJ. Turn-off tail current is observed and the turn-off loss is much higher compared to turn-on loss. Also, the total switching loss is found to be 31.3 mJ which is similar to the 1700 V Si IGBT.

7.2.8 Characterization of 1200 V, 45 A SiC JFET module

The 1200 V, 45 A Infineon SiC JFET module features two pairs of cascodes in series. However, the design approach is novel the LV MOSFET is p-type and the gate of the JFET is not tied to ground as in the classic case. The JFET and MOSFET in each pair have an antiparallel diode across them, as shown in Fig. 7.31. These antiparallel diodes serve as freewheeling agents for the current when the JFET modules are tested under load. Waveforms of the turn-on for $R_g = 0\ \Omega$ is shown in Fig. 7.32. The turn-on current spike appeared due to the capacitance charging of the complementary device.

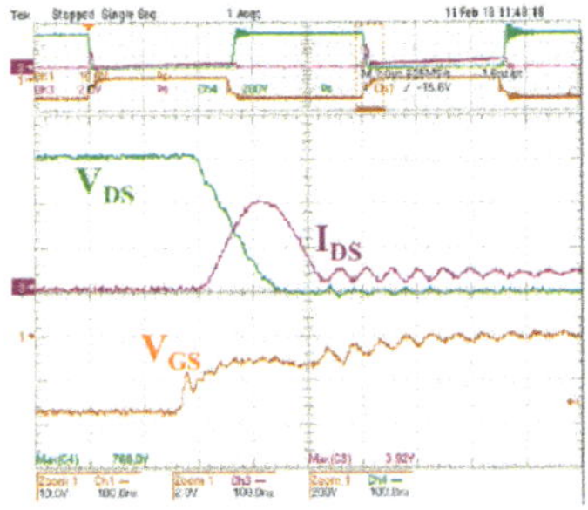

Figure 7.32 Zoomed switching waveforms: $V_{dc} = 600$ V, $R_g = 0\ \Omega$. Scale: (V_{ds}: 200 V/div, I_{DS}: 40 A/div, V_{gs}: 10 V/div). With zero external gate resistance the turn-on and turn-off loss of JFET switching are 1.3 and 1.9 mJ, respectively.

7.2.9 Review of commercially available gate drivers

A few of the commercially available GD ICs are studied. The high-speed switching imposes critical requirements on a SiC GD ICs.

- Wide output supply range: SiC devices specify asymmetrical gate-source voltage in the ranges of +25/ − 10 V. The gate drive circuit must be capable of providing the full range of voltage swing.
- Negative gate voltage: In order to safely turn-off a SiC device, a negative gate voltage needs to be applied across the gate-source for turning it off. In addition, it enables a fast turn-off of the device.
- Extended Common Mode Transient Immunity (CMTI): CMTI is probably the most critical factor while designing a GD IC. While the GD circuit experiences high switching speeds, and thus high *dv/dt*, having a good CMTI becomes critically important.
- Delay Matching: Delay matching is an important factor while choosing the GDs. Unmatched delay especially in a half-bridge circuit might lead to shoot through, and the GDs might exhibit unexpected behavior even in similar conditions.
- Under Voltage Lockout: Undervoltage lockout is a phenomenon in which the GD turns itself off when the voltage falls below a particular value. While it is not as critical as some of the other factors, having an undervoltage lockout protection helps in avoiding potential damage to the converter system.
- Fast desat: An SC protection (desat protection) is preferred in order to avoid SCs and abnormal circuit behavior. The desat trips the signals when it encounters a HV drop across the device during its on state.
- Voltage Drop: Some ICs have an inherent voltage drop. This voltage drop should be kept as low as possible in order to improve the efficiency. A higher voltage drop necessitates the requirement of a higher input voltage which is generally not preferred.
- Optical Isolation: Some GDs offer isolation along with the gate driving capability. This is not necessary if an isolator or an optocoupler is used. However, this helps in the footprint reduction on the PCB and can be used in power dense designs.

A comparison amongst some of the commercially available GDs is shown in Table 7.1. The comparison is made between the optocouplers which are optically

Table 7.1 Comparison between a few commercially available optically isolated gate drivers

Gate driver/ functions	Output supply range (V)	Current sink (A)	CMTI (kV/ μs)	Delay matching (ns)	UVLO	Desat	Voltage drop (V)
Agilent HCPL-316J	15–30	2.3	30	350	Yes	Yes	2.5
Silicon Labs Si826x	6.5–30	1.8	50	25	Yes	No	0.25
Fairchild FOD8342	−0.5–35	3	50	90	Yes	No	0.1
Broadcom ACPL-P343	15–30	2.5	50	100	Yes	No	0.2
Vishay VO3150A	0–35	0.5	35	–	Yes	No	2.1
TI ISO5852S	15–30	5	100	30	Yes	Yes	0.5

CMTI, common mode transient immunity; UVLO, under voltage lockout.

isolated. It should be noted that the list is not exhaustive and gives a basic idea regarding the factors that need to be considered when choosing a GD IC.

7.3 Gate drivers for GaN devices (up to 650 V)

The gate drive design is especially critical while implementing converters using eGaN HEMTs (high-electron mobility transistors). This is due to the low absolute maximum gate voltage (6–7 V) and low device threshold voltage (1–2 V) compared with Si power MOSFETs. The recommended turn-on voltage is close to the absolute maximum rating with a very small headroom of 1–1.5 V. Moreover, the fast switching speeds achievable with GaN transistors introduce further design complexities.

7.3.1 GD specifications and design considerations, challenges, and implementation

The gate-drive circuit for GaN devices is specified by various parameters like operating frequency, turn-on and turn-off voltages, gate-voltage rise-time, power consumption, and whether the device is being switched as a high-side or low-side device. The operating frequency is dictated by the target application though the use of GaN devices allows operation at significantly higher frequencies compared to Si devices. However, this high-frequency operation accompanies various design challenges which must be understood and solved before using the device in an application. Hence, this section describes the challenges associated with gate-drive design for GaN devices as well as possible solutions.

7.3.1.1 Gate-loop inductance

The parasitic inductance in the gate-loop, along with the turn-on gate resistance and the C_{iss} of the device form a resonant LCR circuit. If this circuit has an underdamped response, the overshoot in the gate voltage can violate the absolute maximum limit for the device gate, due to the low headroom in the gate-voltage. Therefore, the criterion given in Eq. (7.1) [9,10] should be met, to avoid overshoot.

$$R_g C_{\text{gs}} \geq 4 \frac{L_{\text{par}}}{R_g}$$

Thus, a lower parasitic inductance ensures that the above criterion is met for a greater range of R_g. Here, R_g refers to the sum of the gate drive transistor resistance, external gate resistor, and the internal gate resistance (R_{Gint}). This becomes a layout issue.

7.3.1.2 Spurious turn-on

At the end of the dead-time period, when the complementary device in a half-bridge is turned on, the C_{gd} of the device experiences a *dv/dt*. As a result, a charging current flows through this capacitance. There are two paths for this current to flow; through the turn-off path of the gate-loop and charging the C_{gs} of the device. If the C_{gs} of device is charged to a value above the threshold of the device, it can lead to spurious turn-on of the device, leading to a partial shoot-through condition. This problem is accentuated due to the high *dV/dt* switching with GaN transistors. This condition to avoid such a condition is described by the following equation:

$$C_{gs}\frac{dV}{dt}R_g\left(1-e^{-\alpha(t_r)}\right)<V_{th}$$

where

$$\alpha=\frac{1}{R_g(C_{gs}+C_{gd})}$$

and t_r is the *dV/dt* transition time. Note that the parasitic inductance of the gate loop has been neglected here. The effect of the gate-loop parasitic inductance has been mathematically derived in literature [11]. Presence of the parasitic inductance leads to ringing in the gate during the period t_r. Moreover, it increases the impedance of the gate loop, leading to an increase in charging current for C_{gs}. Thus, a reduction in the parasitic inductance is helpful in this case as well. In order to prevent a spurious turn-on, the total gate resistance, R_g, in the turn-off path can be reduced, based on Eq. (7.2). However, this can lead to reduced damping in the gate loop during turn-off such that the condition in Eq. (7.1) is not met. In this case, an undershoot may be observed during the turn-off transition in the gate voltage, similar to the overshoot in gate voltage during turn-on. However, the headroom in the gate voltage in the negative direction is usually higher in which case the undershoot might not be an issue. Another approach which can be employed is the application of negative turn-off voltage. The negative voltage adds in magnitude to the right-hand side of Eq. (7.2), increasing the immunity to spurious turn-on. However, this increases the off-state reverse conduction drop for the device. This can lead to increased device loss. The amount of negative turn-off bias required can be reduced with improved layout and selection of turn-off resistance.

The issue of spurious turn-on is not present with devices where the ratio of C_{gd}/C_{gs} is less than 1. This is usually the case for LV devices (up to 100–200 V rating). In such cases, the above design considerations need not be used.

7.3.1.3 Common-source inductance

This refers to the inductance in the source of the device, which is common to the power loop and the gate loop. This inductance experiences the high *di/dt* from the power loop, resulting in an induced voltage which opposes the applied gate voltage, slowing down the turn-on transition and increasing the switching losses. Moreover, at the end of the dead-time period, when the complimentary device in a half-bridge is turned-on, the free-wheeling current in the device is commutated to the complimentary device, leading to an induced voltage which adds to the gate voltage in the negative direction. This is helpful in reducing the probability of a spurious turn-on at the turn-on of the complimentary device. However, since the common source inductance, device C_{gs} and the turn-off path gate resistance form an LCR circuit; if the damping provided by the resistance is insufficient, the voltage across C_{gs} can ring in the positive direction as well, which can lead to a spurious turn-on.

In order to avoid this spurious turn-on issue, it is recommended to minimize the common-source inductance, which can be tackled through careful layout, as explained later. Another approach may be to increase the damping by increasing the resistance in the turn-off path. However, this can in turn increase the probability of spurious turn-on due to the C_{gd} capacitance (miller capacitance) charging current at the turn-on of the complimentary device. Hence, this is not the recommended approach.

7.3.1.4 Common-mode current

Consider a half-bridge circuit. The switching node is the same as the source of the high-side device. With an isolated gate-drive design, the desired gate voltages for driving the high-side device are derived from an isolated power supply. Thus, the switching action causes the entire high-side power supply (both positive and negative terminals) to swing between V_{DC} and the ground with the same transition time. This *dV/dt* is experienced by the isolation capacitances between the input and output terminals of the power supply, resulting in a common-mode current flow for the duration of the transient, through the isolation capacitances to the input-side of the GD. A similar phenomenon occurs with the isolation capacitance of the gate drive, which is characterized by the CMTI rating of the IC. The flow of common-mode current leads to disturbance in the gate-drive input-side ground. Depending on the amplitude of the common-mode current, this can cause false switching of the device since the gate-drive IC provides output pulses based on the signal level at the input-side of the driver with respect to the corresponding ground reference. With the high *dV/dt* switching of GaN devices, the problems associated with common-mode current are increased.

In order to ensure reliable switching operation, the impedance in the common-mode current path must be increased, to reduce the amplitude of the current. This is achieved by using a reduced isolation capacitance power supply for the high-side device gate-drive. The isolated dc–dc converter modules from Recom Power and Murata Power offer an isolation capacitance in the range of 1.5–10 pF, which is usually sufficient for GaN-based switching applications. However, in order to further increase the common-mode impedance, a CMC can be added at the input of the isolated gate-drive power-supply. The CMC acts only in the common-mode

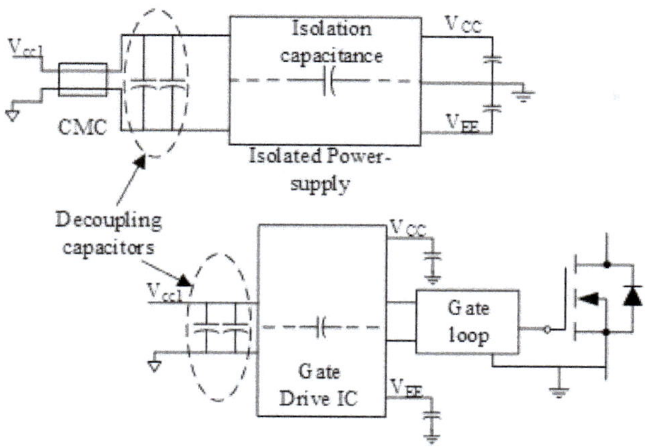

Figure 7.33 Isolated gate-drive design.

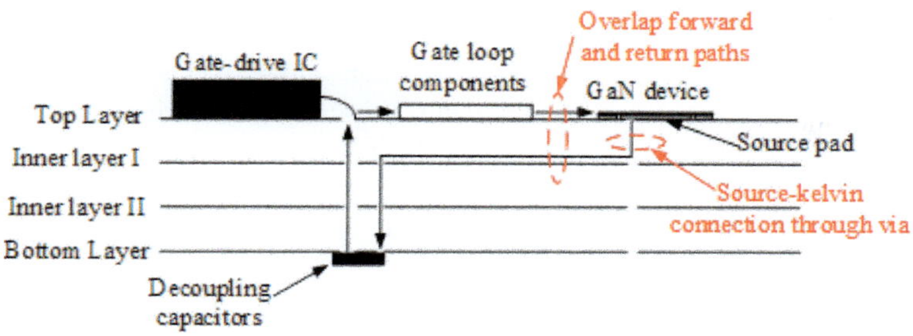

Figure 7.34 Gate-loop layout recommendations.

circuit but acts like a short in the differential-mode circuit. This type of gate drive design is illustrated in Fig. 7.33.

Note that an isolated gate-drive design for high-side switching is recommended over a bootstrap based design, especially for HV applications (above 100 V). However, a low junction capacitance diode with a fast recovery time must be used. Moreover, for low-side device freewheeling, the negative voltage on the switching node can lead to the bootstrap capacitor charging to a value higher than the required gate voltage. Hence, regulation of the voltage after the bootstrap diode is required to ensure correct gate turn-on voltage is applied. This can be done using an low-dropout regulator (LDO) or using a Zener diode clamp.

7.3.2 Layout recommendations

The layout of the gate-loop is essential to tackle many of the design challenges described previously. Fig. 7.34 shows some layout ideas [8,12,13]. First, the forward and return paths for the gate-loop are overlapped. This allows flux cancellation in the two paths. Moreover, utilizing the closest inner layer for the return path helps to reduce the area of the gate-loop. This helps to reduce the parasitic inductance of the gate-loop. Second, the connection from the source pad is made through a via, at the closest edge of the pad. This helps reduce the common-source inductance between the power loop and the gate loop.

The desired power supply voltages can generate using an LDO. However, by keeping low-equivalent series inductance (ESL) decoupling capacitors close to the power supply pins of the IC, as shown in Fig. 7.34, the LDOs can be kept away from the gate-loop components, thus allowing the forward path length of the gate-loop to be reduced. This further helps in reducing the gate-loop inductance.

7.3.3 Gate-drive design for GaN four-quadrant switch (FQS)

A four-quadrant switch (FQS) is a device which can conduct current in both directions of the device as well as block voltages of either polarity during off-state. Such a device has been developed by Transphorm, USA, as shown in Fig. 7.35A.

As an equivalent representation, the device can be considered as two bidirectional conducting devices connected in a common-drain configuration, shown in Fig. 7.35C. Here each device is in a cascode structure with a depletion-mode GaN HEMT connected in series with an enhancement-mode LV Si MOSFET. Since there are two such devices, two independent Si gates need to be driven. The gate-drive voltage requirements are like that of the LV Si MOSFET. Hence, in theory, traditional Si MOSFET drivers may be used to drive the device. However, due to the depletion mode GaN device, the gate drive requirements need more attention [14].

At the turn-on of the LV MOSFET, the gate-source voltage of the depletion-mode GaN transistor is reduced below the threshold voltage, turning-on the GaN device. This turn-on transition occurs at a high *dV/dt* and di/dt. This introduces the issues related to spurious turn-on, increased common-mode currents (and associated ground bouncing), and common-source inductance, as explained previously. Thus, gate drive design considerations similar to an eGAN HEMT must be used for the cascode device as well.

An important point to note for the design of a half-bridge circuit using the FQS is that each device has two independent source connections. This means that gate drive for each device requires two isolated ground connections. In case of the half-bridge circuit, the source connection for the bottom switch of the high-side device and the top switch of the low-side device are common and hence can be generated from a single isolated dc/dc converter module. The isolation capacitance for this module must be low to minimize ground bouncing on the input-side of the GD, as explained earlier.

7.3.4 Commercially available gate driver ICs and trends

A list of commercially available GDs is presented in Table 7.2. It must be mentioned that companies like TI, Peregrine semiconductor provide dual GDs for both high-side and low-side devices in a Bootstrap configuration. Silicon Labs provides

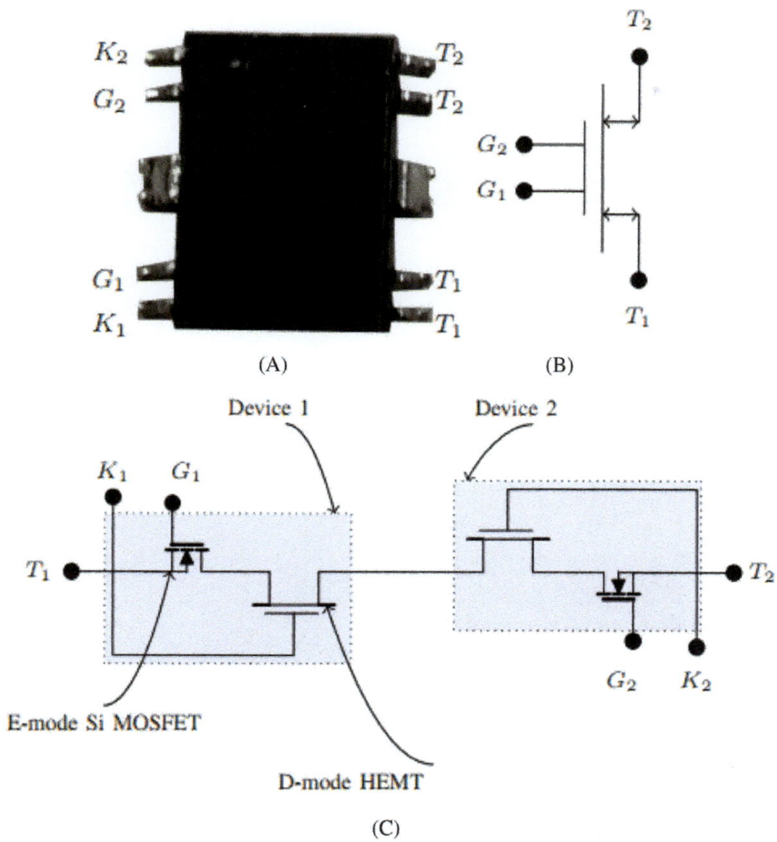

Figure 7.35 (A) Four-quadrant switch in SOIC-16 package, (B) pin configuration of the four-quadrant switch, (C) four-quadrant switch realization using a depletion mode gallium nitride high electron mobility FQS and enhancement mode low voltage silicon MOSFETs—the single chip GaN FQS has been schematically broken in two to facilitate explanation of device operation. *FQS*, four-quadrant switch.

isolated dual GDs, providing isolation between input–output as well as output–output.

Moreover, except for Silicon Labs, other companies provide nonisolated single-channel GDs. These must use a digital isolator to provide the signal level isolation. The CMTI field corresponding to these GDs will depend on the CMTI of the digital isolator used with them. It must be pointed out that many of these GDs cannot be used to provide negative gate-bias for turn-off since the logic ground at the input is the same as the driver ground.

Please note that some other important parameters like propagation delay and rise-time have not been listed in the table but must be considered in the final design.

Table 7.2 **Commercially available GaN gate drivers and relevant parameters**

Part no. (manufacturer)	Max source/ sink current (A)	Single/dual	UVLO	Max vin (V)	Isolated/ nonisolated	Separate pull-up/ pull-down	CMTI
LMG1205 (TI)	1.2/5	Dual (bootstrap)	4.5 V (max)	100	–	Yes	–
LM5113-Q1 (TI)	1.2/5	Dual (bootstrap)	4.5 V (max)	100	–	Yes	–
LM5113 (TI)	1.2/5	Dual (bootstrap)	4.5 V (max)	100	–	Yes	–
UCC27517A (TI)	4/4	Single	4.65 V (max)	–	Nonisolated	No	–
UCC27611 (TI)	4/6	Single	4.15 V (max)	–	Nonisolated	Yes	–
UCC27511 (TI)	4/8	Single	4.65 V (max)	–	Nonisolated	Yes	–
LM5114 (TI)	1.3/7.6	Single	4 V (max)	–	Nonisolated	Yes	–
UCC27524A (TI)	5/5	Dual (indep. channels)	4.65 V (max)	–	Nonisolated	No	–
2EDN7523R (Infineon)	5/5	Dual (indep. channels)	4.2 V (typ.)	–	Nonisolated	No	–
2EDN7524R (Infineon)	5/5	Dual (indep. channels)	4.2 V (typ.)	–	Nonisolated	No	–
2EDN7523F (Infineon)	5/5	Dual (indep. channels)	4.2 V (typ.)	–	Nonisolated	No	–
2EDN7424R (Infineon)	4/4	Dual (indep. channels)	4.2 V (typ.)	–	Nonisolated	No	–
2EDN7424F (Infineon)	4/4	Dual (indep. channels)	4.2 V (typ.)	–	Nonisolated	No	–
2EDN7524G (Infineon)	5/5	Dual (indep. channels)	4.2 V (typ.)	–	Nonisolated	No	–
2EDN7524F (Infineon)	5/5	Dual (indep. channels)	4.2 V (typ.)	–	Nonisolated	No	–

2EDN7523G (Infineon)	5/5	Dual (indep. channels)	4.2 V (typ.)	–	Nonisolated	No	–
1EDN7511B (Infineon)	4/8	Single	4.2 V (typ.)	–	Nonisolated	Yes	–
1EDN8511B (Infineon)	4/8	Single	4.2 V (typ.)	–	Nonisolated	Yes	–
1EDN7512G (Infineon)	4/8	Single	4.2 V (typ.)	–	Nonisolated	Yes	–
Si827x (Silicon Labs)	1.8/4	Single and Dual (Isolated channels)	3.5 and 5.5 V (typ.)	–	Isolated (2.5 kV_{rms})	Only for Si8271	>200 kV/μs
Si826x (Silicon Labs)	4/4	Single	5.5 V (typ.)	–	Isolated (3.75 kV_{rms})	No	50 kV/μs
PE29100 (Peregrine Semiconductor)	2/4	Dual (Bootstrap)	3.6 V (typ.)	100	–	Yes	–
PE29102 (Peregrine Semiconductor)	2/4	Dual (Bootstrap)	3.6 V (typ.)	60	–	Yes	–

CMTI, common mode transient immunity.

7.4 Qualification of gate drivers

After designing GDs for MV SiC applications, it is necessary to test and verify correct operation. The verification of GDs for operation in MV converters is called qualification. In the following chapter, we are going to take a look at the needed steps to achieve qualification of a MV GD.

In this section, stress has been laid on methodical procedures for testing and qualifying GDs for field operation. The first step for qualification is the sequential selection of different topologies to identify different aspects of the GD. The second step involves investigation of the GD for sustaining faults occurring in converters during operation. When the GD passes all of the above tests, it is qualified for field operation.

7.4.1 Gate driver operation for controlling MOSFET turn-on/turn-off

The chosen gate drive waveforms may help in understanding the turn-on and turn-off for the GDs. The turn-on voltage applied for SiC MOSFETs is 20 V max, and the turn-off voltage is chosen to be close to −5 V. We can test this phenomenon in double pulse switching of the MOSFET.

7.4.1.1 Common mode issue in gate drivers

The common mode issue arises due to coupling capacitance in the isolation transformer and to some extent with the coupling capacitance in the gate driving IC as shown in Fig. 7.36. The common mode can be identified by using the GD in high side configuration during the DPT as represented in Fig. 7.37.

During the turn-on of the switch, the source terminal floats at almost dc link voltage. This dc link voltage also appears at the isolation transformer, and the isolator gate driving IC in the GD circuit. We can measure the common mode current

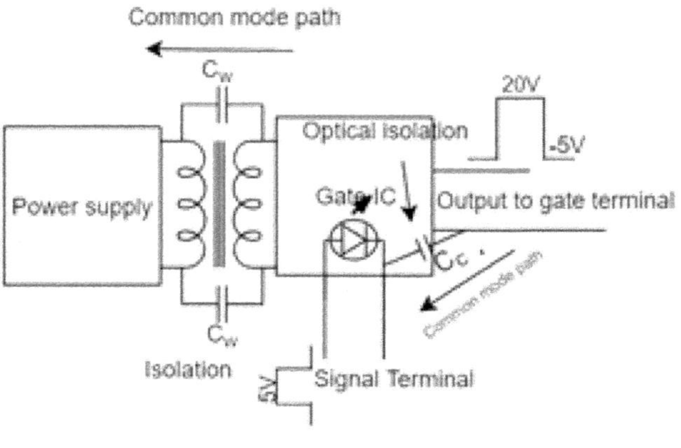

Figure 7.36 Common mode path in the gate driver.

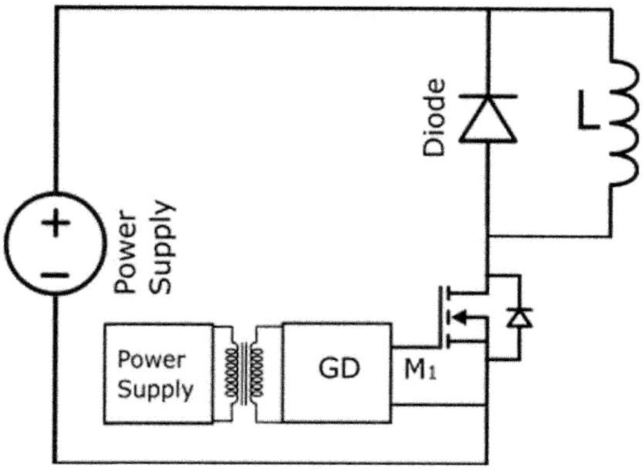

Figure 7.37 Double pulse test setup.

flowing through the input of the GD power supply and the signal ICs. This will help in improving the mitigation of the EMC issues in the GD.

7.4.1.2 Isolation or creepage issue in the driver

The application of GDs at HV requires the design constraint due to appropriate implementation of the creepage and isolation between the HV and signal voltage sides. The DPT will help in identifying these issues prevalent in GDs. The two pulse operation of switches will not help in identifying the thermal issues in the GDs. Therefore, it is required to use continuous run testing of the GDs.

7.4.1.3 Cross talk between top and bottom gate drivers

The turn-off in bottom GD leads to rise of the drain-source voltage across it at the same instant the top device is turning on. The dv/dt caused due to the V_{ds} to 0 transition on top devices causes a common mode current to flow through the C_{ds} of bottom device and thus turns it on. This leads to an overshoot in the current flowing through the legs and may cause SC failure or losses in the converter.

7.4.1.4 Shoot through due to inappropriate dead time

Inappropriate dead time between the operations of complementary switches may lead to shoot through when both of the switches turn-on. To identify the mentioned issues in the designed GD, some new converter topologies are required for testing. To identify the mentioned issues in the designed GD, some new converter topologies are required for testing.

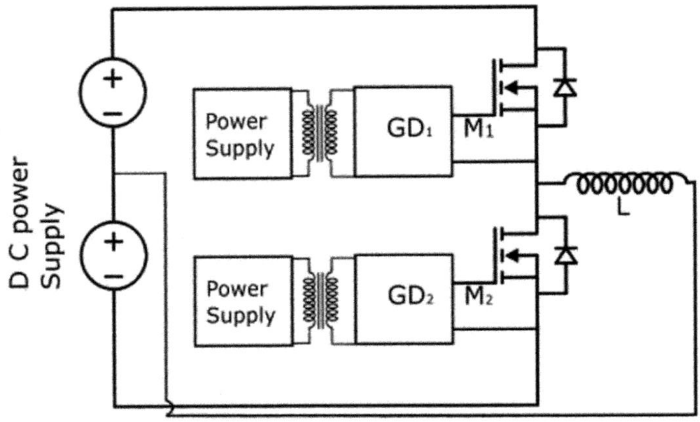

Figure 7.38 Half bridge test setup.

7.4.1.5 VSC pole two level topology

The VSC pole two-level topology is one of the simplest circuits used to qualify a GD for cross talk occurring between the top and bottom switches. It is a single leg configuration as shown in Fig. 7.38. Voltage swing occurring on each of the switches is equivalent to the dc link voltage.

7.4.2 Steps of gate driver qualification

The different aspects of GDs are qualified by using different topologies. The sequence of GD testing is given by following topologies:

7.4.2.1 Double pulse test (DPT) converter

DPT circuit is widely used for the characterization of the solid state devices. It is a basic inductive, clamped, switching circuit that has a FWD in parallel with the inductor. It is used for characterizing the dynamic operation of the switching circuit. Apart from device, the circuit can be used for testing certain aspects of the GDs. What can be tested and qualified by the DPT circuit are as follows:

7.4.2.2 Boost-converter operation

Boost-converter operation helps in identifying the thermal reliability of the GD under test by a continuous run test. The advantage of boost converter is that, it contains the switching component with source linked to ground the terminal of the converter as shown in Fig. 7.39. Therefore, it helps in avoiding any performance issues due to isolation during GD operation. The result is the separation of the thermal-related issues from isolation-related issues. Finally, the testing delivers the qualification of GD for continuous operation without any isolation. GDs designed for the converter operation should have the ability to sustain the isolation and common

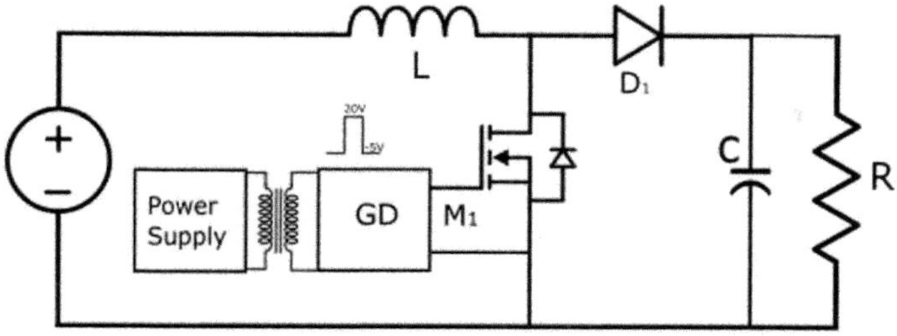

Figure 7.39 Boost converter test setup.

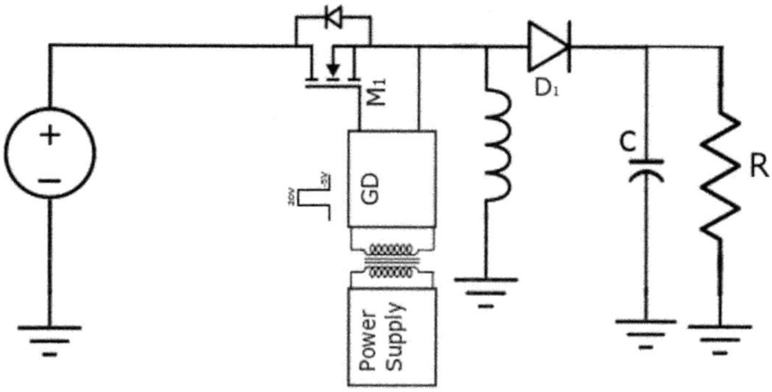

Figure 7.40 Buck-Boost converter test setup.

mode issues due to high *dv/dt*. Therefore, it is required to move to the next stage for GD testing for isolation during continuous operation.

7.4.2.3 Buck-boost converter operation

The common mode voltage and isolation testing of GD for the highest possible voltage swing is tested in the buck-boost converter operation. As shown in Fig. 7.40, the device used in buck-boost converter configuration sees maximum voltage between drain and source terminals during switching operation. Testing of GDs in this topology will help in analyzing the common mode current sustaining capability of the GD. The previously mentioned topologies will only help in testing a GD when it is operating in single. In practical applications, especially in MV operation, converter topologies having more than one switch are being used. The general topology consists of modular multilevel and three level converters. Three level high frequency converters are becoming popular with advancement in MV SiC devices. These topologies involve the operation of more than one GD. This leads to the emergence of new issues caused due to the effect of one GD on other. Most basic topologies have more than one GD in half bridge configuration, which

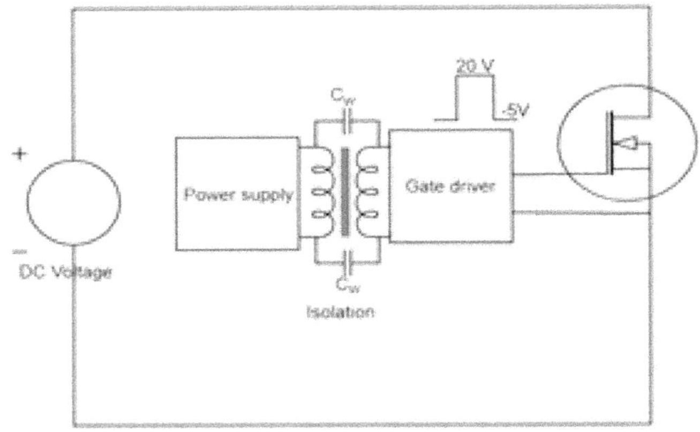

Figure 7.41 Short-circuit test setup.

involves a single leg. Therefore, half bridge can be used as the topology to understand mutual effects of GD operation.

7.4.3 Short-circuit testing of gate driver for high-voltage switch

When the SC happens in the converter, the corresponding GD of the faulty switch has to generate a signal to turn-off the device. The signal should be as fast as possible, in order to avoid further damage to the converter due to persistence of the SC. Fig. 7.41 shows the SC test setup for the GD. It consists of direct attachment of drain and source of the device to the voltage source. When the switch is turned on by the GD, the desat protection system should sense the overcurrent and generate a faulty signal by the gate driving IC. The corresponding time between the fault and tripping can be observed on a scope.

7.4.4 GD characterization for the current switch operation and test circuit

The current switch has found its application in various current source converters. These converters generally implement thyristors. The advantage of current switch converter are that they employ more rugged switches and are very good for the SC protection as well are useful for zero current switching (ZCS) and zero voltage switching (ZVS) operation. The current switch has a diode in series which is generally preferred as the junction barrier Schottky (JBS) in nature to allow low reverse recovery. The packaging of current switch is itself a huge challenge. First of all, it requires to be dealing with peak positive and negative HV stresses. To understand the different issues for the functioning of current switch, it is required to understand the switch by using the different experimental circuits. To estimate the performance of a current switch, two types of test are performed.

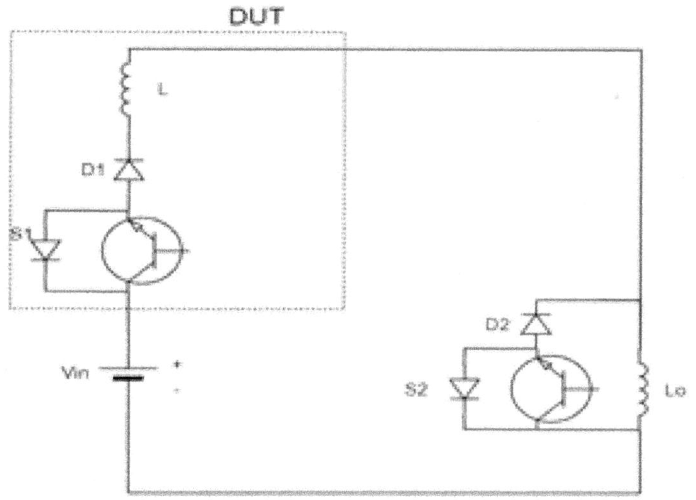

Figure 7.42 DPT test setup for current switch. *DPT*, double pulse test.

7.4.4.1 Double pulse test

To comprehensively characterize the current switch modules, a DPT circuit is built as shown in Fig. 7.42. Special care is taken to minimize the effect of parasitic elements. The loop inductance is minimized by using a sandwiched bus bar arrangement. Emitter anode configuration is used to minimize the package inductance. The separation between top and bottom sides of the metalized ceramic (0.635 mm) should be minimal for achieving a good thermal performance but may result in an increase in package capacitance.

7.4.4.2 Continuous operation test

Continuous pulse test involves the running of current switch at continuous rate in any of the basic topology involving buck converter, boost converter, and buck-boost converter. Buck-boost converter is chosen for the continuous pulse test because it has the capability to produce maximum voltage swing. Therefore, the GD will be tested with the switch operation for the maximum *dv/dt*. Fig. 7.43 shows the schematic of the continuous operation of the current switch using buck-boost converter. The GDs used for current switch operation are same if the MOSFETs or IGBT is used in the circuit. They will be different and more bulky when current controlled switches are used in the operation.

7.5 Gate drivers for HV SiC devices

The recent developments in SiC MOSFET and SiC IGBT device technology have paved way for simple nonseries and noncascaded MV converter topologies. These

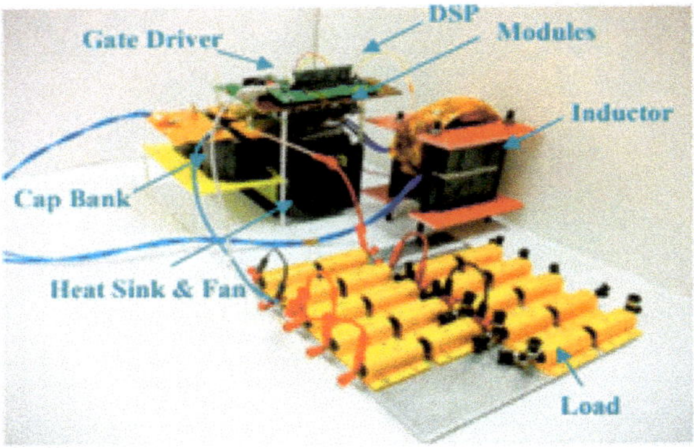

Figure 7.43 Buck-boost converter test setup for current switch.

devices can operate at higher voltages and can switch at higher frequencies compared to their Si counterparts. These characteristics enable applications such as high-speed MV drives, MV dc microgrids, and compact grid-connected converters for renewable energy. The challenges for a 15 kV SiC IGBT and 10 kV SiC MOSFET GD design are the HV isolation and high *dv/dt* (30 kV/s) electromagnetic interference (EMI) immunity. A suitable such basic MV GD with less than 1 pF isolation capacitance power-supply is presented previously in [15]. The driving and desat function is supported by a 2.5 Amp AVAGO driver with integrated Vce-desat detection and fault status feedback. It has recommended nominal blanking time of 2.8 μs and extra 2 μs fall time. The typical protection time provided by this driver is 5 μs. The mentioned devices are still in research and development stage and costly. The SC protection of these devices has currently not been implemented as it is very challenging due to their characteristics. Higher blocking capability for same buffer thickness and better thermal conductivity of these WBG SiC devices have enabled high power density of the chip. But the reduction in chip size has also reduced SC withstand time (SCWT) [16].

7.5.1 GD specifications and design considerations

HV SiC devices produce high dv/dt during switching, typically of the order of 100 kV/μs [15], which gets reflected at the isolation power supply terminals, as shown in Fig. 7.44. A typical switching waveform of 15 kV SiC IGBT is shown in Fig. 7.45. The coupling capacitance C_p across the power isolation stage gets coupled with this dv/dt, which results in a transient leakage current being drawn from the primary side. Due to high dv/dt, this leakage current may surpass the limitations set by industry standard. To avoid high leakage current, C_p should be small, typically below 5 pF. In addition, the isolation transformer must pass industry standard tests for the qualification. High isolation requirement for the transformer results in relatively higher C_p, making it difficult to design the transformer. More details about the isolation stage requirement are discussed in [17].

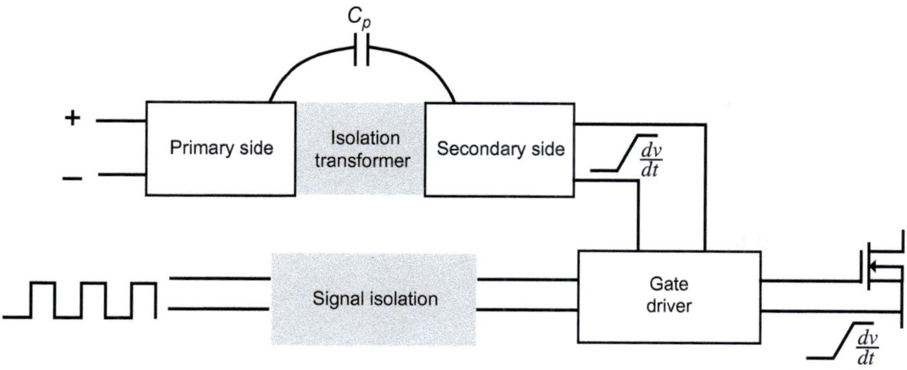

Figure 7.44 Isolation requirements for high voltage gate drivers.

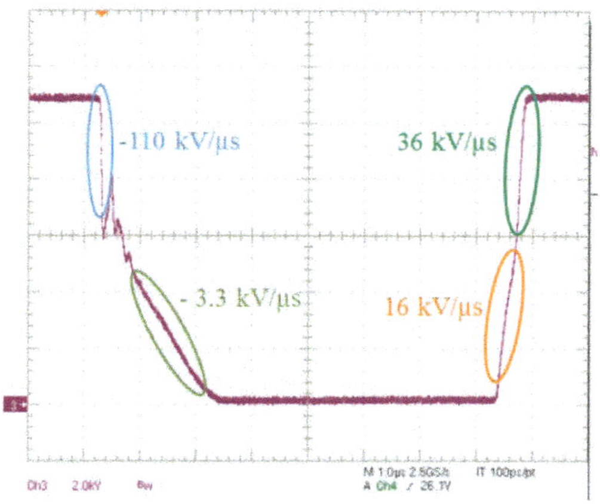

Figure 7.45 Turn-on and turn-off switching voltage transitions with *dv/dt* values at 11 kV, for the 15 kV N-IGBT [15].

The GD channel requires signal isolation from the controller board that is typically termed as CMTI. Optical fibers are usually preferred to isolate the PWM gate pulses. Fig. 7.46 shows a GD developed for 15 kV SiC IGBTs [15] with the specifications listed in Table 7.3. Design of this driver will be discussed in the following sections.

7.5.2 GD power supply

In power electronic converters, the PWM signals and the associated power supplies driving the gate of the power devices whose emitters/sources are at switching nodes have to be floating with respect to the system ground. Such gate drives are often referred to as high side drivers. At HVs, the PWM signals are isolated by using the optical fiber to communicate the gating signals. The gate drive power supplies are

Figure 7.46 High voltage gate driver for 15 kV SiC IGBTs [15].

Table 7.3 Specifications of gate driver for high voltage SiC devices [18]

Parameter	Value
Turn-on voltage	18–20 V
Turn-off voltage	−5 V
Supply input voltage	18–20 V
Switching frequency	Up to 20 kHz
Turn-on gate resistance	10–100 Ω
Turn-off gate resistance	10–33 Ω
Isolation voltage	Up to 20 kV
dv/dt capability	≥50 kV/μs
Isolation transformer coupling Capacitance	≤5 pF (1 Hz–100 MHz)

derived using an isolated dc–dc converter. The isolation stage of the dc–dc converter has to be designed with low coupling capacitance, and also it should not show significant degradation due to HV and high frequency stress.

The HV SiC power devices typically require +20 and −5 V gate voltages to turn-on and turn-off, respectively. A negative gate drive is required to overcome the possibility of gate turn-on due to the effect of collector-gate Miller capacitance on device during turn-off. An isolated dc–dc converter is used for generating these voltages. The schematic of the dc–dc power supply is shown in Fig. 7.47. A push–pull PWM controller LM5030, in open-loop, has been used to generate the complementary pulses to drive the MOSFET Hbridge BD6231F-E2.

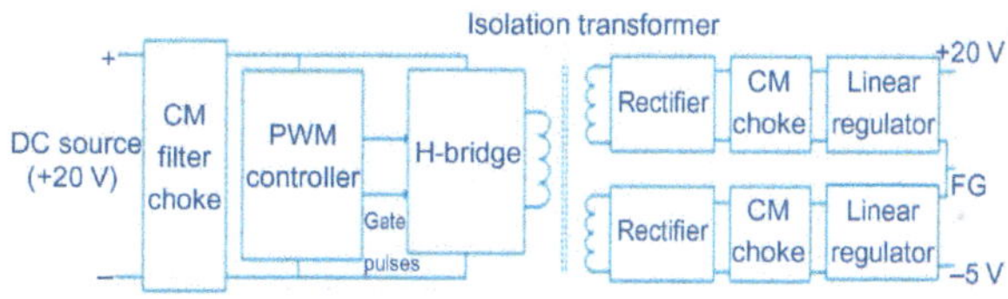

Figure 7.47 Schematic of isolated gate driver power supply (FG) [18]. *FG*, floating ground.

Figure 7.48 Amorphous core for the isolation transformer with $C_p < 2$ pF.

The isolation transformer has two secondary windings; the winding ratios have been selected so as to generate +20 and −5 V upon rectification. To maintain the HV isolation of the transformer, no voltage feedback control has been used. However, the output voltages are regulated using linear regulators at the output stage of the gate drive power supply. In Fig. 7.47, floating ground corresponding to the floating ground which may be connected to the floating emitter/source of power device the gate drive power supply is connected to. CMC s are used both at the input and output of the GD power supply to limit the circulating common mode currents. The fundamental component of this power supply is the isolation stage; the design of the isolation stage is discussed in next section. Very low-ESL capacitors have been used both on primary and secondary of the power supply.

Fig. 7.48 shows the amorphous core used in the isolated power supply. Fig. 7.49 shows the relation of coupling capacitance in isolation transformer to the operating frequency. The coupling capacitance is less than 1.32 pF up to 100 MHz. Photograph of the power supply is shown in Fig. 7.50.

7.5.3 Intelligent gate driver

Due to criticality of application and high stress on the device, it is important to acquire operating data from the device under test for diagnosis or predicting failure beforehand. Some of these functions have been previously implemented for LV Si

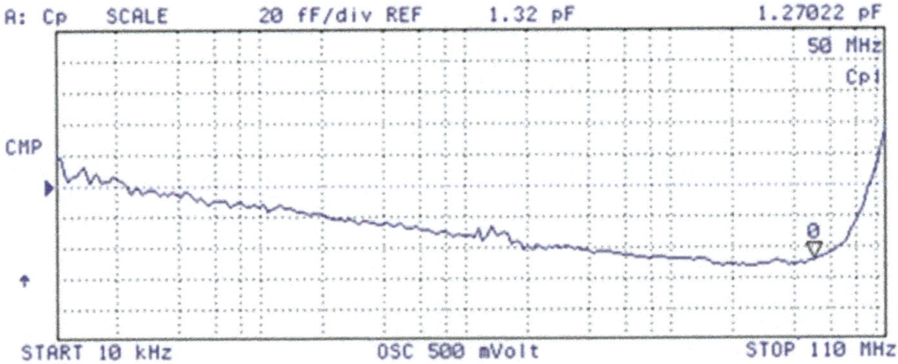

Figure 7.49 C_p vs frequency measurement of the amorphous core.

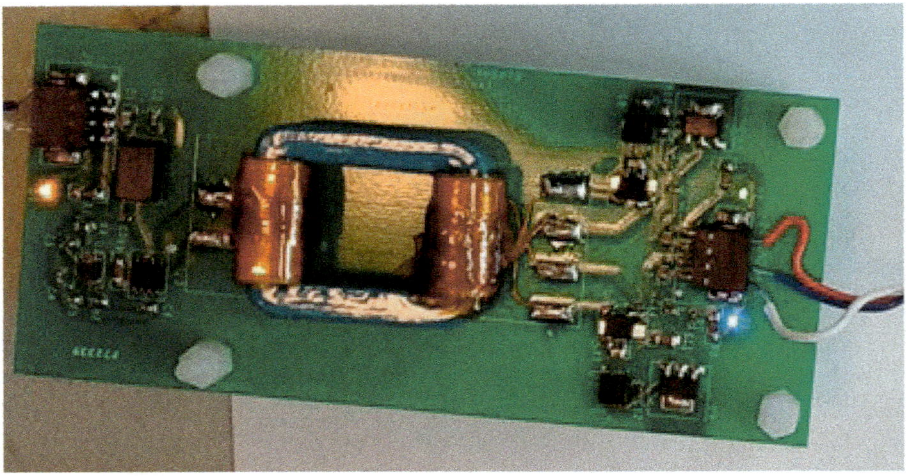

Figure 7.50 Photograph of the isolated power supply for high voltage gate drivers [18].

devices. However, this EMI hardened intelligent MV GD (IMGD), based on a complex programmable logic device (CPLD) digital logic, provides optically isolated data-logging under given high *dv/dt* and isolation voltage. Even in the harsh environment, it measures near-chip module temperature (T_{mod}), drain on-voltage [V_{ds}(on)] and device current (I_d). The fault signal and measured data are optically sent to the control side field programmable gate array (FPGA)-based IMGD interface board for decoding, data-logging, and diagnosis. The purpose of the intelligence addition is generally optimization in terms of the switching loss and the EMI. It also helps for matched series/parallel operation of the devices for scalability. The CPLD provides flexibility in configuring the signals and protections based on custom application. Together with a local clock, it helps in timing the events such as

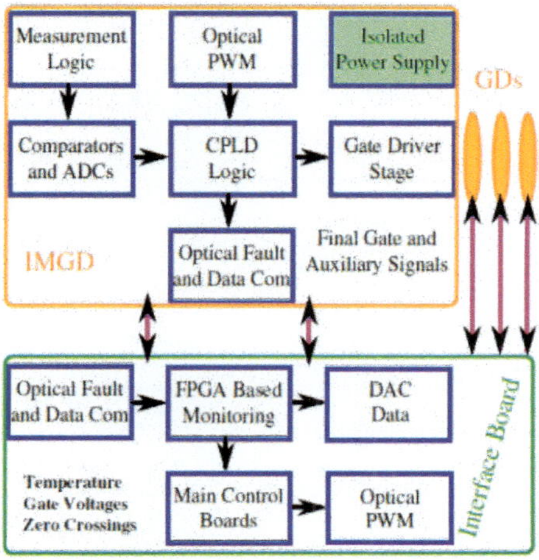

Figure 7.51 Block diagram of intelligent MV gate driver [16]. *MV*, medium voltage.

advanced gating and protections. It interfaces analog to digital converters (ADCs) with the optical signals.

7.5.3.1 Block diagram

Fig. 7.51 shows the block diagram of the IMGD. It mainly consists of the sensing and measurements, comparators, ADCs, communication, active-gating, and the driver stages. A 44pin ALTERA CPLD EPM3032A is used as the local brain. The simple 6pin, 40 MHz serial 8 bit, 3MSPS ADCs are used to convert the critical measurements. The PWM and fault signals are routed through CPLD. The local over-temperature and over-current (OC) fault signals are generated using ultrafast comparators. This board has two optical receivers and four optical transmitters which are configured by CPLD as PWM, clock, fault, and serial ADC signals. These IMGDs communicate with an interface board which can do the data logging and independent abnormal condition controls. The FPGA at the interface board can implement the communication decoding and other digital functions such as status display. The main control boards based on DSP or FPGA, implement the actual control algorithm for the converter systems such as grid-tie front end converter and dc–dc dual active bridge. They also have separate provisions for overvoltage and OC protections of the converters. They need feedback sensors such as voltage and current sensors for closed-loop control. The PWM cables are optical. The feedback sensors must have very small coupling capacitance and good isolation level for EMI resistant functioning.

Fig. 7.52 shows the populated IMGD board with all mentioned functions. This is the first version board with extra components than the requirement. In the final version this will be optimized. It has different analog and digital functional sections

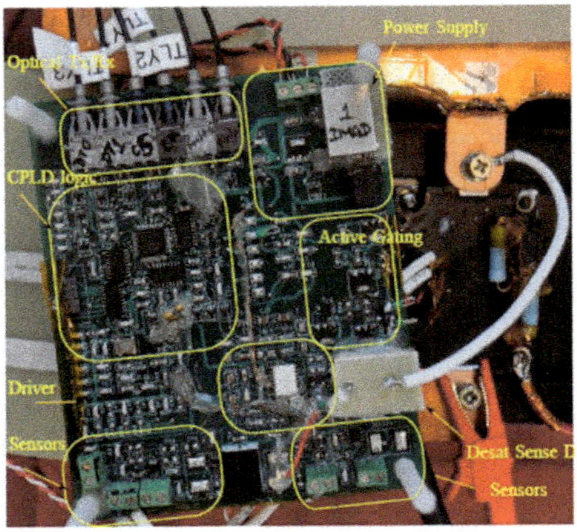

Figure 7.52 Photograph of the intelligent gate driver circuit board [16].

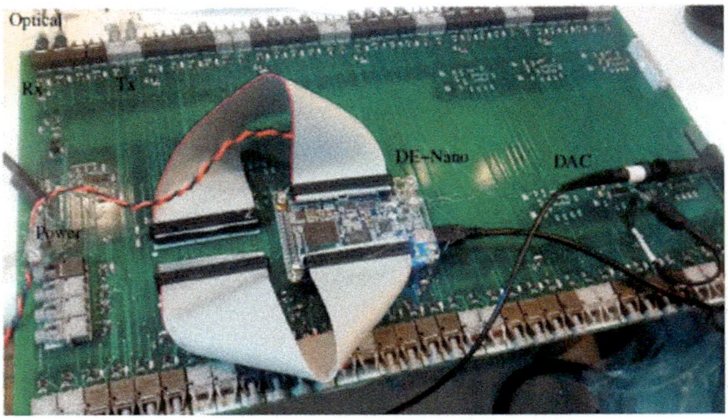

Figure 7.53 Photograph of the IMGD interface board [16]. *IMGD*, intelligent medium voltage gate driver.

which are separated from one another for better EMI performance. Fig. 7.53 shows the populated interface board which communicates with the IMGDs of a MV converter. It has on-board digital to analog converters (DACs) and FPGA board for implementing monitoring and data-logging function. Each IMGD has 4Tx and 2Rx used, respectively, as Fault, DT0, DT1, CS, CK, and PWM. The interface has the opposite type optical Tx/Rx for all signals and 12 IMGDs. The interface FPGA transmits a common clock for IMGD ADC operation and communication. The IMGD generates CS signal for framing of serial data and each bit is read synchronous to the transmitted clock of interface. The critical measurements can be converted by DACs for display and data logging. Based on any abnormality or any fault signal from any particular IMGD, all the IMGDs can be turned off together for safety.

7.5.3.2 Short-circuit protection scheme

SC protection of HV SiC power devices requires its characteristics during the fault and the SCWT based on its package. Hardware test benches for SC characterization of HV SiC devices have been discussed in [19,20]. A detailed discussion on SC characterization of 10 kV, 10 A SiC MOSFET is reported in [21]. The SC withstanding time of this MOSFET is reported to be 8.6 μs at 6 kV blocking voltage with +18/−5 V gate voltage. An SC protection scheme of SiC bipolar junction transistors (BJTs), MOSFETs, JFETs and IGBTs is reported in [19] with trip time less than 1 μs, but only up to blocking voltage of 1.2 kV. In [22], a GD with SC protection is designed for SiC MOSFET modules, but it is evaluated up to 800 V dc bus. Rogowski current sensor based SC protection scheme is proposed in [23] for 1.7 kV SiC MOSFETs and tested up to 1000 V dc bus. An SC protection based on overcurrent level in the ohmic region of 15 kV SiC MOSFET is designed in [24], but no experimental result is provided to demonstrate the protection in the saturation region of the MOSFET. It is required to ensure the protection, when the MOSFET goes into saturation during the SC, and the fault clears off before the SCWT of the device. No SC protection scheme for 10 kV SiC MOSFETs is reported with experimental results beyond 1200 V blocking voltage.

The SC protection block is shown in Fig. 7.54 [25]. The desat diode is required to block the same voltage across the MOSFET during the off-state. Also, the diode must be sufficiently fast to respond to the fault. A resistor divider attenuates the desat sense voltage during the ON-state of the MOSFET to V_{desat}, which is compared against the fixed threshold voltage V_{REF} by a high speed comparator. A zener diode with voltage rating slightly higher that V_{REF} is inserted across R_2 to protect against unexpected voltage overshoot. The blanking capacitor C_{blank} provides the desired blanking time to avoid spurious trip. The comparator sends the fault-trip signal, when V_{desat} surpasses V_{REF}. Receiving this signal, the soft turn-off block and

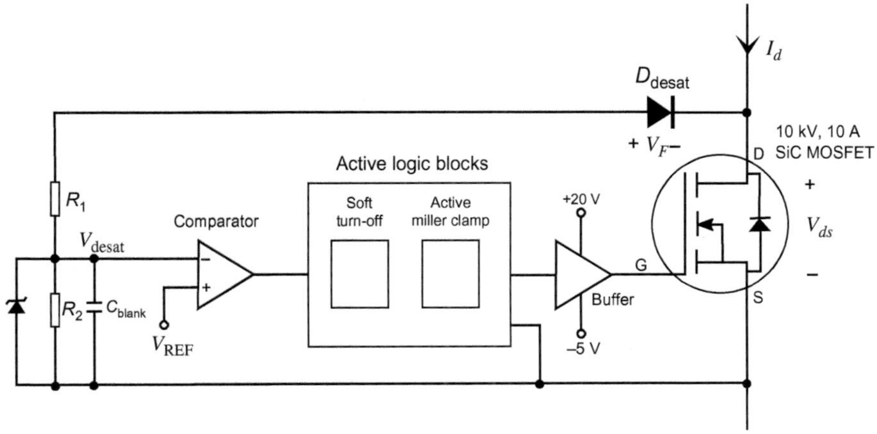

Figure 7.54 Short-circuit protection block in the intelligent gate driver. Other features of the gate driver are not shown here.

active mirror clamp circuit shuts down the gate pulse in two stages, ensuring small voltage overshoot caused by turn-off *di/dt*.

The sensed desat voltage is given by

$$V_{\text{desat}} = \frac{R_2}{R_1 + R_2}(V_F + I_d R_{\text{ds}}) \tag{7.4}$$

V_{desat} must be smaller than V_{REF} during normal operation. The MOSFET current I_d can go up to 10 A in continuous mode. Junction temperature of the SiC MOSFET in power converter typically goes above 100°C, and in some cases, it can be 150°C to obtain maximum power density. On-resistance of this 10 kV MOSFET, R_{ds} is 0.410 Ω at the room temperature and increases to 1.020 Ω at 150°C [26]. Increase in R_{ds} with temperature should also be accounted while deciding the threshold voltage V_{REF}. At rated current, the on-state voltage drop of the MOSFET turns out to be 10.2 V at 150°C. The resistors R_1 and R_2 are selected appropriately to match V_{REF}. Assuming 40% derating of the voltage, the 10 kV MOSFET is tested up to 6 kV blocking voltage. A HV SiC PiN diode is employed to block 6 kV during the off-state of the MOSFET.

The GD is tested on hard switch SC fault up to 6 kV blocking voltage. Then, its continuous operation is validated using single pulse test setup with MOSFET current up to 15 A and junction temperature up to 150°C. Fig. 7.55 shows photograph of the 10 kV MOSFET in die-level package and module package. Capability of the GD is further tested in 5 kV boost converter with the MOSFET switching at 10 kHz frequency.

(A) (B)

Figure 7.55 Photograph of 10 kV, 10 A 4H-SiC MOSFET (A) die package (B) module with isolated base plate.

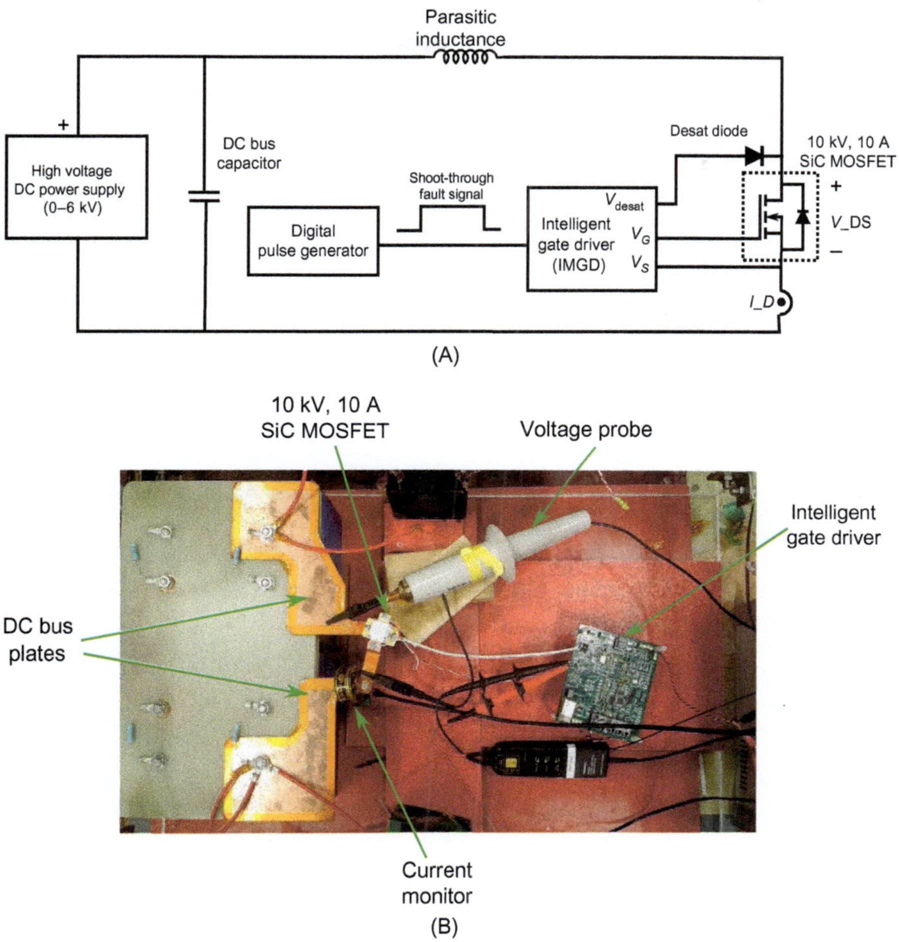

Figure 7.56 Experimental test setup to create hard switch short-circuit fault (A) schematic (B) photograph.

7.5.3.3 Hard switch short-circuit fault test setup

For SC test, the dc bus is designed for minimal parasitic inductance, required for small voltage overshoot across the MOSFET during the fault. The device voltage is measured with a 75 MHz HV passive probe P6015A. The transient current during the fault is monitored with a wideband current monitor 3972. To emulate hard switch SC fault, the MOSFET is connected directly across the dc bus, as shown in Fig. 7.56A. The dc bus is charged to a desired voltage level, and a HIGH gate pulse is sent to the GD for a duration of 4 μs. The fault duration is chosen less than SCWT to protect the device in case the protection circuit fails to work. Fig. 7.56B shows photograph of the experimental test setup.

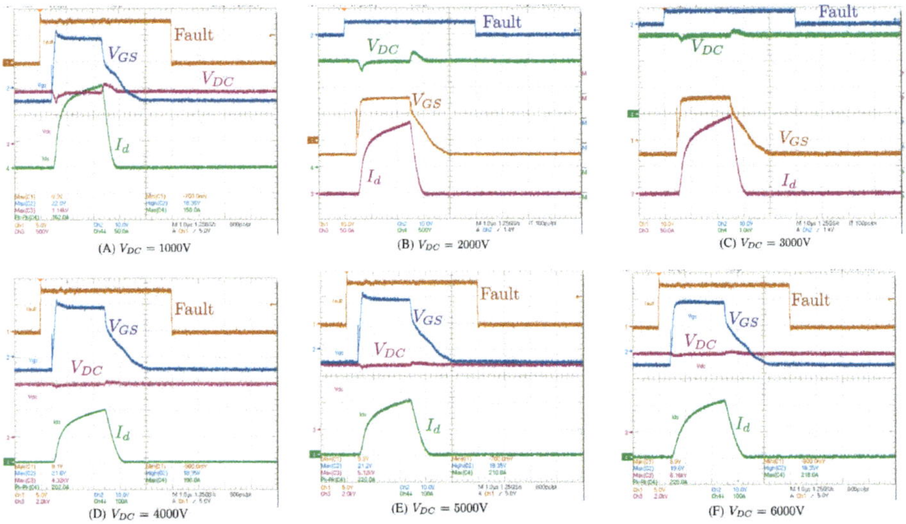

Figure 7.57 Experimental results for short-circuit protection at different dc bus voltage. An intentional FAULT is created for 4 μs. The gate driver responds within 2.4 μs after the short-circuit fault signal is detected.
V_{DC}: (A), (B) 500 V/div; (C) 1 kV/div (D), (E), (F) 2 kV/div I_d: (A), (B), (C) 50 A/div, (C), (D), (E) 100 A/div
V_{gs}: 10 V/div; time scale: 1 μs/div; active FAULT duration: 4 μs [25].

The SC test is performed at dc bus voltage from 1000 to 6000 V in the step of 1000 V. The experimental results are shown in Fig. 7.57. The gate voltage is kept at +20 V as HIGH and −5 V as LOW. During the fault, the MOSFET goes into saturation, and the current goes up to 220 A before it gets quenched by turning off the gate pulse. The blanking time is 1.8 μs, and it takes additional 0.6 μs for the SC current to come to down to zero. The two-stage turn-off ensures almost minimal voltage overshoot across the MOSFET, limiting to 150 V in this hardware setup.

7.5.3.4 Single pulse test setup

The GD in a power converter must be able to protect during the SC fault, and ensure uninterrupted operation in absence of the fault. The 10 kV MOSFET is tested using single pulse test circuit, as shown in Fig. 7.58 up to 15 A of the peak MOSFET current at 25 and 150°C junction temperature. Fig. 7.59 shows the experimental results for the single pulse test at 6 kV dc bus voltage. Oscillations in V_{ds} and I_d at the turn-on instant can be accounted by the parasitic of the circuit. As shown in the experimental setup, there is no overlap between the positive and the negative dc bus plates, causing large parasitic inductance in the dc bus. The overlap was avoided due to precaution considering that the insulating material between the plates may get damaged at HV. A suitable insulating material can be employed to achieve the dc plates overlap as large as possible, which will minimize the

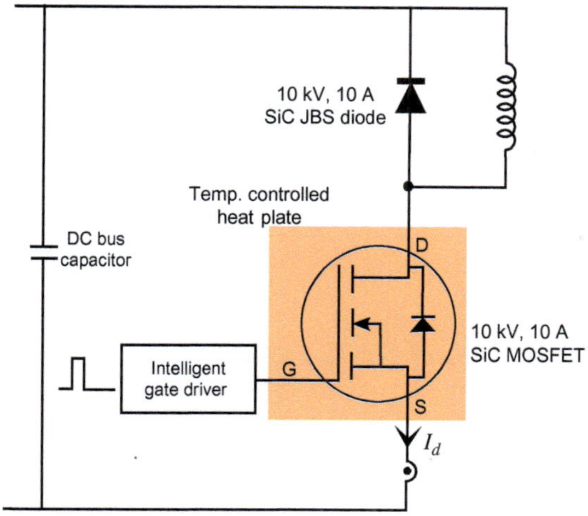

(A)

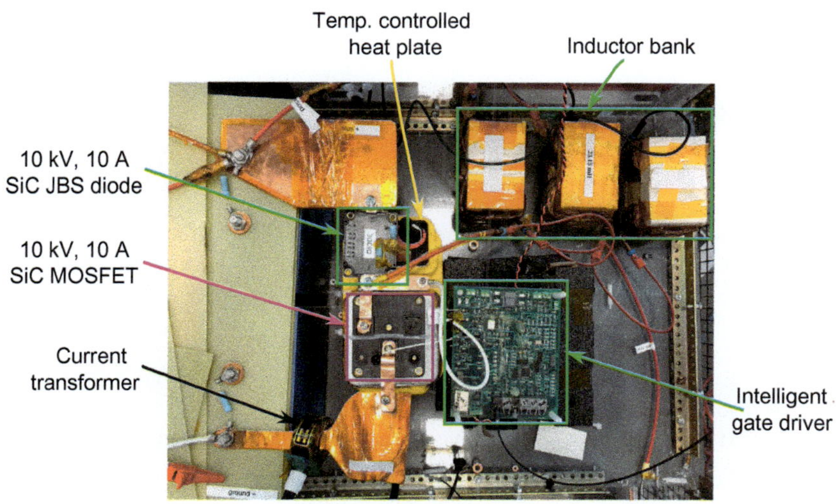

(B)

Figure 7.58 Single pulse test setup (A) schematic (B) photograph. High voltage probes are not visible in the photograph.

oscillation in the waveform. However, objective of the single pulse test is to check the upper overcurrent limit of the SC protection block, when the MOSFET is carrying continuous current. The driver should not trip when the MOSFET is carrying current up to its rated value of 10 A even at high junction temperature. At 150°C junction temperature, R_{ds} of the MOSFET increases, raising the voltage V_{desat} at input of the desat comparator, which may surpass the threshold V_{REF}, and result in *false* trip. From Fig. 7.59, the GD does not trip at 15 A and 150°C, showing that the

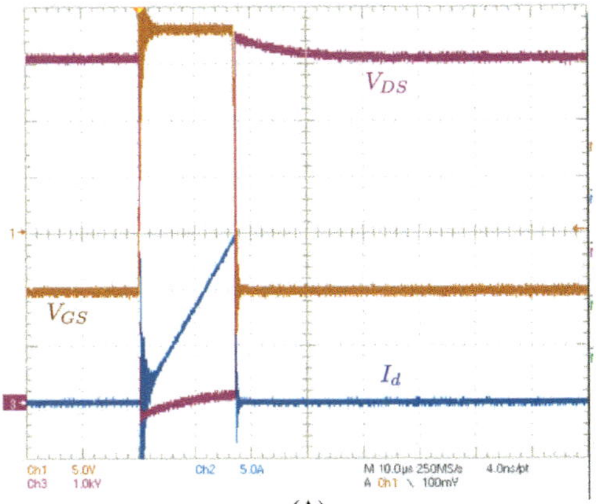

(A)

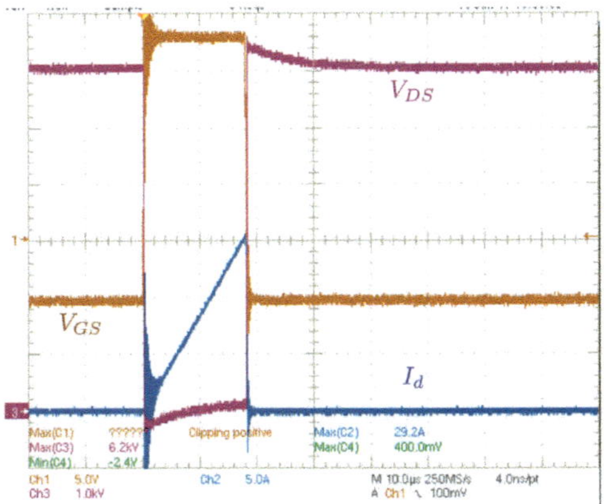

(B)

Figure 7.59 Experimental results for single pulse operation at 6 kV dc bus with short-circuit protection enabled at (A) $T_J = 25°C$ and (B) $T_J = 150°C$. The gate driver does not trip up to 15 A current even at 150°C, ensuring uninterrupted continuous operation required for 10 kV, 10 A SiC MOSFETs.
V_{gs}: 5 V/div; V_{ds}: 1 kV/div; I_d: 5 A/div; time scale: 10 μs/div [25].

overcurrent trip level is higher than the rated current. It qualifies the GD to be applicable in the continuous conduction operation.

7.5.3.5 Boost converter test setup

Having qualified in the hard switch SC fault test and the single pulse test, the GD is further subjected to test under continuous load current condition. A boost converter is built in the laboratory using 10 kV MOSFET to obtain 5000 V dc output voltage. Design of the boost converter at MV level brings challenges in the inductor design. The high frequency inductor, as shown in Fig. 7.60A, will have inherent parasitic capacitance, resulting from interwinding layers and the contact of the winding to the core. The high *dv/dt* associated with HV SiC MOSFETs causes flow of high

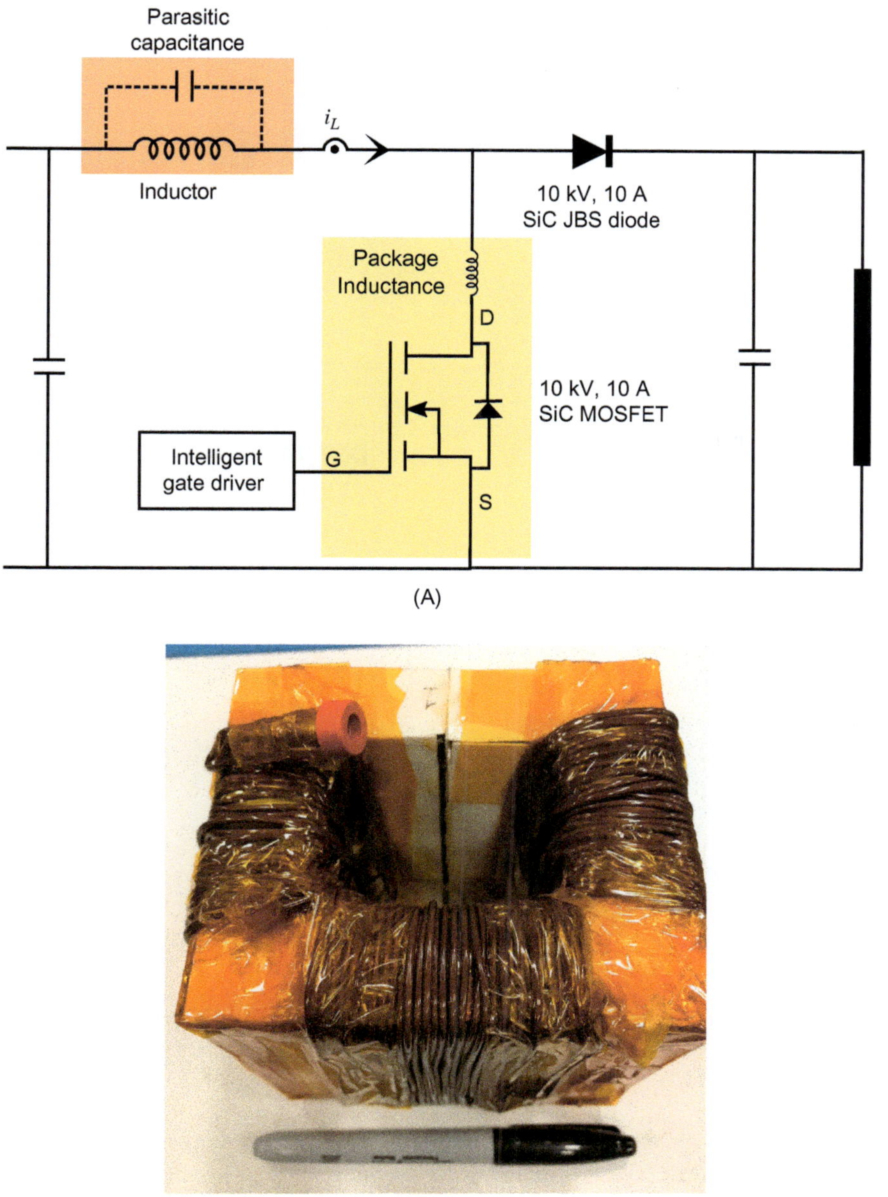

Figure 7.60 Boost converter test setup: (A) the schematic showing parasitic capacitance of the inductor and (B) high frequency inductor with minimal parasitic capacitance, designed in the laboratory.

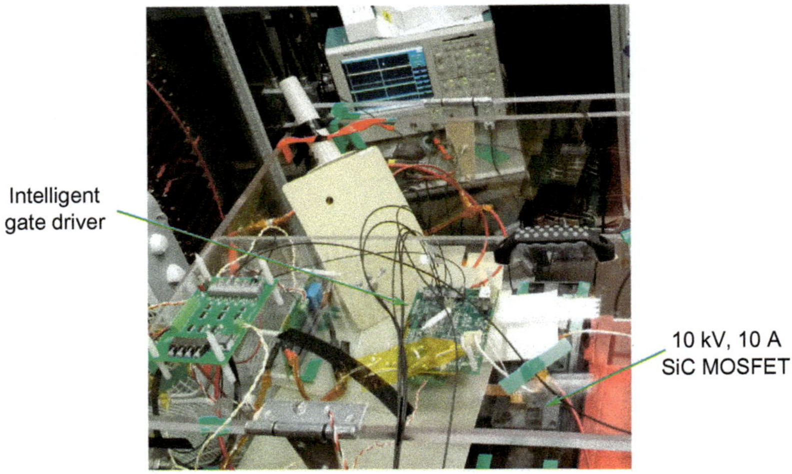

(A)

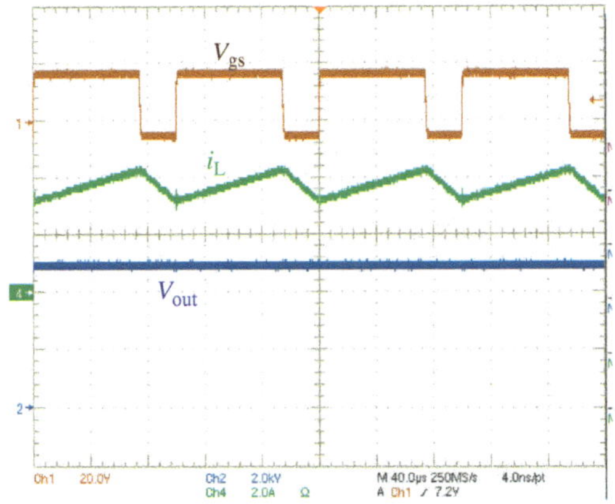

(B)

Figure 7.61 Boost converter test with the intelligent gate driver at 5 kV output voltage, 5 A peak inductor current and 10 kHz switching frequency: (A) photograph of the setup, and (B) experimental results. V_{out}: 2 kV/div; V_{gs}: 20 V/div; i_L: 2 A/div [25].

very high frequency current through the inductor, which in-turn interacts with the inductance of the MOSFET package. It results in resonance in the circuit and produces significant ringing in both voltage and current waveform. This ringing gets superimposed on the waveform to be observed, creating problem in recording clean waveform. A high frequency inductor is designed in the laboratory, shown in Fig. 7.60B, using ferrite core and single layer winding. Its parasitic capacitance is measured to be less than 100 pF. Two of these inductors are connected in series in the boost converter setup to further reduce the parasitic capacitance.

The MOSFET is switched at 10 kHz frequency. Peak of the inductor current i_L is limited to 5 A due to saturation current limit of the inductor. Fig. 7.61A shows hardware setup of the boost converter. The experimental results are shown in Fig. 7.61B. Due to low parasitic capacitance of the inductor, the waveform is relatively clean. Capability of the GD to drive the MOSFET in the given continuous operating conditions is validated by these results.

References

[1] S. Hazra, K. Vechalapu, S. Madhusoodhanan, S. Bhattacharya, K. Hatua, Gate driver design considerations for silicon carbide MOSFETs including series connected devices, in: 2017 IEEE Energy Conversion Congress and Exposition (ECCE), 2017, pp. 1402–1409.

[2] J. Wang, T. Zhao, J. Li, A.Q. Huang, R. Callanan, F. Husna, et al., Characterization, modeling, and application of 10-kV SiC MOSFET, IEEE Trans. Electron Devices 55 (2008) 1798–1806.

[3] K. Vechalapu, S. Bhattacharya, E.V. Brunt, S.H. Ryu, D. Grider, J.W. Palmour, Comparative evaluation of 15-kV SiC MOSFET and 15-kV SiC IGBT for medium-voltage converter under the same dv/dt conditions, IEEE J. Emerg. Sel. Top. Power Electron. 5 (2017) 469–489.

[4] S. Hazra, S. Madhusoodhanan, G.K. Moghaddam, K. Hatua, S. Bhattacharya, Design considerations and performance evaluation of 1200-V 100-A SiC MOSFET-based two-level voltage source converter, IEEE Trans. Ind. Appl. 52 (2016) 4257–4268.

[5] Q. Zhang, R. Callanan, M.K. Das, S.H. Ryu, A.K. Agarwal, J.W. Palmour, SiC power devices for microgrids, IEEE Trans. Power Electron. 25 (2010) 2889–2896.

[6] Z. Zhang, W. Zhang, F. Wang, L.M. Tolbert, B.J. Blalock, Analysis of the switching speed limitation of wide band-gap devices in a phase-leg configuration, in: 2012 IEEE Energy Conversion Congress and Exposition (ECCE), 2012, pp. 3950–3955.

[7] S. Hazra, A. De, L. Cheng, J. Palmour, M. Schupbach, B.A. Hull, et al., High switching performance of 1700-V, 50-A SiC power MOSFET over Si IGBT/BiMOSFET for advanced power conversion applications, IEEE Trans. Power Electron. 31 (2016) 4742–4754.

[8] Y. Xi, M. Chen, K. Nielson, R. Bell, Optimization of the drive circuit for enhancement mode power GaN FETs in dc–dc converters, in: 2012 Twenty-Seventh Annual IEEE Applied Power Electronics Conference and Exposition (APEC), 2012, pp. 2467–2471.

[9] A. Lidow, J. Strydom, eGaN FET drivers and layout considerations, White Paper: WP008, 2016.

[10] A. Lidow, J. Strydom, M. de Rooij, D. Reusch, GaN Transistors for Efficient Power Conversion (2nd edn.), Wiley, 2014.

[11] T. Iwaki, S. Ishiwaki, T. Sawada, M. Yamamoto, Mathematical analysis of GaN high electron mobility transistor false turn-on phenomenon, Electron. Lett. 53 (2017) 1327–1329.

[12] H. Wang, R. Xie, C. Liu, J. Wei, G. Tang, K.J. Chen, Maximizing the performance of 650 V p-GaN gate HEMTs: dynamic RON characterization and gate-drive design considerations, in: 2016 IEEE Energy Conversion Congress and Exposition (ECCE), 2016, pp. 1–6.

[13] X. Zhang, N. Haryani, Z. Shen, R. Burgos, D. Boroyevich, Ultra-low inductance phase leg design for GaN-based three-phase motor drive systems, 2015 IEEE 3rd Workshop on Wide Bandgap Power Devices and Applications (WiPDA), 2015, 119–124.

[14] W. Zhang, X. Huang, F.C. Lee, Q. Li, Gate drive design considerations for high voltage cascode GaN HEMT, in: 2014 IEEE Applied Power Electronics Conference and Exposition – APEC, 2014, pp. 1484–1489.
[15] A. Kadavelugu, S. Bhattacharya, Design considerations and development of gate driver for 15 kV SiC IGBT, in: 2014 IEEE Applied Power Electronics Conference and Exposition – APEC, 2014, pp. 1494–1501.
[16] A. Tripathi, K. Mainali, S. Madhusoodhanan, A. Yadav, K. Vechalapu, S. Bhattacharya, A MV intelligent gate driver for 15 kV SiC IGBT and 10 kV SiC MOSFET, in: 2016 IEEE Applied Power Electronics Conference and Exposition (APEC), 2016, pp. 2076–2082.
[17] T. Batra, G. Gohil, A.K. Sesham, N. Rodriguez, S. Bhattacharya, Isolation design considerations for power supply of medium voltage silicon carbide gate drivers, in: 2017 IEEE Energy Conversion Congress and Exposition (ECCE), 2017, pp. 2552–2559.
[18] K. Mainali, S. Madhusoodhanan, A. Tripathi, K. Vechalapu, A. De, S. Bhattacharya, Design and evaluation of isolated gate driver power supply for medium voltage converter applications, in: 2016 IEEE Applied Power Electronics Conference and Exposition (APEC), 2016, pp. 1632–1639.
[19] D.P. Sadik, J. Colmenares, G. Tolstoy, D. Peftitsis, M. Bakowski, J. Rabkowski, et al., Short-circuit protection circuits for silicon-carbide power transistors, IEEE Trans. Ind. Electron. 63 (2016) 1995–2004.
[20] E.P. Eni, T. Kerekes, C. Uhrenfeldt, R. Teodorescu, S. Munk-Nielsen, Design of low impedance busbar for 10 kV, 100A 4H-SiC MOSFET short-circuit tester using axial capacitors, in: 2015 IEEE 6th International Symposium on Power Electronics for Distributed Generation Systems (PEDG), 2015, pp. 1–5.
[21] E.P. Eni, S. Bczkowski, S. Munk-Nielsen, T. Kerekes, R. Teodorescu, Short-circuit characterization of 10 kV 10A 4H-SiC MOSFET, in: 2016 IEEE Applied Power Electronics Conference and Exposition (APEC), 2016, pp. 974–978.
[22] K. Fink, A. Volke, W. Wei, E. Wiesner, E. Thal, Gate-driver with full protection for SiC-MOSFET modules, in: PCIM Asia 2016; International Exhibition and Conference for Power Electronics, Intelligent Motion, Renewable Energy and Energy Management, 2016, pp. 1–7.
[23] J. Wang, Z. Shen, C. DiMarino, R. Burgos, D. Boroyevich, Gate driver design for 1.7 kV SiC MOSFET module with Rogowski current sensor for short-circuit protection, in: 2016 IEEE Applied Power Electronics Conference and Exposition (APEC), 2016, pp. 516–523.
[24] X. Zhang, H. Li, J.A. Brothers, J. Wang, L. Fu, M. Perales, et al., A 15 kV SiC MOSFET gate drive with power over fiber based isolated power supply and comprehensive protection functions, in: 2016 IEEE Applied Power Electronics Conference and Exposition (APEC), 2016, pp. 1967–1973.
[25] A. Kumar, A. Ravichandran, S. Singh, S. Shah, S. Bhattacharya, An intelligent medium voltage gate driver with enhanced short circuit protection scheme for 10 kV 4H-SiC MOSFETs, in: 2017 IEEE Energy Conversion Congress and Exposition (ECCE), 2017, pp. 2560–2566.
[26] J.W. Palmour, L. Cheng, V. Pala, E.V. Brunt, D.J. Lichtenwalner, G.Y. Wang, et al., Silicon carbide power MOSFETs: Breakthrough performance from 900 V up to 15 kV, in: 2014 IEEE 26th International Symposium on Power Semiconductor Devices IC's (ISPSD), 2014, pp. 79–82.

Applications of GaN power devices

8

Fred C. Lee, Qiang Li, Chao Fei and Yuchen Yang
Center for Power Electronics Systems, Virginia Tech, Blacksburg, VA, United States

With recent advances made in gallium nitride (GaN) power devices, this new generation of switches can be operated with significantly higher frequency compared to their silicon counterparts. It has been demonstrated that an increase of switching frequency by a factor of 10–20 is possible along with improved efficiency. Fig. 8.1 shows some possible applications for GaN devices. For high-voltage GaN (>600 V), they have great potential to penetrate the market of off-line power supply. For low voltage GaN (<100 V), they also could have huge impact on the design of brick DC/DC converter, point-of-load converter and high frequency wireless power transfer system.

It is interesting to note that converter topologies for GaN-based design are converging, in contrast with the silicon-based design. For example, bridgeless totem-pole configuration for power factor correction (PFC) and LLC resonant converters for DC/DC systems are deemed the preferred choices. The standardization of converter topologies will have a significant impact on the development of the standardized modular building blocks toward system-level integration for a broader range of applications.

In this chapter, a number of designs have been demonstrated in applications such as front-end PFC, high-voltage DC/DC converters, and battery chargers. All of these designs demonstrate a significant improvement in power density by a factor of 5–10. It is notable that not only is this not achieved at the expense of efficiency but also in some cases better efficiency is achieved together with significantly higher power density. Furthermore, in most of these designs, magnetics are distributed in the form of matrix transformers or integrated inductors. For operating frequencies in the mega-hertz range, the magnetics are integrated in the printed circuit board (PCB) with significantly improved manufacturability and reduced cost.

8.1 Hard switching vs soft switching [1]

In this section, the detailed loss analysis is presented based on buck converter with experimental and simulation results. The most powerful tool to analyze the device loss is an accurate loss model. Simulation models for silicon MOSFET are commonly used to investigate the device performance. One of the advantages of the simulation model is that the voltage and current information of every node in

Wide Bandgap Semiconductor Power Devices. DOI: https://doi.org/10.1016/B978-0-08-102306-8.00009-5

the simulation can be easily derived. With great effort, the authors cooperate with Transphorm Inc. and develop a 600-V cascode GaN HEMT SPICE simulation model [2]. The accuracy of this model is validated by numerous experiments. The simulation waveforms match with the experimental results very well including the dv/dt, di/dt, magnitude and frequency of voltage/current ringing during transition. Thus, this simulation model can be used to help analyze the transistor loss.

The key parameters for the hardware setup are listed in Table 8.1. It is common sense that common source inductance which is defined as the inductance shared by the power loop and driving loop impacts the switching loss significantly. In addition, the power loop inductance plays an important role on switching performance [3–5]. Therefore, the layout of the PCB is designed to eliminate the common

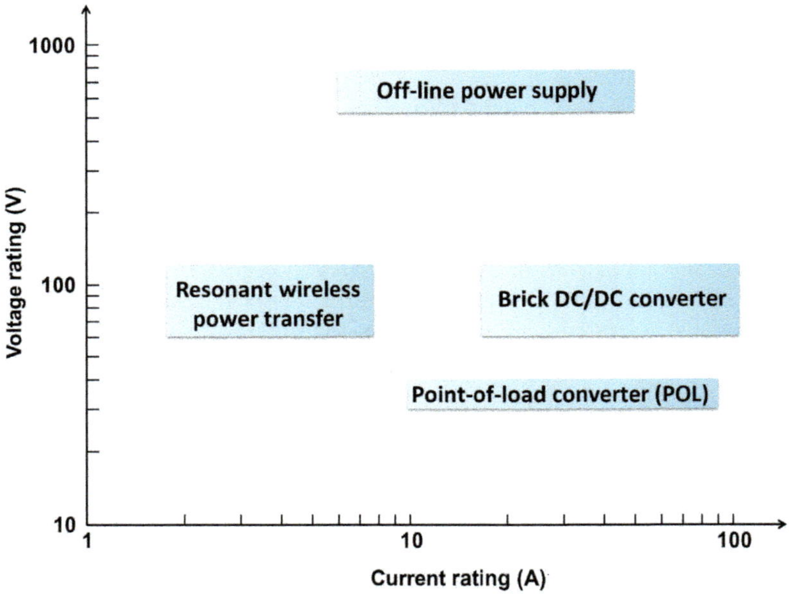

Figure 8.1 Possible applications for GaN devices.

Table 8.1 Parameters of buck converter

Parameters	Value	Parameters	Value
Input voltage	380 V	Output voltage	200 V
Switching frequency	500 kHz	Maximum output current	6 A
Inductor core material	3F35	Inductor current ripple	3 A
Top switch	TPH2006[a]	Bottom switch	TPH2010[b]

[a]TPH2006 is from Transphorm Inc. $V_{ds\_max} = 600$ V, $R_{ds\_on} = 0.15$ Ω.
[b]TPH2010 is Schottky diode from Transphorm Inc. $V_{BR} = 600$ V, $V_F = 1.3$ V@6 A.

source inductance (excluding package parasitic inductance) and minimize power loop inductance. The package for the cascode GaN HEMT also introduces nH level parasitic inductance. The package and layout inductance is extracted from Ansoft Q3D FEA simulation, and it is then applied in SPICE simulation.

In the first setup, a Schottky diode is applied as the bottom switch to eliminate the reverse recovery impact. The green line in Fig. 8.2 shows the converter efficiency. Then the loss breakdown at 6 A output condition based on the device simulation model is shown as green bar in Fig. 8.3. The inductor loss is measured by

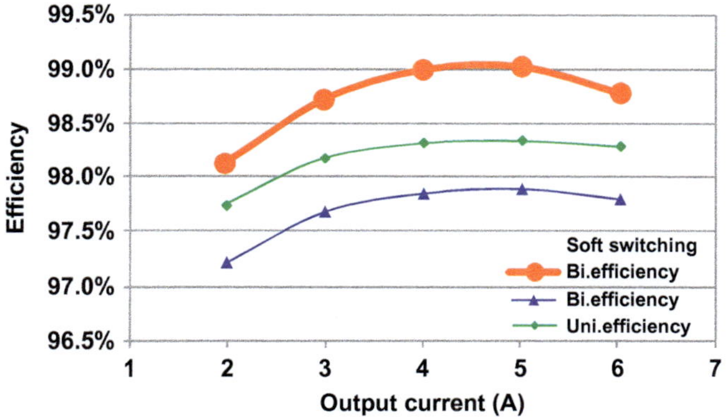

Figure 8.2 Converter efficiency (green line: CCM hard-switching, GaN Schottky diode as bottom switch; blue line: CCM hard-switching, cascode GaN HEMT as bottom switch; red line: CRM soft-switching, cascode GaN HEMT as top and bottom switches). *CCM*, continuous current mode; *GaN*, gallium nitride; *CRM*, critical mode.

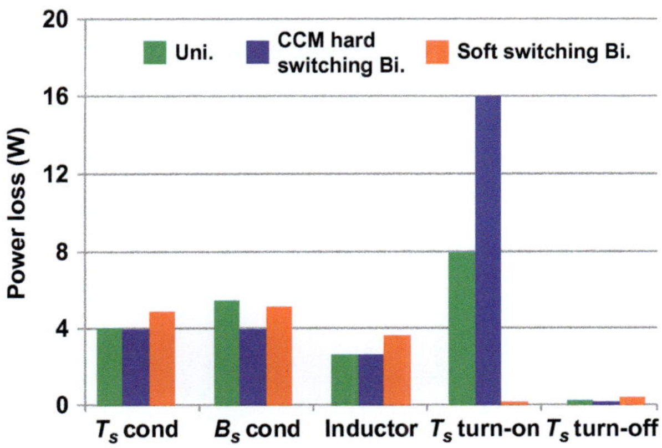

Figure 8.3 Buck converter loss breakdown at 6 A output current condition (green bar: CCM hard-switching, GaN Schottky diode as bottom switch; blue bar: CCM hard-switching, cascode GaN HEMT as bottom switch; red bar: CRM soft-switching, cascode GaN HEMT as top and bottom switches). *CCM*, continuous current mode; *GaN*, gallium nitride; *CRM*, critical mode.

Mu's method [6]. Other loss terms are negligible compared to these five loss bars. The turn-on loss is 40 times larger than the turn-off loss under hard-switching conditions, which is quite different from the low voltage condition [3–5].

The bidirectional buck/boost converter, as shown in Fig. 8.4, is widely used in industrial applications. When power flows from V_a to V_b, the converter operates as buck converter. When power flows from V_b to V_a, the converter operates as boost converter. Compared to the aforementioned unidirectional buck converter, a cascode GaN HEMT is applied as bottom switch. The dead time is fine tuned to optimize the converter efficiency overall load range. The blue line in Fig. 8.2 shows the converter efficiency. It clearly shows that the bidirectional buck converter efficiency is lower that the unidirectional one. Then, the blue bar in Fig. 8.3 shows the loss breakdown at 6 A output condition based on the simulation model. The turn-on loss increases to 16 W, twice larger than the unidirectional case. The other loss bars don't change much.

Operating the buck converter at critical mode (CRM), the bottom switch is turned off with zero current, thus eliminating the reverse recovery effect. Based on the input–output condition, the top switch can achieve zero-voltage turn on which further removes the junction capacitor charge effect. Fig. 8.5 shows key waveforms of top switch turn-on transition waveforms in CRM buck converter. The turn-off transition is similar with continuous current mode (CCM) hard-switching condition. At T_1 instant, the buck inductor resonates with junction capacitors of top and bottom switches. v_{ds_GaN} naturally decreases to zero at T_2. The remaining negative inductor current continues to discharge the drain-source voltage of silicon MOSFET in the cascode structure, as well as gate-source voltage of the GaN HEMT. At T_3 instant, gate signal is applied. Zero-voltage-switching (ZVS) operation is achieved as long as the inductor current is still negative at T_3.

The red line in Fig. 8.2 shows the converter efficiency with soft-switching. The red bar in Fig. 8.3 shows the loss breakdown at 6 A output condition, and it clearly shows that the turn-on loss is minimized and only introduces a little more conduction loss with CRM operation. The increase of conduction loss at CRM is due to the increase of root mean square (rms) value of inductor and switch current. The rms value of triangular waveform (CRM) is definitely larger than square waveform (CCM) with same average value.

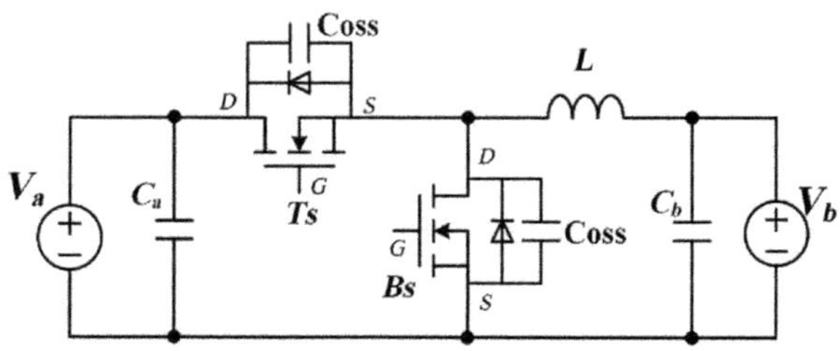

Figure 8.4 Bidirectional buck/boost converter.

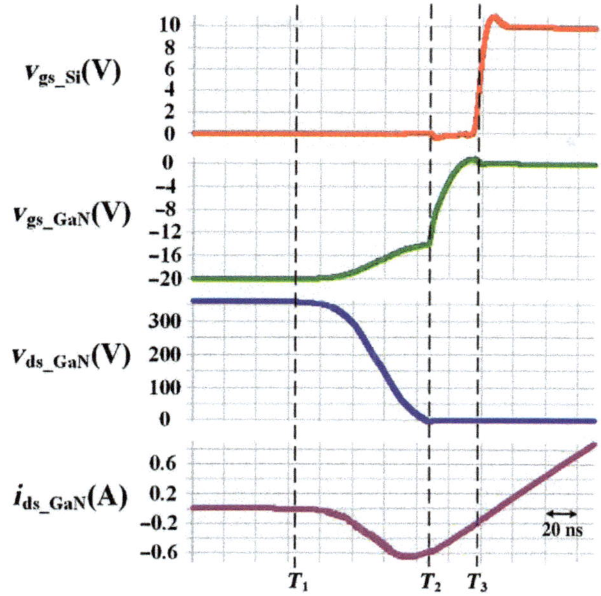

Figure 8.5 Top switch turn-on transition with cascode GaN HEMT as bottom switch at CRM.

In a word, in high-voltage low-current buck converter with cascode GaN HEMT, usually below 10 A, turn-on loss dominates in hard-switching conditions, and turn-off loss is very small due to the intrinsic current source driving mechanism. Actually, this is also true in other topologies, such as Boost, Buck-Boost, etc. Applying a Schottky diode as the bottom switch at CCM hard-switching operation is more efficient due to a small reverse recover charge. While with soft-switching operations, turn-on loss is minimized and introduces a little more conduction loss. It should be noticed that the turn-off loss remains low even the turn-off current doubles. This characteristic makes the cascode GaN HEMT very suitable for high-frequency operation as long as zero-voltage turn on is achieved.

Fig. 8.6 shows efficiency comparison between the cascode GaN HEMT and the silicon MOSFET at CRM soft-switching condition. It clearly shows that the cascode GaN HEMT gains 0.7%–0.8% efficiency improvements at CRM soft-switching condition. Fig. 8.7 shows the loss breakdown at 6 A output condition. The cascode GaN HEMT has much smaller turn-off loss due to current source turn-off mechanism which is unique to the cascode structure. Also, the cascode GaN HEMT has smaller conduction loss because smaller junction capacitor requires smaller circulating energy to achieve ZVS operation.

In summary, the loss analyses of the cascode GaN HEMT based on buck converter are presented. At hard-switching conditions, the turn-on loss dominates due to the reverse recovery charge and junction capacitor charge impact. However, the turn-off loss is negligible due to the intrinsic current source driving mechanism.

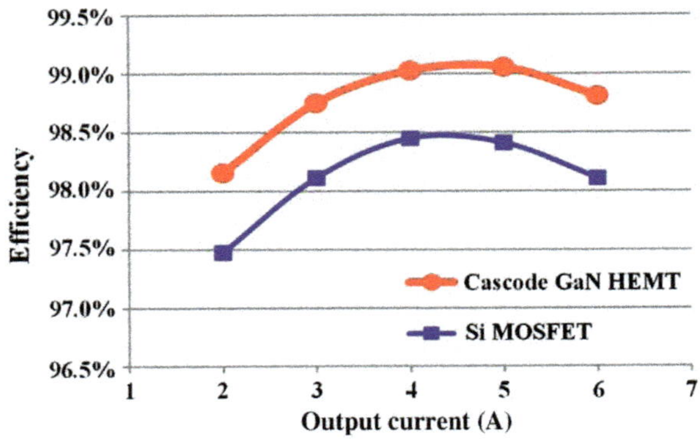

Figure 8.6 CRM buck converter efficiency comparison.

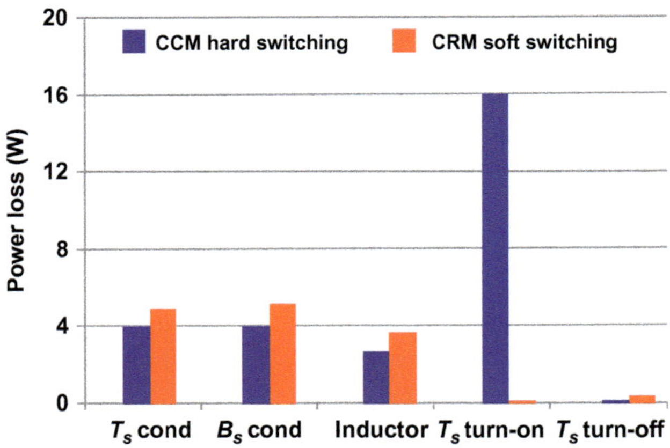

Figure 8.7 CRM buck converter loss breakdown at 6 A output current condition.

This characteristic makes the cascode GaN HEMT very suitable for high frequency operation as long as ZVS turn on is achieved.

8.2 Bidirectional buck/boost converter [7]

The bidirectional buck/boost converter, as shown in Fig. 8.4, is widely used in the power electronics system due to its simplicity and high efficiency, such as the on board charger/discharger for plug-in hybrid electric vehicles [8–10], and the interfaced converter for the energy storage systems [11,12]. Conventional silicon device based bidirectional buck/boost converters are usually intended to be operated in discontinuous current mode (DCM) in order to alleviate reverse recovery issues and to use a small inductor. However, the DCM operation greatly increases turn-off loss because the main switch is turned off at least twice of the load current. As a result, the switching frequency can barely be pushed to hundreds of kilo-hertz due to power loss considerations.

With GaN devices, the switching frequency has been continuously pushed up to several MHz to both reduce the size of the passive components and to increase power density [13–15]. Refs. [1,2,16] have thoroughly illustrated the switching loss mechanism in a high-voltage cascode GaN switch and derived two important switching characteristics. One is that the turn-on switching loss is dominant due to the reverse recovery charge or junction capacitor charge of the free-wheeling device at hard-switching condition. The other notable characteristic is that the turn-off loss is negligible because of the intrinsic current source driving mechanism in cascode structure. These important switching characteristics imply that ZVS is still desired for GaN devices in high frequency applications, while the turn-off current is no longer a big concern for cascode GaN devices.

Critical current mode (CRM) operation is the most simple and effective way to achieve ZVS and is widely used in medium–low power applications [1,13,15]. CRM operation introduces a large current ripple, which is at least twice of the load current. It is necessary to interleave multiple phases to cancel the switching frequency current ripple and lead to a smaller electromagnetic interference (EMI) filter. However, the ZVS range of the CRM DC–DC converter is determined by the input and output voltage, which will significantly impact switching loss at high frequency. Furthermore, the resonant period formed by the inductor and device junction capacitors is too long at high frequency, which leads to a large circulating energy.

The concept of the coupled inductor has been applied successfully in interleaved voltage regular modules for the improvement of the efficiency and transient response [17]. However, the impact of the coupled inductor on the behavior of the converter during the resonant period, which is unique to CRM operation, has not yet been analyzed.

8.2.1 Coupled inductor at CRM

The interleaved bidirectional buck/boost converter with inverse coupled inductor is shown in Fig. 8.8. For simplicity, the two self-inductances are considered to be the same ($L_1 = L_2 = L$). The inverse coupled mutual inductance M is expressed as $M = k \cdot L$, where k is the coupling coefficient.

The key waveforms is shown in Fig. 8.9, where D is the duty cycle. For the buck direction, $D = V_b/V_a$. There are six time intervals in one switching cycle. The time intervals t_0–t_1, t_1–t_2, t_3–t_4, and t_4–t_5 are the same with a conventional two-phase buck converter with a coupled inductor. The analysis of the inverse coupled inductor in CCM operation is illustrated in [17]. For CRM operation, there are two more resonant periods in each switching cycle than in CCM operation. The time intervals t_2–t_3 and t_5–t_6 are the resonant periods that are different from the previous analysis. The equivalent inductance is summarized in Table 8.2.

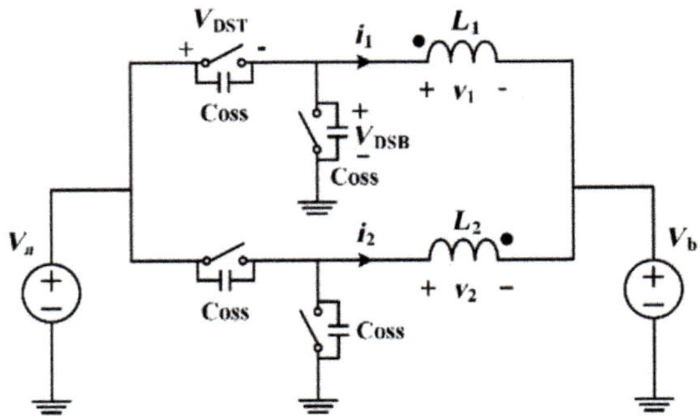

Figure 8.8 Interleaved buck/boost converter with inverse coupled inductor.

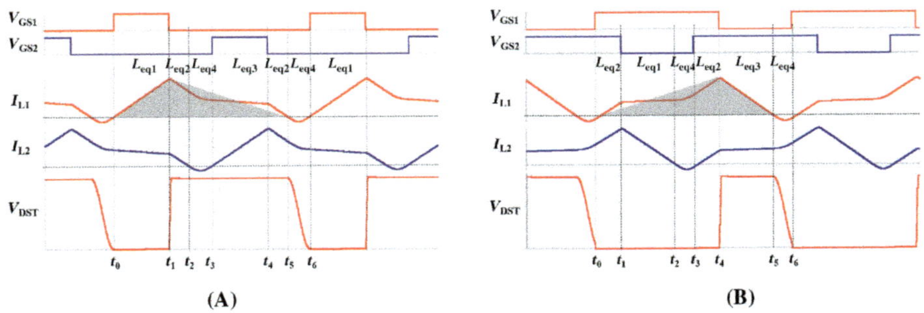

Figure 8.9 Key waveforms of CRM with coupled inductor considering resonant period. (A) $D<0.5$, (B) $D>0.5$.

Table 8.2 Coupled inductor equivalent inductance [14]

L_{eq1}	L_{eq2}	L_{eq3}	L_{eq4}
$(L^2-M^2)/\left(L+\left(\frac{D}{D'}\right)M\right)$	(L + M)	$(L^2-M^2)/\left(L+\left(\frac{D'}{D}\right)M\right)$	$L-(M^2/L)$

As the equivalent resonant inductance with a coupled inductor is $L_{eq4}=L-\left(M^2/L\right)$ which can be rewritten as $L_{eq4}=\mathrm{L}(1-\alpha^2)$, the relation between L_{eq4} and L_{nc} can be expressed as

$$\begin{cases} L_{eq4}=L_{nc}\cdot\left(1+\dfrac{D}{D'}k\right) D<0.5 \\ L_{eq4}=L_{nc}\cdot\left(1+\dfrac{D'}{D}k\right) D>0.5 \end{cases}$$

The inverse coupling coefficient k is negative which means the equivalent resonant inductance of a coupled inductor is always smaller than a noncoupled inductor. Accordingly, the resonant period with coupled inductor is also reduced compared to noncoupled case. The portion of transferring energy increases with the reduction of the resonant period in CRM, which leads to a reduction of conduction loss.

For a bidirectional buck/boost converter with a noncoupled inductor operating in CRM, one of the directions operates at $D < 0.5$ and switches at the valley point, while the other direction operates at $D > 0.5$ and can achieve ZVS easily with extra circulating energy. Even valley switching minimizes the turn-on switching loss, the remaining energy stored in the junction capacitor is still considerable at MHz switching frequency. A similar argument can also be applied to the circulating energy as this part of the power loss is a linear function of the switching frequency.

However, the inverse coupled inductor can modify the converter behavior during the resonant period. As shown in Table 8.3, the resonant amplitude with an inverse coupled inductor is different from the resonant amplitude with a noncoupled inductor. Fig. 8.10 summarizes the benefit of the ZVS range extension and circulating energy reduction with an inverse coupled inductor. The solid line is the ZVS boundary condition, which is the most desired operation point. Regions I and III represent valley switching for the buck direction, while they represent ZVS with circulating energy for the boost direction. Regions II and IV represent ZVS with circulating energy for the buck direction but valley switching for the boost direction. If the operation point is close to the ZVS boundary line, there is smaller turn-on switching loss in one direction and smaller circulating energy in the other direction.

8.2.2 Bidirectional buck/boost converter

To validate the benefit of the converter operating in CRM with inverse coupled inductor, a 380–150-V two-phase interleaved CRM buck/boost converter is built. The circuit is shown in Fig. 8.8. The output current is 8 A for the buck direction and 3 A for the boost direction. The converter can operate with both an inverse coupled inductor and a noncoupled inductor for comparison. The 600-V GaN HEMT from Transphorm is used as the active switch, since the turn-off switching loss is negligible, which means it is suitable for CRM operation and can be pushed to very high frequency. The switching frequency is set to be 1 MHz at full load condition to reduce the size of the passive components. An UI-shaped core with 3F45 material is used in the prototype, and the air gap is fine tune to achieve a certain coupling coefficient. Two ER23 core with 3F45 material are used as the noncoupled inductors for comparison. The prototype is shown in Fig. 8.11. The coupled inductor saves 50% footprint and 25% volume compared with two noncoupled inductors.

The loss breakdown of two-phase buck direction at full load is shown in Fig. 8.12. It clearly shows that the coupled inductor eliminates the turn-on switching loss by ZVS range extension. Moreover, coupled inductor can slightly reduce the conduction loss due to small reduction of rms current. Core loss is also reduced with coupled inductor since core volume can be shrink due to certain DC flux

Table 8.3 **Comparison between noncoupled inductor and inverse coupled inductor**

	Resonant inductance		Resonant amplitude		
	Noncoupled	Inverse coupled	Noncoupled	Inverse coupled	
				$V_b < 0.5V_a$	$V_b > 0.5V_a$
Buck	L	$L-(M^2/L)$	V_b	$V_b \cdot (1-(M/L))$	$V_b+(M/L)\cdot(V_a-V_b)$
boost	L	$L-(M^2/L)$	V_a-V_b	$V_a-V_b\cdot(1-(M/L))$	$(V_a-V_b)\cdot(1-(M/L))$

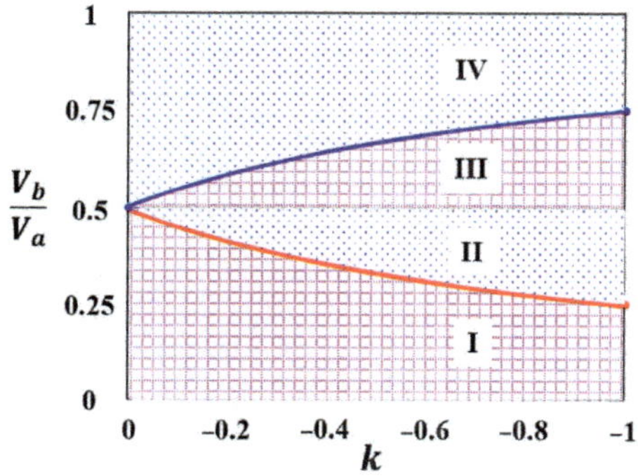

Figure 8.10 Coupled inductor modifies CRM operation. *CRM*, critical mode.

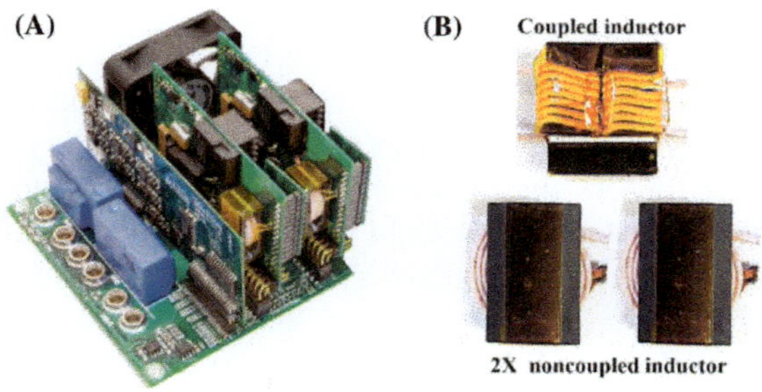

Figure 8.11 Prototype and inductor comparison. (A) Two-phase interleaved buck/boost converter prototype; (B) Comparison of coupled and noncoupled inductor.

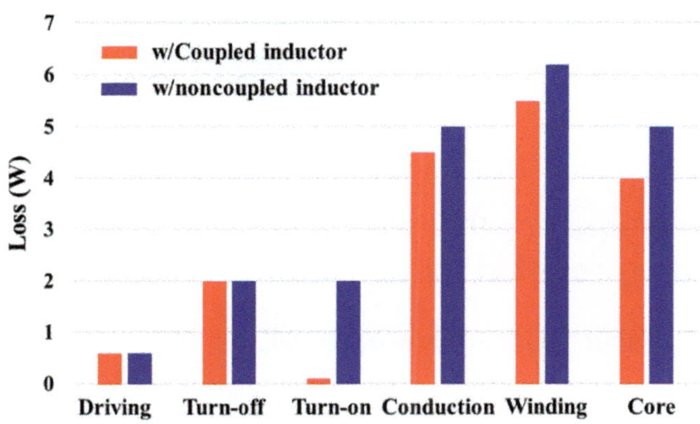

Figure 8.12 Loss breakdown of two-phase buck direction at full load.

cancellation. Overall, coupled inductor saves about 4 W power loss and improves efficiency by 0.3% at full load condition. It should be pointed out that the inductor is not optimized in terms of losses, and therefore, there is still room to improve the efficiency.

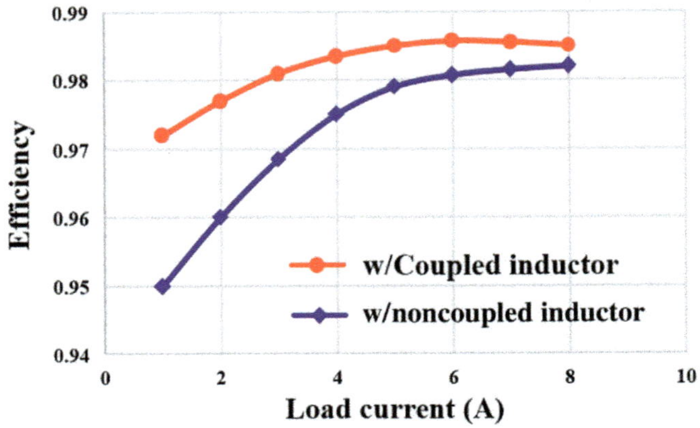

Figure 8.13 Efficiency comparison over wide load range of buck direction.

The full load range efficiency of buck direction is shown in Fig. 8.13. The efficiency of boost direction is similar. The switching frequency increases when the load current reduces due to the nature of CRM operation. The switching related loss becomes the dominant part at light load condition. Therefore, the coupled inductor saves more loss at light load and the efficiency improvement is over 2%.

This section aims to analyze the benefits of the inverse coupled inductors in CRM operation based on a 1 MHz interleaved buck/boost converter with GaN devices. The converter behavior during the resonant period is improved with an inverse coupled inductor. There is a reduction of the resonant period, an extension of the ZVS range when $D < 0.5$ and a reduction of the circulating energy when $D > 0.5$. All of these characteristics are beneficial for high frequency operation in terms of conduction and switching loss reduction. The coupled inductor prototype efficiency is 98.5% at 1 MHz, which is 0.3% higher than the efficiency of the non-coupled inductor, which validate the theoretical analysis.

8.3 High frequency PFC with PCB winding coupled inductor

8.3.1 GaN-based MHz totem-pole PFC

With the advent of 600 V GaN power semiconductor devices, the totem-pole bridgeless PFC rectifier [18,19], which was a nearly abandoned topology, is suddenly become a popular front end candidate for applications like 2-stage high-end adaptor, server and telecommunication power supply, and on-board battery charger. This is mostly attributed to the significant performance improvement of the GaN HEMT compared to Si MOSFET, particularly better figure-of-merit and significantly smaller body diode reverse recovery effect. The circuit topology of totem-pole PFC converter is shown in Fig. 8.14.

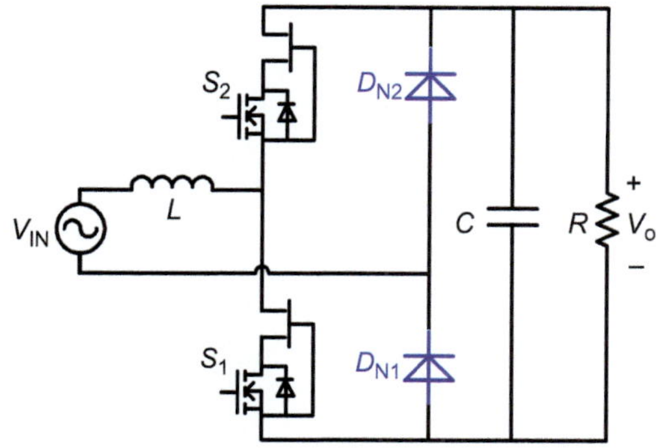

Figure 8.14 Totem-pole bridgeless PFC converter. *PFC*, power factor correction.

GaN-based hard-switching totem-pole PFC rectifier is demonstrated in literature [20]. As the reverse recovery charge of the GaN HEMT is much smaller than the Si MOSFET, hard-switching operation in totem-pole bridge configuration turned to be practical. By limiting switching frequency around or below 100 kHz, the efficiency could be above 98% for a 1-kW level single-phase PFC rectifier. Even the simple topology and high efficiency are attractive, the system level benefit is limited because the switching frequency is still similar to Si-based PFC rectifier.

Based on previous study, soft switching truly benefits the cascode GaN HEMT. As the cascode GaN HEMT has high turn-on loss and extremely small turn-off loss due to the current-source turn-off mechanism, CRM operation is very suitable. A GaN-based CRM boost PFC rectifier is first demonstrated which shows the high-frequency capability of the GaN HEMT and significant system benefits as the volume of the boost inductor and the DM filter is dramatically reduced [13,21].

With a similar system-level vision, the cascode GaN HEMT is applied in the totem-pole PFC rectifier while pushing frequency to above 1 MHz. Several important high-frequency issues, which used to be less significant at low-frequency, are emphasized, and the corresponding solutions are proposed and experimentally verified. They are including ZVS extension in order to solve switching loss caused by non-ZVS valley switching; variable on-time control to improve the power factor, particularly the zero-crossing distortion caused by traditional constant on-time control; and interleaving control for input current ripple cancellation [22].

8.3.1.1 ZVS extension

The CRM PFC rectifier utilizes the resonance between the inductor and the device junction capacitors to achieve ZVS or valley-switching. For boost-type CRM PFC rectifiers, ZVS can be achieved only when the input voltage is lower than one-half of the output voltage, assuming a negligible damping effect, which is often true with good design and limited resonant cycles. Thus, when the input voltage is higher than one-half of the output voltage, the drain-source voltage can only resonate to a valley point which is equal to $(2V_{\mathrm{in}} - V_o)$, so $(0.5CV^2)$ loss occurs at the following turn-on instant. As this loss is directly related to the switching frequency, when the frequency is pushed to the multi-MHz level, the non-ZVS loss is significant and dominant in the total converter loss, as shown in Fig. 8.15.

In order to solve this issue, the ZVS extension strategy explained in [23,24] is used. The concept is to modify the operation from CRM to quasi-square-wave mode. Hence instead of turning off the synchronous rectifier (SR) right before the inductor current crosses zero, a short delay time is purposely added so that there is enough initial energy stored in the inductor to help achieve ZVS after the SR is turned off. The circuit topology is shown in Fig. 8.16. The experimental waveforms are shown in Fig. 8.17, comparing non-ZVS with ZVS extension. The saved switching loss is significant because the total efficiency is increased by 0.3%–1% from full load to half load.

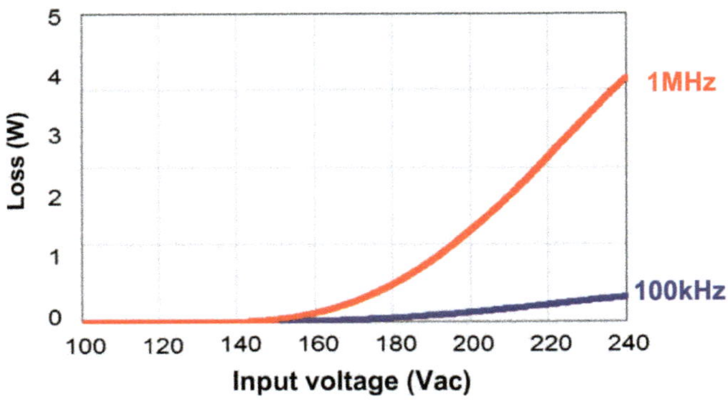

Figure 8.15 Line-cycle averaged non-ZVS loss vs input voltage. *ZVS*, zero-voltage-switching.

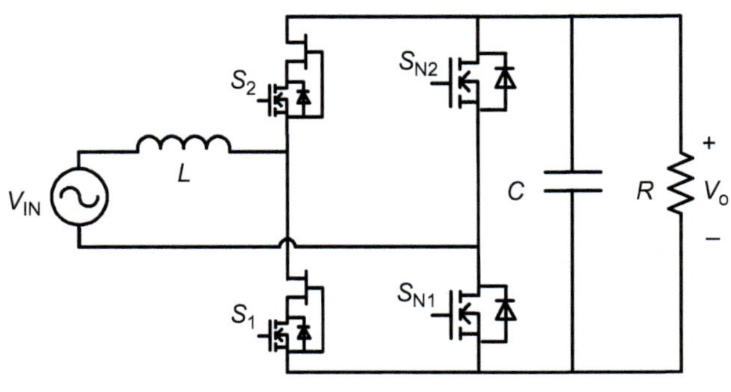

Figure 8.16 Totem-pole PFC with ZVS extension.

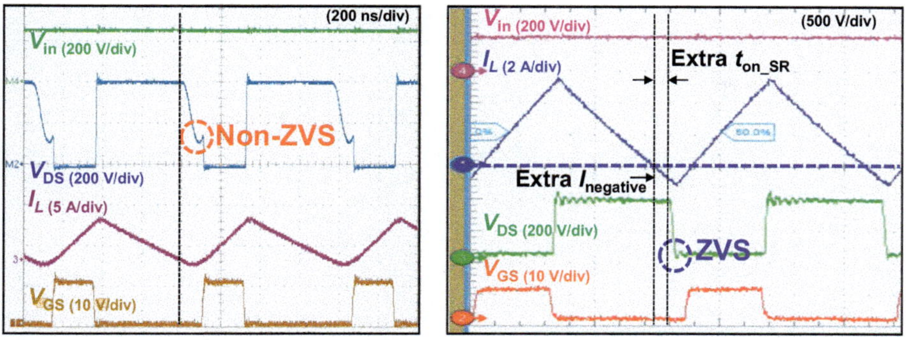

Figure 8.17 Experimental waveform. (A) Non-ZVS operation; (B) ZVS achieved after ZVS extension. *ZVS*, zero-voltage-switching.

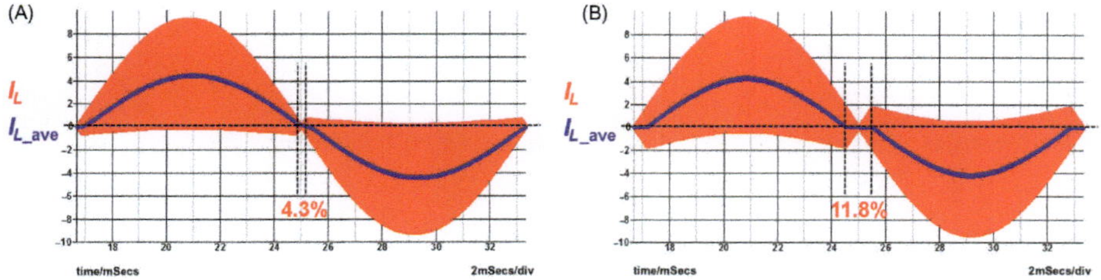

Figure 8.18 Frequency impact on power factor and harmonics (A) 100 kHz constant on-time CRM PFC and (B) 1 MHz constant on-time CRM PFC. *CRM*, critical mode; *PFC*, power factor correction.

8.3.1.2 Variable on-time control

The second high-frequency issue is related to the power quality and harmonics emission. Ideally, the CRM-mode PFC offers unity power factor with voltage mode (constant on-time) control. Since the on-time is constant, the envelope of the inductor peak current follows the shape of the input voltage. Then, if ignoring the negative current, the peak current of the inductor is always twice the average current, which means the input current always follows the shape of the input voltage. However, when the frequency is increased to the MHz range, the negative current during the resonant period is not negligible; thus, there is a notable difference between the shape of the peak inductor current and the average inductor current. In addition, there is also a nonenergy transfer time around the time the line voltage crosses zero in which the average inductor current is zero. Both of these lead to increased harmonics and a poor power factor, as shown in Fig. 8.18.

Variable on-time control is introduced in [25]. A similar concept is used here but with improved and more accurate implementation by using digital control in order

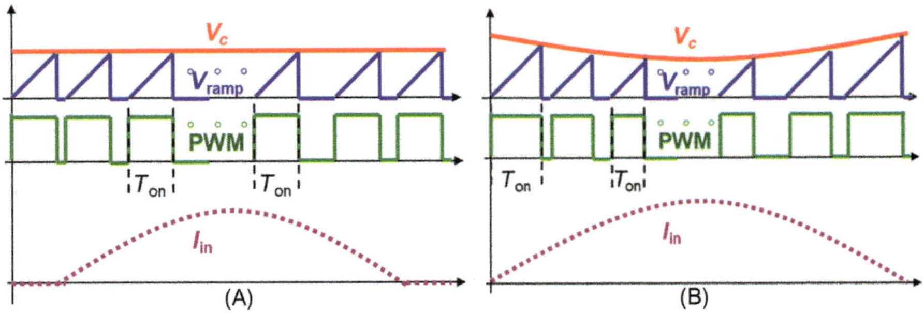

Figure 8.19 Concept diagram of CRM PFC. (A) Constant on-time control. (B) Variable on-time control. *CRM*, critical mode; *PFC*, power factor correction.

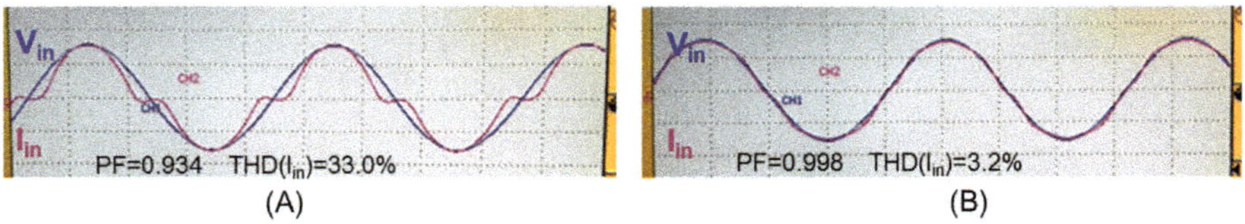

Figure 8.20 Experimental verification. (A) Constant on-time control. (B) Variable on-time control.

to solve the problems of increased harmonics and a poor power factor. The concept is illustrated in Fig. 8.19. By increasing the on time near the zero crossing, the input current is again able to achieve good power factor. Fig. 8.20 shows the experimental verification.

8.3.1.3 Dual-phase interleaving and ripple cancellation

Another drawback of the CRM PFC rectifier is the high current ripple, which leads to not only higher conduction loss but also higher DM noise than the CCM PFC rectifier. To deal with this issue, a two-phase interleaving structure is used to effectively reduce the DM noise by taking advantage of the ripple cancellation effect. The circuit topology is shown in Fig. 8.21.

Interleaving control is very critical to achieving good interleaving and maintaining a small enough phase error. The waveforms in Fig. 8.22 show that good interleaving is achieved. Therefore, even though the current ripple in each phase is always more than two times higher than the average phase input current, the total input current ripple is significantly reduced by interleaving. The interleaving control is usually not an issue for frequencies below 100 kHz, but it becomes a challenge for multi-MHz variable-frequency CRM PFCs. Issues related to MHz-level high-frequency interleaving control and digital implementation are addressed in [26].

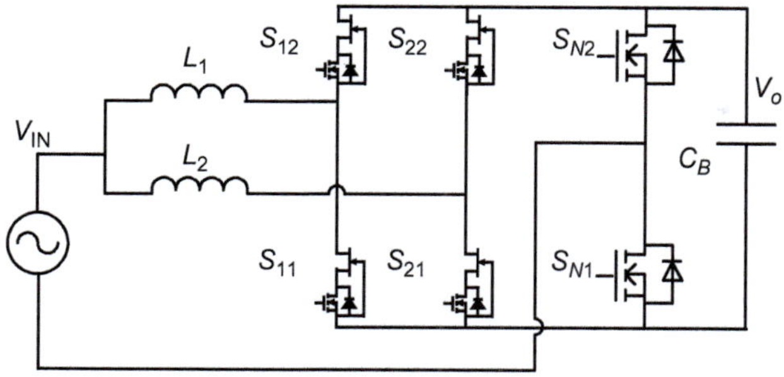

Figure 8.21 Two-phase interleaved totem-pole PFC converter. *PFC*, power factor correction.

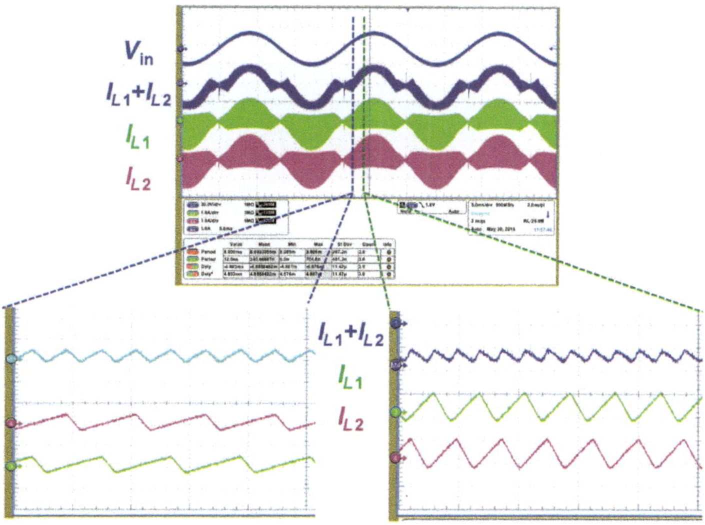

Figure 8.22 Ripple cancellation effect with two-phase interleaving.

8.3.2 PCB winding integrated coupled inductor

The concept of coupled inductor has been widely used in multiphase VRM to reduce loss and improve transient performance [27]. This concept has been extended to two-phase interleaved totem-pole PFC rectifier. One special feature of the coupled inductor in this application is that the effective inductance value will change with duty cycle, in a manner that when duty cycle is approaching 0.5, L value will increase. Subsequently, the switching frequency will decrease as shown in Fig. 8.23. The net benefit is a reduction of the switching losses by 35%.

Previously, the two phase PFC converter has two independent noncoupled inductors. Table 8.4 shows the loss breakdown for two noncoupled inductors. The

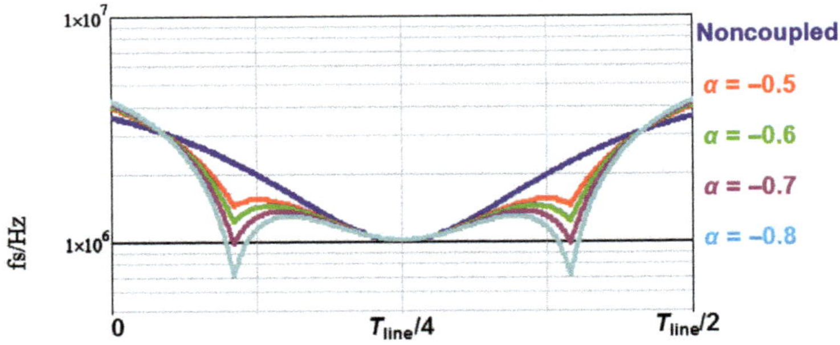

Figure 8.23 Switching frequency variance during half line cycle.

Table 8.4 Loss breakdown for litz wire inductor

DC winding loss (W)	AC winding loss (W)	Core loss (W)	Total loss (W)
0.7	1.6	2.3	4.6

Figure 8.24 Coupled inductor structure I in UI core.

Table 8.5 Loss breakdown for inductor structure I

DC winding loss (W)	AC winding loss (W)	Core loss (W)	Total loss (W)
0.5	6.2	1.3	8.0

noncoupled inductor uses two ER23 core. The winding is 250/46 litz wire with 10 turns for each inductor.

The noncoupled inductor design provided a baseline of inductor size and loss. The design of coupled inductor is shown below. Based on the previous analysis, in order to reduce the average switching frequency, strong coupling is preferred. In this paper, $\alpha = -0.7$ is chosen. However, the typical EI core structure cannot achieve such high coupling coefficient. Therefore, the first attempt is UI core (Structure I).

PCB winding has been used in many applications because it is easy to manufacture and easy to control the parasitic. Fig. 8.24 shows the structure of coupled inductor with PCB winding in a UI core. L_1 is winded on the left leg with five-layer PCB, each layer has two turns. L_2 is winded on the right leg with 10 turns PCB winding. Table 8.5 shows the simulated loss breakdown of this coupled inductor in UI core structure. It can be seen that this coupled inductor has much higher

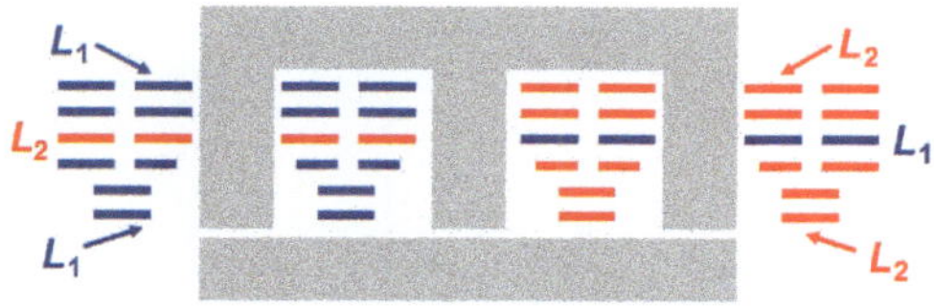

Figure 8.25 Coupled inductor structure II.

Table 8.6 Loss breakdown for inductor structure II

DC winding loss (W)	AC winding loss (W)	Core loss (W)	Total loss (W)
0.5	1.9	1.9	4.3

AC winding loss than the noncoupled inductor. Therefore, the total loss of this coupled inductor is much higher than the noncoupled inductor. This is because of the large eddy current loss in the PCB winding.

In order to reduce the winding loss, a new structure of coupled inductor is developed (Structure II). As shown in Fig. 8.25, the core structure is EI core. In addition, the interleave concept is applied to this structure. Instead of winding L_1 only on left leg and L_2 only on right leg, L_1 and L_2 have switched on the third layer. Now there are eight turns of L_1 and two turns of L_2 on the left leg, eight turns of L_2 and two turns of L_1 on the right leg. The winding can be tapered to avoid fringing flux. In this structure, there is total six layers of PCB winding. The bottom two layers has only one turn. Table 8.6 shows the simulated loss breakdown of this new inductor structure. With interleaving effect, the AC winding loss has been greatly reduced. The total loss of coupled inductor Structure II is smaller than the noncoupled inductor.

According to conventional wisdom, the PCB-based inductor design is inferior to the conventional litz wire wrapped around a core. However, with proposed design, similar total loss is achieved by the proposed PCB based coupled inductor.

A 1.2-kW dual-phase interleaved MHz totem-pole PFC rectifier is built with 99% peak efficiency and 700 W/in.^{3} power density (without bulk cap), as shown in Fig. 8.26, in which the inductor is significantly smaller compared to state-of-the-art industrial practice.

The impact of GaN devices on power electronics goes beyond efficiency and power density improvement. Even though GaN is still in an early stage of development, it is presumably a game-changing device with a scale of impact yet to be defined. Certain design trade off previously inconceivable can be realized with not only significant performance enhancement but also drastic reduction of the labor contents in the manufacturing.

8.3.3 Balance technique to reduce common mode noise

One of the important concerns with using the GaN devices is the potential high EMI noises resulting from high *di*/*dt* and *dv*/*dt* during switching. While the concern

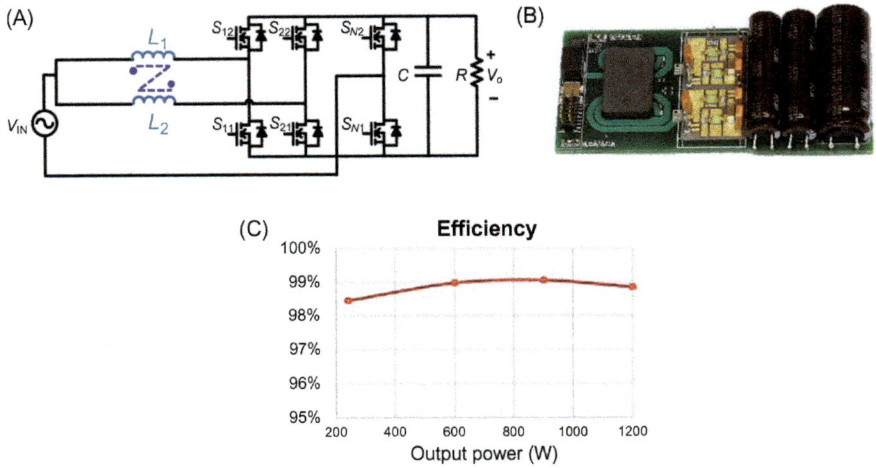

Figure 8.26 GaN-based MHz totem-pole PFC. (A) Topology. (B) Prototype. (C) Measured efficiency. *GaN*, gallium nitride.

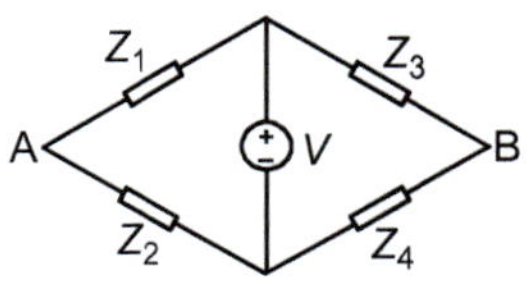

Figure 8.27 Wheatstone bridge circuit.

may be a genuine one with the conventional design practice, the use of PCB integrated magnetics offers the opportunity for significant reduction of common mode (CM) noises. This is achieved by incorporating CPES developed balancing technique relatively easy in the PCB winding structure.

The idea of balance technique is to form a Wheatstone bridge in the circuit, as shown in Fig. 8.27. In this Wheatstone bridge circuit, if $Z_1/Z_2 = Z_3/Z_4$, the voltage of point A is equal to the voltage of point B. Therefore, there would be no current flow between point A and point B.

In order to use the balance principle for CM noise reduction, two additional inductors L_3 and L_4 are employed and are coupled with L_1 and L_2, respectively. Fig. 8.28 shows the circuit topology and the integrated magnetic structure. By use the superposition theory to Fig. 8.29A, the equivalent circuit for the noise source V_{s1}, is shown in Fig. 8.29B. The number of turns of L_1 and L_2 is N_1. The number of turns of L_3 and L_4 is N_2. Thus, we have the balance condition for source V_1 is $N_1/N_2 = C_b/C_d$.

Similarly, the balance condition for noise source V_{N2} is also $N_1/N_2 = C_b/C_d$. The CM noise of this PFC converter can be minimized as long as this balance condition is achieved.

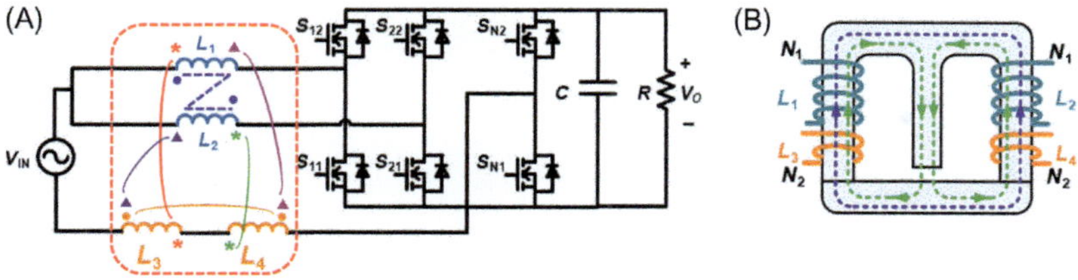

Figure 8.28 Improved balance technique for interleaved totem-pole PFC converter with coupled inductor. (A) Circuit structure. (B) Magnetic structure. *PFC*, power factor correction.

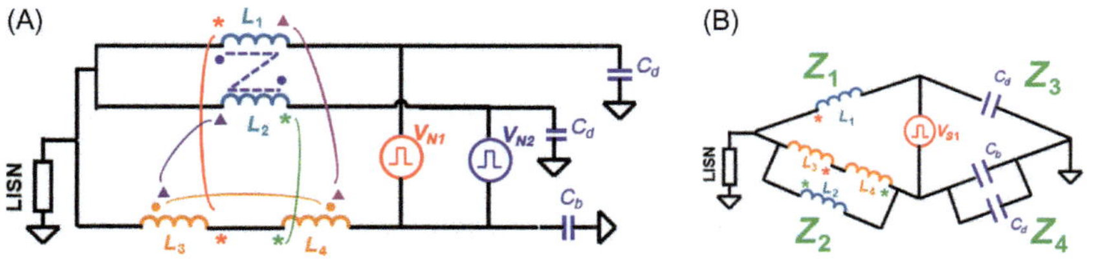

Figure 8.29 Equivalent circuit of CM noise. (A) CM noise model. (B) Effect of one voltage source. *CM*, common mode.

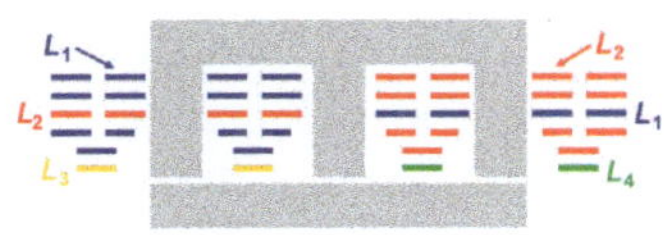

Figure 8.30 Coupled inductor with balance technique.

With the PCB winding, balance technique can be applied to reduce CM noise. The balance winding can be easily implemented by replace the bottom layer of PCB winding with the one-turn balance inductor L_3 and L_4, as shown in Fig. 8.30.

Fig. 8.31 shows that with balance technique, CM noise can be effectively reduced.

8.4 400 V/12 V DCX for server applications

8.4.1 Introduction to datacenter architecture with 400 V bus

Of all the power supplies for industrial applications, those for the data center servers are the most performance driven, energy and cost conscious due to the large electricity consumption. The total power consumption of today's data centers is becoming noticeable. In 2014, data centers in the United States consumed an estimated 70 billion kW h, representing about 1.8% of total US electricity

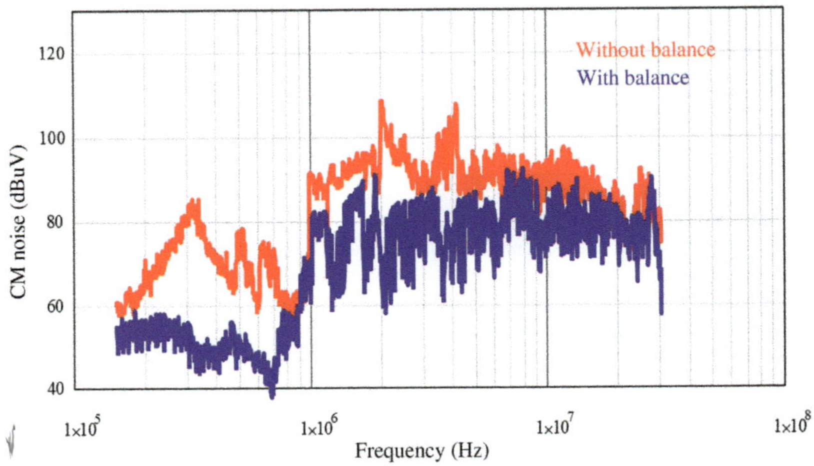

Figure 8.31 CM noise reduction result with balance technique. *CM*, common mode.

consumption [28]. Moreover, with the increase in cloud computing and big data, energy use of data centers is expected to continue rapidly increasing in the near future.

The current practice of data center power architecture is illustrated in Fig. 8.32A, where all major processor/memory devices are powered from a 12-V bus. The i^2R loss for a 12-V bus is excessive, and there are many energy conversion stages, which reduces the total system efficiency. To mitigate the heavy bus-bar loss and reduce energy conversion stages in the power distribution path, industry leaders such as Google, Facebook, Cisco, and IBM are already implementing a new data center design with a higher voltage distribution bus, such as 48 or 400 V instead of 12 V [29–32]. Power architecture with 48 V bus, shown in Fig. 8.32B, is advantageous than the current design practice with the elimination of online UPS and cables and harness. Recently, the International Electronics Manufacturing Initiative at the behest of IBM and other server manufacturers undertook a project to develop an industry standard for DC/DC converters [33]. This converter is used to step down 380 V directly to 12 V and is placed directly on the motherboard. This new power architecture with 400 V bus, shown in Fig. 8.32C, is deemed superior to the current practice and could further replace the power architecture with 48 V bus.

Isolated high output current DC/DC converters are critical for future data center power architecture. Present DC/DC converters, operating at 50–100 kHz with a power density less than 50 W/in.^{3}, cannot be placed on the motherboard to fulfill the requirement of the future data centers. The trend for datacenters or distributed power systems is the pursuit of higher efficiency and higher power density. The LLC resonant converters are suitable for both high-efficiency and high-power-density design [34–38]. The LLC converters can achieve ZVS for zero to full load range with a low turn-off current for the primary switches, and at the same time,

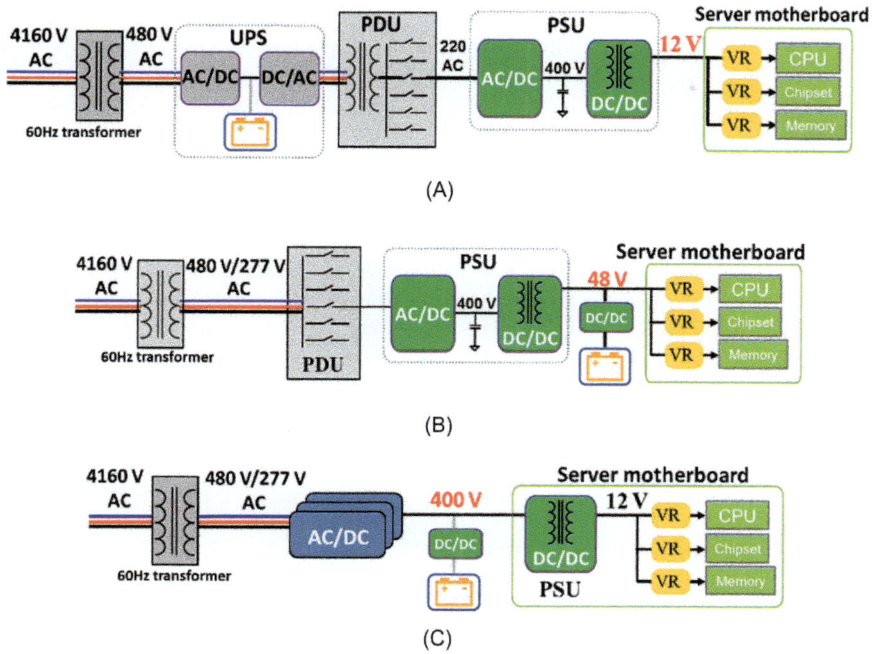

Figure 8.32 Comparison of data center power architectures. (A) Current architecture with 12 V bus. (B) Alternative architecture with 48 V bus. (C) Future architecture with 400 V bus.

ZCS for the SRs. Such soft-switching properties also reduce electromagnetic interference [38]. In comparison with soft-switching PWM converters, LLC converters can achieve a higher switching frequency with better efficiency, resulting in a higher power density and lower total cost [35,37]. The fast development of wide bandgap devices and novel magnetic materials provides opportunities to push switching frequency higher [1,39,40]. MHz LLC converters with GaN devices have been designed for different applications and significantly improved power density has been demonstrated [41–43].

The design of isolated high output current DC/DC converters is very challenging due to the large conduction losses from SR devices and transformer windings, as well as the huge AC termination loss when connecting SRs and secondary windings. To reduce the primary AC winding loss, a hybrid transformer structure, including litz wires primary winding and PCB secondary winding, are employed; to reduce the secondary termination loss, the SRs and output capacitors are mounted on the same PCB layer of the secondary winding [43,44]. Although such a structure in [43,44] can help reduce conduction loss and termination loss, it is too complex for manufacturing and mass production. The matrix transformer can help increase the output current capability by distributing the secondary current with multiple cores [45,46]. The concept of flux cancellation is proposed to reduce core size and loss [40], but the transformer is implemented with a very expensive 12-layer PCB, which also has large distributed interwinding capacitance and, hence, large CM noise current for high input voltage applications. One alternative solution over the 12-layer PCB is to use a simple four-layer PCB to implement a matrix transformer for a 380 V/12 V LLC converter and integrate SRs and output capacitors as part of the secondary winding in order to eliminate AC termination loss [41]. When using four-layer PCB windings, two shielding layers could be placed in between primary and secondary windings. Each shielding layer is connected to the primary ground. Therefore, the CM noise current can only circulate in the primary side [47]. The 380-V/12-V LLC converter design is further optimized with a design optimization procedure, and a peak efficiency of around 97% is demonstrated [42]. But these designs still have the problem of complex structure with multiple cores, and there is room for efficiency improvement. Further efforts have been made on optimization and magnetic integration of matrix transformer in order to improve both efficiency and power density [48].

8.4.2 400 V/12 V LLC converter with matrix transformer

For applications that require low-voltage and high-current outputs, such as computer servers, there are several important design considerations for the LLC converter:

1. The SR devices have limited current capability due to packaging and thermal constrain. For server applications, one should consider paralleling 4–8 SRs to reduce conduction loss as illustrated in Fig. 8.33A. Both static and dynamic current sharing, when paralleling a large number of SRs, are difficult to achieve.
2. The large sum of high frequency and high *di/dt* AC currents must flow through common termination points between the transformer and the SRs, which is marked using red dots in Fig. 8.33A. This will result in large termination losses.
3. It is difficult to place the large number of SRs close to the termination, which will result in large leakage inductances at the transformer's secondary windings, as well as large winding losses.

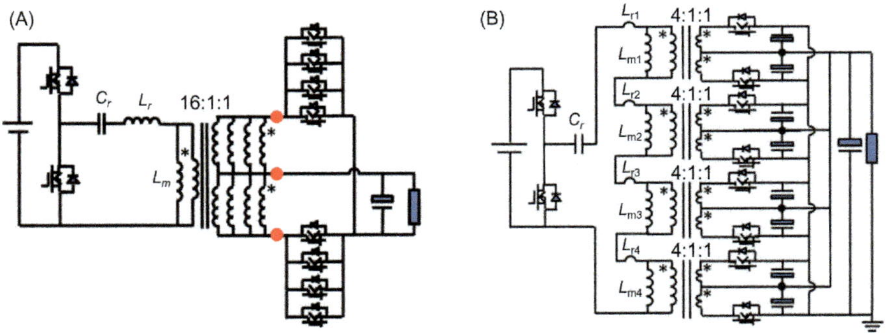

Figure 8.33 LLC converters. (A) With conventional transformer. (B) With matrix transformer.

Such LLC converter design, even though it has a lower core loss, will suffer large winding losses and termination loss. On the other hand, the transformer in commercial power supplies normally uses litz wires as the primary winding and copper foils as the secondary winding. This is very bulky, labor-intensive, and expensive.

A planar transformer with PCB winding can be adopted to achieve automatic manufacturing and high power density. The matrix transformer, which is defined as an array of elemental transformers interwired to form a single transformer, can be employed to evenly distribute the large secondary current to different SRs [45]. An LLC converter with a matrix transformer is shown in Fig. 8.33B. The traditional single core structure was divided into a four-core structure, with the primary windings in series and the secondary windings in parallel. Since the primary current for the four elemental transformers is the same due to a series connection, the secondary current is perfectly balanced.

The matrix transformer in Fig. 8.33B can be formed by using four sets of identical UI cores. The primary winding winds one pillar of each core as shown in Fig. 8.34A, where only the primary winding layer is shown for simplicity. The matrix transformer in Fig. 8.34A has an increased core loss with four sets of UI cores compared to the transformer in Fig. 8.33A with only one ER-core. To reduce the core size, as well as core loss in Fig. 8.34A, the primary winding can be modified as shown in Fig. 8.34B to achieve flux cancellation [40].

The termination points are very critical for efficiency. With a 12-layer PCB as the transformer windings, vias have to be used to connect the secondary winding and the SRs [40], which would cause extra losses since all the AC current would go through these vias. It is proposed in [41] to integrate the SRs and output capacitors into the secondary winding as shown in Fig. 8.35A and employ a simple four-layer PCB winding as shown in Fig. 8.35B, where the top and bottom layers are two sets of secondary windings for center-tap structure and the middle two layers are the primary windings. With this approach, the termination points, where all currents are summed, occur on the DC side; thus, there would be no AC termination loss. Transformer winding losses are significantly reduced as are the leakage inductances.

A design optimization procedure is then provided in [42], and the number of cores is doubled as well as the number of SRs to reduce the winding losses and

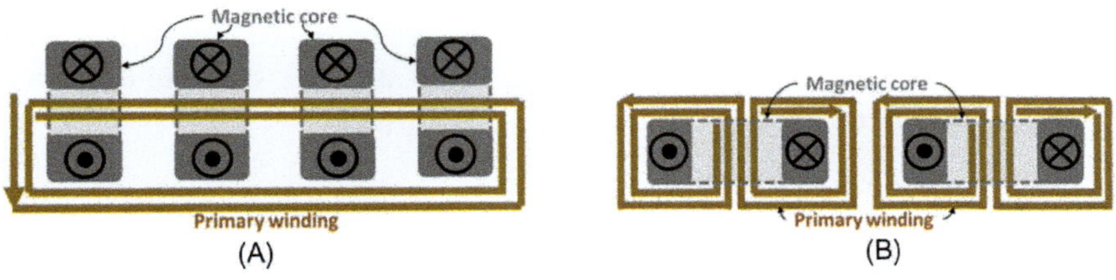

Figure 8.34 Primary winding pattern for matrix transformer. (A) Original matrix transformer. (B) Matrix transformer with flux cancellation.

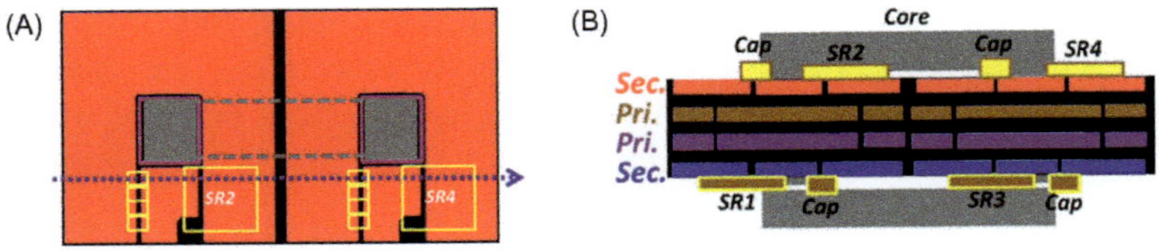

Figure 8.35 Four-layer PCB winding transformer with integrated SRs and output capacitors. (A) Top view. (B) Cross-sectional view. *PCB*, printed circuit board; *SR*, synchronous rectifier.

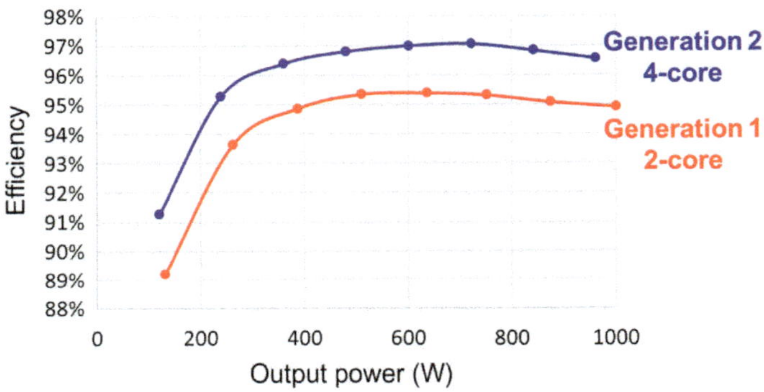

Figure 8.36 Efficiency of LLC converters with matrix transformer.

SRs' conduction loss. Refs. [41,42] provide two designs of 1 kW 380 V/12 V LLC converters with matrix transformer operating at 1 MHz, whose efficiency curves are shown in Fig. 8.36. From the Gen. 1 design with two magnetic cores [41] to the Gen. 2 design with four magnetic cores [42], the peak efficiency is improved from 95.5% to 97.1%, with the same power density of around 700 W/in.^{3}.

Despite all the efforts, the LLC converters with matrix transformer still have some challenges to overcome and require further efficiency improvement. To achieve a high output current, multiple cores are still used even with flux cancellation, which is complex for the manufacturing. Further magnetic integration is indispensable to overcome this challenge. Furthermore, to replace the present DC/DC converters operating at 50–100 kHz, further efficiency improvement for the high frequency DC/DC converters is required. The following sections will present a detailed design methodology and a novel matrix transformer structure to overcome the design challenge.

8.4.3 Integrated planar matrix transformer

To overcome the challenge of multiple cores and further improve efficiency, two matrix transformer structures, both of which can integrate four elemental transformers into one magnetic core, are proposed as follows, and comparison between them has been made.

The original matrix transformer with flux cancellation is shown in Fig. 8.37. Four elemental transformers are integrated into two magnetic cores. Only the primary winding layer is shown for simplicity. Magnetic Core 1 has two pillars numbered as 1 and 2, and magnetic Core 2 has another two pillars numbered as 3 and 4. There is a flux of Φ_B from Pillar 1 to Pillar 2, and another flux of Φ_B from Pillar 3 to Pillar 4.

The proposed matrix transformer Structure 1 is derived as follows: Core 2 in Fig. 8.37 is moved just below Core 1 as shown in Fig. 8.38A; now, there are still two magnetic cores in Fig. 8.38A; then the magnetic plates for the two cores are replaced by the integrated plates as shown in Fig. 8.38B to integrate the two magnetic components into one. With the proposed Structure 1, although the flux pattern and core loss remains the same as the original two-core structure, the challenge of multiple cores in previous designs has been overcome.

The proposed matrix transformer Structure 2 with reduced flux density is derived as follows: Core 2 in Fig. 8.37 is moved just below Core 1, and then rotated by 180° as shown in Fig. 8.39A; now, there are still two magnetic cores in Fig. 8.39A and the flux pattern remains the same as the that in Fig. 8.37; then, the magnetic plates for the two cores are replaced by the integrated plates as shown in Fig. 8.39B to integrate the two magnetic components into one. With the proposed Structure 2, although the flux within the pillars remains the same, the flux density in the magnetic plates is reduced by half. This is very beneficial for high frequency ferrite

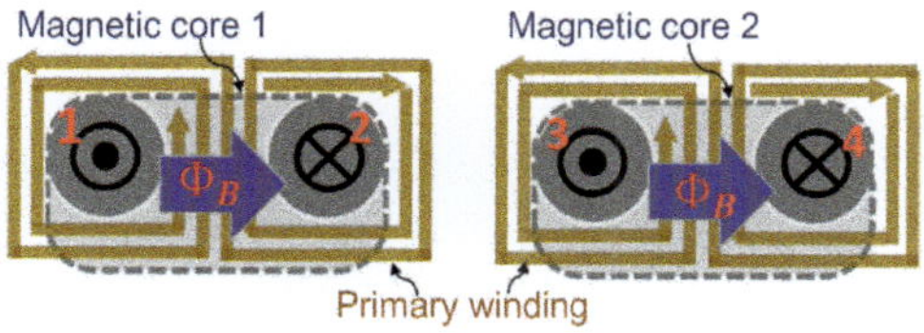

Figure 8.37 The original matrix transformer with flux cancellation.

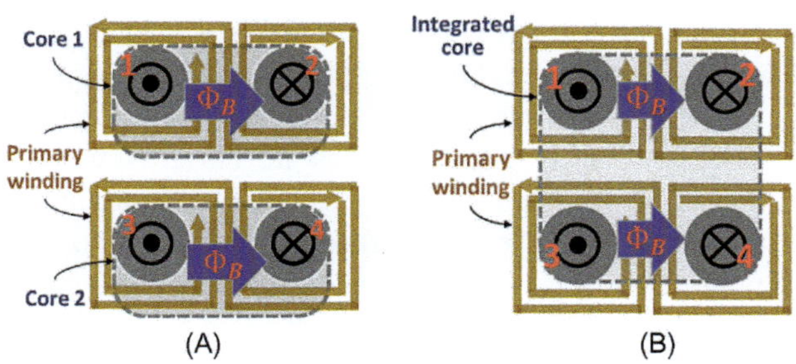

Figure 8.38 Proposed matrix transformer Structure 1. (A) Before integration. (B) After integration.

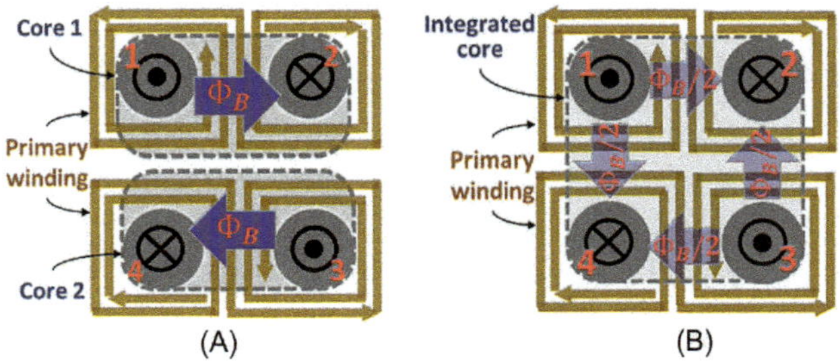

Figure 8.39 Proposed matrix transformer Structure 2 with reduced flux density. (A) Before integration. (B) After integration.

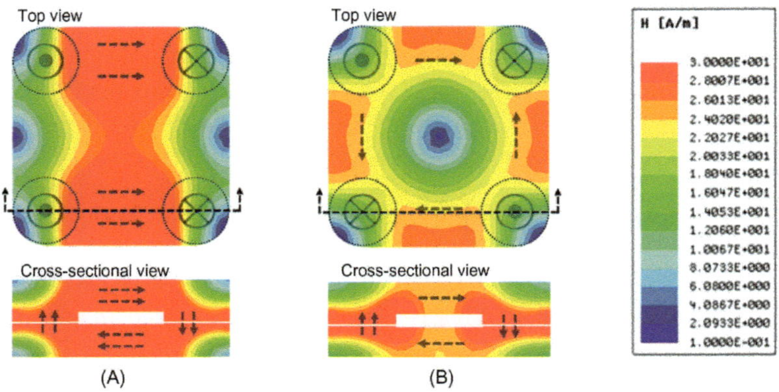

Figure 8.40 Comparison of flux distribution for two proposed matrix transformer structures. (A) Structure 1. (B) Structure 2.

materials since the core loss is a considerable portion in the total loss and is mainly determined by the flux density.

The flux distribution of two proposed matrix transformer structures are compared by simulation as shown in Fig. 8.40. Two structures have identical geometry for the magnetic core, but different winding arrangements as described in Figs. 8.38 and 8.39. Since the proposed Structure 2 has a more evenly distributed flux, the core loss is reduced by around 40% compared to Structure 1 according to the simulation results.

Both of the two proposed matrix transformer structures consist of four identical elemental transformers, so their winding losses are close. However, the proposed Structure 2 has a significantly reduced core loss due to a better arrangement of the four elemental transformers. Furthermore, since the four elemental transformers are coupled in the proposed Structure 2, it is more robust to the tolerance between the four elemental transformers. Overall, Structure 2 is considered superior to Structure 1 due to lower core loss and more robustness.

The detailed winding arrangement for the proposed matrix transformer Structure 2 is illustrated in Fig. 8.41, where the yellow arrows indicate current

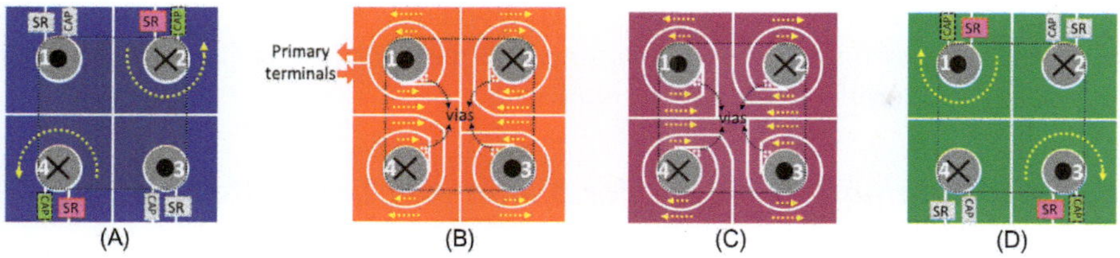

Figure 8.41 Winding arrangement for the proposed matrix transformer Structure 2. (A) Layer 1 for secondary. (B) Layer 2 for primary. (C) Layer 3 for primary. (D) Layer 4 for secondary.

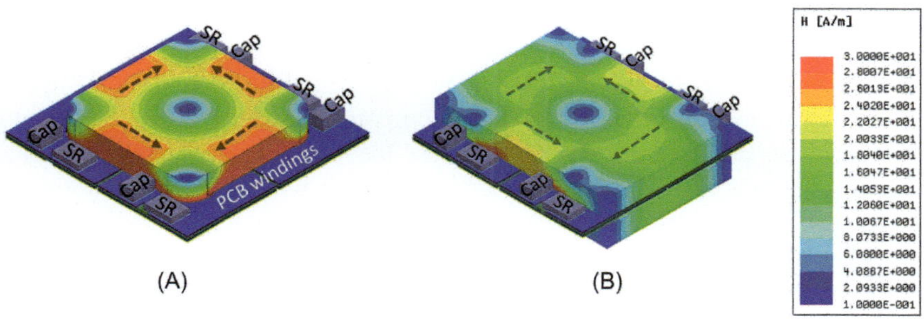

Figure 8.42 Improvement of proposed Structure 2 to further reduce core loss. (A) Before improvement. (B) After improvement.

direction in the positive current cycle. The top layer (Layer 1) and bottom layer (Layer 4) are the secondary windings. SRs and output capacitors are integrated into the secondary windings to eliminate AC termination loss. No vias are needed to connect the secondary PCB layers due to the simple four-layer PCB winding implementation. The middle two layers (Layers 2 and 3) are the primary windings. Although the primary windings are connected through vias, the vias related loss is relatively small since the primary current is small. Another benefit of this winding structure is that the two primary terminals, as shown in Fig. 8.41B, are very close to each other, which means there is negligible leakage inductance and termination loss related to the primary terminals. Although the proposed design employs buried vias, the overall PCB cost is still lower than the 12-layer PCB implementation in [40], and interwinding capacitance has been reduced a lot with four-layer PCB winding.

The proposed Structure 2 can be improved to further reduce the core loss without sacrificing power density. Fig. 8.42A shows the 3D view of the original Structure 2, including the flux distribution on the surface of the integrated magnetic core. Since the SRs and output capacitors are on the two edges of the PCB windings and there are no components on the other two edges; the top and bottom magnetic plates can be expanded to the other two edges of the PCB windings as shown in Fig. 8.42B. By doing this, the flux density in the magnetic plates can be further reduced, so is the core loss. The improved Structure 2 has the same dimension and power density as that of the original one, but the core loss is further reduced by around 30% according to the simulation results.

8.4.4 Shielding techniques for planar matrix transformer

The PCB winding matrix transformers have fully interleaved windings, which leads to a large distributed interwinding capacitance and, hence, large CM noise current. A lot of effort has been done to suppress the CM noise in isolated DC/DC converters, which is mainly caused by the displacement current flowing through the parasitic capacitances between the high *dv/dt* nodes and ground. The interwinding capacitance model and the CM noise model have been developed to predict the noise spectrum [49–51], but none of them have analyzed the matrix transformer structures. Passive cancellation approaches or balance techniques can be employed to reduce the CM noise [52–55], but they require additional passive components and are very sensitive to the tolerance in components and circuitry. The shielding technique is an effective method to suppress the CM noise [56–59], and can be embedded automatically in PCB fabrication process if PCB winding is employed [47], but all these shielding methods would cause extra losses and reduce efficiency. In order to promote the adoption of the shielding technique, it would be very meaningful to develop the interwinding capacitance and CM noise model for the matrix transformers, and come up with a novel shielding technique for the matrix transformer without sacrificing efficiency.

But the matrix transformer has a large CM noise current due to fully interleaved PCB windings. The shielding technique is an effective method to suppress CM noise. It is achieved by inserting shielding layers in between the primary and the secondary windings. There are several methods to achieve the shielding technique. In [56], both the primary and secondary windings have multiple turns, and two layers of identical single-turn copper foil are employed as the shielding; the shielding layer next to the primary windings is connected to the primary ground, and the other layer is connected to the secondary ground. Hence, the noise induced by the primary side will flow to the shielding next to it and circulate back to the primary ground, so as the noise induced in secondary side. And there is no CM noise between the two shielding layers, since they are identical and have the same voltage potential distribution. In [57], both the primary and secondary windings have multiple turns, and one layer of single-turn copper foil is employed as the shielding, and this shielding layer is connected to the primary and made not fully cover the secondary windings purposely. In the covered area, the CM current flows from the secondary windings to the shielding, and then to the primary ground. In the noncovered area, the CM current flows from the primary windings to the secondary windings. By balancing these two currents, the CM noise can be minimized. In [58,59], only one layer of shielding is employed and the shielding is made identical to the secondary winding. The shielding is connected to the primary ground so the noise generated by the primary side will flow back to the primary ground through the shielding. This method is extended PCB winding in [47].

Specifically for a four-layer PCB winding matrix transformer, shielding could be achieved by simply placing two shielding layers in between the primary and secondary windings as shown in Fig. 8.43. Each shielding layer is connected to the primary ground. Therefore, the CM noise current induced by the primary winding will flow to the shielding and circulate back to the primary ground. The shielding layers are made identical to the secondary windings, both are single-turn windings, so they have the same voltage potential distribution. Therefore, even there is a parasitic capacitance between the shielding layers and the secondary windings, there is no CM current between them since the voltage potential difference across this parasitic capacitance is zero.

The shielding technique in Fig. 8.43 is achieved by simply placing two shielding layers in between primary and secondary windings. In a fully interleaved transformer, the space between the primary and secondary winding has the highest magnetomotive force; when placing the PCB layers as shielding into this space, there would be an eddy current induced in the shielding layers. Decreasing the thickness of the shielding layers can limit the eddy current loss, however, due to the cost concern, the thinnest copper thickness is 0.5 oz for standard PCB manufacturing. It has been demonstrated that using 0.5 oz copper thickness as shielding layers can achieve 20 dB CM noise attenuation, but at cost of 0.2% efficiency decrease [47].

In order to get the benefit of CM noise attenuation without simultaneously sacrificing the efficiency, it is proposed to utilize half of the shielding as the primary windings as shown in Fig. 8.44 [60]. It is defined that the voltage excitation applied to the primary windings is V_{Pri}, then the floating nodes of Shielding 2, 4, 6, 8 have a voltage potential of $+V_{Sh} = V_{Pri}/16$, while the floating nodes of Shielding 1, 3, 5, 7 have a voltage potential of $-V_{Sh}$. By connecting all the floating nodes of Shielding 2, 4, 6, 8 to a common node *S*, which is marked as green dot in Fig. 8.44, a parallel connection of these shielding serves as one additional turn for the primary winding. Doing this does not impact the function of shielding, since the voltage potential on the shielding in Fig. 8.44 remains the same as the secondary windings.

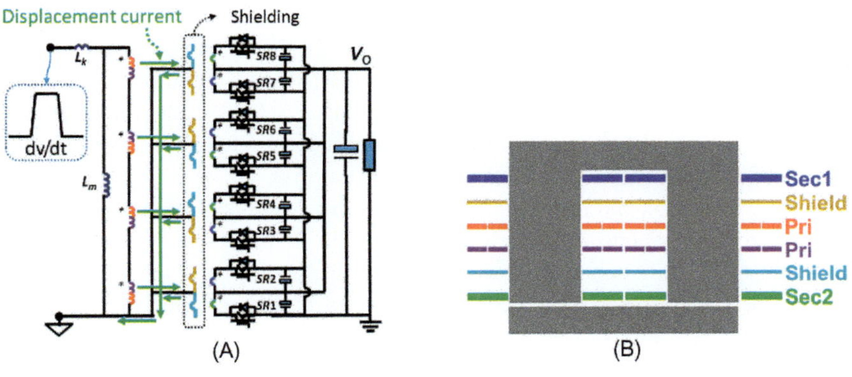

Figure 8.43 PCB winding matrix transformer with two shielding layers. (A) Schematics. (B) Cross-sectional view. *PCB*, printed circuit board.

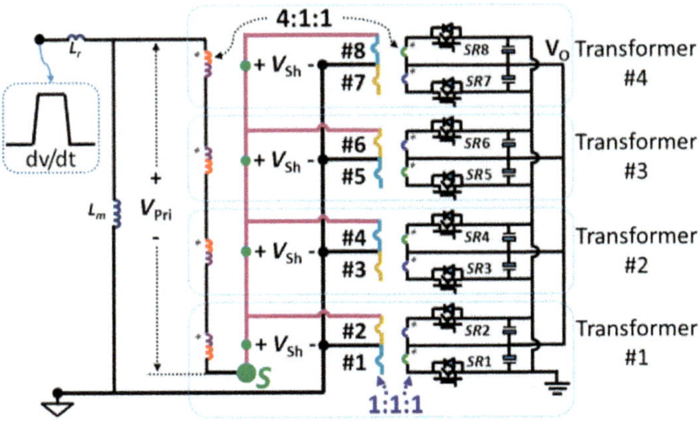

Figure 8.44 Schematics of matrix transformer with the proposed shielding.

So, there is zero potential difference between the shielding winding and the secondary winding, thus, no CM current. Although the voltage potential on the primary winding is increased by $+V_{Sh}$, the CM noise current induced by the primary windings still circulates in the primary side since the shielding is connected to the primary ground.

When utilizing half of the shielding as one additional turn for the primary winding, the equivalent primary to secondary turn ratio is changed from 16:1 to 17:1. By doing so, the primary current is reduced by 5.9% for a given output voltage V_O and current, which can help reduce the primary MOSFET conduction loss by 11%. The reduction of winding losses needs to be verified by the finite element analysis (FEA). In server applications, the turn ratio for most transformer designs of the LLC converter is between 16:1 and 18:1 [44,61,62]. When using a matrix transformer consisting of four elemental transformers, the turn ratio must be 16:1 since it can only be an integer multiplied by 4. But with the proposal shielding, the turn ratio can be 17:1 by using the shielding as an additional turn.

The PCB winding implementation of the proposed shielding is shown in Fig. 8.45. The ground connection and voltage pulsation for the secondary windings are also included. The shielding is made identical to the secondary windings. To achieve the proposed shielding structure, the windings for the shielding have to be connected through additional PCB traces around the output terminals to the primary ground and node S, which would increase the footprint, cause interference between the primary windings and output terminals, and thus increase CM noise.

To solve this problem, the shielding layout can be rotated by 270° as shown in Fig. 8.46, while still achieving the same effect. The top of Fig. 8.46 shows the shielding at Layer 5 and the secondary winding at Layer 6 for elemental transformer 1. Two terminals of secondary windings are marked as A and B, and those of shielding are marked as A' and B'. B is connected to the secondary ground and B' is connected to primary ground. The windings can be stretched along the x-axis to map the voltage potential at each point on the windings to the $U-x$ coordinate at

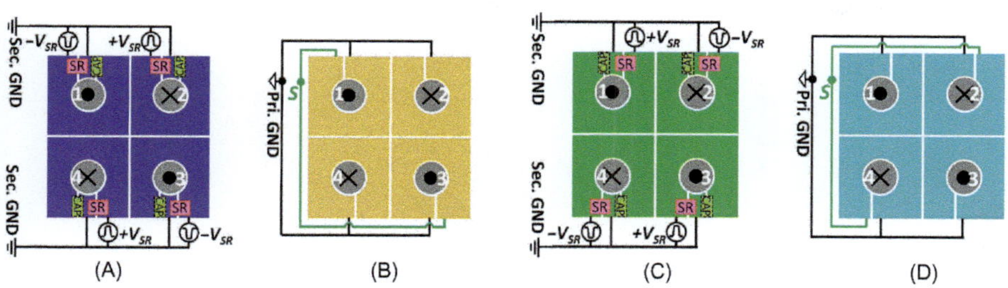

Figure 8.45 PCB winding implementation of the proposed shielding. (A) Layer 1 for secondary. (B) Layer 2 for shielding. (C) Layer 6 for secondary. (D) Layer 5 for shielding. *PCB*, printed circuit board.

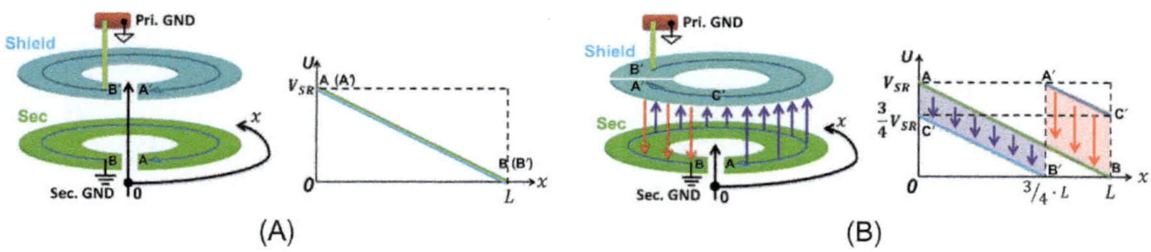

Figure 8.46 Voltage potential on secondary windings and shielding. (A) Shielding identical to secondary winding. (B) Shielding rotated by 270°.

the bottom of Fig. 8.46. In Fig. 8.46A, since the secondary winding and shielding are identical, the voltage potentials of both at the same position on the x-axis are identical, so the two curves on the $U-x$ coordinate overlap each other, and have $U=V_{SR}$ at $x=0$ and $U=0$ at $x=L$. In Fig. 8.46B, the secondary winding remains the same, while the shielding is rotated by 270°. Then the shielding has $U=3/4\cdot V_{SR}$ at $x=0$ and $x=L$. Since the shielding has an opening between terminals A' and B', the shielding has $U=0$ on the left edge of $x=3/4\cdot L$ and $U=V_{SR}$ on the right edge of $x=3/4\cdot /$. It can be seen from Fig. 8.46B that there is a displacement current circulating from the secondary winding to the shielding and then back to the secondary winding, so the net CM current between the secondary winding and shielding is still zero although the shielding is rotated by 270°. Similarly, it can be further proven that the net current is zero even when the shielding is rotated by an arbitrary angle.

By rotating the shielding, the two terminals of the shielding can converge at the center of the matrix transformer as shown in Fig. 8.47. Then, the ground terminals for the shielding can be connected through the PCB trace from the center to the middle of the left edge, so that the PCB trace for the primary does not occupy the output terminals on the top and bottom edges, and the interaction between primary and secondary is minimized.

The PCB winding implementation can be completed by including the traces to connect node S, as shown in Fig. 8.48. Vias are also needed for the connection

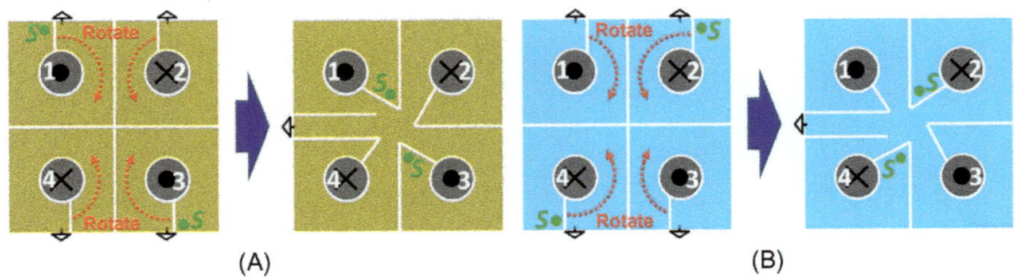

Figure 8.47 Rotating the shielding. (A) Layer 2. (B) Layer 5.

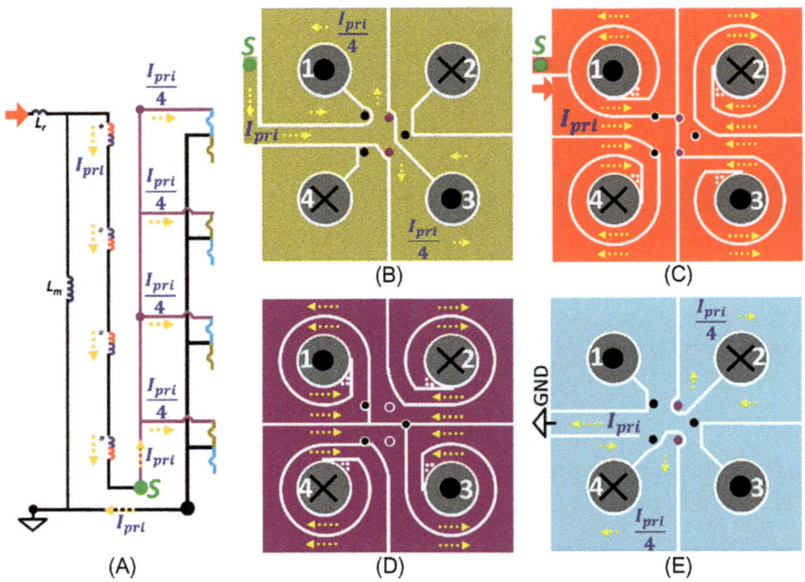

Figure 8.48 PCB winding implementation of the proposed rotated shielding. (A) Schematics. (B) Layer 2 for shielding. (C) Layer 3 for primary. (D) Layer 4 for primary. (E) Layer 5 for shielding. *PCB*, printed circuit board.

between the primary windings and the shielding. With the proposed PCB winding implementation, all the windings and the connection are limited to within the footprint of the matrix transformer. Although vias are added for the connection between shielding, the vias related loss is relatively small since the primary current is less than 5 A and distributed to four sets of shielding.

To verify the winding loss reduction with the proposed shielding, the FEA simulation result is shown in Fig. 8.49, in which the windings are idealized by eliminating vias and termination. The Secondary 1 is set to be conducting a secondary current. The termination and vias effects are analyzed with the method mentioned in [41,42]. The proposed shielding doesn't impact the secondary winding loss. Although the primary winding resistance increases by 8% due to the series of one extra turn, the

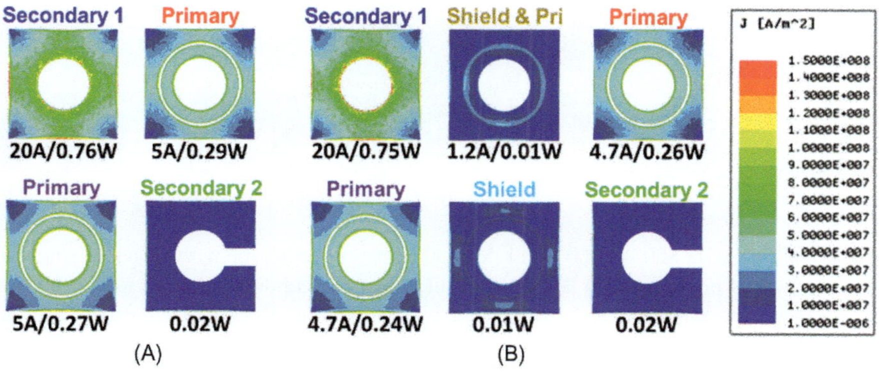

Figure 8.49 Winding loss simulation. (A) Without shielding. (B) With shielding.

primary winding loss is reduced thanks to the primary current reduction. The total winding loss is reduced by 4% with the proposed shielding under full-load condition.

8.4.5 Hardware demonstration

The 1-MHz 800-W 400-V/12-V LLC converter prototype with the proposed matrix transformer (four elemental transformers in one magnetic core) and shielding is shown in Fig. 8.50. It has the same footprint as a quarter-brick, but a much smaller height of 0.27 in., compared to the height of 0.50 in. for a quarter-brick. An output power of 800 W at such a dimension is equivalent to a power density of around 900 W/in.^{3}. Adding the proposed shielding does not increase the volume of the LLC converter. The prototype maintains the same power density as the design without shielding.

The experimental waveforms under the full-load condition are shown in Fig. 8.51A. It shows that both primary switch drain-source voltage V_{ds_Q2} and SR drain-source voltage V_{ds_SR2}, V_{ds_SR4} reduce to zero during dead time, so ZVS is achieved. And resonant current i_{Lr} touches magnetizing current just before dead time, so ZCS for SR is achieved. Moreover, there is no ringing in the V_{ds_SR} due to the minimum secondary leakage. Fig. 8.51B shows the measured CM noise spectrums of the proposed LLC converter. The red curve is the design without shielding and the blue curve is the design with the proposed shielding. The proposed shielding can attenuate the CM noise by around 30 dB, and such attenuation is effective in all frequency spectrums of interest up to 30 MHz. A conventional CM choke cannot achieve such a good attenuation at very high-frequency range due to its parasitic capacitance.

The measured efficiency of the proposed LLC converter with four transformers in one magnetic core is shown in Fig. 8.52. Compared to the design without shielding, the proposed shielding can improve the full-load efficiency from 97.2% to 97.4% and improve the peak efficiency from 97.6% to 97.7%. The design with

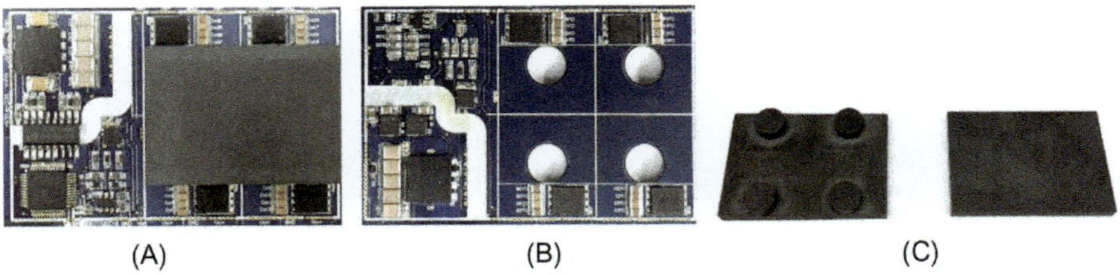

Figure 8.50 1 MHz 800 W 400 V/12 V LLC converter prototype with proposed shielding. (A) Top view of prototype. (B) Bottom view of prototype without magnetic core. (C) Magnetic cores.

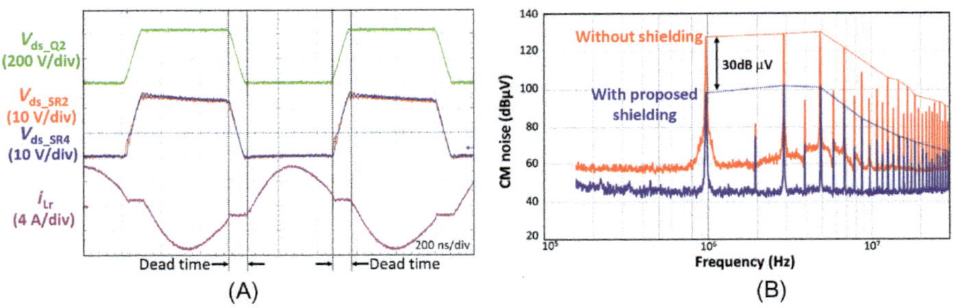

Figure 8.51 Experimental results. (A) Waveforms under the full-load condition at the resonant frequency. (B) Measured CM noise spectrums. *CM*, common mode.

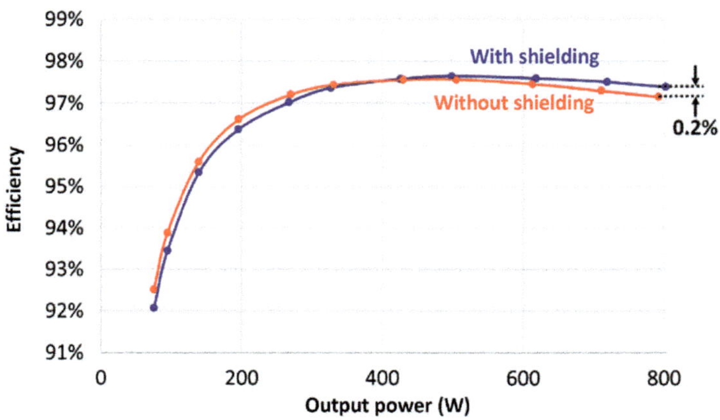

Figure 8.52 Measured efficiency of proposed LLC converter with four transformers in one magnetic core.

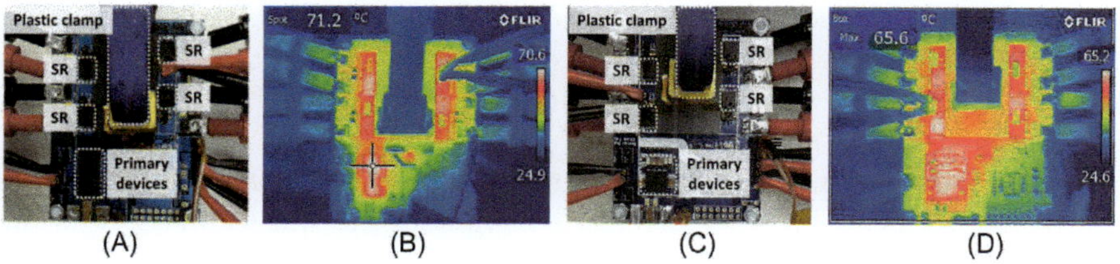

Figure 8.53 Thermal test under full-load. (A) Setup for prototype without shielding. (B) Setup for prototype with shielding. (C) Thermal image without shielding. (D) Thermal image with shielding.

proposed shielding has a slightly lower light load efficiency due to the following reasons: the design with proposed shielding has a higher V_{IN} due to the extra one turn in the primary winding while the same V_O as the design without shielding, resulting in a higher turn-off loss; the primary RMS current is mainly determined by the magnetizing current at the light load condition, which is almost same for both designs, while the design with proposed shielding has a higher primary winding resistance as explained in Section 8.4.4, thus a higher conduction loss at the light load condition; there is extra eddy current loss on those shielding which are not serving as the primary windings.

Fig. 8.53 provides the thermal test results of the proposed LLC converter under the full-load condition, 25°C ambient temperature and 200 LFM (Linear Feet per Minute) fan speed. The proposed design has better thermal behavior due to a higher efficiency, improved layout and two more copper layers to dissipate the heat compared to the design without shielding.

8.4.6 Conclusions

The matrix transformers for high output current LLC converter are investigated in this section. To improve the current design practice, a novel matrix transformer structure is proposed to integrate four elemental transformers into one magnetic core and utilized a simple four-layer PCB as the windings. The proposed design can utilize flux cancellation and reduce flux density in the magnetic plates to reduce core loss and integrate SRs and output capacitors into the secondary winding to minimize leakage and termination loss. The core loss with the proposed matrix transformer is reduced by more than half compared to the state-of-the-art matrix transformer technique. The proposed matrix transformer is superior to the state-of-the-art due to the much reduced core loss, simple four-layer PCB windings and integrated magnetic structure.

However, the PCB winding matrix transformer suffers from a large interwinding capacitance of the PCB windings, which causes a large CM noise. This is more severe when GaN device is applied because it has a higher *dv/dt* than its Si counterpart. Shielding is an effective method to attenuate the CM noise in all frequency spectrums of interest, and it is more suitable for PCB windings since it can automatically be embedded in the fabrication process. But shielding will cause extra losses and decrease efficiency. A novel shielding structure is proposed in this section to utilize half of the shielding as the primary winding while still maintaining the benefit of the CM noise attenuation.

By pushing switching frequency up to MHz with GaN devices, the proposed matrix transformer with shielding can demonstrate the impact of GaN in such important issues as efficiency, power density and manufacturability. With the academic contribution in this section, we can design a converter with 10 times, or even 20 times in switching frequency, comparing to the current practice using silicon devices. Finally, a 1-MHz 380-V/12-V 800-W LLC converter with GaN devices using the proposed matrix transformer structure is demonstrated. The prototype fits in quarter-brick footprint and achieves a peak efficiency of 97.7% and a power density of 900 W/in.[3].

8.5 EMI filter design for high frequency GaN converters

In the previous sections, GaN-based high frequency PFC stage and DC–DC stage have been demonstrated. In order to pass the EMI standard, an EMI filter is needed. Usually, the EMI filter will occupy one-fourth to one-third total volume of the power supply. Fig. 8.54 shows the conventional EMI filter design. It can be seen that the filter topology is a two-stage LC filter structure. As a result, the size of this filter is very large. In addition, two-stage EMI filter requires more inductor and capacitor components, which increases the cost.

With GaN devices, the switching frequency can be pushed 10 times higher than traditional design. This will not only improve the power density of PFC and DC–DC converters, but also help improve EMI filter design. With MHz switching frequency, it is easy to apply the balance and shielding techniques to help reduce the EMI noise generated by converters. In addition, since the switching frequency is 10 time higher, the corner frequency of the EMI filter can be much higher than

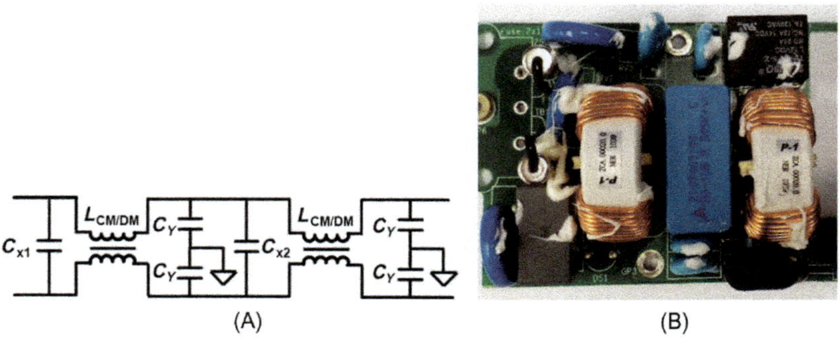

(A) (B)

Figure 8.54 Conventional two-stage EMI filter. (A) Filter topology. (B) Product picture.

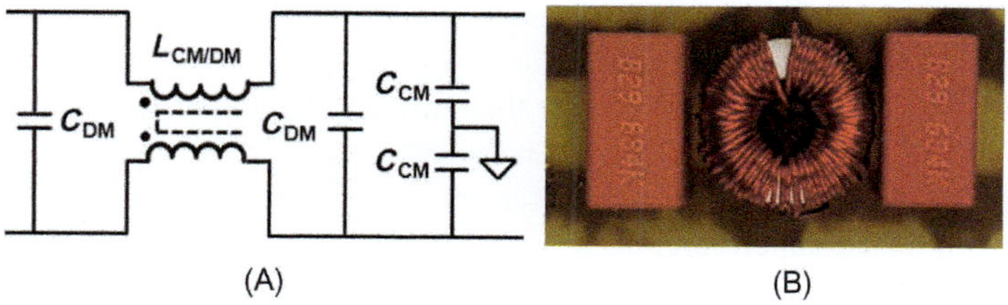

Figure 8.55 Simple one-stage EMI filter. (A) Filter topology. (B) Prototype picture.

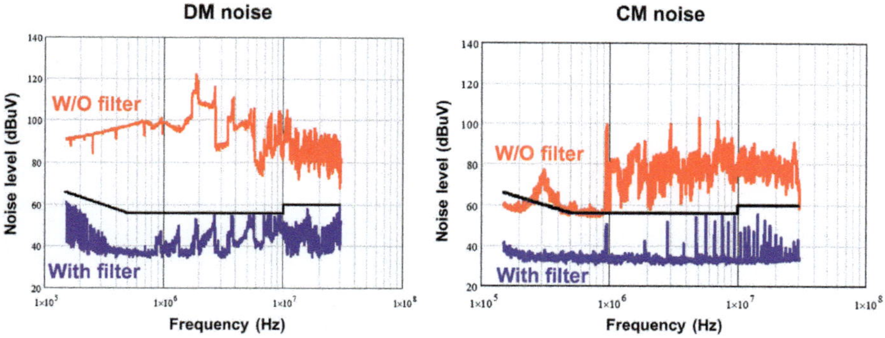

Figure 8.56 EMI test result for a 1-kW 1-MHz front-end converter (PFC + LLC) with a single-stage EMI filter. *PFC*, power factor correction.

traditional design. This provides the opportunity to use a simple one-stage EMI filter structure to achieve the required EMI noise attenuation.

Fig. 8.55 shows the design of one-stage EMI filter. Comparing with the conventional two-stage EMI filter design, the one-stage EMI filter is much simpler. It requires less components and small volume. With one-stage EMI filter, it can achieve 80% volume reduction compared with conventional two-stage EMI filter design.

Fig. 8.56 shows the attenuation result of this one-stage EMI filter. It can be seen that the simple and compact one-stage EMI filter can achieve the required attenuation. With GaN devices, the switching frequency can be pushed above MHz. Therefore, the volume of EMI filter, PFC converter and DC−DC converter can be greatly reduced. The power density of the server power supply can be improved 10 times higher than conventional design.

8.6 Summary

In summary, with recent advances made in GaN power devices, it will make a significant impact on the development of the modular power electronics building blocks toward system-level integration.

It is evident that, for any given design, simply replacing silicon devices with GaN devices will improve efficiency. Furthermore, it is clear that GaN devices can operate at much higher frequencies than their silicon counterparts. Consequently, as much as a 5–10-fold reduction in size and weight is achievable using GaN devices, as has been demonstrated in some applications. It is also clear that GaN devices can operate at a higher temperature. This opens the doors for many new opportunities, such as integrating power electronics with an aircraft engine in the engine compartment, integrating power electronics with an IC engine in a car, or placing PV inverters on a hot rooftop.

Still, to realize the full potential of GaN devices, we should start with challenging our current design practice, which has been taken for granted in the past. Perhaps, certain design tradeoffs previously inconceivable can be realized with not only significant performance enhancements but also a drastic reduction of the labor required for the manufacturing and assembly process, ultimately leading to a reduction in cost. First, we should be mindful of starting our design with the idea of manufacturability. Building upon this idea, a few examples in this chapter book have demonstrated quite remarkable levels of improvement of power density, efficiency, and manufacturability. This is an exciting time, as we are entering the next generation of power electronics.

References

[1] X. Huang, Z. Liu, Q. Li, F.C. Lee, Evaluation and application of 600 V GaN HEMT in cascode structure, IEEE Trans. Power Electron. 29 (5) (2014) 2453–2461.

[2] Z. Liu, X. Huang, F.C. Lee, Q. Li, Package parasitic inductance extraction and simulation model development for the high-voltage cascode GaN HEMT, IEEE Trans. Power Electron. 29 (4) (2014) 1977–1985.

[3] Y. Ren, M. Xu, J. Zhou, F.C. Lee, Analytical loss model of power MOSFET, IEEE Trans. Power Electron. 21 (2) (2006) 310–319.

[4] M. Rodriguez, A. Rodriguez, P.F. Miaja, D.G. Lamar, J.S. Zuniga, An insight into the switching process of power MOSFETs: an improved analytical losses model, IEEE Trans. Power Electron. 25 (6) (2010) 1626–1640.

[5] Z. Zhang, J. Fu, Y. Liu, P.C. Sen, Switching loss analysis considering parasitic loop inductance with current source drivers for buck converters, IEEE Trans. Power Electron. 26 (7) (2011) 1815–1819.

[6] M. Mu, F.C. Lee, Q. Li, D. Gillham, K. Ngo, A high frequency core loss measurement method for arbitrary excitations, in: Proc. IEEE Applied Power Electronics Conference, 2011, pp. 157–162.

[7] X. Huang, F.C. Lee, Q. Li, W. Du, High frequency high efficiency GaN-based interleaved CRM bi-directional buck/boost converter with inverse coupled inductor, IEEE Trans. Power Electron. 31 (6) (2016) 4343–4352.

[8] O. Hegazy, R. Rarrero, J.V. Mierlo, P. Lataire, N. Omar, T. Coosemans, An advanced power electronics interface for electric vehicles applications, IEEE Trans. Power Electron. 28 (12) (2013) 5508–5521.

[9] M.A. Khan, A. Ahmed, I. Husain, Y. Sozer, M. Badawy, Performance analysis of bidirectional DC–DC converters for electric vehicles, IEEE Trans. Ind. Appl. 51 (4) (2015) 3442–3452.
[10] H.F. Ahmed, H. Cha, S. Kim, D. Kim, H. Kim, Wide load range efficiency improvement of a high-power-density bidirectional DC–DC converter using an MR fluid-gap inductor, IEEE Trans. Ind. Appl. 51 (4) (2015) 3216–3226.
[11] W. Zhang, D. Dong, I. Cvetkovic, F.C. Lee, D. Boroyevich, Lithium-based energy storage management for DC distributed renewable energy system, in: Proc. IEEE Energy Conversion Congress & Expo., 2011, pp. 3270–3277.
[12] J. Baek, W. Choi, B. Cho, Digital adaptive frequency modulation for bidirectional DC–DC converter, IEEE Trans. Ind. Electron. 60 (11) (2013) 5167–5176.
[13] Z. Liu, X. Huang, M. Mu, Y. Yang, F.C. Lee, Q. Li, Design and evaluation of GaN-based dual-phase interleaved MHz critical mode PFC converter, in: Proc. IEEE ECCE, 2014, pp. 611–616.
[14] D. Reusch, J. Strydom, Evaluation of gallium nitride transistor in high frequency resonant and soft-switching DC–DC converters, IEEE Trans. Power Electron. 30 (9) (2015) 5151–5158.
[15] X. Huang, Z. Liu, F.C. Lee, Q. Li, Characterization and enhancement of high-voltage cascode GaN devices, IEEE Trans. Electron. Devices 62 (2) (2015) 270–277.
[16] X. Huang, Q. Li, Z. Liu, F.C. Lee, Analytical loss model of high voltage GaN HEMT in cascode configuration, IEEE Trans. Power Electron. 29 (5) (2014) 2208–2219.
[17] P. Wong, Q. Wu, P. Xu, Y. Bo, and F.C. Lee, Investigating coupling inductors in the interleaving QSW VRM, in: Proc. IEEE Applied Power Electronics Conference, 2000, pp. 973–978.
[18] B. Su, J. Zhang, Z. Lu, Totem-pole boost bridgeless PFC rectifier with simple xero-current detection and full-range ZVS operating at the boundary of DCM/CCM, IEEE Trans. Power Electron. 26 (2) (2011) 427–435.
[19] C. Marxgut, F. Krismer, D. Bortis, J.W. Kolar, Ultraflat interleaved triangular current mode (TCM) single-phase PFC rectifier, IEEE Trans. Power Electron. 29 (2) (2014) 873–882.
[20] L. Zhou, Y.-F. Wu, U. Mishra, True bridgeless totem-pole PFC based on GaN HEMTs, PCIM Eur. (2013) 1017–1022.
[21] Y. Yang, Z. Liu, F.C. Lee, Q. Li, Analysis and filter design of differential mode EMI noise for GaN-based interleaved MHz critical mode PFC converter, in: Proc. IEEE ECCE, 2014, pp. 4784–4789.
[22] Z. Liu, F.C. Lee, Q. Li, Y. Yang, Design of GaN-based MHz totem-pole PFC rectifier, IEEE J. Emerg. Sel. Top. Power Electron. 4 (3) (2016) 799–807.
[23] B. Su, Z. Lu, An interleaved totem-pole boost bridgeless rectifier with reduced reverse-recovery problems for power factor correction, IEEE Trans. Power Electron. 25 (6) (2010) 1406–1415.
[24] L. Huang, W. Yao, Z. Lu, Interleaved totem-pole bridgeless PFC rectifier with ZVS and low input current ripple, in: Proc. IEEE Energy Convers. Congr. Expo., 2015, pp. 166–171.
[25] J.W. Kim, S.M. Choi, K.T. Kim, Variable on-time control of the critical conduction mode boost power factor correction converter to improve zero-crossing distortion, in: Proc. IEEE Power Electronics and Drive Systems Conf. (PEDS), Nov. 2005, pp. 1542–1546.
[26] Z. Liu, Z. Huang, F.C. Lee, Q. Li, Digital-based interleaving control for GaN-based MHz CRM totem-pole PFC, in: IEEE Applied Power Electronics Conference and Exposition (APEC), Long Beach, CA, USA, 2016, pp. 1847–1852.

[27] P. Wong, P. Xu, B. Yang, F.C. Lee, Performance improvements of interleaving VRMs with coupling inductors. IEEE Trans. Power Electron. 16 (4) (2001) 499–507.

[28] Ernest Orlando Lawrence Berkeley National Laboratory, United States Data Center Energy Usage Report, June 2016.

[29] A. Pratt, P. Kumar, T.V. Aldridge. Evaluation of 400 V DC distribution in telco and data centers to improve energy efficiency, in: Proc. IEEE INTELEC, 2007, pp. 32–39.

[30] G. AlLee, W. Tschudi, Edison redux: 380 V dc brings reliability and efficiency to sustainable data centers, IEEE Power Energy Mag. 10 (6) (2012) 50–59.

[31] S.M. Lisy, B.J. Sonnenberg, J. Dolan. Case study of deployment of 400 V DC power with 400 V/ − 48 V DC conversion, in: Proc. IEEE INTELEC, 2014, pp. 1–6.

[32] Vicor white paper, From 48 V Direct to Intel VR12.0: Saving 'Big Data' $500000 Per Data Center, Per Year. [online]: ⟨http://www.vicorpower.com/documents/whitepapers/wp_VR12.pdf⟩, July 2012.

[33] International Electronics Manufacturing Initiative, iNEMI Statement of Work (SOW) DC–DC Power Module, Phase 2. [online]: ⟨http://thor.inemi.org/webdownload/2015/SOW_DC-DC_Power_Module_Phase2_050115.pdf⟩, May 2015.

[34] B. Yang, F.C. Lee, A.J. Zhang, G. Huang, LLC resonant converter for front end DC/DC conversion, in: Proc. IEEE APEC, 2002, pp. 1108–1112.

[35] B. Yang, Y. Ren, F.C. Lee, Integrated magnetic for LLC resonant converter, in: Proc. IEEE APEC, 2002, pp. 346–351.

[36] B. Lu, W. Liu, Y. Liang, F.C. Lee, J.D. Van Wyk. Optimal design methodology for LLC resonant converter, in Proc. IEEE APEC, 2006, pp. 533–538.

[37] D. Fu, B. Lu, F.C. Lee. 1MHz high efficiency LLC resonant converters with synchronous rectifier, in Proc. IEEE PESC, 2007, pp. 2404–2410.

[38] F.C. Lee, S. Wang, P. Kong, C. Wang, D. Fu, Power architecture design with improved system efficiency, EMI and power density, in: Proc. IEEE PESC, 2008, pp. 4131–4137.

[39] S. Ji, D. Reusch, F.C. Lee, High-frequency high power density 3-D integrated gallium-nitride-based point of load module design, IEEE Trans. Power Electron. 28 (9) (2013) 4216–4226.

[40] D. Reusch, F.C. Lee, High frequency bus converter with low loss integrated matrix transformer, in: Proc. IEEE APEC, 2012, pp. 1392–1397.

[41] D. Huang, S. Ji, F.C. Lee, LLC resonant converter with matrix transformer, IEEE Trans. Power Electron. 29 (8) (2014) 4339–4347.

[42] M. Mu, F.C. Lee, Design and optimization of a 380 V–12 V high-frequency, high-current LLC converter with GaN devices and planar matrix transformers, IEEE J. Emerg. Sel. Top. Power Electron. 4 (3) (2016) 854–862.

[43] D. Fu, F.C. Lee, and S. Wang, Investigation on transformer design of high frequency high efficiency dc–dc converters, in: Proc. IEEE APEC, 2010, pp. 940–947.

[44] C. Yan, F. Li, J. Zeng, T. Liu, and J. Ying, A novel transformer structure for high power, high frequency converter, in: Proc. IEEE PESC, 2007, pp. 940–947.

[45] E. Herbert, Design and application of matrix transformers and symmetrical converters, in: Presented at High Freq. Power Convers. Conf., Santa Clara, CA, USA, May 1990.

[46] K.D.T. Ngo, E. Alpizar, J.K. Watson, Modeling of losses in a sandwiched-winding matrix transformer, IEEE Trans. Power Electron. 10 (4) (1995) 427–434.

[47] Y. Yang, D. Huang, F.C. Lee, Q. Li, Analysis and reduction of common mode EMI noise for resonant converters, in: Proc. IEEE APEC, 2014, pp. 566–571.

[48] C. Fei, F.C. Lee, Q. Li, High-efficiency high-power-density LLC converter with an integrated planar matrix transformer for high output current applications, IEEE Trans. Ind. Electron. 64 (11) (2017) 9072–9082.
[49] D. Fu, P. Kong, S. Wang, F.C. Lee, M. Xu, Analysis and suppression of conducted EMI emissions for front-end LLC resonant DC/DC converters, in: Proc. IEEE PESC, 2008, pp. 1144–1150.
[50] D. Fu, S. Wang, P. Kong, F.C. Lee, D. Huang, Novel techniques to suppress the common-mode EMI noise caused by transformer parasitic capacitances in dc–dc converters, IEEE Trans. Ind. Electron. 60 (11) (2013) 4968–4977.
[51] L. Xie, X. Ruan, Z. Ye, Equivalent noise source: an effective method for analyzing common-mode noise in isolated power converters, IEEE Trans. Ind. Electron. 63 (5) (2016) 2913–2924.
[52] S. Wang, F.C. Lee, Analysis and applications of parasitic capacitance cancellation techniques for EMI suppression, IEEE Trans. Ind. Electron 57 (9) (2010) 3109–3117.
[53] D. Cochrane, D.Y. Chen, D. Boroyevic, Passive cancellation of common-mode noise in power electronic circuits, IEEE Trans. Power Electron. 18 (3) (2003) 756–763.
[54] P. Kong, S. Wang, F.C. Lee, Z. Wang, Reducing common-mode noise in two-switch forward converter, IEEE Trans. Power Electron. 26 (5) (2011) 1522–1533.
[55] Y. Chu, S. Wang, A generalized common mode current cancellation approach for power converters, IEEE Trans. Ind. Electron 62 (7) (2015) 4130–4140.
[56] L. Pentti, O. Hyvönen, Electrically decoupled integrated transformer having at least one grounded electric shield, U.S. Patent 7733205 B2, Jun. 8, 2010.
[57] S. Lin, M. Zhou, W. Chen, J. Ying, Novel methods to reduce common-mode noise based on noise balance, in: Proc. IEEE PESC, 2006.
[58] C.W. Park, Method and apparatus for substantially reducing electrical earth displacement current flow generated by wound components without requiring additional windings, U.S. Patent 7109836 B2, Sep. 19, 2006.
[59] Y. Yang, D. Huang, F.C. Lee, Q. Li, Transformer shielding technique for common mode noise reduction in isolated converters, in: Proc. IEEE ECCE, 2013, pp. 4149–4153.
[60] C. Fei, Y. Yang, Q. Li, F.C. Lee, Shielding technique for planar matrix transformers to suppress common-mode EMI noise and improve efficiency, IEEE Trans. Ind. Electron. 65 (2) (2018) 1263–1272.
[61] Texas Instruments, LLC Resonant Half-Bridge Converter, 300-W Evaluation Module. [online]: ⟨http://www.ti.com/lit/ug/sluu361/sluu361.pdf⟩, April 2009.
[62] Texas Instruments, Digitally Controlled LLC Resonant Half-Bridge DC–DC Converter, [online]: ⟨http://www.ti.com/lit/ug/sluub97a/sluub97a.pdf⟩, Jan. 2016.

Applications of SiC devices

9

Dirk Kranzer, Andreas Hensel, Jürgen Thoma, Cornelius Armbruster, Patrick Hercegfi and Stefan Schönberger
Fraunhofer Institute for Solar Energy Systems ISE, Freiburg, Germany

9.1 Retrospective

Depending on individual beliefs one can state, that in the early beginning there was a creative will or a single bang. At least, when it came to first tests of SiC prototypes in applications, there was a lot of creative will on the side of the power electronic engineers and additionally there were several bangs-literally-until everything worked fine. It is now slightly more than 10 years ago that first SiC-transistor prototypes were made available. Of course, there were several prototypes years before, but they were mainly hand-crafted pieces of semiconductor physicists to improve the device and its process. The real fun of circuit design started, when the samples were rather stable and accessible for power electronic research engineers in an appropriate volume for first designs. These were mainly the years 2006–09.

SiC-transistors were expensive and for some applications they are still too expensive today. So especially in the beginning, it had to be a high-value application to settle a new device technology. Beside military and harsh-environmental applications, renewable energy systems were the most promising industrial application with high growth rates. Especially for photovoltaic (PV)-inverters, SiC-transistors were desirable devices to increase efficiency and reduce system costs. The power range of the PV-inverters (3–15 kVA) fit to the current range of the first samples of SiC-transistors and modules, whereas inverters for wind turbines were above 800 kVA and totally out of the feasible range for SiC, at that time. Today, the situation is different.

Of course also the automotive industry had a close look on it, but they struggled with the costs and the voltage range of the devices. Around 2007, commercial available 650 V SiC-Schottky diodes in conjunction with silicon insulated gate bipolar transistors (IGBTs) were not common for the automotive industry. It has even been seen critical for future designs, at that time.

Whereas at the same time SiC-Schottky diodes were already used since their first generation in commercial PV-inverters to reduce switching losses and to increase efficiency.

Increased efficiency, reduced volume and improved system performance are general goals for the development of all kind of power electronics. But the question is always, what these improvements are worthwhile to afford the step towards a new

Wide Bandgap Semiconductor Power Devices. DOI: https://doi.org/10.1016/B978-0-08-102306-8.00010-1

technology. Especially for the application in PV-inverters it was possible to take benefits of all these options.

Fig. 9.1 shows the comparison of the weighted European Efficiency of a three-phase 7 kVA inverter with 1200 V IGBTs and with the first available engineering samples of SiC-metal-oxide-semiconductor field-effect transistors (MOSFETs). Just by the exchange of devices, it was possible to increase the efficiency by 2.4%. The switching frequency was 16 kHz, which was the usual switching frequency of PV-inverters of this power range at that time [1].

Ten years ago, the feed-in tariffs in Europe for PV were set high to support the change towards renewable energy generation, so that an improvement in efficiency led to increased energy harvesting and increased financial benefits, as can be seen in Table 9.1. So over the lifetime of an inverter, the additional higher costs for SiC devices could be afforded.

Similar was done with normally-off junction gate field-effect transistors (JFETs) in a 5 kVA single-phase PV-inverter with HERIC topology as seen in Fig. 9.2. The switching frequency was 16 kHz. It was possible to achieve a maximum efficiency of 99%, which was a world record at that time [2].

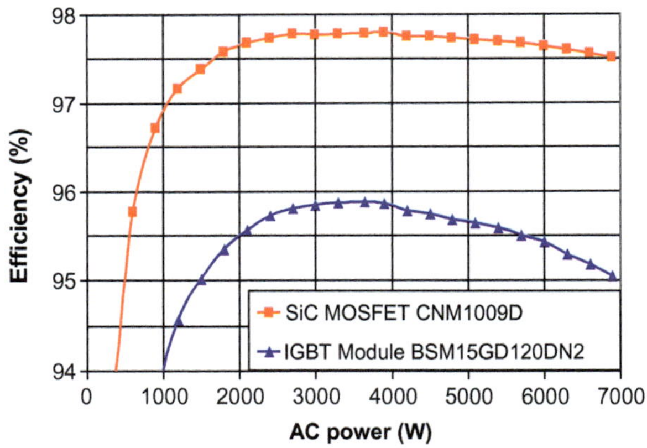

Figure 9.1 Comparison of the efficiency of a 7 kVA three-phase PV-inverter, 1200 V SiC-MOSFET versus 1200 V Si-IGBT [1].

Table 9.1 Annual financial gain for a system established in 2009 for a 7 kVA PV-inverter due to an efficiency improvement of 2.4% based on the local feed-in tariff [1]

	Maximum feed-in tariff (2009)/kWh	Financial gain/a
Freiburg, Germany	0.49 EUR	81 EUR
Almeria, Spain	0.44 EUR	145 EUR
Marseille, France	0.55 EUR	164 EUR

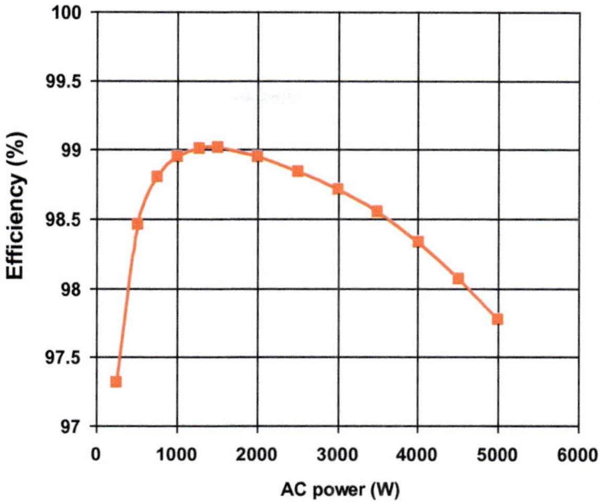

Figure 9.2 Maximum efficiency of 99% for a single-phase PV-inverter with normally-off JFETs [2].

Research on suited applications for SiC-transistors was not only performed at institutes, but also in industry. Thus it was no wonder that soon after the commercialization of the first generation of SiC devices, also the first inverter products followed.

In 2011 SMA released a commercial 20 kVA three-phase PV-inverter (SUNNY TRIPOWER 15000TL/20000TL high efficiency) with three-level topology with SiC-transistors. It had a maximum efficiency of 99% [3].

With an annually declining feed-in tariff, it was clear, that not only the highest efficiency could have been the main target for a cost optimized system design; furthermore the design of the inverter has to be improved to prevail a growing cost pressure. Fig. 9.3 shows the sharing of costs of a 5 kVA PV-inverter, as it was in 2009 with the use of 1200 V IGBTs in a three-phase full-bridge [1].

As can be seen, the cost share of the inductor was nearly as twice as high than the costs of the IGBTs. Also the cost for the passive cooled heatsink was higher than for the semiconductors. Taking this into account, it was clear that there were more cost benefits than only increased income due to higher efficiency and feed-in tariff. As seen in Fig. 9.4 tests were made with switching frequencies up to 144 kHz, which was far beyond the-state-of-the-art at that time. But as also can be seen, the efficiency was too low, as the switching losses were too high for a two-level topology. A high switching frequency like 144 kHz and a high efficiency above 98% is also possible, but only with three-level topologies. This requires twice the amount of devices and it was not economical in the early days of SiC-transistors.

PV-inverters do not only consist of an inverter stage. For string inverters it is usual to have an additional booster stage to fit the MPP-voltage of the PV-string to the required DC-link voltage for grid feeding. Fig. 9.5 shows the comparison in efficiency of a 1200 V IGBT and a 1200 V SiC-BJT. It is a booster for a three-phase inverter, thus 1200 V devices have to be used for a common DC-link range of 650–900 V. The switching frequency was 48 kHz [4].

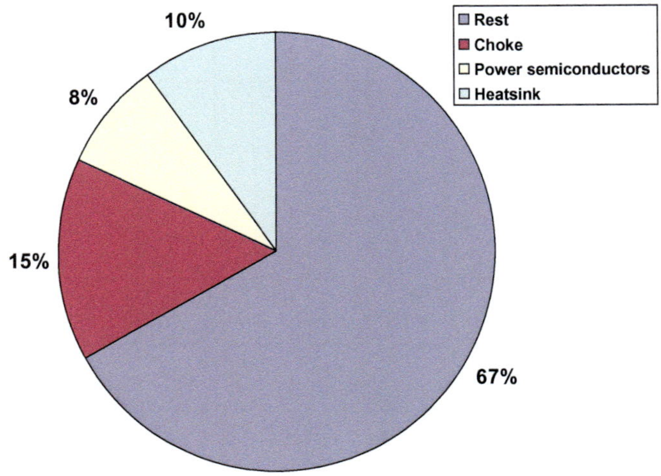

Figure 9.3 Related costs of a 5 kVA three-phase PV-inverter with IGBTs and 16 kHz switching frequency [1].

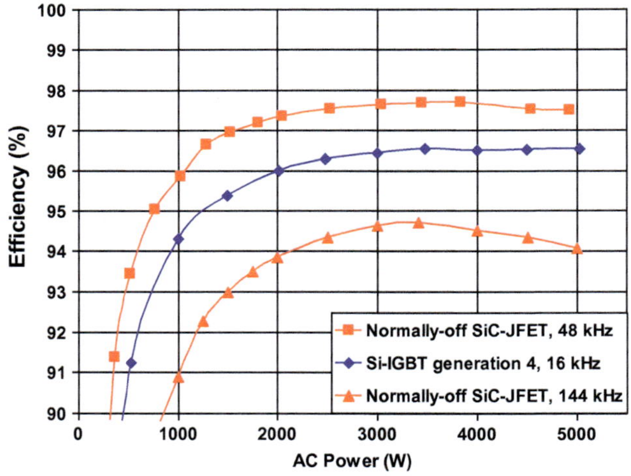

Figure 9.4 Efficiency of a 5 kVA three-phase full-bridge inverter with normally-off SiC-JFETs and Si-IGBTs [2].

The main benefit of unipolar SiC-transistors is the opportunity to use synchronous rectification as shown in Fig. 9.6. The conduction losses can be further decreased, because only during the commutation between the bridge sections the phase current must be conducted by a diode to avoid a short circuit.

Beside reduced switching losses, this is the main advantage of SiC-MOSFETs toward any bipolar device made of Si or SiC. Instead of 12 semiconductor devices, a full-bridge with SiC-MOSFETs consists of only 6 devices.

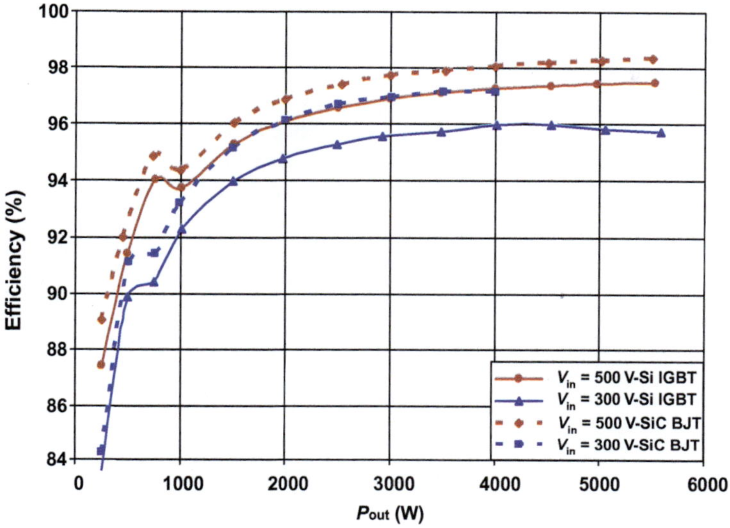

Figure 9.5 Booster efficiency at different input voltages and 700 V output voltage for 1200 V IGBT and 1200 V SiC-BJT at 48 kHz [4].

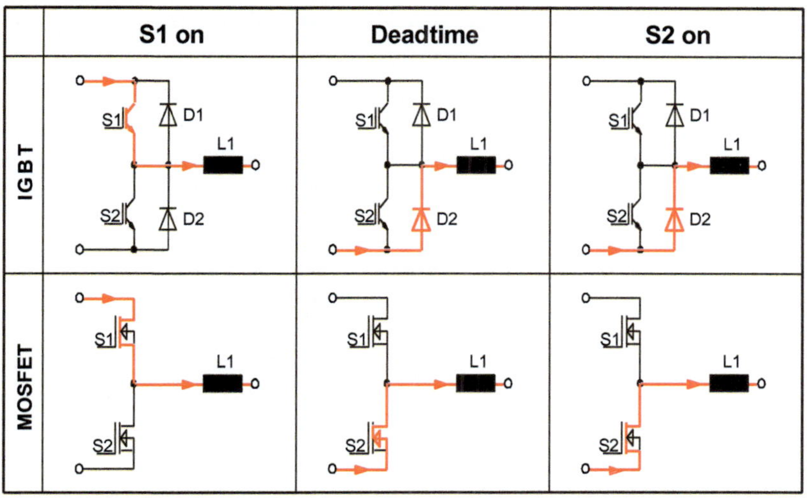

Figure 9.6 Commutation between the bridge sections in case of IGBTs and SiC-MOSFETs [1].

While in the beginning the variety of feasible applications was small, the variety of device types was the highest and more or less all of them were considered to be a possible solution:

- There were highly rugged BJTs, but they were not attractive, as even with SiC the current-gain was too low, so that the base-drive would have become complicated at higher power ranges. Furthermore, there was no option of synchronous rectification. TRANSIC in Sweden was active in this area. They were purchased by FAIRCHILD, but Fairchild later stopped the development.

- There were normally-on JFETs, but they were not attractive, as a pure normally-on device is not suitable for many power electronic applications. They required a cascade configuration to convert them into a normally-off system. INFINEON was the strongest and most persistent representative of this device type. They announced an MOSFET in 2016 and by this also ended the discussion about a JFET, for the time being.
- There were normally-off JFETs with SEMISOUTH as main representative, but the discussion about this device stopped, when the company shut down, respectively their lead investor stopped the investment.
- There were MOSFETs, but they were said not to be reliable due to stability problems with the gate oxide and too expensive regarding their complexity. Furthermore they were said to be too expensive compared to the other device types.

The variety decreased and up until now the MOSFET remained as standard SiC-transistor for power electronics. Some of the other device types are still available, but clearly for some niche applications.

9.2 Application examples with SiC devices

9.2.1 *High efficient 10 kVA uninterruptible power supply inverter with 1200 V MOSFETs*

9.2.1.1 *Introduction*

The proposed system is an inverter for an online uninterruptible power supply (UPS) system (Fig. 9.7). Online UPSes are providing the highest safety against grid

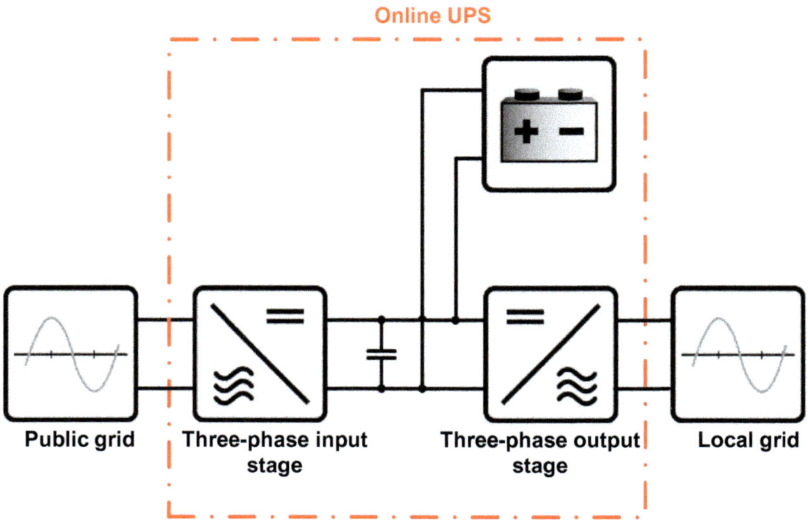

Figure 9.7 Topology of an online UPS system.

failures or power shorts for a protected local grid. Any problem in the public grid will not reach the protected local grid due to its system architecture [5]. Efficiency in these systems are highly important as all the energy consumed in the local grid flows through the two inverter stages with each operating with a specific efficiency. This leads to constant power losses and costs that increases the electricity costs [6].

A demonstrator of this three-phase UPS inverter with a total power of 10 kW was developed. This range of power is matched to the product lineup (single chip) of different SiC device manufacturers. An increased switching frequency of 100 kHz is used to reach a higher power density compared to commercially available products. The main inductors are capable of being directly mounted on the printed circuit board (PCB), which leads to cost savings in mass production. Based on the very good dynamic characteristics of the SiC devices the volume of the cooling system can be reduced, which in total leads to a highly compact demonstrator.

The nominal voltage of the DC link of the system is 820 V and feeds a three-phase 50 Hz grid with 230 V grid phase to neutral. An output power of 10 kVA can be delivered continuously and an overload capability of 120% for 30 minutes and 200% for 30 seconds is implemented. The size of one stage is only 230 mm × 210 mm × 110 mm with only 4.4 kg including all periphery and components. The switching frequency is 100 kHz.

As topology the mixed voltage neutral point clamped (MNPC) or T-type topology (Fig. 9.8) is chosen. This offers good possibilities to demonstrate the advantages of SiC [7]. On the common three-phased grid two transistors with a blocking voltage of 1200 V (T1, T4) as well as two transistors with a blocking voltage of 650 V (T2, T3) per phase-leg are used. The difference in efficiency is more significant by using this type of topology. With modified switching schemes the common mode currents can be reduced [8]. This point is potentially interesting for drive inverters. A photography of the demonstrator is shown in Fig. 9.9.

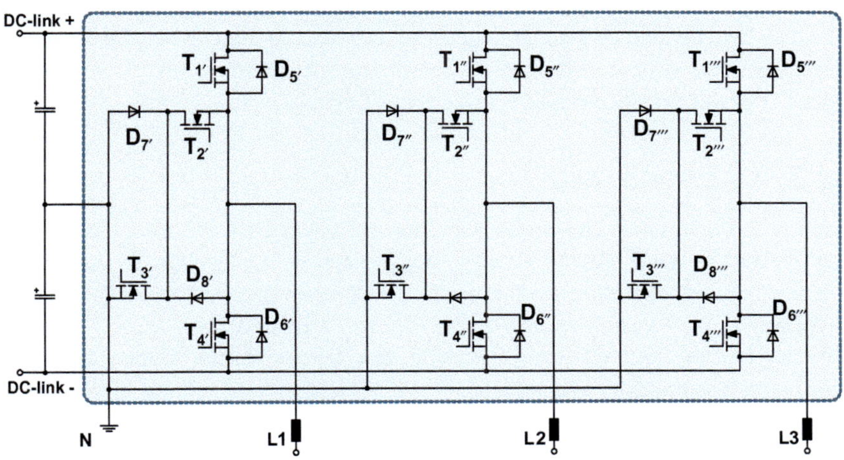

Figure 9.8 Three-phase MNPC topology of the inverter.

Figure 9.9 Photography of the 10 kVA three-phase SiC inverter.

Table 9.2 Measurement parameters

Description of measurement	Efficiency measurement—different DC voltages
Topology	MNPC three-phase
Used equipment	Yokogawa WT3000 (power meter) Regatron (DC supply) Ohmic load
Gate resistor	2.2 Ω
Dead time	500 ns
Switching frequency	100 kHz

9.2.1.2 Results

Table 9.2 shows the measurement parameters for the test setup. The measurement results for two different input voltages can be seen in Fig. 9.10. The results are showing the total system efficiency. A maximum efficiency of 98.7% is reached at part load. For almost the complete curve the efficiency is above 98%.

9.2.1.3 Operation costs

For online UPS systems the main costs are not caused by the direct costs of the power electronic itself. As all energy is always flowing through the inverter the efficiency has big influence on the operation costs of the systems [9]. Calculated for a lifetime of 10 years the part of the efficiency related costs (energy costs) are approx. 80% of the total costs of ownership. For the calculation of the running costs the following assumptions are made:

- Nominal power of the system is 10 kW
- System runs at 365 days per year at 24 h with $P_N/2$
- Efficiency of SiC three-level inverter: 98.4%

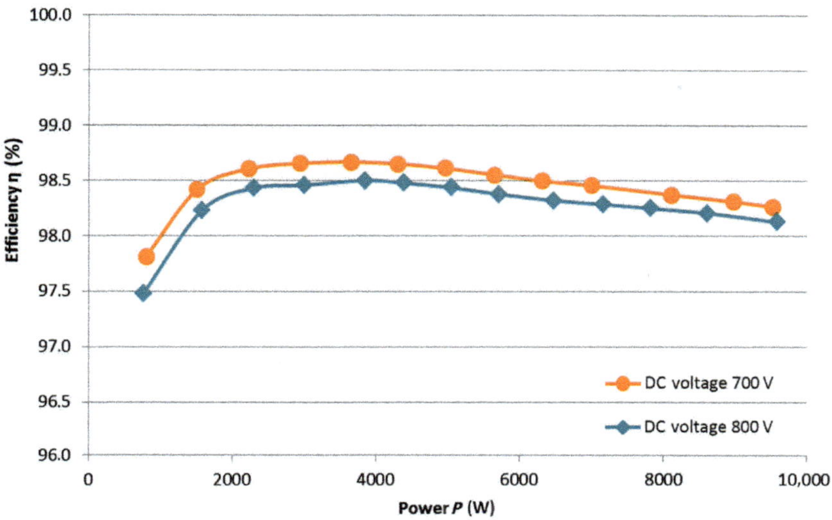

Figure 9.10 Measurement of the efficiency for different loads with 700 and 800 V input voltage.

- Efficiency of Si three-level inverter: 97.4%
- Efficiency of Si two-level inverter: 95.4%
- Constant energy price of 0.17 € per kWh

The relative running costs with an energy price of 0.17 € per kWh are shown in Fig. 9.11. The savings are almost 40% per year when comparing the demonstrator with a conventional three-level Si system.

Per year the difference of 1% point regarding the efficiency is leading to savings of almost 80 € per year. In 10 years without considering deposits and with a constant energy price the savings sum to more than 750 €. Relatively the SiC UPS system has 40% less operation costs per year.

9.2.1.4 *Weight-to-power ratio compared to commercial systems*

In order to evaluate the system in relation to competitor products the weight-to-power ratio of the system is compared to systems in the same power range of five companies. The power density of the demonstrator is by the factor of 6–10 times higher compared to the commercial products. As the demonstrator is designed without housing assumptions were made for a fair comparison (Fig. 9.12).

9.2.1.5 *Summary*

Designed is the demonstrator for a nominal power of 10 kW and an overload capability of 120% for 30 minutes. The demonstrator has a switching frequency of 100 kHz which is approx. 6 times higher than in a commercial product in this power range. The application of SiC devices allows the operation at such a high

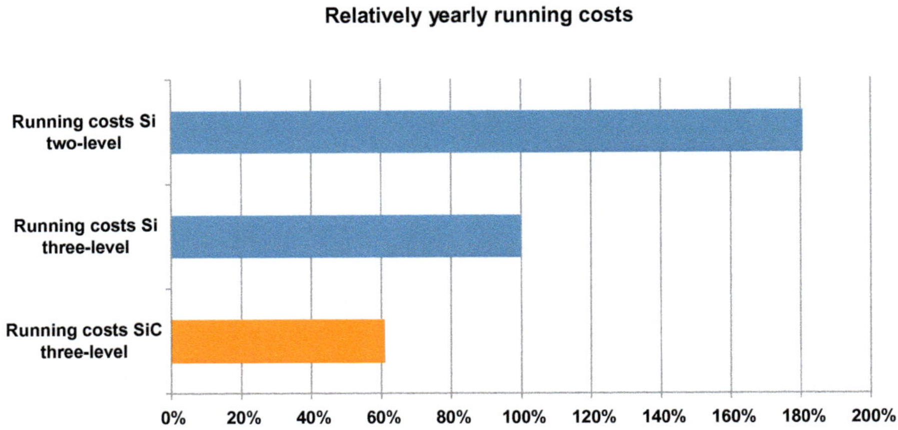

Figure 9.11 Relative yearly running costs at 0.17 € per kWh (operating time 365 days at 24 h at $P_N/2$).

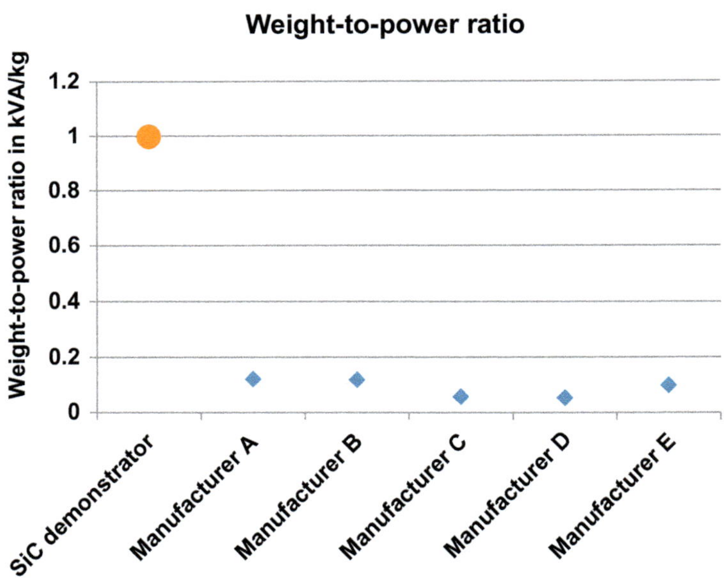

Figure 9.12 Weight-to-power ratio of the system compared to commercial systems, kVA/kg.

switching frequency with still a higher efficiency compared to conventional Si systems. The inverter reaches a maximum efficiency of 98.7%. The gain in efficiency leads to lower running costs in the application of online UPS systems. Compared to conventional Si systems 40% of the running costs per year can be saved.

On system level the smaller passive components and heat sink are leading to a small and highly compact system. The reduced costs of these parts are more than compensating the higher costs of the power parts (SiC device vs Si device).

The results can also be transferred to other application where efficiency and size is an issue and a high current quality is required (e.g., PV-inverters, etc.).

9.2.2 Modular and compact 1 MW SiC-inverter in a single 19″ rack for storage and PV

9.2.2.1 Introduction

The possibilities of SiC devices are shown with a demonstrator of 1 MW inverter for a hybrid energy storage system comprised of a lithium-ion battery supported by a supercapacitor, feeding into the medium voltage grid via the inverter and a transformer. The 1 MVA inverter is directly connected to the battery with a single high current DC bus, while the supercapacitor is connected to the same DC bus but via its own DC/DC converter. The inverter hardware itself can not only be used as a bidirectional battery inverter but also as a PV-inverter in other system setups.

Unlike PV central inverters, however, which are commonly built as monolithic 2–3 MW devices now, storage systems and thus battery inverters are often chosen for a specific load case which, i.e., in industrial applications (PV self-consumption) can change over time due to changes in the machinery pool or its usage patterns. A flexible solution is presented here, modularizing the inverter and allowing resizing of the inverter at any time.

9.2.2.2 Hardware

As housing for the eight inverter units, a single 19″ rack ($60 \times 80 \times 220$ cm^3 including base) is sufficient (Fig. 9.13). The weight without the cooling system but including all inverter units is 590 kg, which results in a total power density of 0.95 kVA/L and 1.7 kVA/kg. This is by a factor of two to four higher than comparable inverters on the market.

In the back of the rack, copper bus bars with a nominal current of 1600 A run from the top—where they connect the DC and the AC side of the inverter units in parallel—to the connection point at the bottom. The coolant distribution is placed next to the bus bars and separated from the units by a sheet metal wall which holds the coolant sockets. Connectors for communication, 230 V auxiliary supply as well as signals to and from the coolant system and the circuit breaker are placed on the front panel and are connected in parallel across all units as well.

All connections for power and coolant are realized by simply pushing the units into the rack, which means units can be easily be replaced (hot swapping).

The inverter units are liquid cooled with water in combination with an external heat exchanger. Air cooling would lead to a derating of the system at extreme conditions like high temperature or high altitude, while this can be avoided by using a

Figure 9.13 Megawatt battery inverter in 19″ technology [10].

separate liquid cooling system that can be dimensioned for the specific conditions on the installation site.

The units are housed in 19″ cases which are 150 mm in height, 490 mm in depth (without connectors), and weigh 42.5 kg. At a nominal power of 125 kVA this results in a high power density of 3.70 kVA/L and 2.94 kVA/kg (Figs. 9.14 and 9.15).

Each of the units has its own DC fuse as well as DC and AC contactors, so it can completely disconnect from the system if necessary.

The AC filter consists of a three-phase choke with cores made from modular iron powder pellets and three power capacitors in star configuration with the star point connected to the negative pole of the DC link.

The recently available line of high current SiC semiconductor modules made it possible to combine high power with high switching frequency without strong negative influence on the efficiency. In this case three 1.2 kV, 300 A half-bridge modules are used in a simple B6 topology with the battery voltage on the input and operated at 40 kHz to make up the core of each of the single-stage inverter units.

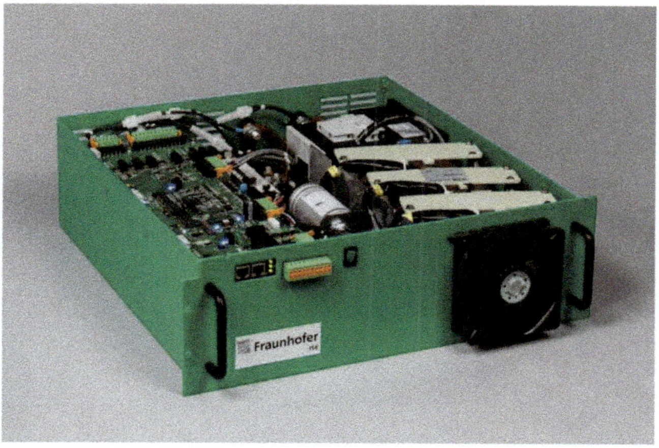

Figure 9.14 Front view of a 125 kVA inverter unit for the 19″ rack with 150 mm height [10].

Figure 9.15 Rear view with high current DC and AC connectors (left) and leakage-free coolant connectors (right) [10].

High switching speeds of up to 7.5 kA/μs and at peak currents of around 300 A require low inductances in the commutation cells to limit voltage overshoots. Laminated bus bars were therefore taken into consideration but calculations showed that in this case a special thick copper PCB can achieve similar results at lower cost (Fig. 9.4).

As this is a three-phase three-wire inverter with symmetrical power output, there is almost no 100 and 150 Hz ripple to be buffered by the DC-link capacitors. The remaining switching ripple was realized by a film capacitor bank with very low equivalent series resistance (ESR) and equivalent series inductance (ESL) values. Due to the low total parasitic inductance, snubber capacitors are not necessary here.

Measurements were taken with high current SiC half-bridge modules from two manufacturers and compared for performance. Different solutions for the SiC module drivers were investigated and compared for a solution that has high current capability at small size and does not interfere with the control board despite the high-voltage rise rates. The final solution is a high current IGBT driver from SEMIKRON, of which the voltage levels have been modified via an external adapter board to fit the requirements of the used SiC MOSFETs.

A custom built aluminum cold plate is used to cool the three semiconductor modules. It is fitted with leakage-free couplings on the back to enable the inverter unit to be snapped in and out of the rack even during operation of the other units.

9.2.2.3 Measurements

Switching curves

One inverter unit was operated with a 256 kW DC supply at 800 V and the nominal power of 125 kW was fed into a 200 kVA three-phase grid simulator set to an RMS phase-to-phase voltage of 400 V while switching curves were recorded for one of the high-side MOSFETs at the maximum inductor current (Figs. 9.16 and 9.17).

The drain current $I_{D,HS}$ was measured via a Rogowski coil around the high-side drain terminal. The voltage $V_{GS,ext,HS}$ was measured at the external gate contacts of the SiC module. However, this does not correspond to the internal gate voltage $V_{GS,HS}$ directly as the half-bridge units have a relatively high internal gate resistance.

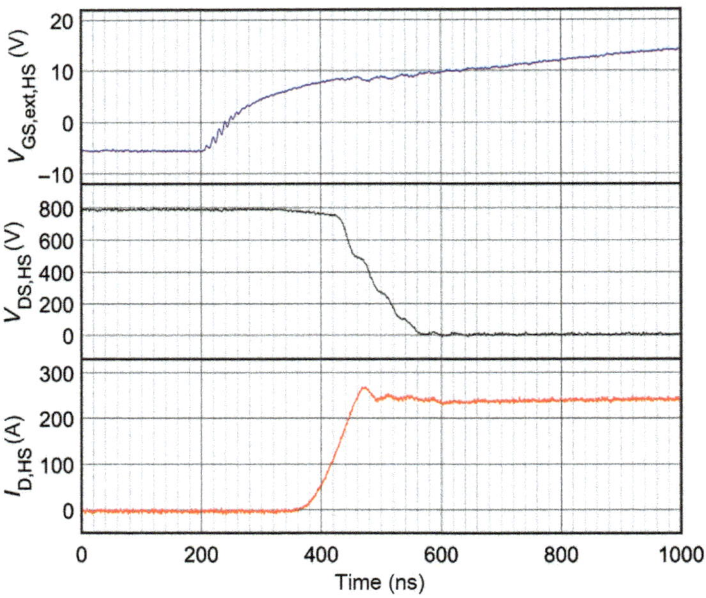

Figure 9.16 High-Side MOSFET turn-on transient at 238 A; V_{DS} fall-time from 90% to 10% is 112 ns [10].

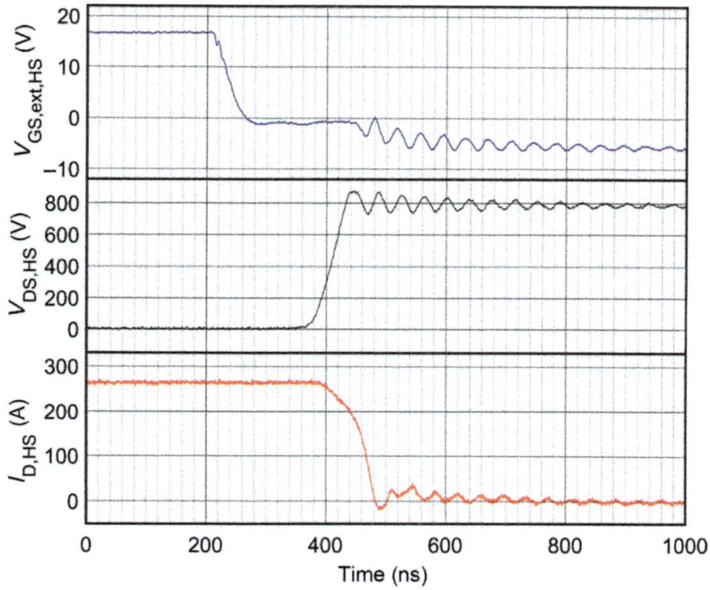

Figure 9.17 High-side MOSFET turn-off transient at 267 A; V_{DS} rise-time from 10% to 90% is 46 ns [10].

Externally a capacitance of 18 nF was added between gate and source to lower the impedance at high frequencies and to reduce the risk of parasitic turn-on. The external gate resistors were chosen to be $R_{G,on} = 5\ \Omega$ and $R_{G,off} = 1.1\ \Omega$.

Efficiency

For efficiency measurements, one inverter unit was operated inside the rack using a 256 kW DC supply and a 200 kVA three-phase grid simulator set to an RMS phase-to-phase voltage of 400 V. The neutral wire was connected to earth and not the inverter. Voltages were measured at the connection point of the inverter system at the bottom of the rack and include connector and bus bar losses. Auxiliary and cooling power was not included in the measurement but is below 100 W per inverter unit.

The maximum battery voltage used in this project is 843 V. At this voltage efficiencies up to 97.8% were measured with standard modulation (reference curve). With the 120 degrees Flat-Top modulation, however, it is boosted to 98.4% (Fig. 9.18).

For the minimum battery voltage used in the project, 685 V, the efficiency was as high as 98.7%.

Measurements were also taken for the minimum DC voltage required by the inverter, which is 591 V. Here efficiencies of up to 98.9% could be reached.

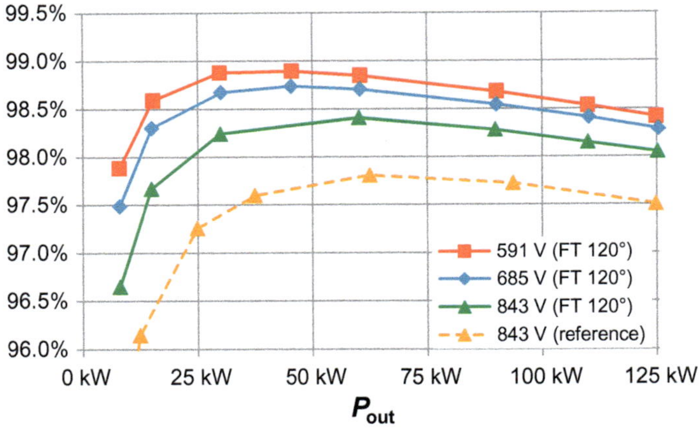

Figure 9.18 Efficiency of one inverter unit at $V_{Grid,RMS} = 400$ V, $f_{sw} = 40$ kHz [10].

9.2.3 Medium voltage inverter with SiC

Power electronics with devices above 6.5 kV have not been common for commercial applications. Nevertheless, they are offering a lot of opportunities for new system solutions. For demonstration of the significant advantages but also to show technical problems and solutions first prototypes with high-voltage SiC devices have been developed.

9.2.3.1 Photovoltaic application with 10 kV SiC MOSFETs

As an example for future applications of medium voltage power electronics with SiC devices in renewable energy systems the structure of PV power plants is discussed in this section and a new system architecture is proposed. Therefore, a demonstrator with 10 kV SiC MOSFETs was built [11].

In today's PV power plants with up to 100 MVA and more, the distribution of electrical energy is done at low voltage level. However these power plants mostly feed into the medium voltage grid.

The DC voltage of the generator strings is usually not higher than 1000 V. The outputs of the generators are collected and lead to an inverter. The inverter generates a three-phase alternating voltage with normally 250–400 V phase to phase. The 50/60 Hz transformer has to convert this voltage for the connection to the medium voltage grid (see Fig. 9.19, left). The PV generators, the inverter, the transformer and an additional switch gear form one subunit. The power of one subunit generally is in the range of 1 MVA. PV power plants consist of a large number of these subunits. It would be preferable to increase the power of the subunits to reduce system costs. Raising the power with unchanging voltage level leads to higher currents. Drawbacks are higher copper diameters for the cables and rising

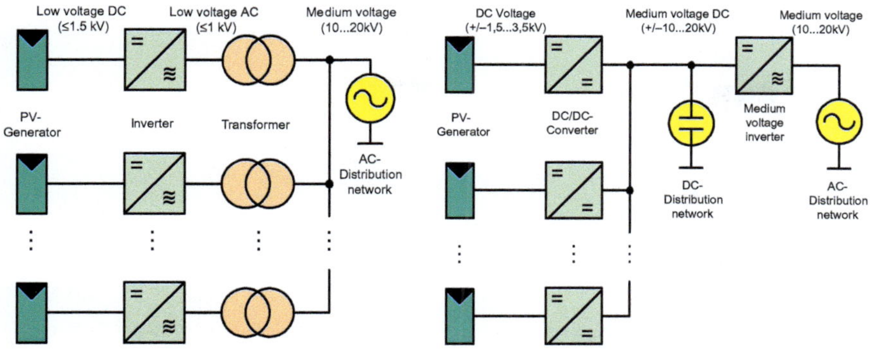

Figure 9.19 Topology of today's PV power plants (left) and conceivable topology of future PV power plants (right) [11].

thermal losses. It is also inefficient to increase the power of the transformers at the existing high transmission ratios because physical limits will be reached.

The only reasonable way to increase the power of one subunit is the raise of the system voltage level. Although the low voltage directives have to be left, this step will lead to many advantages.

Fig. 9.19 (right) shows a concept for a PV power plant topology with a DC-collection at the medium voltage level. The output voltage of the PV modules can be increased in future. DC/DC converters will connect the PV generators to a common DC distribution network. Because of the collection of the generators in the medium voltage level cable diameters can be relatively small. With a medium voltage converter the PV power plant will feed into the grid. In this concept the transformer can be omitted, which can save high costs for core materials and copper. The total number of system components will be reduced. The nominal power of one medium voltage inverter can be multiple megawatts, depending what semiconductor modules will be available in future.

With available 10 kV/10 A MOSFETs integrated 10 kV/10 A SiC JBS diode [12] a DC/DC converter was developed, which could be one part of the proposed PV power plant structure. The converter was built as a boost converter with an input voltage of 3.5 kV and an output voltage of 8.5 kV. PV modules with 3.5 kV output voltage are conceivable in the long term. The rated power of the converter is 28 kW. This rating results from the maximum values of the 10 kV/10 A SiC MOSFETs including only a small safety margin.

Due to the low switching energies of the MOSFETs the switching frequency was chosen to 8 kHz. Compared with conventional medium voltage converters this corresponds to an about 10 times higher value. The higher the switching frequencies the smaller the passive components can be dimensioned. This leads to a reduction of material consumption, volume, and costs for inductor and capacitors. The

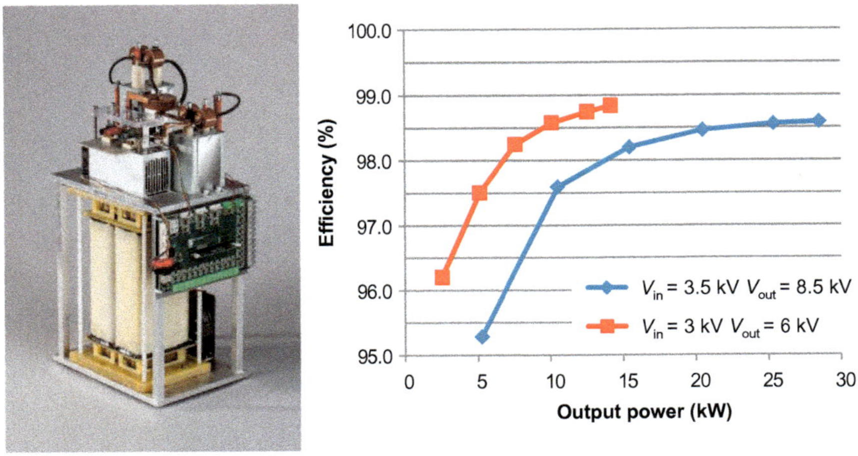

Figure 9.20 Mechanical construction of the 28 kW boost converter (left) and efficiency of the boost converter at the nominal voltage ratings from 3.5 to 8.5 kV and at lower voltages from 3 to 6 kV (right) [11].

high-frequency inductor consists of an amorphous core and has the dimensions of about 330 × 210 × 160 mm and is actively cooled by air.

The MOSFETs are placed on a heat sink. The bottom of the housings is on drain potential with up to 8.5 kV. This implies that the housing has to be electrically isolated to the heat sink. Additionally a sufficient heat flow from the semiconductors to the heat sink is necessary. Therefore a ceramic disk of aluminum nitride (AlN) was inserted. The AlN disk has an excellent thermal conductivity of 200 W/m*K and a breakdown voltage of 15 kV/mm. The transistors are fixed to the AlN disk and the heat sink mechanically with the help of a pressure plate.

Fig. 9.20 (left) shows the mechanical construction. The control circuit board is mounted on the side of the converter. For the control input voltage, output voltage and the current through the low-side MOSFET can be measured. The efficiency of the converter for an operation at a medium voltage resistance is shown in Fig. 9.20 (right). A maximum efficiency of 98.5% was reached.

9.2.3.2 AC grid application with 15 kV SiC devices

Today, active filters for medium voltage grids are mainly low voltage inverters (e.g., 690 V), which are coupled by a 50/60 Hz transformer into the higher voltage level (e.g., 3–30 kV) as can be seen in Fig. 9.21 (left). So nearly regardless of the dynamic performance of the inverter, the bandwidth of the active filter is limited by the cut-off frequency of the transformer. Using silicon IGBT of 3.3 or 6.5 kV to design a directly grid coupled medium voltage inverter will not lead to an increased bandwidth, as the switching frequency would be too low and thus the bandwidth would be in the same range as before. A directly coupled inverter with

SiC-transistors and drastically increased dynamic range would overcome this problem (Fig. 9.21, right).

Following results of a research project of a three-phase medium voltage inverter with 15 kV devices are presented [13,14]. The work has been funded by the German Federal Ministry of Education and Research. For the project 15 kV MOSFETs, IGBTs, and JBS diodes were available [15].

Figs. 9.22 and 9.23 show double pulse measurements with these devices [16]. The measurements show that the overall switching energies of the MOSFET are

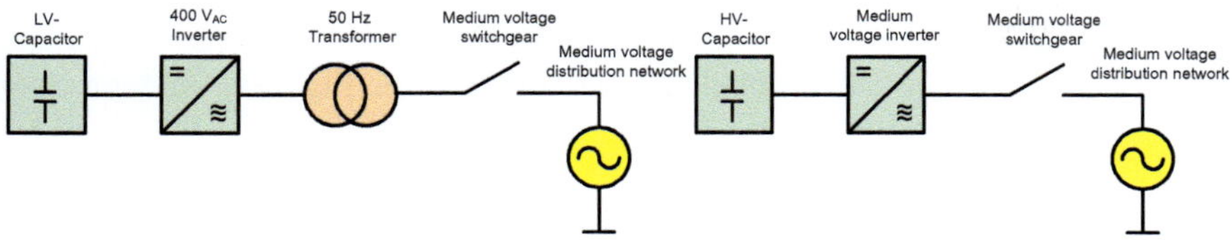

Figure 9.21 Active filter as state-of-the-art with low voltage inverter (left) and active filter with SiC medium voltage inverter (right) [13].

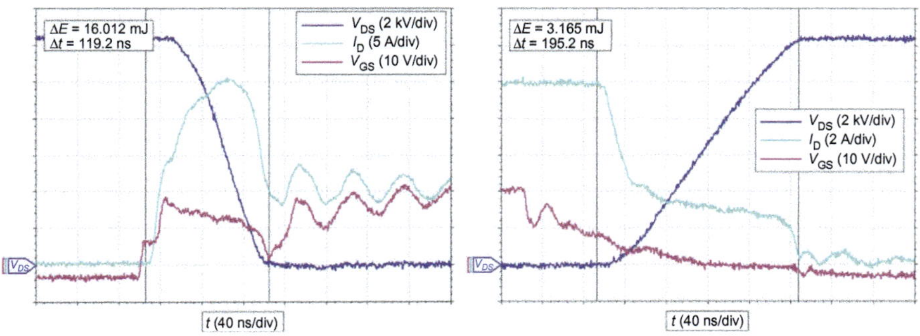

Figure 9.22 Turn-on and turn-off transition of the 15 kV MOSFET (commutation with JBS diode) at $V_{DS} = 12$ kV, $I_D = 10$ A, $R_G = 6.8\ \Omega$ [16].

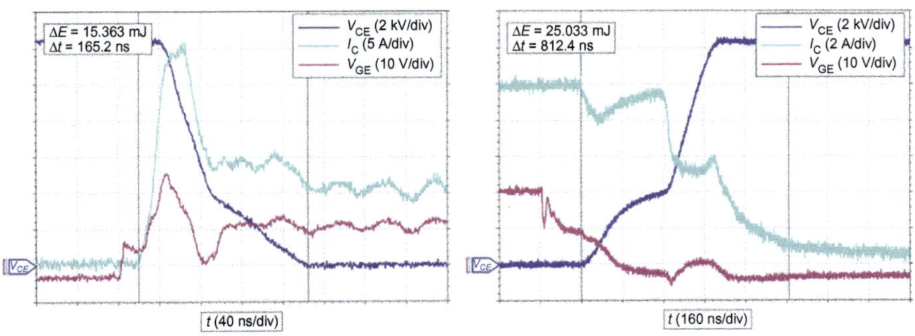

Figure 9.23 Turn-on and turn-off transition of the 15 kV IGBT at $V_{CE} = 12$ kV, $I_C = 10$ A, $R_G = 6.8\ \Omega$ [16].

much lower compared to the IGBT. Furthermore, the IGBT has a very high dv/dt of about 180 kV/μs during the turn-on transition, which cannot be influenced by the gate resistance. The measured dv/dt of the MOSFET is 160 kV/μs, but could be reduced by the use of higher gate resistance. This is why the MOSFET can be preferred in applications with very high switching frequencies and was used in the project.

Another advantage of the MOSFET is that its intrinsic diode can be used for freewheeling. In Fig. 9.24 it can be seen, that the dynamic behavior of the intrinsic diode is comparable to the behavior of the JBS diode (Fig. 9.25).

Due to the maximum blocking voltage of 15 kV of the MOSFETs at least a three-level topology is needed to feed into a 10 kVAC grid. Therefore, the active neutral point clamped (ANPC) topology was selected (see Fig. 9.1). In comparison

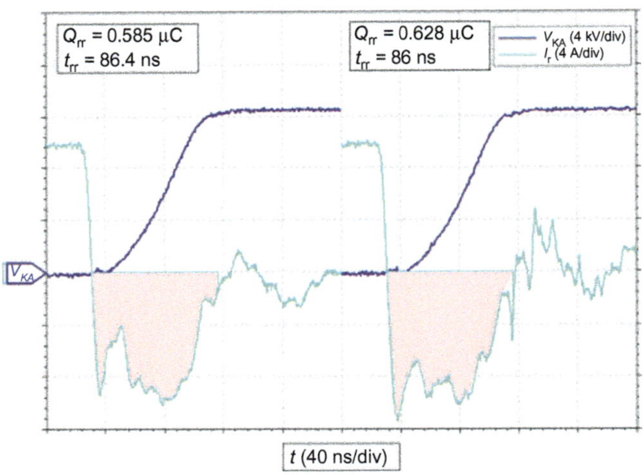

Figure 9.24 Reverse current of the JBS diode (left) and the intrinsic diode (right) at VKA = 12 kV, IF = 10 A [16].

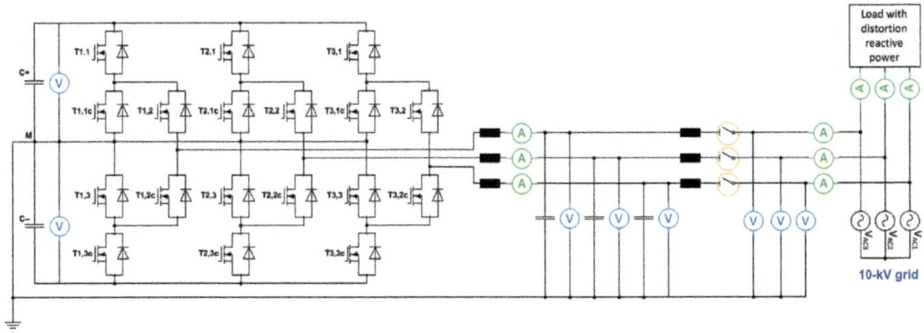

Figure 9.25 Circuit of the inverter with LCL filter, voltage and current measurements, and grid connection [14].

Figure 9.26 Design parameters and specifications of the 10 kV inverter (left) and 20 kV_{DC} stack in ANPC topology as a single-phase leg with 15 kV MOSFETS, drivers and capacitors attached [13].

to the neutral point clamped topology, MOSFETs instead of diodes are used in the clamping path. The use of an active switch in the clamping path leads to more degrees of freedom in the modulation method of the inverter. By using an appropriate pulse width modulation (PWM) strategy, the apparent switching frequency of the bridge voltage can be doubled [17]. Fig. 9.26 (left) shows the specifications and design parameters of the inverter.

One main challenge in the inverter design was the handling of the high dv/dt values of the 15 kV MOSFETs with up to 180 kV/μs. Besides the high requirements for the isolation coordination and materials it is important to minimize all parasitic capacitances near to the fast changing voltage potentials. Otherwise high interference currents will occur (Table 9.3).

Because there are no 15 kV modules available until now, the MOSFET chips are placed in a not insulated prototype housing with drain potential at the base plate. To achieve a sufficient electrical isolation of the housings, good heat dissipation and a compact, low inductive structure, the MOSFETs of one-phase leg were placed on one big ceramic plate (see Fig. 9.26, right). The aluminum nitride based ceramic is mechanical processible, so it is possible to mount the devices by screw attachments on the plate. On the bottom side of the ceramic there a milled slots, in which aluminum fins are stucked in for dissipating the heat with cooling fans. The specific thermal conductivity of the ceramic plate is 100 W/m K. The parasitic capacitance between the MOSFETs and the grounded aluminum fins is approximately inversely proportional to the thickness of the

Table 9.3 Design parameters and specifications of the 10 kV inverter

Parameter	Symbol	Value
Nominal grid voltage	V_{AC_OUT}	10 kV_{AC}
DC-link voltage	V_{DC_link}	20 kV_{DC}
Switching frequency	f_S	16 kHz
Ripple frequency	f_R	32 kHz
Nominal power	S_{OUT}	100 kVA
DC-link capacity	C_{DC_link}	$2 \times 12\ \mu F$
Main inductance	$L_{M,LCL}$	22 mH
Filter capacitor	$C_{F,LCL}$	200 nF
Grid filter inductance	$L_{G,LCL}$	1.3 mH

Figure 9.27 15 kV SiC-MOSFET with attached gate driver and galvanically isolated power supply (left) and three-phase 10 kV inverter system (right) [14].

ceramic plate. With a thickness of about 10 mm the capacitance was kept in the range of a few tens of pF.

To achieve short current paths, the gate drivers are attached directly on top of the MOSFETs (Fig. 9.27, left) and clearance and creepage distances are increased by additional isolation barriers. So the MOSFETs can be arranged close to each other.

Also the source-potentials of the MOSFETs are bouncing with the very high dv/dt-rates. Therefore, the power supply of the gate driver has not only to insulate the voltage but also has to prevent capacitive couplings from the high to the low voltage side. This is achieved with a design shown in Fig. 9.27. The primary windings of the transformer of the gate power supply are coiled through the middle of the toroid core. Thus the isolation is ensured and the parasitic capacitance between high and low voltage side is below 3 pF. The gate signals are transmitted potential-free by optical fibers.

The mechanical construction of the three-phase inverter system is built in a cabinet with the dimensions of $80 \times 80 \times 200\ cm^3$ (Fig. 9.4). On the three upper levels the three phases are located. Each phase level consists of one stack with one ANPC phase leg and six gate driver power supplies above the stack. Also on each phase level there is a part of the DC-link capacitors, the LCL filter, two current and voltage measurements, and a high-voltage relay for grid-coupling. The current measurement is realized with a shunt measurement. The electronic circuit for the shunt is galvanically isolated by a power supply similar to the one used for the gate driver supply. Via delta sigma converters and optical fibers the measured currents are transmitted to the control board. For the voltage measurement RC-compensated voltage dividers were designed.

The lower level of the cabinet consists of several auxiliary power supplies with 12 V and 24 V, connection points for DC-input and AC-output, surge arresters, circuits for capacitor charge and discharge and the control board. The control is implemented on an field programmable gate array (FPGA), which receives and transmits all measure-, control-, and PWM-signals galvanically isolated by fiber optics.

Initially the commissioning was implemented in open-loop control in single-phase operation. For the input a high-voltage-DC power supply was used. At the output a medium voltage resistive load was connected. At a duty cycle of 100% the DC voltage was raised up to 17 kV. At this operation point the nominal output voltage of 5.8 kVRMS was reached, which corresponds to the string voltage of a 10 kV grid. Due to the minimum adjustable value of the available medium voltage resistor of 2.4 kΩ the output current is 2.4 A and the output power is 14 kW (Fig. 9.28, left). All three phases were tested successfully. Fig. 9.28 (right) shows efficiency measurements for different DC voltages.

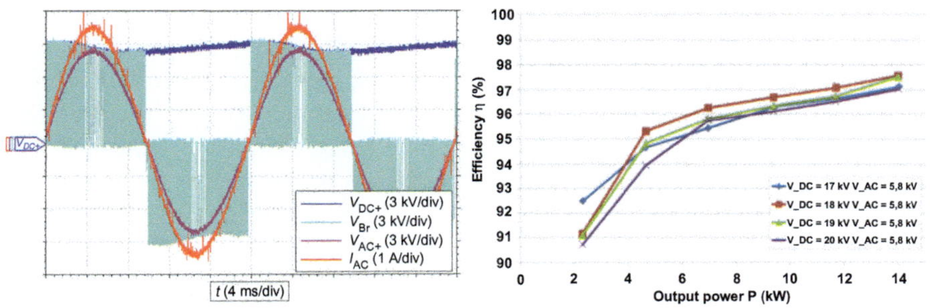

Figure 9.28 Positive DC-link voltage V_{DC+}, bridge voltage V_{Br}, output voltage VAC, and output current IAC in single-phase operation (left) and efficiency of the inverter for different DC voltages [14].

9.3 What will happen?

Regarding the low voltage range, 1200 V SiC-MOSFETs are commercial products since 2011. The early days were stormy, not only for the device designers, but also for the power electronic engineers. They were used to silicon IGBTs and they had to rethink their ways of designing systems to take all benefits of SiC, including redesigns of control electronics, more complex concepts of inductor designs and new cooling methods for a drastically increased power density.

The technology has settled and the further development will be similar to the history of silicon devices. The diameter of the wafers and the quality of the material are continuing to increase. The packaging technology is evolving and enabling to take advantage of the material properties of SiC.

As the costs decreased since the first prototypes, also lower voltage levels are getting more and more economically feasible, tackling with applications for silicon superjunction MOSFETs. WOLFSPEED is selling a 900 V MOSFET, MICROSEMI has a 700 V device available, and ROHM offers a 650 V device. With further expansion in this low voltage area, SiC will become a clear rival for GaN, neither in switching frequency nor in terms of lateral circuit integration, but in the ability to get higher currents by directly paralleling chips in power modules. It is still a long way to go for GaN to improve the assembling technologies and to get in power ranges above 450 V_{DC}/15 kW.

For the application of SiC, the current range is no more limited like in the beginning. The chip-size increased and also the knowledge in the design of SiC power modules. The PV-inverter introduced by GE in 2016 was an amazing step in this direction [18]. It demonstrated the economic feasibility of a SiC-based inverter even in a high power range of 2.5 MVA with AC currents in the range of 2.6 kA and thus a large accumulated chip area.

For the low voltage range, the further improvements are rather evolutionary with a growing availability of devices and modules on the market, whereas for the high voltage range of 3.3 kV and above there is still a lot of scientific work ahead to get to the point of knowledge where we are today in the low voltage range.

Of course, there has been a lot of research on HV-SiC devices even far above 3.3 kV for more than 10 years. Mitsubishi is proving their 3.3 kV MOSFET prototype in a traction inverter in a train as field test [19]. Also Wolfspeed, Rohm, and Hitachi are working on the commercialization of a 3.3 kV SiC-MOSFETs [20–22]. So this voltage class will be soon commercially available. Also for sure 6.5 kV SiC devices will come [23]. This voltage class is already existing for silicon devices, thus there is also an existing market and maybe the power electronics engineers do not know yet how to handle the improved dynamics with SiC, but at least they know how to handle this voltage class. Beyond the border of 6.5 kV there is pure wilderness with a bunch of new chances in power electronic system design, but also with a lot of challenges to settle this technology.

First of all, the chances for power electronic system design with SiC devices above 6.5 kV are revolutionary and the change in technology will come! But the barrier at this point is, that power electronic system engineering for HV-SiC has only just begun and it is still far behind the potential of the HV-SiC devices. Less with low voltage SiC, but mainly with low voltage GaN, power electronic engineering entered the field of high-frequency engineering. With HV-SiC it will be high-frequency and high-voltage engineering. Here, high-frequency engineering is not meant in sense of high switching frequencies. They will be rather low. But the extreme high dv/dt of HV-SiC devices will lead to phenomena of traveling waves and reflections in a highly more extensive way as we are used with the application of 6.5 kV silicon IGBTs.

Fig. 9.29 shows the voltage slope in a three-phase 10 kV inverter with 15 kV SiC-MOSFETs. The *blue line* is the rising voltage at the output terminal of the half-bridge. Here the dv/dt is 160 kV/μs with 12 kV as the maximum voltage of the DC-link capacitor. The *purple line* shows the rising voltage at the input terminal of the inductor. Due to reflections at the end of the line, the voltage peak is at 24 kV and the dv/dt is 320 kV/μs.

HV-SiC devices offer a high potential of possible switching speed, increased switching frequency, and still an improved efficiency. But to seize the full potential of HV-SiC, a strong effort has to be put on the challenges of system design, isolation materials, and high-voltage engineering.

Yet, the related technologies are not capable to use the full potential of HV-SiC and not all common materials can handle the high-voltage stress caused by the high dv/dt. Future improvements in medium voltage power electronics with HV-SiC will be determined by the challenges and improvements on an electrophysical level.

There will be a scientifically challenging time ahead until everything works fine, but it will be worthwhile. Improvements in power electronic were always driven by reduced system costs, increased power density, increased efficiency and increased dynamics. Despite the troubles we will have with HV-SiC in the beginning, the pressure of these benefits will be high enough to push the industrial and academic efforts.

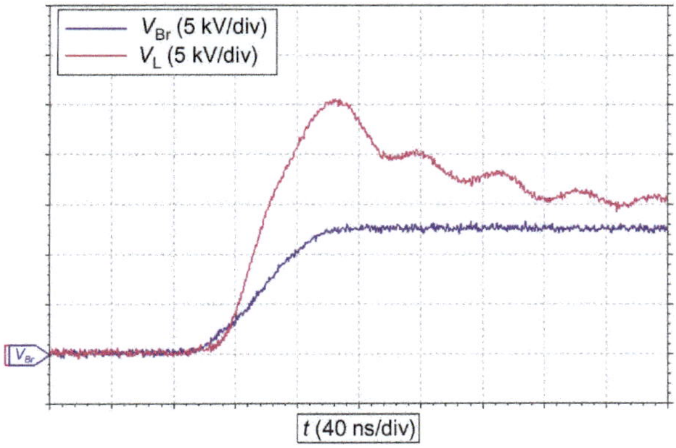

Figure 9.29 Voltage slopes at the output of the bridge and at the input of the inductor of a 10 kV_{AC} inverter with 15 kV SiC-MOSFETs [13].

References

[1] B. Burger, D. Kranzer, O. Stalter, Cost reduction of PV-inverters with SiC-DMOSFETs. In: 5th International Conference on Integrated Power Systems (CIPS), VDE, 2008. S. 1–5.

[2] D. Kranzer, et al., System Improvements of Photovoltaic Inverters with SiC-Transistors, Materials Science Forum, Trans Tech Publications, 2010, pp. S. 1171–1176.

[3] G. Deboy, et al., New SiC JFET boost performance of solar inverters, Power Electron. Eur. (4) (2011) S. 29–33.

[4] A. Hensel, C. Wilhelm, D. Kranzer, Development of a boost converter for PV systems based on SiC BJTs. In: Proceedings of the 2011-14th European Conference on Power Electronics and Applications (EPE 2011), IEEE, 2011. S. 1–7.

[5] M. Milad, M. Darwish, Comparison between Double Conversion Online UPS and Flywheel UPS technologies in terms of efficiency and cost in a medium Data Centre. In: 2015 50th International Universities Power Engineering Conference (UPEC), IEEE, 2015. S. 1–5.

[6] F. Cammarota, S. Sinigallia, High-efficiency on-line double-conversion UPS. In: 29th International Telecommunications Energy Conference, INTELEC 2007. IEEE, 2007. S. 657–662.

[7] C. Schöner, D. Derix, A. Hensel, Comparison and evaluation of different three-level inverter topologies for PV systems. In: 2014 16th European Conference on Power Electronics and Applications (EPE'14-ECCE Europe), IEEE, 2014. S. 1–10.

[8] B. Muralidhara, et al., Experimental measurement and comparison of common mode voltage, shaft voltage and the bearing current in two-level and multilevel inverter fed induction motor, Int. J. Inf. Electron. Eng. 1 (3) (2011) S. 245–250.

[9] S. Buschhorn, K. Vogel, Saving money: SiC in UPS applications. In: PCIM Europe 2014; Proceedings of International Exhibition and Conference for Power Electronics, Intelligent Motion, Renewable Energy and Energy Management, VDE, 2014. S. 1–7.

[10] P. Hercegfi, S. Schoenberger, Modular and Compact 1 MW Inverter in One 19" Rack for Storage and PV. In: PCIM Europe 2017; Proceedings of International Exhibition and Conference for Power Electronics, Intelligent Motion, Renewable Energy and Energy Management; VDE, 2017. S. 1–5.

[11] J. Thoma, D. Chilachava, D. Kranzer, A highly efficient DC-DC-converter for medium-voltage applications. In: 2014 IEEE International Energy Conference (ENERGYCON), IEEE, 2014. S. 127–131.

[12] M.K. Das, et al., 10 kV, 120 A SiC half H-bridge power MOSFET modules suitable for high frequency, medium voltage applications. In: 2011 IEEE Energy Conversion Congress and Exposition (ECCE), IEEE, 2011. S. 2689–2692.

[13] D. Kranzer, et al., Development of a 10 kV three-phase transformerless inverter with 15 kV silicon carbide MOSFETs for grid stabilization and active filtering of harmonics. In: 2017 19th European Conference on Power Electronics and Applications (EPE'17 ECCE Europe), IEEE, 2017. S. P. 1–P. 8.

[14] D. Kranzer, et al., Design and commissioning of a 10 kV three-phase transformerless inverter with 15 kV silicon carbide MOSFETs. In: 20th European Conference on Power Electronics and Applications (EPE'18 ECCE Europe), 2018. S. P. 1–P. 8.

[15] S. Ryu, et al., Ultra high voltage MOS controlled 4H-SiC power switching devices, Semicond. Sci. Technol 30 (8) (2015) S. 084001.

[16] J. Thoma, et al., Characterization of high-voltage-SiC-devices with 15 kV blocking voltage. In: 2016 IEEE International Power Electronics and Motion Control Conference (PEMC), IEEE, 2016. S. 946–951.

[17] D. Floricau, E. Floricau, M. Dumitrescu, Natural doubling of the apparent switching frequency using three-level ANPC converter. In: International School on Nonsinusoidal Currents and Compensation, ISNCC 2008. IEEE, 2008. S. 1–6.

[18] Ge Power Conversion. LV5 + 1500V Solar Inverter, Datasheet, 2016.

[19] K. Hamada, et al., 3.3 kV/1500 A power modules for the world's first all-SiC traction inverter, Jpn J. Appl. Phys. 54 (4S) (2015). S. 04DP07.

[20] T. Sakaguchi, et al., Characterization of 3.3 kV and 6.5 kV SiC MOSFETs. In: PCIM Europe 2017; Proceedings of International Exhibition and Conference for Power Electronics, Intelligent Motion, Renewable Energy and Energy Management, VDE, 2017. S. 1–5.

[21] T. Ishigaki, et al., 3.3 kV/450 A full-SiC nHPD2 (next high power density dual) with smooth switching. In: PCIM Europe 2017; Proceedings of International Exhibition and Conference for Power Electronics, Intelligent Motion, Renewable Energy and Energy Management, VDE, 2017. S. 1–6.

[22] J. Hayes, et al., Dynamic characterization of next generation medium voltage (3.3 kV, 10 kV) silicon carbide power modules. In: PCIM Europe 2017; Proceedings of International Exhibition and Conference for Power Electronics, Intelligent Motion, Renewable Energy and Energy Management, VDE, 2017. S. 1–7.

[23] Mitsubishi Electric Corporation. Mitsubishi Electric's New 6.5 kV Full-SiC Power Semiconductor Module Achieves World's Highest Power Density, Press Release No. 3164, Tokyo, January 31, 2018.

Synopsys

B. Jayant Baliga
North Carolina State University, Raleigh, NC, United States

10

10.1 The silicon IGBT

The silicon insulated gate bipolar transistor (IGBT) was invented and rapidly commercialized at the General Electric Company (GE) in the early 1980s [1]. Its high power handling capability, simple gate control, and ruggedness have made it the technology of choice for all medium and high power applications. The availability of the IGBT transformed power electronics from analog to digital control allowing more precise and efficient energy management. The applications for the IGBT now span all sectors of the economy, as shown in Fig. 10.1, leading to device sales of over $ 4 billion annually. The breakdown in the market for silicon IGBTs is provided in Fig. 10.2.

The energy efficiency gained by using the IGBT enabled advances in power electronics has been quantified over a 25-year time span from 1990 to 2015 using three examples. The first impact is in the transportation sector by the creation of the electronic ignition system to improve the efficiency of gasoline-powered cars and trucks by 10%. Since 76% of fuel is used by cars and trucks around the world with a total annual consumption of about 650 billion gallons, the cumulative gasoline savings over the 25-year time span is 1.48 trillion gallons. The second impact is in the consumer and industrial sector by the creation of the adjustable speed motor drives. Two-thirds of all the 25,000 TW h of electrical energy in the world is used to run motors. Adjustable speed drives have improved the efficiency for motor control by 40%. These drives have a market penetration of 50% due to their higher cost. Using this information, the IGBT-enabled adjustable speed drives have saved 56,910 TW h of electricity over the 25-year time span. The third impact is in the lighting sector by the creation of the compact fluorescent lamp (CFL). The CFL allows reducing power consumed by a 60 W incandescent bulb to only 15 W for generating the same amount of light. More than 20 billion CFLs are now used around the world resulting in a cumulative energy savings of 16,120 TW h over the 25-year time span. These energy savings are summarized in Fig. 10.3 together with the corresponding cost savings for consumers. Just in these three sectors, the IGBT has enabled a cumulative cost savings of $ 23.7 trillion for consumers.

The consumption of 1 gal of gasoline produces 19.4 pounds of carbon dioxide emissions. The generation of 1 kW h of electricity produces 1.1 pounds of carbon dioxide because 70% of electricity is generated using fossil fuels such as coal or natural gas. Using this information, the IGBT-enabled efficiency enhancements have reduced carbon dioxide emissions by 109 trillion pounds during the 25-year

Wide Bandgap Semiconductor Power Devices. DOI: https://doi.org/10.1016/B978-0-08-102306-8.00011-3

Figure 10.1 Applications for IGBTs. *IGBT*, insulated gate bipolar transistor.

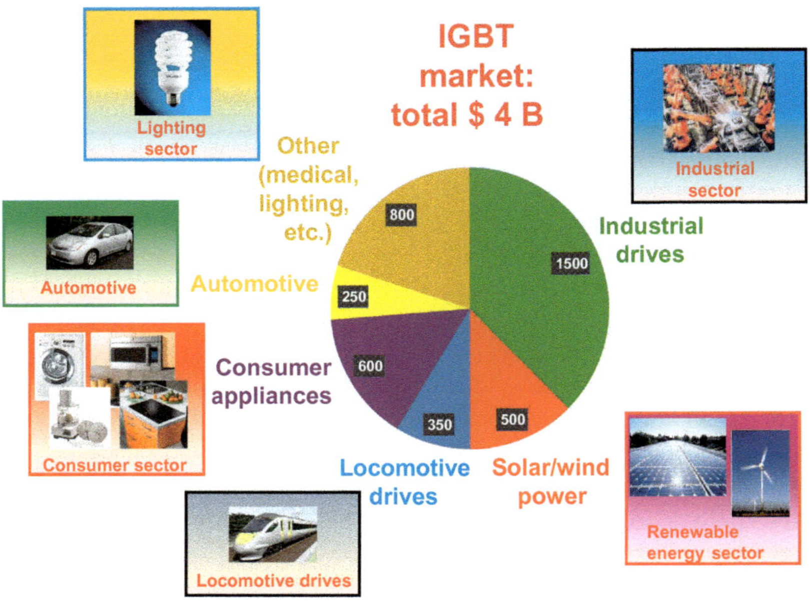

Figure 10.2 Market distribution for IGBTs. *IGBT*, insulated gate bipolar transistor.

Social impact

IGBT enabled application	Cumulative gasoline or energy savings		Consumer cost savings		Utility cost savings	
	US	World	US	World	US	World
Electronic ignition system	318 B gall	1477 B gal	$ 0.654 T	$ 9.125 T	–	–
Adjustable speed motor drive	25,170 TW h	56,910 TW h	$ 2.537 T	$ 11.38 T	$ 0.398 T	$ 1.12 T
Compact fluorescent lamp	1550 TW h	16,120 TW h	$ 0.155 T	$ 3.224 T	$ 0.267 T	$ 2.98 T
Total			$ 3.35 T	$ 23.73 T	$ 0.665 T	$ 4.10 T

Figure 10.3 Energy and cost savings accrued from IGBTs. *IGBT*, insulated gate bipolar transistor.

Social impact

IGBT enabled application	Cumulative gasoline or energy savings		Cumulative carbon dioxide emission reduction	
	US	World	US	World
Electronic ignition system	318 B gal	1477 B gal	6.16 T pounds	28.66 T pounds
Adjustable speed motor drive	25,170 TW h	56,910 TW h	34.25T pounds	62.61 T pounds
Compact fluorescent lamp	1,550 TW h	16,120 TW h	2.09 T pounds	17.74 T pounds
Total			42.5 T pounds	109 T pounds

Figure 10.4 Energy and carbon emission savings accrued from IGBTs. *IGBT*, insulated gate bipolar transistor.

time span as shown in Fig. 10.4. This has offset carbon emissions from all human activity over a period of 2 years.

It can be concluded that the silicon IGBT has had a tremendous impact on society. It has improved the comfort, convenience, and health for billions of people around the globe while reducing environmental impact and global warming. The wide bandgap semiconductor devices discussed in this book were proposed to replace the silicon IGBT to gain further enhancements in energy efficiency. The prospects and challenges to achieve this goal are discussed in this chapter.

10.2 History of wide bandgap semiconductor power devices

The history of evolution of silicon carbide power devices from 1980 to 2000 is illustrated in Fig. 10.5. The proposal to develop power devices from wide bandgap semiconductors can be traced to a theoretical analysis first performed at the General Electric Company in 1979 [2]. This analysis was eventually published in 1982 with quantitative solutions for various semiconductors whose properties were relatively well known at that time [3]. However, the critical electric field for breakdown in semiconductors was not available. Consequently, the analysis related the specific on-resistance of drift regions in unipolar devices (Schottky rectifiers, junction field effect transistors (JFETs), and metal oxide semiconductor field effect transistors (MOSFETs)) to the energy bandgap of the semiconductor. The paper projected that huge reductions in the specific on-resistance were possible by replacing silicon with larger bandgap semiconductors.

The analysis did not include silicon carbide material at that time due to the primitive status of the material. The analysis predicted a reduction in the specific on-resistance of unipolar power devices by a factor of 13.6 by replacing silicon with gallium arsenide (GaAs), the most mature semiconductor technology at that time after silicon due its applications for IR light emitting diodes (LEDs). This prompted an intense effort at GE from 1980 to 1985 to develop the technology for growth of high purity GaAs epitaxial layers and process technology for making high quality Schottky and Ohmic contacts. The first high performance GaAs Schottky rectifiers were announced by GE in 1985 [4,5] which were subsequently commercialized by several companies. These devices represent the first wide bandgap semiconductor power devices providing confirmation of the projections from the theoretic model.

The growth and commercial supply of 6 H and later 4 H silicon carbide (SiC) wafers by CREE Research Inc. in the 1990s made the development of SiC power

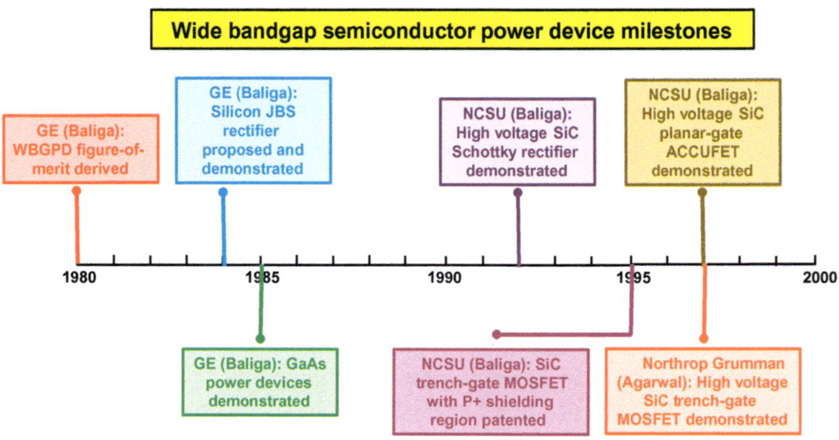

Figure 10.5 History of evolution of wide bandgap semiconductor power devices from 1980 to 2000.

devices feasible. The increase in the diameter of the SiC wafers during this period is shown in Fig. 10.6. The first high voltage SiC power device, demonstrated at North Carolina State University in 1991 and reported in 1992, was a 400-V Schottky rectifier without edge termination [6]—whose breakdown voltage was later enhanced to 1000-V using the argon-implanted termination [7]. This was an important milestone as the first verification of the low specific on-resistance of the drift region in SiC devices. This work also demonstrated the elimination of the reverse recovery current observed in silicon P–i–N rectifiers due to bipolar current flow. The leakage current of these Schottky rectifiers was later successfully reduced by using the junction-barrier concept first proposed and demonstrated for silicon Schottky diodes in 1984 [8,9].

The demonstration of SiC power MOSFETs was delayed by problems with the interface between the semiconductor and the gate oxide [10]. This prompted the development of SiC high voltage JFETs. The normally on nature of these devices was solved using the Baliga-Pair (or Cascode) configuration [11]. The SiC JFETs were quickly overshadowed by the demonstration of shielded planar-gate SiC power MOSFETs in 1997 at North Carolina State University [12]. This structure reduced the oxide electric field using a P^+ shielding region and an accumulation-mode channel to improve the channel mobility. It also employed the double-implanted channel definition approach [13] which is now commonplace for manufacturing SiC power MOSFETs. The fabrication of a 1 kV trench-gate SiC power MOSFET was also reported in 1997 [14]. These devices lacked the shielding of the gate oxide that is necessary for reliable operation. The shielding of the gate oxide at the trench-bottom using a P^+ region was patented in 1995 [11] and later experimentally demonstrated in 2002 [15].

The second era for silicon carbide power devices began in 2000. Major investments in product development occurred in the United States, Europe, and Japan. The technology had sufficiently matured to allow introduction of the first SiC Schottky power rectifier in 2001 as shown in Fig. 10.7. These devices became ideal

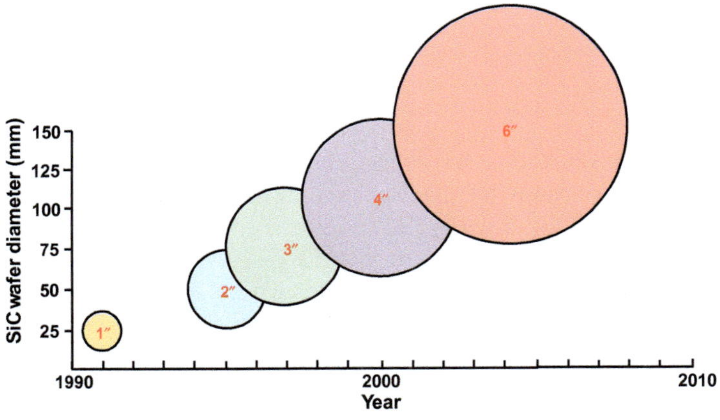

Figure 10.6 History of SiC wafer supply.

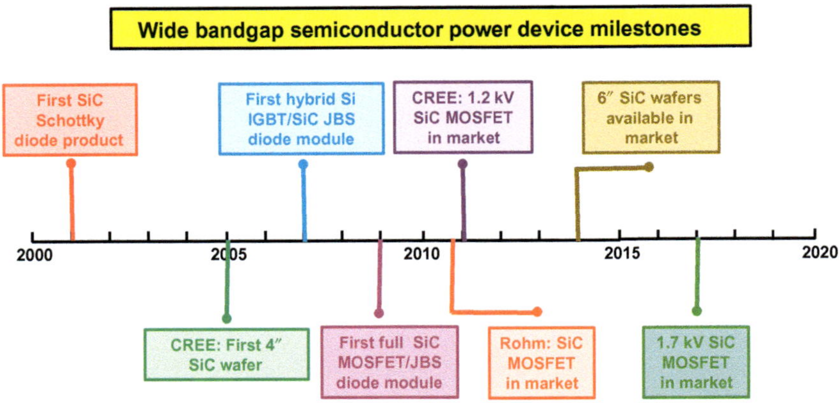

Figure 10.7 History of evolution of wide bandgap semiconductor power devices from 2000 to 2017.

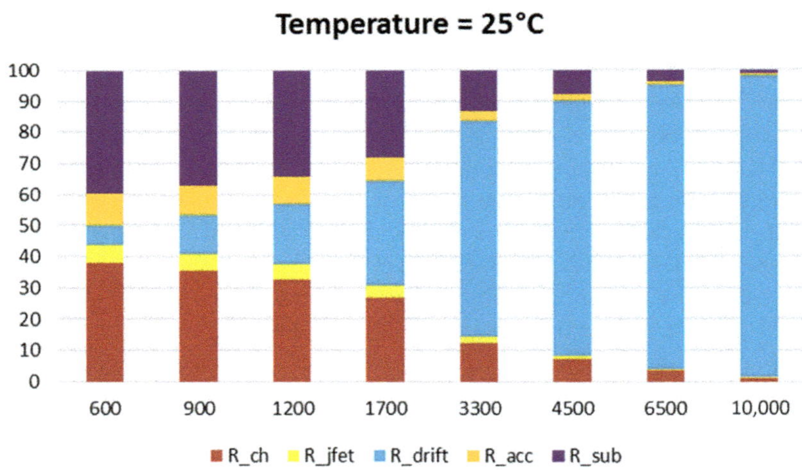

Figure 10.8 Internal resistance contributions for SiC power MOSFETs with various voltage ratings at 25°C.

companions to silicon IGBTs for H-bridge motor-control applications. The hybrid SiC Schottky diode/silicon IGBT module became feasible in 2007. The market for SiC Schottky diodes has grown to nearly $200 million after their reliability was proven.

The introduction of the SiC power MOSFET into the market occurred in 2011 with 1.2-kV rated devices. These devices offered very favorable reduced switching losses when compared with silicon IGBTs in inverter circuits. The ability to increase the circuit operating frequency by using the SiC power MOSFETs reduced the size and cost of the passive elements (inductors, capacitors, and filters) offsetting the higher device cost. The channel resistance contribution in the 1.2 kV SiC power MOSFET is about 30% as shown in Fig. 10.8 at 25°C and reduces to 25% at

150°C as shown in Fig. 10.9. The drift region resistance, especially at the usual junction operation of 150°C, becomes large for higher voltage ratings, while the channel contribution becomes dominant at lower voltage ratings.

The 1.7 kV rated SiC power MOSFET product was introduced into the market in 2015 as shown in Fig. 10.7. The introduction of SiC power MOSFETs with other voltage ratings is projected in Fig. 10.10. The existence of competition from silicon COOLMOS transistors with voltage ratings of 600 V and 900 V delayed the introduction of the 900-V SiC power MOSFET into the market until 2017. By this time, the availability of larger (6 in.) diameter SiC wafers had driven down the cost of the SiC power devices to enable market penetration. SiC power MOSFETs with higher voltage ratings can be expected to become available in the future as shown

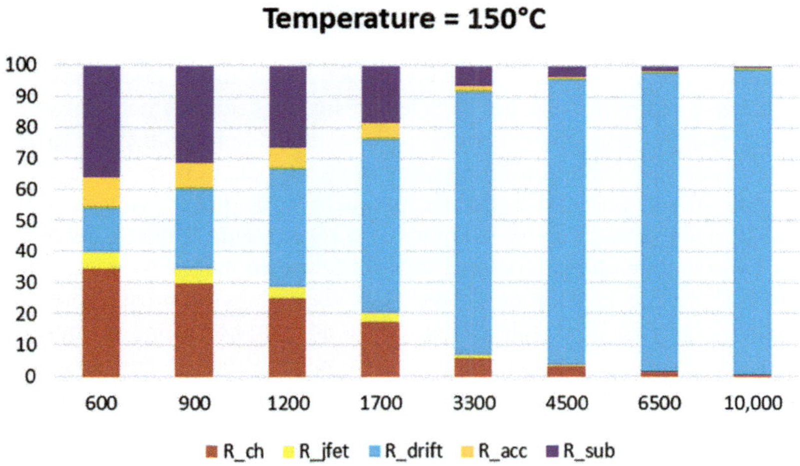

Figure 10.9 Internal resistance contributions for SiC power MOSFETs with various voltage ratings at 150°C.

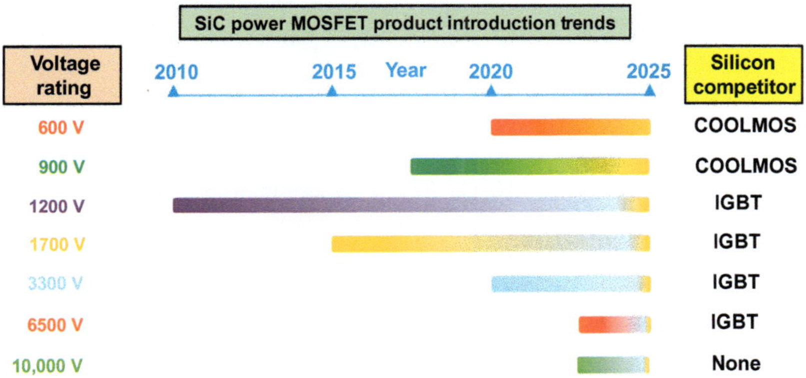

Figure 10.10 SiC power MOSFET market introduction by voltage rating.

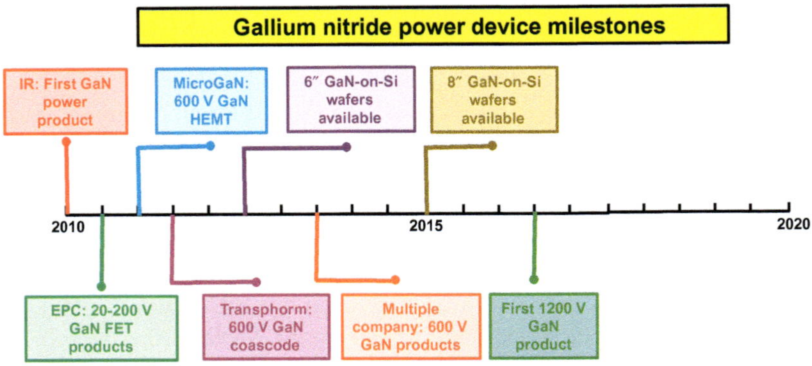

Figure 10.11 History of evolution of Gallium Nitride power devices from 2000 to 2017.

in the figure. The 600 V rating is highly attractive due to the large market volume for the electric and hybrid-electric vehicles applications, but the automotive industry has a long lead time and is very demanding with regard to reliability and ruggedness.

The evolution of gallium nitride power device products is traced in Fig. 10.11. There was initial push to establish products with lower voltage ratings of 20−200 V. This strategy must overcome the significant strides made for silicon devices by using the charge coupling concept to create the split-gate silicon trench MOSFET [16]. The application of the GaN technology to create very compact, lightweight, chargers for mobile devices, such as cell-phones and laptops, has provided market traction for these devices. More recent products have been designed with a rated voltage of 600 V. They have been used to make solar or Photo-Voltaic (PV) inverters for domestic power generation. Many companies are relying on a future burgeoning market in electric and hybrid-electric vehicles which will take many years to materialize. One major factor in favor of the lateral GaN devices is availability of low cost 6 in. and more recently 8 in. GaN-on-Si wafers. Another factor is the commitment of a large foundry (TSMC) to manufacture this technology for the power semiconductor industry.

10.3 Technology trends

The trends for the technology used to manufacture silicon carbide power MOSFET products are shown in Fig. 10.12. The first commercial products utilized the shielded planar-gate structure with typical design rules of 2 μm for the features on the masks. This was reduced to the 1 μm level by 2015 by the conversion of a silicon automotive products foundry to manufacturing silicon carbide power devices. As the demand and volume for the silicon carbide devices grows, it is anticipated that the design rules for features will be reduced to 0.5 μm by 2020 and 0.3 μm by 2025. This is necessary for the development of products with lower voltage ratings (<1.2 kV).

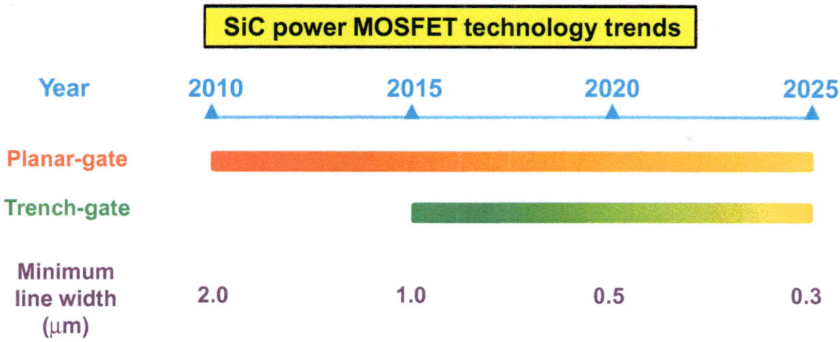

Figure 10.12 Technology trends for manufacturing SiC power MOSFETs.

Another important technological trend is the development of SiC power MOSFETs with trench-gate structures. Silicon power MOSFETs evolved from planar-gate to trench-gate structures in the 1990s to achieve a significant reduction in the specific on-resistance. A similar evolution for SiC power MOSFETs is hindered by the problems with preventing high electric fields in the gate oxide. Despite this issue, some companies have introduced 1.2 kV rated SiC trench-gate MOSFETs in 2015. It may be possible to reduce the specific on-resistance of these devices to the 2 mΩ cm^2 range compared with 4–5 mΩ cm^2 for planar-gate structures.

10.4 Wide bandgap semiconductor device applications

The primary near term applications for silicon carbide power devices have been identified to be: (1) power supplies for data centers; (2) renewable energy sources—solar and wind power; (3) motor drives; (4) rail transportation; and (5) electric and hybrid-electric vehicles. The penetration of the market for these applications is shown in Fig. 10.13 with the device ratings required for each case and the highest circuit operating frequency. The device manufacturers have already released products with voltage ratings of 900 V–1.7 kV to supply low power residential solar inverters and consumer power factor correction (PFC) circuits. Devices with higher voltage ratings ($>$3.3 kV) are under qualification to serve the needs of heating-ventilating and air-conditioning and the wind power applications. The more rigorous qualification required to penetrate the electric vehicle applications is expected to take longer as shown in the figure. The medium voltage motor drives market will develop later due the higher voltage ratings for the devices and its smaller market size.

The near-term applications identified for the gallium nitride power devices are (1) DC–DC converters; (2) solar inverters for residential use; (3) cell phone base-station power supplies; (4) PFC in consumer appliances; and (5) charging stations for electric vehicles. The penetration of the market for these applications is shown

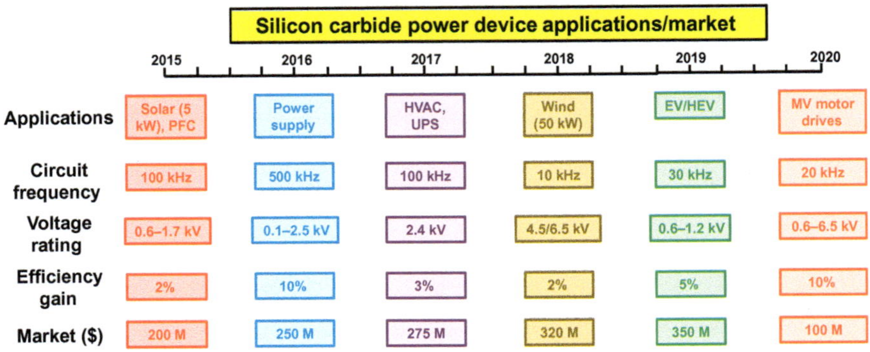

Figure 10.13 SiC device application requirements and market size.

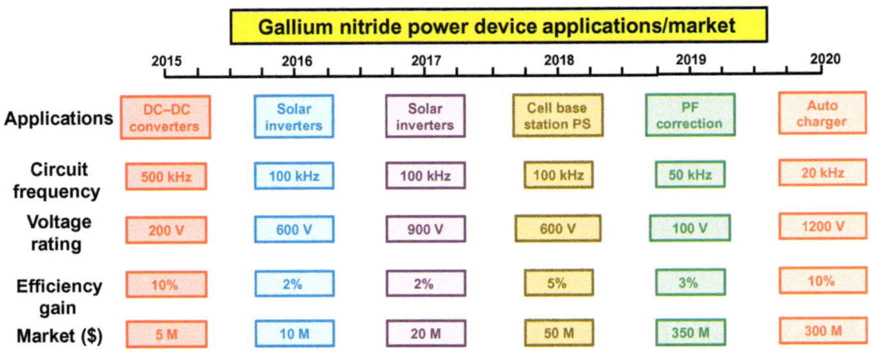

Figure 10.14 GaN device application requirements and market size.

in Fig. 10.14 with the device ratings required for each case and the highest circuit operating frequency. Devices with low voltage ratings (<200 V) have been deployed for low power DC–DC converters used for powering cell phones and laptops. Devices with 600 V ratings have been successfully used to commercialize a compact solar-powered generator for residential use.

Power electronics based up on using silicon IGBTs already has a relatively high efficiency ranging from 85% to 97%. Consequently, although the power losses in the power electronics can be reduced by 50% by replacing silicon IGBTs with SiC power MOSFETs, the efficiency gains are modest—in the range of 2%–10% as shown in Figs. 10.13 and 10.14. The social impact of wide bandgap semiconductor power device technology is therefore modest when compared with the silicon IGBT as shown later in the chapter.

10.5 Wide bandgap semiconductor device market

The market size for the SiC power devices for each of the applications is estimated in Fig. 10.13. The ability to displace silicon IGBTs in these applications is strongly dependent on how quickly the cost of manufacturing the SiC power devices can be scaled down. The cost for manufacturing the SiC power devices is discussed in the next section.

An estimate for the market size for the GaN power devices for each of the applications is estimated in Fig. 10.14. The ability to displace silicon IGBTs in these applications has been accelerated by manufacturing the power devices with GaN grown on silicon wafers.

The growth of the market for wide bandgap semiconductor power devices has been projected for various applications [17] as shown in Fig. 10.15. The near-term market applications are in power supplies and solar (PV) inverters. A major growth in the market is anticipated with the acceptance of these devices for electric and hybrid-electric vehicles.

The factors that will impact the penetration of wide bandgap semiconductor power devices in the applications are summarized in Fig. 10.16 for short-term

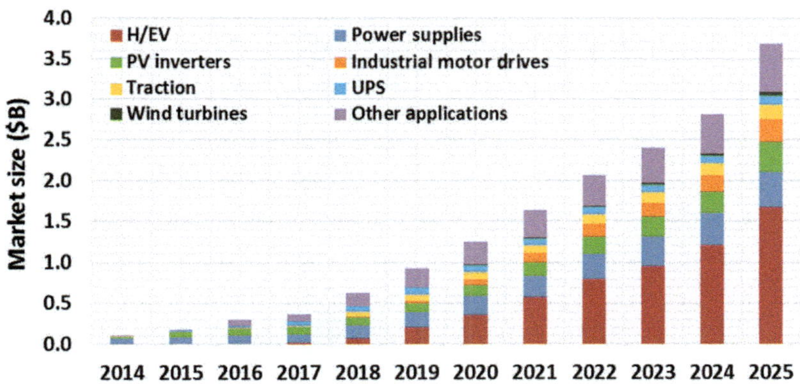

Figure 10.15 Projected market growth for wide bandgap semiconductor power devices.

Market criteria for wide bandgap semiconductor device: short term applications

	Power supply	UPS	Solar	Wind	Motor drives	Rail	EV
Efficiency	High	Medium	High	High	High	Medium	High
Weight/size	Medium	Low	Low	Medium	Medium	High	High
Operating temperature	Low	Medium	Low	Low	High	Medium	High
Switching frequency	Low	Low	Low	Low	Medium	Medium	High
Cost	Medium	High	High	Medium	Medium	Low	High

Figure 10.16 Market factors determining acceptance of wide bandgap semiconductor power devices in the short term.

Market criteria for wide bandgap semiconductor device: long-term applications

	Sensors	Heating process	Medical	Well logging	Wireless charging
Efficiency	High	High	Medium	Medium	High
Weight/size	Low	Medium	High	Medium	Medium
Operating temperature	High	Low	Low	High	Low
Switching frequency	Low	High	High	Low	Medium

Figure 10.17 Market factors determining acceptance of wide bandgap semiconductor power devices in the long term.

applications and in Fig. 10.17 for long-term applications. Achieving greater efficiency is a factor for all the applications. In many cases, such as electric vehicles, weight and size reductions are compelling attributes of the new technology. The ability to operate the power electronics at higher frequencies by using wide bandgap semiconductor power devices enables reduction of weight, size, and cost of passive elements such as transformers, inductors, and capacitors. Operating these devices at higher temperatures than feasible with silicon devices can provide cost savings in the cooling systems for hybrid electric vehicles.

10.6 Silicon carbide power MOSFET price projections

The cost (or price) of the wide bandgap semiconductor power devices will ultimately become a deciding factor for their implementation in the applications. The cost of SiC and GaN devices is at present three to five times larger than that of the silicon COOLMOS or IGBT devices they will be replacing. The price target for silicon carbide power MOSFETs with various voltage ratings is given in Fig. 10.18. A value of less than $0.20/Amp is required for 1.2-kV rated devices to displace the silicon IGBT. Lower values are required for the 600- and 900-V rated products while that for higher voltage rated devices grows from $ 0.70 to $ 2.00, to $ 5.00, to $ 15.00 for the 3.3, 6.5, 10, and 15 kV devices, respectfully.

The strategy for achieving the lower price targets for 1.2 kV SiC power MOSFETs is illustrated in Fig. 10.19 by using a 2.5×2.5 mm chip size as an example [18]. The case 1 represents manufacturing the devices in a captive 4 in. (100 mm) SiC wafer facility as has been historically done by several companies. This results in a relatively high price of $0.54/Amp which is five-times the price for silicon IGBTs. The price can be reduced to $0.26/Amp in case 2 by using a commercial foundry converted from antiquated silicon products to making SiC power MOSFETs. Further, price reduction to $0.21/Amp is feasible in case 3 by moving the production to 6 in. (150 mm) SiC wafers. The desired target of

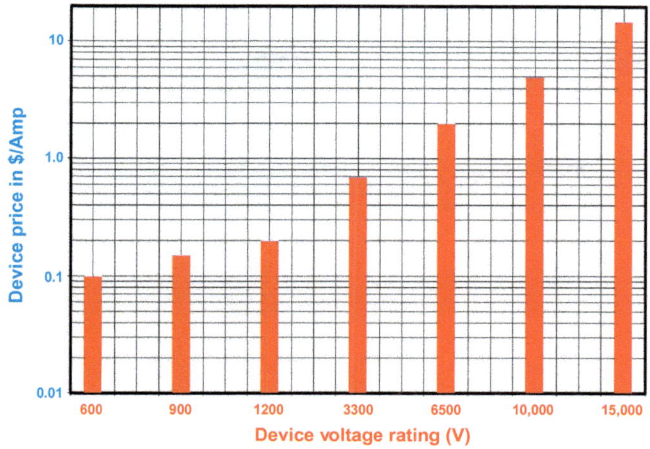

Figure 10.18 Price targets for SiC power MOSFETs with various voltage ratings.

1.2 kV SiC power MOSFET pricing model (2.5 × 2.5 mm chip)

	Case 1	Case 2	Case 3	Case 4
Foundry	Dedicated	Commercial	Commercial	Commercial
Wafer size	100 mm	100 mm	150 mm	150 mm
Substrate+ Epi cost	$ 1200	$ 1200	$ 3000	$ 800
Processing cost	$ 1800	$ 700	$ 700	$ 500
Yield	60%	80%	80%	80%
Cost per chip	$ 5.46	$ 2.60	$ 2.09	$ 0.73
Cost per Amp	$ 0.27	$ 0.13	$ 0.105	$ 0.037
Price per Amp	$ 0.54	$ 0.26	$ 0.21	$ 0.074

Figure 10.19 Strategy for reduction of SiC power MOSFETs price. The price is based up on a 50% gross margin.

achieving SiC power MOSFET prices at or below that of silicon IGBTs can be reached by projections at high-production volumes in case 4.

10.7 Social impact of wide bandgap semiconductor devices

The enormous impact of the silicon IGBT on society was documented in Section 10.1. Its availability in the 1980s allowed achieving an increase in efficiency for motor drives by 40% and for lighting by 75%. As a consequence of transforming power electronics from analog control to digital control, the efficiency

Potential energy savings and carbon emission reductions			
	Electricity (TW h)	Gasoline (B gal)	CO_2 (B pounds)
Data centers	22	–	20
Solar/wind	15	–	15
Motor drives	20	–	20
Rail	2	–	2
Electric vehicles	5	1.0	20
Total	**64**	**1.0**	**77**

Figure 10.20 Annual energy and carbon dioxide emission savings from deployment of wide bandgap semiconductor power devices.

for applications today is typically above 90%. This leaves a much smaller room for improvement in efficiency with wide bandgap semiconductor power devices.

The Power America Institute and Department of Energy websites project an energy savings potential from using wide bandgap semiconductor power devices in industrial motor drives as equivalent to powering 1 million homes. The US Energy Information Administration states that the annual electricity consumption by an average home in the United States was 10,766 kW h in 2016. The annual electricity savings from wide bandgap semiconductor power devices used in industrial motor drives is then 108 TW h. The projected savings in electricity in consumer electronics and data centers is estimated to be equivalent to powering 1.3 million homes which translates to 140 TW h. The projected savings in electricity in conversion of solar and wind energy is estimated to be equivalent to powering 0.7 million homes which translates to 76 TW h.

The potential energy savings have also been estimated by scientists from the Oak Ridge National Laboratory [17]. Their numbers are based up on detailed analysis of each of the applications listed in Fig. 10.20. These estimates are considerably lower than those in the previous paragraph. A total annual electricity savings from all of the application is found to be 64 TW h. They also estimate an annual gasoline energy savings of 1 billion gallons based up on application of wide bandgap semiconductor power devices in electric vehicles.

Approximately 1.1 pounds of carbon dioxide is emitted into the earth's atmosphere for each kW h of generated electricity, and 19.4 pounds of carbon dioxide is generated by burning one gallon of gasoline. Using these values, the total reduction of annual carbon dioxide emissions by deployment of wide bandgap semiconductor power devices is estimated in Fig. 10.20 to be 77 billion pounds.

10.8 Summary

The development of silicon carbide and gallium nitride power devices has been motivated by the opportunity to create unipolar devices with high blocking voltage capability. Silicon carbide JBS rectifiers and power MOSFETs with excellent on-state voltage drop have been demonstrated with blocking voltages up to 5000 V. These devices offer low switching losses when compared with the silicon IGBT enabling increasing the circuit operating frequency which reduces the size of passive components and filters in applications. Gallium nitride high electron mobility transistor (HEMT) devices have been commercialized with blocking voltages of 100−600 V. The applications of the SiC and GaN devices are constrained by their higher cost when compared with silicon devices. The lower switching losses of these devices can be utilized to increase the circuit operating frequency leading to smaller, less-expensive passive elements to offset the higher device cost. The deployment of wide bandgap semiconductor power devices will benefit society by reducing electricity and gasoline consumption leading to mitigating carbon dioxide emissions.

References

[1] B.J. Baliga, The IGBT Device: Physics, Design, and Applications of the Insulated Gate Bipolar Transistor, Elsevier Press, 2015.

[2] B.J. Baliga, Optimum Semiconductors for High Voltage Vertical Channel Field Effect Transistors, GE Class 2 CRD Report no. 79CRD186, November 1979.

[3] B.J. Baliga, Semiconductors for high voltage vertical channel field effect transistors, J. Appl. Phys. 53 (1982) 1759−1764.

[4] B.J. Baliga, et al., Gallium arsenide Schottky power rectifiers, IEEE Trans. Electron Devices Ed-32 (1985) 1130−1134.

[5] P.M. Campbell, et al., Trapezoidal groove Schottky gate vertical channel GaAs FET, in: IEEE International Electron Devices Meeting, Abstract 7.3, 1984, pp. 186−189.

[6] M. Bhatnagar, P.M. McLarty, B.J. Baliga, Silicon carbide high voltage (400 V) Schottky barrier diodes, IEEE Electron Device Lett. EDL-13 (1992) 501−503.

[7] D. Alok, B.J. Baliga, P.K. McLarty, A simple edge termination for silicon carbide with nearly ideal breakdown voltage, IEEE Electron Device Lett. EDL-15 (1994) 394−395.

[8] B.J. Baliga, The pinch rectifier: a low forward voltage drop high speed power diode, IEEE Electron Device Lett. 5 (1984) 194−196.

[9] F. Dahlquist, et al., "A 2.8 kV JBS diode with low leakage", silicon carbide and related materials – 1999, Mater. Sci. Forum 338−342 (2000) 1179−1182.

[10] B.J. Baliga, Critical nature of oxide/interface quality for SiC power devices, Microelectron. Eng. 28 (1995) 177−184.

[11] B.J. Baliga, Silicon Carbide Switching Device with Rectifying Gate, U.S. Patent 5,396, 085, Issued March 7, 1995.

[12] P.M. Shenoy, B.J. Baliga, High voltage planar 6H-SiC ACCUFET, IEEE Electron Device Lett. 18 (1997) 589−591.

[13] B.J. Baliga, M. Bhatnagar, Method of Fabricating Silicon Carbide Field Effect Transistor, U.S. Patent 5,322,802, Issued June 21, 1994.

[14] A.K. Agarwal, et al., 1.1 kV 4H-SiC SiC power UMOSFETs, IEEE Electron Device Lett. 18 (1997) 586−588.

[15] Y. Li, J.A. Cooper, M.A. Capano, High voltage (3 kV) UMOSFETs in 4H-SiC, IEEE Trans. Electron Devices 49 (2002) 972−975.

[16] B.J. Baliga, Power Semiconductor Devices having improved High Frequency Switching and Breakdown Characteristics, U.S. Patent 5,998,833, Issued December 7, 1999.
[17] K. Armstrong, S. Das, L. Marlino, Wide Bandgap Semiconductor Opportunities in Power Electronics, Oak Ridge National Laboratory Report, December 2016.
[18] A. Agarwal, et al, Wide Bandgap Power Devices and Applications: The U.S. Initiative, in: European Solid State Device Research Conference, 2016, pp. 206–209.

Index

Note: Page numbers followed by "*f*" and "*t*" refer to figures and tables, respectively.

T

CPI Antony Rowe
Chippenham, UK
2018-12-06 01:35